Aquaculture Law and Policy

Aquaculture is a rapidly expanding industry predicted to play an increasingly important role in meeting demand for fish. The essays in this volume focus on getting a "good governance" grip on the aquaculture industry, which is facing major environmental and social challenges.

The book highlights the numerous law and policy issues that may need to be addressed in the search for effective regulation of aquaculture. Those issues include, among others: the equitable assignment of property rights; the design of effective dispute resolution mechanisms; adoption of a proper taxation system for aquaculture; resolution of aboriginal offshore title and rights claims; recognition of international trade law restrictions such as labeling limitations and food safety requirements; and determination of whether genetically modified fish should be allowed and, if so, under what controls.

The main themes running through the book are the need to understand and address major limitations in existing aquaculture laws and policies, particularly the "free-market" leasing/licensing approach and lack of integrated coastal planning processes; and the need to rethink national aquaculture laws and institutional arrangements in light of international environmental principles, such as integration, precaution, public participation, community-based management, intergenerational equity and indigenous rights developments.

While previous books on this topic have tended to be descriptive and narrowly focused on just one country, *Aquaculture Law and Policy* attempts to be prescriptive and comparative in its analyses, suggesting ways in which aquaculture legislation might be reformed in light of sustainable development principles and drawing on global and other national experiences.

David L. VanderZwaag, Canada Research Chair in Ocean Law and Governance at Dalhousie Law School, is a member of the Commission on Environmental Law (CEL) and the World Conservation Union (IUCN), and presently co-chairs the IUCN Specialist Group on Oceans, Coasts and Coral Reefs. **Gloria Chao** practices environmental and energy law with the Vancouver office of Blake, Cassels & Graydon LLP. Prior to joining Blakes, Gloria obtained a Master of Laws degree in marine environmental law and completed a clerkship with the Supreme Court of Canada.

Routledge advances in maritime research
Edited by H. D. Smith
Cardiff University, UK

The oceans and seas of the world are at a critical juncture in their history. The pressures of development brought about by the globalization of the world economy continue to intensify through the major sectors of ocean use. In parallel, marine management and policy issues become larger, more numerous and more urgent. The response of this series is to provide in-depth analysis of ocean development, management and policy from a multi-disciplinary perspective, encompassing a wide range of aspects of interrelationships between the oceans and seas on the one hand, and maritime human activities on the other. Several strands run through the series:

- Studies of the development and management of major ocean industries and uses including shipping and ports; strategic uses; mineral and energy resources; fisheries and aquaculture; the leisure industries; waste disposal and pollution; science and education; and conservation.
- Inter- and multidisciplinary perspectives provided by the natural sciences, geography, economics, sociology, politics, law and history.
- Responses to the need to devise integrated ocean policies and management measures which cover the deep oceans, the bordering seas and coastal zones.
- Regional studies at a variety of geographical scales from large ocean regions to regional seas.

The series is of interest to all concerned professionally with the oceans and seas, ranging from scientists and engineers to surveyors, planners, lawyers and policy-makers working in the public, private and voluntary sectors. It is also of wider public interest to all those interested in or having a stake in the world ocean and its bordering seas.

1 **Development and Social Change in the Pacific Islands**
   Edited by A. D. Couper

2 **Marine Management in Disputed Areas**
   The case of the Barents Sea
   Robin Churchill and Geir Ulfstein

Aquaculture Law and Policy

Towards principled access and operations

Edited by David L. VanderZwaag and Gloria Chao

Routledge
Taylor & Francis Group

LONDON AND NEW YORK

First published 2006
by Routledge

Published 2017 by Routledge
2 Park Square, Milton Park, Abingdon, Oxon OX14 4RN
711 Third Avenue, New York, NY 10017, USA

Routledge is an imprint of the Taylor & Francis Group, an informa business

First issued in paperback 2012

Copyright © 2006 Selection and editorial matter, David L. VanderZwaag
and Gloria Chao; individual chapters, the contributors

Typeset in Garamond by Wearset Ltd, Boldon, Tyne and Wear

The Open Access version of this book, available at
www.tandfebooks.com, has been made available under
a Creative Commons Attribution-Non Commercial-No
Derivatives 4.0 license.

British Library Cataloguing in Publication Data
A catalogue record for this book is available from the British Library

Library of Congress Cataloging in Publication Data
A catalog record for this book has been requested

ISBN13: 978-0-415-70201-0 (hbk)
ISBN13: 978-0-415-65357-2 (pbk)

Contents

Illustrations

Figures

Tables

Notes on contributors

Richard Apostle is Professor of Sociology and Social Anthropology at Dalhousie University, Halifax, Canada. He is continuing aquaculture research and writing with Professors Hovgaard and Phyne. He is currently completing a book concerning municipal change and local democracy in the Faroes.

Gloria Chao practices environmental law with Blake, Cassels & Graydon, Vancouver, Canada. Ms. Chao obtained a BCom from the University of British Columbia, her LLB and LLL degrees from the University of Ottawa, and her LLM degree, specializing in marine environmental and maritime law, from Dalhousie University. In 2002, Ms. Chao completed a clerkship with the Honourable Mr. Justice Louis LeBel of the Supreme Court of Canada.

Aldo Chircop, JSD, is Associate Professor of Law, Dalhousie Law School, Halifax, Canada, with cross-appointments to the Faculty of Graduate Studies (Marine Affairs Programme) and Faculty of Arts and Social Sciences (International Development Studies Programme). On leave of absence from Dalhousie University (2003–2005), he held the Canadian Chair in Marine Environmental Protection at the World Maritime University, Malmö, Sweden. Dr. Chircop's research interests are in comparative ocean law and policy, law of the sea, international marine environmental law, maritime law and integrated coastal management. He is co-editor of the *Ocean Yearbook* (University of Chicago Press and Transnational Press) and co-author of *Maritime Law* (Toronto: Irwin Law, 2003).

Mark Covan, LLM, is Counsel with the Federal Prosecution Service, Department of Justice, Canada, and a former Research Associate with the AquaNet Law and Policy Project, Marine and Environmental Law Programme, Dalhousie University, Halifax, Canada.

Richard Devlin, LLM, is Professor of Law, University Research Professor, and Associate Dean, Graduate Studies and Research, Dalhousie University, Halifax, Canada. He has published over forty articles in various

Canadian, American and British journals, and is editor of *Canadian Perspectives on Legal Theory*, and co-editor of *Critical Disability Theory: Essays in Philosophy, Politics, Policy and Law*.

Jeremy Firestone, JD, PhD, is Assistant Professor, College of Marine Studies, University of Delaware, Newark, USA. He has practiced as a lawyer for the United States Environmental Protection Agency and the State of Michigan on environmental protection and natural resources issues. His current research is focused on fish and wildlife issues and governance, including invasive species, ship–whale interactions, and offshore wind power development. He is a member of a multidisciplinary research team developing an operational framework for aquaculture in the US exclusive economic zone.

Richard Finn obtained his LLB from Dalhousie Law School, Halifax, Canada, in 2003 where he participated in the School's Marine and Environmental Law Program (MELP), graduating with a specialization in marine law. Richard also holds undergraduate and graduate degrees in biological science. His interest in coastal development and management issues led him to become involved in the national AquaNet policy initiative in the summer of 2002.

Diana Ginn, LLM, is Associate Professor of Law, Dalhousie University, Halifax, Canada. She is a Member of the Nova Scotia Barristers' Society, and a former Member of the Law Society of Upper Canada and the Law Society of the Northwest Territories. She teaches and writes in a variety of areas, including aboriginal title, administrative law, property law, and gender and the law.

Arthur J. Hanson, LOC, PhD, is Distinguished Fellow with the International Institute of Sustainable Development, having served as President and CEO of IISD from 1992 to 1998. Dr. Hanson was named an Oceans Ambassador for the Minister of Fisheries and Oceans from June 2000 to June 2004. In August 2003, he was appointed an Officer in the Order of Canada. He is currently Lead Expert and Member of the China Council for International Cooperation on Environment and Development, a member of the Canadian Foundation for Innovation, and a Mentor in the Trudeau Foundation. Dr. Hanson is engaged in research on Biotechnology and Sustainable Development through the Canadian Biotechnology Advisory Committee.

Marcus Haward, PhD, is Associate Professor and Head of School in the School of Government at the University of Tasmania, Hobart, Australia. He is also Program Leader, Policy Program, Antarctic Climate and Ecosystems Cooperative Research Centre. He has research interests in fisheries management and oceans policy, and has published extensively in each of these areas. He is a member of the Australia Canada Ocean

Research Network (ACORN), with research interests focusing on Australian and Canadian experiences in oceans policy development and governance.

Nathanael Hishamunda is Fishery Planning Officer with the FAO (Food and Agriculture Organization of the United Nations) in Rome, Italy. He holds an MSc in Aquaculture Economics and a PhD in Agricultural Economics with a Minor in Agriculture Policy, both from Auburn University, Alabama, USA.

Gestur Hovgaard is Senior Researcher, Center for Local and Regional Development, Klaksvík, the Faroes. In 2002, he was awarded a post-doctoral fellowship at the Faculty of Arts and Sciences, Dalhousie University, Halifax, Canada. Dr. Hovgaard's interests are in the fields of economic sociology and political economy.

William Howarth is Professor of Environmental Law at the University of Kent at Canterbury, United Kingdom. He is General Editor of the journal *Water Law* and has published extensively on water and environmental law. His international work has involved advising the governments of Turkey, Romania, Thailand and Bulgaria on water and fisheries legislation, and he has also acted as a consultant to the Food and Agriculture Organization of the United Nations and national bodies including the National Rivers Authority, the Environment Agency and the former Department of the Environment, Transport and the Regions.

Tamara Lorincz graduated with a combined MBA and LLB degree from Dalhousie University, Halifax, Canada, in 2003, with a specialization in marine and environmental law. Ms. Lorincz holds a BA from the University of Alberta, Edmonton, Canada, and a BCom from the University of Victoria, Victoria, Canada. She is a legal researcher and was recently nominated to the national board of the Sierra Legal Defence Fund. Currently, Ms. Lorincz is the Coordinator of the Nova Scotia Environmental Network, a provincial non-profit organization.

Moira L. McConnell, BA, LLB, PhD, is Professor of Law at Dalhousie Law School, Halifax, Canada, and is the Director of its Marine and Environmental Law Institute. Dr. McConnell's teaching and research interests are in the fields of marine environmental protection, shipping, law of the sea, biosecurity, integrated coastal and ocean management, business and environmental law/management, public and corporate governance, regulatory design, dispute resolution processes and international trade and environment. She is a special adviser on maritime issues to UN agencies (the International Maritime Organization and the International Labour Organization) and is a member of the IUCN's Commission on Environmental Law. She is a co-editor of the interdisciplinary *Ocean Yearbook* and has published widely in the fields of international and environmental law, regulatory system design, maritime law and policy, social justice and human rights.

Ted L. McDorman, LLM, is Professor, Faculty of Law, University of Victoria, Victoria, Canada. Professor McDorman has written extensively on ocean law, policy and management issues. Since 2000, he has been "editor-in-chief" of *Ocean Development and International Law: The Journal of Marine Affairs*. From June 2002 to 2004, he served as Academic-in-Residence, Bureau of Legal Affairs, Department of Foreign Affairs and International Trade, Ottawa. Canada.

Douglas Moodie, LLM, is former Assistant Professor of Law, Dalhousie University, Halifax, Canada. He is a Member of both the Nova Scotia Barristers' Society and the Law Society of Upper Canada and is currently a Senior Solicitor with the Nova Scotia Department of Justice. He is a part-time lecturer in the Political Science Department of Saint Mary's University Halifax. His research interests include environmental and First Nations law.

Ronalda Murphy, LLM, is Associate Professor of Law, Dalhousie University, Halifax, Canada. She teaches a wide variety of courses, including Canadian constitutional law and comparative constitutional law. Professor Murphy lectured full-time at the University of the Witswatersrand in Johannesburg, South Africa, and worked on the transition from apartheid to constitutional democracy from 1991 to 1994. Her publications are in the fields of constitutional law and theory, evidence law, aboriginal law, feminist legal theory and postmodernism.

John Phyne is Professor of Sociology and Coordinator of the Interdisciplinary Studies in Aquatic Resources Program at St. Francis Xavier University, Antigonish, Canada. His research focuses upon the socio-economic and regulatory dimensions of the salmon aquaculture industry. He is the author of *Disputed Waters: Rural Social Change and Conflicts Associated with the Irish Salmon Farming Industry, 1987–1995*.

Hamish Rennie, PhD, is Senior Lecturer, Department of Geography, Tourism and Environmental Planning, University of Waikato, Hamilton, New Zealand, specializing in fisheries/aquacultural geography and coastal planning. Prior to joining the University of Waikato in 1995, Dr. Rennie spent approximately twelve years working in policy roles for various central government departments. He helped write parts of New Zealand's aquaculture and resource management legislation and its Coastal Policy Statement, and has worked for applicants and as a commissioner hearing applications and proposed plan changes. He is a Member of the New Zealand Planning Institute and the New Zealand Association of Resource Management.

Douglas Sanders is Professor Emeritus, Faculty of Law, University of British Columbia, Vancouver, Canada, and an LLM program professor,

Chulalongkorn University, Bangkok, Thailand. He has worked on indigenous legal issues since the mid-1960s, including the fishing rights case *Regina* v. *Jack*, which was later influential in the leading fishing rights decision *Sparrow* v. *The Queen*. He has done comparative work on indigenous and tribal issues in a number of countries, including Australia, India, Japan, Malaysia, Norway and the United States.

Phillip M. Saunders is Dean and Associate Professor, Faculty of Law, Dalhousie University, Halifax, Canada, with a cross-appointment to the School for Resource and Environmental Studies. He is a member of the Nova Scotia Bar. His teaching and research interests include international marine and environmental law, maritime boundary delimitation and fisheries law. Professor Saunders was formerly with the International Centre for Ocean Development, as Senior Policy Adviser and as Field Representative, South Pacific.

Torsten Ström, LLM, of the Bar of Ontario, is Senior Counsel with the Trade Law Bureau at the Department of Justice/Department of International Trade. He has previously published articles and given presentations on the legal aspects of the trade and environment interface.

David VanderZwaag, PhD, is Professor of Law and Canada Research Chair in Ocean Law and Governance at Dalhousie University, Halifax, Canada. He was a co-founder of Dalhousie's interdisciplinary program in Marine Affairs. He is a member of the Commission on Environmental Law, the World Conservation Union (IUCN), and co-chair of the Specialist Group on Oceans, Coasts and Coral Reefs.

Faye Woodman is Professor of Law, Dalhousie University, Halifax, Canada. Her teaching and research areas are primarily in the areas of property (broadly conceived) and taxation. The latter interest dates back to a stint at the Department of Finance, Ottawa, Canada, working on tax policy. She is most recently the co-editor of and a contributor to a text on the law of trusts.

Foreword

When I was first exposed to aquaculture as an adviser to the Nova Scotia Department of Fisheries, the province had recently passed an Act respecting aquaculture and it was my job to brief the federal government. Off to Ottawa I went, where I spent a very unpleasant day hearing from the Department of Fisheries and Oceans lawyers why the province had no constitutional jurisdiction to do what the Act says we may.

Twenty years later, things are a bit better but we continue to struggle with the ongoing challenge of providing a workable and cooperative governance structure for an industry whose development, and the issues associated therewith, have dramatically outpaced our law and policy capacity. Aquaculture is heralded by its proponents as part of the solution to the world's food shortage and as an alternative to the dwindling wild fisheries. Aquaculture opponents cite pollution, food health, animal welfare, conflict of use and plain old NIMBY ("not in my backyard") as reasons for expending phenomenal energy opposing any form of aquaculture.

This book is a thoughtful and comprehensive synthesis and analysis of the plethora of social, economic, environmental and legal issues that must be addressed in order to ensure that the world's growing aquaculture industry is sustainable into the future. There are no perfect resolutions to the tensions surrounding aquaculture. Every country will have nuances in its governance regimes, and every aquaculture site will require different considerations, depending on the ecosystem in which it resides.

This book provides an excellent source document for anyone interested in understanding the issues facing one of the most rapidly growing and complex areas of the global food basket. The authors also have the courage to provide some well-reasoned recommendations that should move through currently acrimonious debates to a more constructive dialogue.

Peter C. Underwood
Deputy Minister
Nova Scotia Department of Natural Resources

Acknowledgments

This volume, the result of a four-year research project centered at Dalhousie Law School, would not have been possible without funding support from AquaNet, Canada's Research Network in Aquaculture. Special thanks goes to the funders of AquaNet, the Natural Sciences and Engineering Research Council (NSERC), the Canadian Institutes of Health Research (CIHR) and the Social Sciences and Humanities Research Council (SSHRC), in partnership with Industry Canada.

Many of the chapters of this volume have benefitted from discussions and critiques at the "Canadian Aquaculture Law and Policy Workshop: Towards Principled Access and Operations," held in Halifax, Nova Scotia, 26–28 February 2003. Workshop sponsorship by AquaNet, the Nova Scotia Department of Agriculture and Fisheries, and the Office of the Commissioner for Aquaculture Development, Fisheries and Oceans Canada, is gratefully acknowledged. The Workshop Steering Committee consisted of Yves Bastien, Commissioner for Aquaculture Development, Fisheries and Oceans Canada; Mark Burgham, Director Aquaculture Policy, Office of Sustainable Aquaculture, Fisheries and Oceans Canada; Norman Dale, Northern Coastal Information and Research Program, University of Northern British Columbia; Ralph Matthews, Professor of Sociology, University of British Columbia; David Rideout, Executive Director, Canadian Aquaculture Industry Alliance; Peter Underwood, Deputy Minister, Nova Scotia Department of Agriculture and Fisheries; and Richard Wex, Director-General, Office of Sustainable Aquaculture, Fisheries and Oceans Canada.

Understanding of aboriginal title and rights issues in aquaculture was greatly enhanced by participation of a number of First Nations organizations and the sharing of aboriginal perspectives at the Workshop. Those participants, among others, included Martin Weinstein, 'Namgis First Nation Treaty Office; Patricia Doyle-Bedwell, Dalhousie Law School; Hugh Braker, Tseshaht First Nation of the Nuu-chan-multh People (West Coast Vancouver Island); and Arnie Narcisse, BC Aboriginal Fisheries Commission.

Thanks are also owed to Annick Van Houtte, Legal Officer with FAO's Development Law Service. She helped organize meetings of the editors with various FAO specialists in fisheries and aquaculture and provided a wealth of information about aquaculture legislative approaches and challenges around the globe.

Appreciation is also extended to Dr. Alex Brown, Head, Environmental Affairs, Undersecretariat for Fisheries, Chile. He not only provided researchers with an overview of aquaculture law and policy developments in Chile but was instrumental in convening a further workshop (Iquique, Chile, 1–3 September 2004) under Asia-Pacific Economic Cooperation (APEC) auspices, comparing the implementation of environmental principles and policies in aquaculture.

Numerous persons at Dalhousie Law School helped make this volume a reality. Molly Ross patiently undertook meticulous word processing. Ann Morrison, Law School Librarian, provided space in the library for student researchers and enthusiastically added the project's webpage to the library's website. Former Dean Dawn Russell ensured adequate staff support and facilitated a positive research environment. Over twenty research assistants drawn from students in the LLB program assisted chapter authors, and many of the students are specifically recognized in the subsequent chapters.

Susan Rolston, Seawinds Consulting Services, was tireless in providing editorial assistance and in formatting the book for publication.

The Canada Foundation for Innovation (CFI) was also instrumental in supporting research for this volume. Infrastructure funding was provided by CFI for the establishment of the Ocean Law and Governance Research Centre at Dalhousie Law School.

Each of the authors deserves thanks for being patient during the rather lengthy peer review and production processes. Most of the chapters were originally written in 2003, and the authors have undertaken various updating efforts to at least the end of 2004 and in some cases into 2005.

David VanderZwaag and Gloria Chao

Abbreviations

AAFRD	Alberta Agriculture, Food and Rural Development
ABARE	Australian Bureau of Agricultural and Resource Economics
ABMA	aquaculture bay management area
ACAP	Atlantic Coastal Action Program (Canada)
ADR	alternative dispute resolution
AECC	Aquaculture Environmental Coordinating Committee (New Brunswick)
AFFA	Agriculture, Fisheries and Forestry Australia
AFMA	Australian Fisheries Management Authority
AIA	advance informed agreement
ALC	Aquaculture Licensing Committee (Newfoundland and Labrador)
ALMB	Aquaculture Leasing Management Board (Prince Edward Island)
AMA	aquaculture management area
AOP	Aquaculture Occupation Permit (New Brunswick)
APF	Aquaculture Policy Framework
ASEC	Aquaculture Site Evaluation Committee (New Brunswick)
ASMFC	Atlantic States Marine Fisheries Commission (United States)
BC	British Columbia
BCAL	British Columbia Assets and Land Corporation
BCMAL	British Columbia Ministry of Agriculture and Lands
BIEAP	Burry Inlet Environmental Action Program
BMA	bay management agreement
BMP	best management practice
BNA Act	*British North America Act*
BSE	bovine spongiform encephalopathy
BST	bovine somatotropin
CAAPF	Concentrated Aquatic Animal Production Facility
CAAR	Coastal Alliance for Aquaculture Reform
CAC	Codex Alimentarius Commission

CAIA	Canadian Aquaculture Industry Alliance
CAIS	Canadian Agriculture Income Stabilization Program
CAP	Common Agricultural Policy (of the European Union)
CBAC	Canadian Biotechnology Advisory Committee
CBD	Convention on Biological Diversity
CCFAM	Canadian Council of Fisheries and Aquaculture Ministers
CCFL	Codex Committee on Food Labeling
CCPC	Canadian-controlled private corporation
CCSBT	Commission for the Conservation of Southern Bluefin Tuna
CEA	*Clean Environment Act* (New Brunswick)
CEAA	*Canadian Environmental Assessment Act*
CEPA	*Canadian Environmental Protection Act*
CFIA	Canadian Food Inspection Agency
CFIP	Canadian Farm Income Program
CFP	Common Fisheries Policy (of the European Union)
CLC	*Convention on Civil Liability for Oil Pollution Damage*
CMA	coastal management area
CSI	Environment and Development in Coastal Regions and in Small Islands (UNESCO)
CWA	"Clean Water Act" (*Federal Water Pollution Control Act*) (United States)
CZMA	*Coastal Zone Management Act* (United States)
DAF	Department of Agriculture and Fisheries (Nova Scotia)
DAFA	Department of Agriculture, Fisheries and Aquaculture (New Brunswick)
DFO	Fisheries and Oceans Canada (Department of Fisheries and Oceans)
DG – SANCO	Directorate-General – Health and Consumer Protection (European Commission)
DM	Deputy Minister
DNV	Det Norske Veritas
DPS	distinct population segment
DR	dispute resolution
DSB	Dispute Settlement Body (of the World Trade Organization)
EAO	Environmental Assessment Office (British Columbia)
EAP	environmental assessment process
EC	European Commission
EEC	European Economic Community
EEZ	exclusive economic zone
EFSA	European Food Safety Authority
EFTA	European Free Trade Association
EIA	environmental impact assessment
ENGO	environmental non-governmental organization

EPA	*Environmental Protection Act* (Prince Edward Island; Newfoundland and Labrador); Environmental Protection Agency (United States)
EPBC Act	*Environment Protection and Biodiversity Conservation Act* (Australia)
ERDF	European Regional Development Fund
ERMA	Environmental Risk Management Authority (New Zealand)
ESCOR	United Nations Committee on Economic, Social and Cultural Rights
ESF	European Social Fund
ESSIM	Eastern Scotian Shelf Integrated Management Initiative
FAA	Framework Agreement on Aquaculture
FADS	Federal Aquaculture Development Strategy (Canada)
FAO	Food and Agriculture Organization (of the United Nations)
FCRA	*Fisheries and Coastal Resources Act* (Nova Scotia)
FDA	Food and Drug Administration (United States)
FIFRA	*Federal Insecticide, Fungicide, and Rodenticide Act* (United States)
FMC	fishery management council
FOS	Fish Farmers' Sales Organisation (Norway)
FRDC	Fisheries Research and Development Corporation
FREMP	Fraser River Estuary Management Program
FSO	Fish Sales Organization (Norway)
FVO	Food Veterinary Office (European Union)
FWS	Fish and Wildlife Service (United States)
GAA	Global Aquaculture Alliance
GATT	General Agreement on Tariffs and Trade
GCSAI	Guysborough County Sustainable Aquaculture Initiative
GE	genetically engineered
GESAMP	Joint Group of Experts on the Scientific Aspects of Marine Environmental Protection
GH	growth hormone
GIS	geographic information systems
GM	genetically modified
GMAO	genetically modified aquatic organism
GMO	genetically modified organism
GST	goods and services tax (Canada)
HACCP	Hazard Analysis and Critical Control Point
HNS	hazardous and noxious substances
HSNO	*Hazardous Substances and New Organisms Act* (New Zealand)
HST	harmonized sales tax (Canada: Atlantic provinces)
IAAC	income-averaging annuity contracts
ICA	Interdepartmental Committee on Aquaculture (Canada)
ICES	International Council for the Exploration of the Sea
IISD	International Institute for Sustainable Development

IM	integrated management
IOPFC	*International Convention on the Establishment of an International Fund for Compensation for Oil Pollution Damage*
ISA	infectious salmon anemia
ITOPF	International Tanker Owners Pollution Federation
ITQ	individual transferable quota
JSA	Joint Subcommittee on Aquaculture (United States)
KAAC	Kimberley Aquaculture Aboriginal Corporation
LIFDC	low-income food deficit country
LMO	living modified organism
LOMA	large ocean management area
LRC	Leasing Referral Committee (Prince Edward Island)
LUCO	Land Use Coordination Office (British Columbia)
LWBC	Land and Water British Columbia Inc.
MAC	management advisory committee
MAF	Ministry of Agriculture and Fisheries (New Zealand)
MAFF	Ministry of Agriculture, Food and Fisheries (British Columbia)
MFA68	*Marine Farming Act 1968* (New Zealand)
MFA71	*Marine Farming Act 1971* (New Zealand)
MFDP	Marine Finfish Development Plan (Nova Scotia)
MFN	most favored nation
MIP	minimum import price
MOE	Minister of Environment
MOU	memorandum of understanding
MPA	marine protected area
mt	metric ton
NAA	*National Aquaculture Act* (United States)
NAC	National Aquaculture Council (Australia)
NACA	Network of Aquaculture Centres in Asia-Pacific
NADC	National Aquaculture Development Committee (Australia)
NAS	National Academies of Science (United States)
NASCO	North Atlantic Salmon Conservation Organization
NEPA	*National Environmental Policy Act* (United States)
NHK	Norwegian Federation of Fish and Aquaculture Industries
NISA	Net Income Stabilization Account (Canada)
NL	Newfoundland and Labrador
NLDFA	Newfoundland and Labrador Department of Fisheries and Aquaculture
nm	nautical mile
NMFS	National Marine Fisheries Service (United States)
NOAA	National Oceanic and Atmospheric Administration (United States)
NOK	Norwegian krone
nprPPM	non-product-related processing and production method

NRTA	Natural Resources Transfer Agreement
NS	Nova Scotia
NSDAF	Nova Scotia Department of Agriculture and Fisheries
NSEA	*Nova Scotia Environment Act*
NSN	New Substances Notification
NWPA	*Navigable Waters Protection Act* (Canada)
OCAD	Office of the Commissioner for Aquaculture Development (Canada)
OCS	Offshore Constitutional Settlement (Australia)
OCSLA	*Outer Continental Shelf Lands Act* (United States)
ODAS	ocean data acquisition systems
OECD	Organisation for Economic Co-operation and Development
OMAF	Ontario Ministry of Agriculture and Food
OMRN	Ocean Management Research Network (Canada)
OOA	Office of Offshore Aquaculture (United States; proposed)
OTA	Office of Technology Assessment (United States)
OVPIC	Office of Veterinary and Phytosanitary Inspection and Control (European Union)
PBA	*Planning and Building Act* (Norway)
PCB	polychlorinated biphenyl
PEI	Prince Edward Island
PFRCC	Pacific Fisheries Resource Conservation Council
PMSEIC	Prime Minister's Science, Engineering and Innovation Council (Australia)
PRC	People's Republic of China
PRT	project review team
QMS	quota management system
RA	responsible authority
RADAC	Regional Aquaculture Development Advisory Committee (Nova Scotia)
RHA	*Rivers and Harbors Act* (United States)
RMA	*Resource Management Act* (New Zealand)
ROFA	*Rock Oyster Farming Act* (New Zealand)
SAP	site allocation policy
SAR	*Salmon Aquaculture Review*
SARA	*Species at Risk Act* (Canada)
SCFA	Standing Committee on Fisheries and Aquaculture (Australia)
SEA	strategic environmental assessment
SLA	*Submerged Lands Act* (United States)
SMEs	small and medium-sized enterprises
SOLAS	*International Convention for the Safety of Life at Sea*
SPS	Sanitary and Phytosanitary Measures
SSHRC	Social Sciences and Humanities Research Council of Canada

SSOPF	Ship-Source Oil Pollution Fund
TACC	total allowable commercial catch
TBT	Technical Barriers to Trade
UNCED	United Nations Conference on Environment and Development
UNCLOS	United Nations Convention on the Law of the Sea
UNEP	United Nations Environment Programme
UNESCO	United Nations Educational, Scientific and Cultural Organization
USDA	United States Department of Agriculture
USPIRG	United States Public Interest Research Group
VEC	valued ecosystem component
WCVI	West Coast Vancouver Island
WDE	Washington Department of Ecology
WGA	Working Group on Aquaculture (Australia)
WPCHB	Washington Pollution Control Hearing Board
WTO	World Trade Organization

Introduction

Aquaculture law and policy: struggling in the wake of the blue revolution

David L. VanderZwaag

Rapid growth of aquaculture developments around the globe has been likened to a blue revolution.[1] Aquaculture contributions of fish, crustaceans and mollusks have grown from 3.9 percent of total world production by weight in 1970 to 29.9 percent in 2002.[2] Aquaculture has grown at an average rate of 8.9 percent per year since 1970, compared to 1.2 percent for capture fisheries.[3] Aquaculture is predicted to play an increasingly important role in meeting food fish demands in light of world population growth, rising per capita incomes, urbanization trends[4] and the depletion of many wild fish stocks.[5]

In the wake of aquaculture expansions, a host of environmental, social, economic, health and ethical issues are raised. For example, marine finfish farming carries numerous potential environmental effects, including eutrophication, sedimentation and stimulation of harmful algal blooms from nutrient and organic matter enrichment caused by uneaten food pellets and feces.[6] Spread of diseases and parasites from farmed fish to wild stocks[7] and potential adverse effects of escaped fish on wild fish through competition and interbreeding are further concerns.[8] The environmental impacts of chemicals, such as drugs used to treat sea lice and antibiotics targeting infectious diseases, are largely unknown.[9]

Social conflicts are also prevalent. Traditional fishers and boaters may be unhappy at the prospect of losing access to marine spaces. Coastal residents and tourism operators may be upset over aesthetic interferences. Who in society should be given priority for aquaculture site access, particularly in offshore marine areas where common property regimes have dominated, is another contested question.[10] Lower prices for farmed fish may also give rise to animosity from fishers faced with going out of business because of the aquacultural market glut.[11]

Economic questions, having social and cultural dimensions, also abound. How far should governments limit corporate control and ownership of aquaculture operations? How should taxation of aquaculture profits be handled, and what level of fees should be imposed for leasing and licensing privileges?

Complex health issues surround aquaculture products. Elevated levels of organic contaminants, such as dioxins and PCBs in farmed salmon, originating from contaminated feeds, have spawned both advisories against consuming large amounts of farmed fish[12] and advocacies regarding the overall positive health benefits of consuming farmed salmon.[13] The potential human health impacts of using antibiotics in aquaculture have received little study.[14]

Ethical dimensions also loom over aquaculture developments. Persons having ecocentric viewpoints may be distraught over the very notion of "farming" the wild seas, and in particular the morality of taking fish from the ocean food chain to supply fishmeal for intensive aquaculture operations.[15] The prospect of genetic modification of fish to maximize aquaculture profits – for example, from faster-growing fish – perhaps most starkly raises the ethical policy choices.[16] Whether to limit aquaculture to native species is a further question.[17] Fish welfare and avoidance of cruelty also lurk as issues.[18]

The governance of aquaculture, especially in the marine and coastal context, has largely lagged behind the pace of development pressures. Many countries, not having specifically tailored aquaculture legislation, have struggled to control aquaculture access and operations through dated and marginally relevant legislation such as old fisheries acts, navigable waters protection laws and general environmental protection statutes.[19] Environmental impact assessment processes are often not applied to established aquaculture operations and proposed shellfish projects[20] even though they may carry possibly significant environmental consequences.[21] Effective regulation has also foundered as a result of limited governance capacity and lack of political will in light of the short-term benefits offered by rapid aquaculture development.[22] Regional agreements and arrangements have tended to focus on transboundary fisheries and pollution challenges[23] and have generally neglected transboundary threats of aquaculture such as transborder disease transfers and the impacts of escaped fish.[24]

Further developing and strengthening national aquaculture governance is likely to involve substantial social and political struggles. With no global treaty specific to aquaculture, countries have wide latitude to establish their own environmental and health standards,[25] the setting of which is almost certain to spark interest and value conflicts among multiple actors interacting to reflect their viewpoints.[26] Working out the appropriate mix of the three main modes of governance[27] – hierarchical governance, self-governance and co-governance – and sorting out the roles of science, public opinion and bureaucratic expertise in decision-making are likely to be other areas of tension.[28]

To assist countries with law and policy assessments and reforms in support of ecologically and socially sustainable aquaculture,[29] various principles have emerged from international agreements,[30] declarations[31] and codes.[32] Those principles include, among others, the precautionary approach, public participation, integration, intergenerational and intragenerational equity, the ecosystem approach and the "polluter pays" principle.[33]

Perhaps one of the greatest struggles in coming decades will be over how to translate principled coastal/ocean governance into practice.[34] Many of the principles are still subject to considerable interpretive controversies,[35] and governance implications and details have yet to be fully fleshed out.[36] Multiple meanings of principles[37] and the interrelationship between/among principles are added challenges.[38]

The need for ensuring legal frameworks supportive of sustainable aquaculture has been widely recognized. For example, the Food and Agricultural Organization (FAO) Code of Conduct for Responsible Fisheries urges states to establish and maintain an appropriate legal and administrative framework that facilitates the development of responsible aquaculture.[39] The Bangkok Strategy for Aquaculture Development, adopted at an international conference in February 2000 involving some 540 participants from sixty-six countries and more than 200 governmental and non-governmental organizations, urges countries to develop "comprehensive and enforceable laws, regulations and administrative procedures that encourage sustainable aquaculture" and to clarify "legal frameworks and policy objectives regarding access and user rights for farmers."[40]

This book highlights the numerous law and policy issues that must be addressed in the search for effective regulation of aquaculture. Those issues include, among others, the equitable assignment of property rights; the design of effective dispute resolution mechanisms; clarification of what maritime laws apply to aquaculture (for example, are injured aquaculture workers covered by special maritime compensation legislation or traditional workers' compensation legislation?); adoption of a proper taxation system for aquaculture; resolution of aboriginal offshore title and rights claims; recognition of international trade law restrictions, such as labeling limitations and food safety requirements; and determination of whether genetically modified fish should be allowed and, if so, under what controls.

This volume explores principled aquaculture law and policy approaches and challenges through a five-part format. Part I includes two chapters providing global overviews. Chapter 1, by William Howarth, discusses aquaculture regulatory challenges and diversities. Chapter 2, by Nathanael Hishamunda, highlights the two categories of legal priorities, addressing the "supply side" of aquaculture, for example, through secure access rights and clean water protections – and the "demand side" – for example, through aquatic food safety standards (Nathanael Hishamunda, Chapter 2).

The five chapters in Part II describe Canadian experiences and challenges in implementing sustainability principles in relation to aquaculture. Chapter 3, by David VanderZwaag, Gloria Chao and Mark Covan, gauges how adequately the federal government, British Columbia and the four Atlantic provinces (New Brunswick, Nova Scotia, Prince Edward Island and Newfoundland and Labrador) have implemented four key principles: integration, precaution, environmental impact assessment and public participation (including social equity). The chapter concludes by suggesting ways forward for ensuring more principled decision-making in Canadian coastal/ocean governance, including the need to strengthen public participation in environmental impact assessment of aquaculture proposals and to fully incorporate sustainability principles into provincial aquaculture laws, particularly the presently ignored principle of social equity. In Chapter 4, Phillip Saunders and Richard Finn summarize Canada's limitations in applying a principled approach to property rights in aquaculture and highlight a critical flaw in ensuring secure access to marine aquaculture sites, namely the lack of federal legislation legitimizing interferences with public rights to fish. Chapter 5, by Moira McConnell, explores the nature of aquaculture siting conflicts and, drawing from Norwegian and Canadian experiences, concludes that a strategic sea-use planning process is necessary for aquaculture to achieve its full potential. In Chapter 6, Aldo Chircop examines the largely unexplored relationship between mariculture activities and maritime law, also referred to as admiralty law. The legal status of maritime installations, mariculture workers and safety standards applicable to offshore work sites is reviewed. Chapter 7, by Faye Woodman, discusses the appropriateness of the agricultural model of taxation to aquaculture operations and looks at non-income taxes, including sales taxes and property taxes, and the appropriateness of aquaculture leaseholder fees and license fees.

Part III covers the largely unaddressed issues of how indigenous rights in the offshore may interact with aquaculture access and operations. In Chapter 8, Diana Ginn focuses on whether the doctrine of aboriginal title, developed by the courts in relation to land, could also apply to submerged marine areas, and she speculates on how the concept of aboriginal title might be reconciled with existing common law rights, such as navigation and fishing rights, and international rights such as the right of innocent passage for shipping. In Chapter 9, Ronalda Murphy, Richard Devlin and Tamara Lorincz review Canadian jurisprudence on the duty to consult with aboriginal communities and the potential impacts of the duty on proposed aquaculture developments. Chapter 10, by Douglas Sanders, surveys international developments relating to indigenous rights to lands, fisheries, the foreshore and the offshore and briefly reviews the still incomplete attempts of Australia, Canada and New Zealand to resolve indigenous rights that complicate the granting of secure tenures.

Part IV highlights some of the international trade law and policy dimensions in aquaculture. In Chapter 11, Ted McDorman and Torsten Ström examine how far countries may utilize trade measures to protect the health and safety of citizens from imported aquaculture products without violating international trade rules. Chapter 12, by John Phyne, Richard Apostle and Gestur Hovgaard, tracks the evolution of European Union food safety policies and emphasizes the potential consequences for salmon exports from coastal communities, using case studies from Norway and the Faroe Islands. Douglas Moodie, in Chapter 13, summarizes the international legal framework for addressing genetically modified aquatic organisms and reviews the adequacy of Canadian regulatory approaches.

Part V provides a comparative perspective on how three developed countries have approached aquaculture governance. The legal frameworks applicable to mariculture in the United States (Jeremy Firestone, Chapter 14), Australia (Marcus Haward, Chapter 15) and New Zealand (Hamish Rennie, Chapter 16) are reviewed and critiqued.

Through global and national comparative perspectives, with an emphasis on Canadian approaches and challenges in moving towards principled aquaculture governance for coastal and offshore areas, this monograph is aimed at partially filling the relative vacuum in aquaculture law studies. Existing books with an aquaculture law and policy focus are either quite dated[41] or largely focused on a single country.[42]

The overall reality that emerges from the multiple chapters that follow is that Canada and other countries are still struggling to modernize law and policies in the wake of fast-paced aquaculture developments.[43] Major challenges include, among others, the need to overcome a "free-market" mentality to aquaculture leasing and licensing;[44] to place aquaculture developments within the context of integrated coastal/ocean management processes;[45] to develop clear regulatory standards for aquaculture operations to ensure environmental and social health;[46] and to forge political, bureaucratic and industry awareness and willingness to change existing laws in light of sustainability principles.[47]

One of the greatest challenges may be to reach societal consensus on what appropriate aquaculture policies should be[48] – for example, whether transgenic fish should be allowed to be produced and marketed[49] and whether the farming in open cages or pens of non-native carnivorous fishes, requiring high feed inputs, releasing wastes and carrying disease, parasite and escape threats, should be encouraged.[50] Lessening ecosystem impacts may occur through better management practices for non-enclosed systems,[51] but the transition to enclosed and recirculating systems seems likely to hasten in light of growing demands to follow precautionary and ecosystem approaches and to minimize impacts of every kind.[52]

Building a boat while navigating rough seas is an image portrayed in the concluding chapter of this volume by Arthur Hanson. That picture amply captures the law and policy struggles facing many countries in the wake of the blue revolution.

Notes

1 "Special Report: Fish Farming: The Promise of a Blue Revolution," (2003) 368 (8336) *The Economist* 20.
2 FAO Fisheries Department, *The State of World Fisheries and Aquaculture* (Rome: Food and Agriculture Organization of the United Nations, 2004) at 14.
3 *Ibid.*
4 Cécile Brugère and Neil Ridler, "Global Aquaculture Outlook in the Next Decades: An Analysis of National Aquaculture Production Forecasts to 2030," FAO Fisheries Circular 1001 (Rome: FAO, 2004) at 8.

5 It is estimated that in 2003 about half (52 percent) of the fish stocks monitored were fully exploited while approximately one-quarter were overexploited, depleted or recovering from depletion. FAO, *supra* note 2 at 32.

6 See B. T. Hargrove, "Far-Field Environmental Effects of Marine Finfish Aquaculture," in *A Scientific Review of the Potential Environmental Effects of Aquaculture in Aquatic Ecosystems*, Vol. I (2003) Canadian Technical Report on Fisheries and Aquatic Sciences 2450 [Scientific Review Vol. I] and D. J. Wildish, M. Dowd, T. F. Sutherland and C. D. Levings, "Near-Field Organic Enrichment from Marine Finfish Aquaculture," in *A Scientific Review of the Potential Environmental Effects of Aquaculture in Aquatic Ecosystems*, Vol. III (2004) Canadian Technical Report on Fisheries and Aquatic Sciences 2450 [Scientific Review Vol. III].

7 See Martin Kikosck, Mark A. Lewis and John P. Volpe, "Transmission Dynamics of Parasitic Sea Lice from Farm to Wild Salmon," (2005) 272 *Proceedings of the Royal Society B: Biological Sciences* 689. See also, C. E. Nash, P. R. Burbridge and J. K. Volkman, eds., *Guidelines for Ecological Risk Assessment of Marine Fish Aquaculture*, U.S. Dept. of Commerce, NOAA Tech. Memo NMFS–NWFSC–71 (2005) at Appendix D.

8 See Rosamond Naylor, Kjetil Hindar, Ian A. Fleming, Rebecca Goldburg, Susan Williams, John Volpe, Fred Whoriskey, Josh Eagle, Dennis Kelso and Marc Mangel, "Fugitive Salmon: Assessing the Risks of Escaped Fish from Net-Pen Aquaculture," (2005) 55(5) *BioScience* 427; and ICES Mariculture Committee, "Report of the Working Group on Environmental Interactions of Mariculture (WGEIM), 5–9 April 2004, Galway, Ireland" at Annex 6.

9 See, for example, L. E. Burridge, "Chemical Use in Marine Finfish Aquaculture in Canada: A Review of Current Practices and Possible Environmental Effects," in Scientific Review Vol. I, *supra* note 6.

10 For a discussion of the difficult problems surrounding "common pool" resources, see Jan Willem Van Der Schans, "Governing Aquaculture: Dynamics and Diversity in Introducing Salmon Aquaculture Farming in Scotland," in Jan Kooiman, Martijn van Vliet and Svein Jentoft, eds., *Creative Governance: Opportunities for Fisheries in Europe* (Aldershot, UK: Ashgate Publishing, 1999) at 102–103.

11 Michael L. Weber, *What Price Farmed Fish: A Review of the Environmental and Social Costs of Farming Carnivorous Fish* (Providence, RI: Sea Webb Aquaculture Clearinghouse, 2003) at 11–12.

12 See Ronald A. Hites, Jeffrey A. Foran, David O. Carpenter, M. Coreen Hamilton, Barbara A. Knuth and Steven J. Schwager, "Global Assessment of Organic Contaminants in Farmed Salmon," (2004) 303 (5655) *Science* 226.

13 See Health Canada, "Food Safety and PCBs Found in Fish" (12 January 2004). Online. Available http://www.hc-sc.gc.ca/ahc-asc/media/nr-cp/2004/2004_pcb-bpc_e.html (accessed 23 September 2005).

14 Weber, *supra* note 11 at 28.

15 While fishmeal use in aquaculture during 1994 was estimated to be 17 percent of the world's production, by 2000 aquaculture consumed about 35 percent of the world's fishmeal. Christopher L. Delgado, Nikolas Wada, Mark. W. Rosegrant, Siet Meijer and Mahfuzuddin Ahmed, *Fish to 2020: Supply and Demand in Changing Global Markets* (Washington, DC: International Food Policy Research Institute, 2003) at 83.

16 See Douglas J. R. Moodie, "The Cautious 'Frankenfish': Environmental Protection and other Canadian Regulatory Issues Relating to Transgenic Fish," (2004) 1 *Macquarie Journal of International and Comparative Environmental Law* 49.

17 See FAO Committee on Fisheries, "Report of the Second Session of the Sub-Committee on Aquaculture, Trondheim, Norway, 7–11 August 2003," *FAO Fisheries Report* No. 716 (Rome: FAO, 2003) at 11.

18 See Roger S. V. Pullin and U. Rashid Sumaila, "Aquaculture," in Jan Kooiman, Maarten Bavinck, Svein Jentoft and Roger Pullin, eds., *Fish for Life: Interactive Governance for Fisheries* (Amsterdam: Amsterdam University Press, 2005) at 103.

19 For a good review of aquaculture regulatory approaches in European and Scandinavian countries, see the special issues of the *Journal of Applied Ichthyology* (2000, Vol. 16).

20 Secretariat of the Convention on Biological Diversity, "Solutions for Sustainable Mariculture: Avoiding the Adverse Effect of Mariculture on Biological Diversity," *CBD Technical Series* No. 12 (2004) at 38.

21 See, for example, S. M. Bower and S. E. McGladdery, "Disease Interactions between Wild and Cultured Shellfish," *Scientific Review of the Potential Environmental Effect of Aquaculture in Aquatic Ecosystems*, Vol. II, Canadian Technical Report on Fisheries and Aquatic Sciences 2450; and P. Crawford, M. Dowd, J. Grant, B. Hargrave and S. McGladdery, "Ecosystems Level Effects of Marine Bivalve Aquaculture," in Scientific Review Vol. I, *supra* note 6.

22 Delgado *et al.*, *supra* note 15 at 78.

23 For a review of regional agreements/arrangements addressing marine pollution, see Elizabeth A. Kirk, "Protection of the Marine Environment from Land-Based Activities," (2003) 14 *Yearbook of International Environmental Law* 287.

24 One of the exceptions is the North Atlantic Salmon Conservation Organization (NASCO), which has developed Protocols for the Introduction and Transfer of Salmonids and various guidelines including Guidelines on Containment of Farm Salmon and Guidelines for Action on Transgenic Salmonids. See Resolution of the Parties to the Convention for the Conservation of Salmon in the North Atlantic Ocean to Minimise Impacts from Aquaculture, Introductions and Transfers, and Transgenics in Wild Salmon Stocks (the Williamsburg Resolution, adopted at the Twentieth Annual Meeting of NASCO in June 2003 and amended at the Twenty-first Meeting of NASCO in June 2004) CNL (04) 54.

25 For example, the Sub-Committee on Aquaculture of the FAO Committee on Fisheries has noted that import and export standards for aquaculture products vary between countries and regions and has urged the FAO to assist in the harmonization of health and safety standards through the Codex Alimentarius Commission process. FAO Committee on Fisheries, *supra* note 17 at para. 47.

26 See generally, Jan Kooiman *et al. supra* note 18.

27 See Jan Kooiman and Maarten Bavinck, "The Governance Perspective," in Jan Kooiman *et al.*, *supra* note 18 at 21–22.

28 For a discussion of the loss of confidence in science in the postmodern era and the difficult task facing decision-makers of defining and weighing competing or conflicting social interests, see Nicholas de Sadeleer, *Environmental Principles: From Political Slogans to Legal Rules* (Oxford: Oxford University Press, 2002) at 247–251.

29 On the need for aquaculture to be ecologically and socially responsible and to embrace appropriate technologies and support coastal community well-being, see Barry A. Costa-Pierce, "Ecology as the Paradigm for the Future of Aquaculture," in Barry A. Costa-Pierce, ed., *Ecological Aquaculture: The Evolution of the Blue Revolution* (Oxford: Blackwell Science, 2002) at 364–365.

30 For example, almost every post-1980s international environmental agreement, including the 1992 *Convention on Biological Diversity* and the *UN Framework Convention on Climate Change*, articulate a version of the precautionary principle/approach. See Carolyn Raffensperger and Joel A. Tickner, eds., *Protecting Public Health and the Environment: Implementing the Precautionary Principle* (Washington, DC: Island Press, 1999) at Appendix B.

31 For example, the Rio Declaration on Environment and Development, 14 June 1992, 31 *I.L.M.* 874 (1992), sets out twenty-seven principles to guide states

towards achieving sustainable development. For a review of the relevance of the principles in the ocean context, see Jon M. Van Dyke, "The Rio Principles and Our Responsibilities of Ocean Stewardship," (1996) 31 *Ocean and Coastal Management* 1.

32 For example, the FAO Code of Conduct for Responsible Fisheries, meant to apply both to capture fisheries and to aquaculture, sets out various general principles in Art. 6, including the precautionary approach, public participation and notions of the ecosystem approach, and adds principles specific to aquaculture in Art. 9, such as the need to establish effective environmental impact assessment procedures specific to aquaculture. Food and Agriculture Organization of the United Nations, *Code of Conduct for Responsible Fisheries* (Rome: FAO, 1995). The Global Aquaculture Alliance (GAA) has adopted Guiding Principles for Responsible Aquaculture and various Codes of Practice for Responsible Shrimp Farming in order to help implement the principles. The guiding principles are available online. Available http://www.gaalliance.org/prin.html (accessed 23 September 2005).

33 For discussions regarding the normative importance of principles in directing and spurring public policy shifts, see de Sadeleer, *supra* note 28 at 249–251; Philippe Sands, "International Law in the Field of Sustainable Development: Emerging Legal Principles," in Winfried Lang, ed., *Sustainable Development and International Law* (London: Graham & Trotman/Martinus Nijhoff, 1995); and Alhaji B. M. Marang, "From Rio to Johannesburg: Reflections on the Role of International Legal Norms in Sustainable Development," (2003) 16 *Georgetown International Environmental Law Review* 21.

34 For a recent review of how the principles of ecosystem health, social justice, livelihood and employment, and food security and safety are still filtering into governance practices, see Maarten Bavinck and Ratana Chuenpagdee, "Current Principles," in Jan Kooiman *et al.*, *supra* note 18 at 245–263.

35 One of the most contested principles is the precautionary principle, where debates continue not only over terminology (principle versus approach) but also over how strong precautionary measures should be and whether scientific risk assessment should be required to trigger precautionary actions. Whether the legal burden of proof should be placed upon development proponents is another contentious question. See David L. VanderZwaag, Susanna D. Fuller and Ransom A. Myers, "Canada and the Precautionary Principle/Approach in Ocean and Coastal Management: Wading and Wandering in Tricky Currents," (2002–2003) 34 *Ottawa Law Review* 117.

36 This seems especially true of the principle of ecosystem-based management (often used interchangeably with the term "ecosystem approach"), where a "fragmented array" of international articulations have occurred and key questions remain open, such as how healthy ecosystems are to be defined and what should be ecosystem conservation objectives and indicators. See Donald R. Rothwell and David L. VanderZwaag, "The Sea Change towards Principled Oceans Governance," in Donald R. Rothwell and David L. VanderZwaag, *Towards Principled Oceans Governance: Australian and Canadian Approaches and Challenges* (London: Routledge, 2006). For some of the "articulations," see FAO Fisheries Department, "The Ecosystem Approach to Fisheries," *FAO Technical Guidelines for Responsible Fisheries* No. 4, Suppl. 2 (Rome: FAO, 2003); Decision VII/11 "Ecosystem Approach," Conference of the Parties to the Convention on Biological Diversity. Online. Available http://www.biodiv.org/decisions/default. asp?lg=0&m=cop-07&d=11 (accessed 23 September 2005); and K. L. McLeod, L. Lubchenco, S. R. Palumbi and A. A. Rosenburg, "Scientific Consensus Statement on Marine Ecosystem-Based Management" (Communication

Partnership for Science and the Sea 2005). Online. Available http://compasson-line.org/?q=EBM (accessed 23 September 2005).

37 For example, the principle of integration has various shades of meaning includ-ing, among others, "integrated culturing," where various species, such as fish, shellfish and seaweeds, are raised together in order to reduce wastes and maxi-mize production and "integrated coastal zone management." See M. Troell, C. Halling, A. Neori, T. Chopin, A. H. Buschmann, N. Kautsky and C. Yarish, "Integrated Mariculture: Asking the Right Questions," (2003) 226 *Aquaculture* 69; A. Neori, T. Chopin, M. Troell, A. H. Buschmann, G. P. Kraemer, C. Halling, M. Shpigel and C. Yarish, "Integrated Aquaculture: Rationale, Evolu-tion and State of the Art Emphasizing Seaweed Biofiltration in Modern Aqua-culture," (2004) 231 *Aquaculture* 361; and S. M. Stead, G. Burnell and P. Goulletquer, "Aquaculture and Its Role in Integrated Coastal Zone Manage-ment," (2002) 10 *Aquaculture International* 447.

38 For a discussion of potential synergies among principles, see David Van-derZwaag, "The Precautionary Principle and Marine Environmental Protection: Slippery Shores, Rough Seas, and Rising Normative Tides," (2002) 33 *Ocean Development and International Law* 165 at 174–175.

39 Code of Conduct, *supra* note 32, Art. 9.1.1.

40 Network of Aquaculture Centres in Asia-Pacific (NACA) and FAO, *Aquaculture Development beyond 2000: The Bangkok Declaration and Strategy, Conference on Aquaculture in the Third Millennium, 20–25 February 2000, Bangkok, Thailand* (Bangkok: NACA, and Rome: FAO, 2000).

41 See Thomas E. Kane, *Aquaculture and the Law* (Miami: University of Miami Sea Grant Program, 1970); and Bruce H. Wildsmith, *Aquaculture: The Legal Frame-work* (Toronto: Emond-Montgomery, 1982).

42 See William Howarth, *The Law of Aquaculture: The Law Relating to the Farming of Fish and Shellfish in Britain* (Oxford: Fishing News Books, 1990); and Henry D. McCoy II, *American and International Aquaculture Law: A Comprehensive Legal Treatise and Handbook Covering Aquaculture Law, Business and Finance of Fishes, Shellfish, and Aquatic Plants* (Peterstown, WV: Supranational Publishing Company, 2000) (focusing mainly on the United States).

43 For a review of needed legislative improvements in Madagascar, Malawi, Mozambique, Nigeria and Zambia, see R. D. Percy and N. Hishamunda, *Promo-tion of Sustainable Commercial Aquaculture in Sub-Saharan Africa*, vol. 3: *Legal, Reg-ulatory and Institutional Framework*, FAO Fisheries Technical Paper No. 408/3 (Rome: FAO, 2001).

44 For example, as discussed in Chapter 3 by VanderZwaag, Chao and Covan, Canada has not sought to control corporate ownership for aquaculture sites and has not through existing aquaculture laws ensured priority access to coastal communities and residents.

45 See FAO, Fishery Department Planning Service, Fisheries Department, *Integra-tion of Fisheries into Coastal Area Management*, FAO Technical Guidelines for Responsible Fisheries No. 3 (Rome: FAO, 1996).

46 The general lack of regulatory standards based upon indicators of environmen-tally sound aquaculture has been noted and the example of Japan given, where some regulatory standards have been set, such as a limitation on the quantity of sulfide in the mud under fish cages and the requirement that benthos, like lug-worms, be present in the mud. See Secretariat of the Convention on Biological Diversity, *supra* note 20 at 40.

47 An example of the challenge can be seen in Canada, where the Canadian govern-ment has not shown a willingness to pass federal aquaculture legislation even though such legislation seems essential in light of provincial offshore jurisdic-

tional limitations and the need for federal legislation to authorize interferences with public rights to fish. See Chapters 3 (VanderZwaag, Chao and Covan) and 4 (Saunders) in this volume.

48 A component of the challenge may be determining what approach or combination of approaches should be followed to develop policies: for example, developing an overall aquaculture policy through public consultation, applying strategic environmental assessment to a proposed aquaculture policy and holding a referendum on a proposed policy.

49 See Chapter 13 (Moodie) in this volume.

50 The great challenge is that many of these species currently farmed also have the greatest market value, and substantial investments in equipment have already been made. See Jane Lubchenco, "The Blue Revolution: A Global Ecological Perspective," (2003) 34(4) *World Aquaculture* 8 at 10.

51 Better management practices may include, for example, better siting to ensure proper water circulation, using polyculture, improving utilization of feeds and reformulating feeds. Secretariat of the Convention on Biological Diversity, *supra* note 20 at 26.

52 For advocacy for such a transition, see Secretariat of the Convention on Biological Diversity, *supra* note 20 at 26.

Part I

Aquaculture law and policy at the millennium

Global trends and challenges

1 Global challenges in the regulation of aquaculture

William Howarth

Introduction: food from land

Perhaps the greatest and most enduring challenge for humanity is that of feeding itself. Since the dawn of time, human beings have struggled to find a balance between the extremes of surplus and shortage found in nature. That struggle continues today, with an alarmingly large proportion of the world still suffering food shortages.

Insofar as security of food supplies is the greatest challenge for humanity, *agriculture* might be seen as its greatest achievement. On the timescale of humanity, we have progressed relatively recently from being hunters and gatherers of our food, dependent entirely upon the bounties and vagaries of nature, to forming societies that know how to feed themselves by growing the crops and keeping the livestock that provide the bulk of our food supplies. Security of food supplies has provided the bedrock upon which all subsequent technological and other development has been made possible.

The triumph of agriculture, however, deserves a moment's reflection. How did this come about? Certainly by thousands of years of trial and error by individuals and communities, which long predated modern "scientific" knowledge about the most productive methods of cultivating plants and rearing animals. The techniques of agriculture are an ancient heritage of hard-learned lessons from persons who faced starvation because of their mistakes and cherished their successes for good reason.

Equally, agriculture is a lesson in the way that human beings must organize their activities for mutual benefit. There could have been no incentive for people to plant crops, without the expectation that after months of painstaking work tending them, they would be assured the end product at the time of harvesting. Agriculture is possible only on the basis of a system of land tenure that allows its participants a sufficient degree of ownership over the products of their labor. Where there is a surplus of suitable land this may not be problematic, but in almost all societies of which we have knowledge, the need for defined land rights is a precursor to almost any kind of productive agricultural activity. The achievements of agriculture rest upon the ability to define and protect property interests as much as they have depended upon the development of practical farming techniques. Without the former, the latter would have been futile.

Having noted these fundamental points about the dependence of agriculture upon well-defined property rights, this chapter will contrast the rather embryonic state of legal regimes for aquaculture, where tensions between public and private rights have not yet been satisfactorily resolved. Focusing on the role of regulation in realizing sustainable development, the chapter draws on examples from Scotland that may have broader relevance to the international community. Within the United Kingdom, Scotland has experienced the most precocious and extensive aquaculture development, particularly in the farming of salmon. Many of the environmental and regulatory issues that have been identified have direct counterparts in other jurisdictions that have engaged in similar kinds of development. Hence, some general lessons are to be learned about the need for regulatory integration, precaution and community-based management. The purpose of the chapter is to outline the main environmental concerns and to indicate *possible* regulatory responses, alongside issues of implementation and enforcement, in the broader context of securing a more sustainable aquaculture industry.

Food from waters

Turning from the production of food from land to the production of food from waters, the primitiveness of our present situation is apparent. Technological advancement has preceded, rather than followed, organizational development. Sufficient fishery know-how has been acquired to denude the world's oceans of almost every fish that it is commercially worth catching. What has only recently come to be generally appreciated is that the first come, first served approach to fishery resources may not be genuinely conducive to the long-term interests of humanity. Moreover, to the extent that it is not conducive to those interests, humanity is only in the relatively early stages of establishing a system of rights that is appropriate for protecting the continuing productivity of the sea against the ravages of excessive exploitation. In short, common resources provided by nature need to be subjected to legal regimes that allocate rights of exploitation to ensure that sustainable yields are not exceeded and that those yields are equitably shared for the general benefit.

The warning was issued in 1968 with the publication of Garrett Hardin's famous essay "The Tragedy of the Commons," in which he envisaged that "the inherent logic of the commons remorselessly generates tragedy" by maximizing individual gain to the collective detriment so that "freedom in a commons brings all to ruin."[1] The validity of this thesis is hotly disputed, but in the case of fisheries, common property in natural resources has enabled a ruinous decline in global fisheries to take place. It would be good to think that humanity has the good sense to address this by mutual self-regulation, but many would take the view that the response has been too little too late.

Although it is encouraging to see recent international cooperation on fisheries, and particularly the need to address conservation of migratory stocks of fish which move between waters within the jurisdiction of different states,[2] much more needs to be done. International fisheries regulation is on the cusp of moving from pure hunter-gathering into a new era of managed international natural resource regulation. The fact remains that most terrestrial agriculture passed through this point of development several thousand years ago.

The birth of aquaculture

Where does aquaculture feature on this scale of development? Certainly it has undergone massive scientific and technological advancement in an astonishingly short space of time. Although awareness that containment of fish within ponds was a means of securing supplies in times of shortage can be traced back to ancient times,[3] the capacity to breed and culture fish, and more significantly to enhance natural productivity, is a modern discovery. Within a few decades, developed countries have surmounted the problems of rearing fish in marine waters and have transformed salmon from a luxury product to a relatively cheap source of food, at least in developed countries. The cultivation of shellfish has developed unrecognizably in the space of a few years and, again, products have become available with a security that would not have been previously imagined. The prospects for the future look to be even more encouraging, with the increasing sophistication of culture techniques and installations, and the prospect of farming an increasing range of species.

The remarkable success story evidenced by the recent development of aquaculture provides convincing evidence that this is the route for future development. The shortcomings of capture fisheries and the progressive decline in catches, despite increased fishing effort, are capable of being perceived as the last throes of an industry that is in the process of being gradually superseded. The vision for the future is of seas that will be "farmed," with progressive reductions in reliance upon purely natural fish stocks. This prospect is likely to be rather too futuristic for some, at the present at least, but if the analogy with agriculture is a fair one, it may not be entirely fanciful.

The recent achievements in aquaculture have been nothing short of miraculous. Compared to the thousands of years that farming took to evolve to its present state, the farming of fish and shellfish has developed "overnight." However, while the achievement should not be diminished, neither should it be overlooked that it has come at a cost. The dazzling technical achievement of aquaculture has been followed by a more reflective period of assessing the impacts and costs, and perhaps a skeptical feeling in the minds of some spectators that the achievement is simply too good to be true.

The skeptical observer does not need to look too hard for a downside. Aquaculture inevitably involves an intrusion into the natural environment that leads into a catalogue of environmental quality and ecological adversities. Any perception that aquaculture involves "something for nothing" may be swiftly dismissed. The evidence substantiating a range of *actual* environmental, economic and social adversities is considerable, and the evidence for a range of other *potential* harms is at least plausible, if presently lacking in substantiation by research. The second-wave issues are about how the balance of benefits and costs is most appropriately drawn.

The role of regulation

Lawyers always have to tread carefully when wading into this kind of exercise. Many of the issues are most centrally placed within the remit of environmental scientists, who must make expert determinations as to the extent and gravity of environmental impacts. Insofar as the benefits and impacts of aquaculture development can be properly assessed, they provide boundaries that need to be reflected in regulation. The task of the lawyer is that of trying to reflect the balance of costs and benefits in regulation that facilitates the development of the industry while providing the safeguards that are needed against unacceptable environmental and other kinds of intrusion. The knee-jerk reaction to any kind of adversity, that "there ought to be a law against it," should be avoided. Even where regulation provides an appropriate response to a problem, it must be appreciated that laws are capable of being formulated in an infinite variety of ways. The test of a good law is how well it reflects the balance of interests involved in defining the boundary between the permissible and the impermissible.

Clearly, there is conviction that the free-for-all in respect of capture fishery resources that has prevailed until fairly recent times should not serve as a model for a free-for-all in respect of aquaculture development. The present focus is upon regulation that facilitates the new industry, to allow it to provide the substantial benefits to humanity that are within its potential, but ensures that its environmental and social impacts are brought within acceptable bounds.

Sustainable development

Inevitably, since the Rio Conference on Environment and Development in 1992 it is necessary to look to the regulatory touchstone provided by the imperative of "sustainable development." The Brundtland Report of the World Commission on Environment and Development defines it as "development which meets the needs of the present without compromising the ability of future generations to meet their own needs."[4]

The critics of, and skeptics towards, sustainable development have drawn attention to the open-ended and obscure nature of the international imperative, and their observations are difficult to refute. Nonetheless, in the absence of anything better, we are stuck with the concept of sustainable development as a guide to the way in which we should regulate all kinds of developmental activity to ensure that they take place at an acceptable environmental cost. What is useful about the concept is that it recognizes that development has a cost, as well as a benefit, and that this cost should be minimized and not allowed to exceed the value of environmental goods that are lost as a result of development.

Elaboration of sustainable development

The concept of sustainable development, however, should always be seen as a starting point rather than a point of destination. The concept needs to be reinterpreted by each nation in the specific context of its developmental needs. Within nations, the reinterpretation needs to be taken a stage further in identifying what the concept means for each sector of activity, including fishery and aquaculture development.

Some important international work has been done in reinterpreting sustainable development in relation to aquaculture activities. The Food and Agriculture Organization's *Code of Conduct for Responsible Fisheries*[5] (FAO Code) is a valuable, though largely voluntary, global statement of principles addressed to governments, organizations and individuals engaged in fishery activities, which seeks to encourage adherence to principles and standards conducive to sustainability in the conservation, management and development of fisheries.

Article 9 of the FAO Code, concerned specifically with aquaculture development, encourages states to promote sustainability in aquaculture by the use of strategic planning to take account of, and minimize, adverse environmental and social impacts and to adopt a cooperative approach towards transboundary impacts. At a national level, this involves conservation of genetic diversity and integrity of ecosystems by encouragement of appropriate practices and procedures. At the operational level, promotion of responsible aquaculture should involve encouragement, and, in some respects, the regulation of chemical inputs that are hazardous to human health and damaging to the state of the environment. Not least important are the need to prevent aquaculture practices that are actually harmful to the aquaculture industry itself, such as practices that give rise to disease transmission, and the need to ensure food safety in aquaculture products. The potentially catastrophic implications of failure to ensure protection against these matters will be self-evident to all who are involved.

Further elaboration of the practical implications of the general principles set out in the FAO Code is provided in FAO's Technical Guidelines for Responsible Fisheries.[6] In addition, further international guidance on sustainability in aquaculture is provided by the Holmenkollen Guidelines for Sustainable Aquaculture,[7] which are addressed to governmental authorities and those actively involved in aquaculture, and identify key issues needing to be addressed in securing greater sustainability. Also, the Bangkok Declaration and Strategy for Aquaculture Development beyond 2000[8] offers insightful observations as to key policies that states should pursue to enhance sustainability in aquaculture.

However, it is in the nature of international guidelines that they must be couched as generally formulated approaches to the environmental, ecological and social problems, approaches that may be misread as indicating all-purpose solutions for a diversity of national and local situations and activities. This diversity in practice is clearly a practical strength of aquaculture, providing widely different responses to meet different local needs. Hence, internationally established principles and proposals must always have their limitations insofar as they are drafted remotely from the actual circumstances and activities that need to be influenced or regulated. The international guidance always needs to be reinterpreted in the light of local conditions and in the context of the particular activities that are being undertaken.

Developmental relativity

Arguably, the valuable work undertaken by international organizations might be seen as posing, rather than answering, the pertinent questions that individual states must address in organizing their relevant activities. The point needs to be stressed that no two nations are at an identical state of development and the environmental balance sheet needs to be the subject of national introspection. As a generality, sustainable development implies that more developed countries should place a greater weight upon environmental goods than less developed countries, but how this balance sheet should be drawn up is a matter for each state to determine.

I was fortunate to be engaged to work with the legal staff of the Food and Agriculture Organization a couple of years ago on a project looking at the regulation of shrimp farming in developing countries.[9] The team of us who were involved in the project sought information about the state of regulation of aquaculture in some twenty different countries in Asia, East Africa and South America to identify some pointers towards best practice in the sector and to offer some suggestions as to regulatory and administrative measures that might be seen as desirable in developing countries.

The shrimp legislation project was revealing, showing the spectrum of approaches adopted by a range of countries that, within limits, might be seen as being at comparable states of development. Each of them had a pressing need for the economic benefits provided by shrimp farming, but also had legislation reflecting varying levels of concern as to the extent to which economic benefits should be allowed to damage their natural environments. It was encouraging to see that some states had taken account of the need for sustainable development in formulating policies and legislative measures that sought to facilitate shrimp farming development, but in a way that minimized environmental and ecological harm. Even where seriously limited resources were available to particular states, it was seen to be appropriate to devote these resources to mitigating environmental impacts.

In developed countries, such as Canada or the United Kingdom, environmental expectations should be all the higher. Generally, sustainable development envisages that strong economies are expected to take the lead on matters of environmental performance. Adverse environmental and ecological impacts, which might be justified by extreme developmental needs in developing countries, should not be acceptable in developed countries, where the threat of poverty and starvation cannot be provided as a justification for environmental damage. Hence, developed countries should be aiming for the highest environmental standards.

The environmental checklist

The "checklist" of possible environment impacts that may arise from aquaculture and need to be addressed is not too difficult to ascertain by examining the issues covered by the international initiatives. Key general issues include environmentally or ecologically destructive land use; unacceptable impacts on water resources, either through abstraction or through the release of contaminated discharges; damaging ecological impacts upon natural resources through the excessive removal of specimens from the wild or the release of farmed species into the wild; the transmission of disease to farmed stocks and natural ecosystems by the unrestrained movement of farmed stocks; and a range of intrusions upon water and land uses which conflict with legitimate interests of others.[10]

Again the international literature needs to be read with caution. Although useful as a checklist of the problems that *may* need to be addressed, it needs to be equally appreciated that aquaculture activity is remarkably diverse. Small-scale, low-intensity fish farming operations in developing countries are capable of having almost negligible environmental or ecological impacts, and need to be regulated accordingly. Where economic circumstances preclude the use of pesticides and chemicals in aquaculture, or the importation of non-native species, it may not be necessary to devise elaborate systems of regulatory control to address a problem that does not exist in practice.

However, the intensive and large-scale approach to aquaculture taken in developed countries is likely to place it in a different regulatory category. The onus is upon each state to ascertain to what extent the *potential* environmental problems associated with aquaculture are *actually* evidenced as consequences of the kinds of operations that take place within its jurisdiction. Large-scale sea cage installations, perhaps farming non-native species and subjecting stocks to sophisticated programs of medication, give rise to a wide range of potential environmental and ecological risks. Clearly, the extent to which *risks* are capable of being converted into *harms* requires extensive investigation and research to be undertaken. Where research of this kind has not been undertaken, legislation may be justified on preventive or precautionary grounds.

The regulatory response

The central issue is the role of the law in seeking to provide for a balance between aquaculture development and environmental protection. On this, a personal conviction has to be expressed that the best kind of regulation is *self-regulation*. A belief that something should be done because it is *right* to do it is always more effective than a mandatory legal prohibition backed by a sanction. In the field of aquaculture, the paths of self-regulation and self-interest follow close parallels. It is not in the interests of fish farmers to contaminate the waters upon which their livelihood depends, to allow their stock to become tainted by pesticides or other chemicals, or to alienate their consumers by engaging in anything less than environmentally exemplary performance. In all the main respects, commercial good sense and environmental good sense run along the same lines.

It is evident that the interests of the aquaculture industry as a whole may not be the same as the interests of particular fish farmers who must compete for commercial advantage against their fellow producers. The incentive to secure greater profitability is capable of motivating individual fish farmers to "cut corners" at the expense of the environment, but the obvious first-resort mechanism to address this is through measures imposed by the industry upon its members.

The role of producer groups in securing cohesion within the aquaculture industry is fundamental. Uniformly high standards may be facilitated through self-education and through product and practice standards, perhaps by the establishment of codes of practice that determine what activities all participants should adhere to.[11] The incentive for adherence to those standards by particular fish farmers is that their product will be seen to have the certifiably high quality stipulated by the industry as a whole. The counterpart of this is that producer groups should have the power to exclude from their membership any producer who is not willing to adhere to the standards set by the industry for itself. Given recent experiences of "food scares" in diverse sectors, the desirability of any food producers' group taking measures to safeguard the quality of its products seems justifiable. It takes only one fish farmer misusing chemicals in treating stock to blight the industry as a whole, and the power to dissociate the non-compliant farmer from the quality assurance standards promulgated by the industry as a whole seems justifiable.

A good illustration of the benefits that may be secured by cohesion within the aquaculture industry is to be seen in the "Scottish Quality Salmon" quality assurance scheme. This scheme requires members to be certified as complying with both product standards and environmental management schemes that are independently accredited to international standards, such as ISO 14001.[12] Hence, the benefits to producers of a recognized status being attached to a product entail a commitment to adhere to high environmental standards throughout the production process. In addition, the Scottish aquaculture industry is in the process of developing a Code of Best Practice (covering disease control, welfare, and health and safety), and within that a Code of Best Environmental Practice. It was intended that every licensed aquaculture operator would formally subscribe to these codes by 2004, and compliance is to be externally monitored and audited.[13]

Alongside action taken at an industry-wide level, significant benefits can be achieved by local cooperation between individual fish farmers. An example of such cooperation in Scotland is the increasing coordination of restocking, medication and fallowing operations. It is recognized that breaking potential disease cycles, by minimizing the risk of cross-infection and preventing "souring" of the seabed beneath cages requires areas to be left fallow for periods. However, for this to be effective, close liaison between neighboring fish farms is needed. The indications are that voluntary management agreements have been successfully reached between neighboring fish farmers to reduce the possibilities of adverse interaction. These inter-company agreements have produced mutual benefits to operators and have reduced environmental impacts.[14]

Education and the formulation of standards or codes of practice should involve close collaboration between producers' groups and government. A code of practice promulgated by government has an additional status in that it can be used as a prerequisite for other kinds of action. For example, an undertaking that a prospective fish farmer will adhere to the terms of the code might be required as a condition for having a license for a particular operation or as a prerequisite for some kind of grant or fiscal advantage being made available to a particular fish farmer. Hence, it is possible to interlink a code with other kinds of benefit to enhance its importance.

Self-regulation by producers' groups or in cooperation with government, though important, is not without its limitations. The issue remains as to what legal response is needed at the point where self-regulation runs out, or perhaps where legislation is seen as an important means of bolstering support by self-regulation. When it comes to regulation, properly so called, it is not a simple matter of prohibiting, and providing sanctions for, each activity that is perceived to be problematic and not adequately addressable by self-regulation. In each instance, the task is that of identifying a legal response that is appropriate and proportionate as the most likely mechanism to secure the desired policy objective. The diversity of possible regulatory approaches deserves careful consideration.

Regulatory diversity

From the outset, it must be appreciated that law is a remarkably flexible control mechanism. In each instance, a wide range of regulatory approaches is possible, and the determination of which of these is appropriate requires a range of factors to be weighed in the balance. Legislation may be used comprehensively to prohibit an activity; to restrict the manner in which the activity is undertaken; to license, permit or authorize the activity to enable it to be undertaken in limited circumstances (perhaps subject to locally formulated conditions); to require the registration of an installation where an activity is conducted; or positively to compel (rather than prohibit) an activity to be undertaken. Moreover, legal foundations are frequently needed for information-based measures that facilitate research, monitoring, investigation, and educational measures that allow for the financial or technical support in relation to an activity.

This wide spectrum of potential legal responses needs to be stressed because of the misplaced tendency to suppose that the use of the law means, necessarily and exclusively, the banning and punishment of unacceptable practices. In fact, it facilitates a diverse range of possible requirements, backed by a variety of sanctions and incentives, and important issues arise in weighing up the advantages and disadvantages of the different legal responses that are possible.

In light of the potential flexibility of the legal response, the task for national authorities is that of relating the gravity of environmental problems generated by aquaculture to the appropriate level of legal coercion that is needed to address them. Broadly, it may be supposed that the most damaging environmental practices need to be made subject to the greatest degree of legal stringency, assuming that a legal response is found to be necessary to address the problem in the first place. At the other extreme, a range of issues may be better addressed by non-mandatory mechanisms. Education and training are arguably as valuable as law in directing fish farmers towards good practice, and economic incentives that are made conditional upon good environmental performance may provide a greater incentive for avoiding environmental damage than the threat of legal proceedings and punishment.

Initial controls

Categorized according to the point of impact, legal controls upon aquaculture may be subdivided into "initial" and "continuing" forms of control. Initial controls are concerned with the procedures that determine whether an aquaculture installation may be established at a particular location, whereas continuing controls are those measures that apply to subsequent activities that take place at the installation following its establishment. For good reasons, the questions, first, where a fish farm should be located, and second, how it should conduct its operations, are best approached sequentially.[15]

The issue of rights to land covered by water is of considerable legal complexity, particularly where marine waters are at issue. Insofar as the seabed is not the subject of private ownership, it will be necessary to acquire a lease or similar authorization from the public body with responsibility for seabed operations. "Development licensing" is capable of taking diverse legal forms, but essentially these all involve a system of public land-use control that seeks to subsume particular lease determinations to broader strategic policies for development of areas or sectors of activity.[16]

The mechanisms for regulating developments of the seabed have recently been reconsidered in Scotland. Presently, it is necessary for a prospective fish farmer to obtain both development consent and a license from the Crown Estate Commissioners, who administer the Crown Estate, including the seabed.[17] In the future, it is envisaged that responsibility for development control will pass to local authorities, which will also have responsibility for environmental impact assessment where this is required. The involvement of the Crown Estate Commissioners will continue, in that a lease for the site will be needed from the Commissioners. However, the move represents a shifting of control away from a body that, in the past, has been seen to be too closely associated with the commercial aspects of authorizing a development, insofar as the Commissioners would be entitled to rent under any lease that was granted. Under interim arrangements, local planning authorities will advise the Commissioners on fish farm applications.[18]

It is for government, or perhaps regional or local licensing authorities, to determine strategic policies for land use, but this needs to be done as a result of open deliberation and widespread consultation about the issues to be taken into account. Aquaculture frequently has special requirements for suitable marine areas, perhaps because of the need to ensure sufficient water flows but to avoid excessive exposure under storm conditions. The difficulty is often that the areas that are sought by prospective fish farmers are also most valued for traditional fishing, recreation or conservation, or need to be kept free of obstructions for the purposes of navigation. Insofar as it is possible to do so, an integrated development planning exercise will be needed for marine areas to identify those areas that are most suitable for the different kinds of activity and minimize the conflicts that are likely to arise between activities that compete for water space.

An illustration of a strategic approach to aquaculture planning is to be seen in the Scottish Executive Environment and Rural Affairs Department's recently revised *Locational Guidelines for the Authorisation of Marine Fish Farms in Scottish Waters*.[19] The purpose of these Guidelines is to provide guidance on the factors to be taken into account in considering proposals for fish farms and to encourage local authorities to prepare non-statutory marine fish farming framework plans, as a guide to the location of future fish farms. Drawing upon a range of recent ecological information, the Guidelines reaffirm a presumption against fish farm developments on the north and east coasts of Scotland. Elsewhere, three categories of area are envisaged on the basis of nutrient loading and benthic impact: first, where fish farms will only be acceptable in "exceptional circumstances"; second, where areas are at the limits of their carrying capacity, so that further development will result in these areas being placed in the first category; and third, where there is a "better prospect" of satisfying nutrient loading and benthic impact requirements. In summary, only those areas in the third category are likely to be acceptable for new fish farm development. Even then, for applications in category three areas it is stated that the circumstances will need to be carefully considered before permission will be given, and proposals are likely to succeed only where they involve a package of measures designed to reduce environmental impact. Such measures, for example, will include feeding techniques that minimize food wastage and integrated sea lice management practices that reduce the use of chemical treatments.

Although strategic zoning approaches of this kind are helpful in guiding development towards areas that are seen to be least problematic, actual determinations of whether a particular project should be allowed in a particular zone need to be decided against a background of specific information about the nature of the project and its foreseeable impacts. Although the developer must provide much of this information, the determination process should seek to maximize involvement in the decision-making process by encouraging representations to be made by relevant public bodies, organizations and members of the public. Where representations of this kind are made, it should be the responsibility of the decision-making body to take these into account in the final decision.

For example, in Scotland the consultation process surrounding an application for development consent for a fish farm is an important mechanism for securing the opinion of the statutory nature conservation and landscape protection body, Scottish Natural Heritage. Scottish Natural Heritage is well placed to advise on a range of implications for flora and fauna of fish farm applications. Notably in this respect, substantial areas of land are subject to conservation designation in accordance with national provisions or legislation adopted at European Community level, and it is important that aquaculture should be steered away from these areas. Another relevant aspect is the need to minimize adverse impacts on predator species of seabirds and mammals. In this respect, Scottish Natural Heritage is able to advise on avoidance of such conflicts, though it is also notable that the industry and conservation organizations have agreed to a Code of Practice on the Interaction of Fish Farming and Predatory Wildlife.[20]

In particular determinations as to whether an aquaculture installation should be allowed at a particular location, environmental impact assessment has proved to be an invaluable tool in many jurisdictions. Essentially, the process requires the prospective developer to provide information about the intended project and to anticipate the range of impacts that it will have upon the surrounding environmental media and ecosystems, and on other activities that are conducted in the vicinity. The final determination is made on the basis of this information, and should seek to incorporate mechanisms to ensure that the adverse impacts are minimized to an extent that is acceptable, or that the application is declined where sufficient mitigation is not possible. Hence, an eventual authorization is likely to be hedged around with a list of conditions that seek to minimize adverse environmental consequences.

In the United Kingdom, regulation is provided for under *Environmental Impact Assessment (Fish Farming in Marine Waters) Regulations 1999*[21] and associated guidance, which implements European Community requirements for environmental impact assessment.[22] The regulations apply where any part of the development is within a "sensitive area," where the development is designed to hold a biomass of 100 tonnes or greater, or where the proposed development will extend to 0.1 hectare or more. Nonetheless, there are concerns that environmental impact assessment is not functioning as well as it might in Scotland because of the relatively variable standard of environmental statements submitted in relation to fish farm applications and their failure to identify cumulative impacts of developments and discharges.[23]

Development licensing, in conjunction with environmental assessment, is, therefore, a very flexible legal tool, which may be applied with different degrees of rigor and sophistication depending upon national and local needs. In an extreme case, development licensing may allow the most comprehensive investigation of all environmental and social impacts to which a project may give rise. The licensing authority will have before it all the information that it needs to ascertain whether the project is consistent with the requirements of sustainable development. Accordingly, it may impose conditions of a preventive or precautionary kind where it feels that there are special risks that need to be secured against. Not least important is that development licensing procedures provide a public forum for considering the potential adverse environmental and social effects of a proposed aquaculture installation. These adversities must be assessed against the economic and developmental benefits that the project will be likely to secure. All parties with an interest in a aquaculture proposal have the opportunity to express their views, and the determining body should strive to take these views into account in the final decision.

Continuing controls

Development licensing and environmental assessment serve as invaluable mechanisms for evaluating projects before the commencement of an aquaculture operation. However, they have limitations in regulating the various subsequent activities that may take place at the site, and that need to be regulated by various kinds of continuing control. The kinds of continuing control that are needed will depend upon the nature of the installation and the location, but typically the following might be regulated by separate licensing requirements.

Water-use licensing will be required, for example, where a land-based operation is involved, to allow a sufficient quantity of water, of sufficient quality, to be drawn from available sources. Again, this kind of licensing is capable of being controversial in giving rise to conflicts with other water users who seek to use the same water for other purposes such as agriculture, industry or drinking water supply. License applications will need to be openly considered so that scarce water resources are seen to be fairly allocated according to competing needs.

Wastewater discharge licensing is necessary when it is recognized that fish farms have the capacity to produce large amounts of waste products of various kinds that are capable of having seriously detrimental effects upon environmental quality and surrounding ecosystems. Equally, the discharge of poor-quality effluent from an aquaculture installation may have a damaging impact upon other installations in the locality through transmission of contamination or disease, particularly where fish farms share a common watercourse or marine area. In relation to terrestrial fish ponds, with a discrete point of outflow, monitoring is relatively straightforward, but in relation to cage culture in open marine waters, monitoring of effluent, and setting license conditions, may be more problematic because it is not always clear what the "discharge" refers to.

In Scotland, the difficulties of setting discharge consents for operations that involve treatment of fish for veterinary purposes, and inevitably involve natural dispersion of the chemicals involved, have been recognized by the environmental regulatory authority, the Scottish Environment Protection Agency. The approach adopted in Scotland is to adopt environmental quality standards for the receiving waters, which are set at a level to ensure that concentrations of compounds are below that at which adverse ecological effects are detectable.[24] Similarly, the problem of waste accumulating as bottom sediments below fish cages, and consequent "site souring," has necessitated a modified approach towards the setting of fish farm consents.[25] Control of this problem requires a limitation of the biomass of fish cages to match the dispersive characteristics of the site, so that assimilative capacity is not exceeded. This process forms part of the assessment carried out by the Agency when determining discharge consent applications or reviewing existing consents.[26]

Chemical use licensing is another approach to the control of adverse impacts upon water quality. Various chemicals used in aquaculture for veterinary, pest control or disinfection purposes, may need to be specifically controlled to avoid contamination of surrounding waters, by more direct mechanisms than wastewater discharge licensing. Hence, for control of particularly hazardous kinds of chemicals, restrictions may usefully be specifically imposed upon those chemicals. However, even where a risk is identified, a spectrum of legal responses is possible depending upon the gravity of the risk. In an extreme case, it is possible to impose total prohibition upon the possession of a particular chemical for use in aquaculture by creating appropriate criminal offenses, extending perhaps to the importing, distribution or sale of the chemical for aquaculture purposes. For less hazardous chemicals, it is possible to enact less severe control measures by making use subject to licensing requirements, veterinary prescription requirements or the adherence to requirements relating application of the substance, such as the certification of users or a code of practice governing use.

Fish movement licensing may be an appropriate response to a range of concerns that arise from aquaculture operations. First, the collection of stock, particularly shellfish stock, from the wild may have a detrimental impact on natural stocks and ecosystems. Second, the escape of a farmed species into the wild may adversely impact upon local ecosystems through adverse genetic impacts upon native stocks or through invasive habitat competition. This may be seen as especially problematic where non-native species are involved and adverse ecological impacts are particularly difficult to assess. Finally, there is the significant threat of disease transmission, which is capable of having devastating economic consequences for the aquaculture industry as a whole. Equally important is the possibility of disease being transmitted from farmed to wild species.

For this range of reasons, control over fish movements is generally regarded as a legal essential for all but the most local, low-intensity aquaculture operations. Inevitably, controls must take the form of prohibitions or restrictions upon certain categories of fish movements, which will need to be made the subject of criminal offenses. General prohibitions or restrictions upon movement will, however, need to be subject to a facility for allowing particular movements to be individually licensed where, after veterinary and ecological investigation, sufficient safeguards can be provided against the potential hazards that have been noted.

Fish movement licensing is invaluable as a means for securing prior assessment of risks and mitigation measures. Equally, it is of general benefit to the aquaculture industry in that it enables all significant movements to be recorded and information communicated to the relevant authority as a license condition. The monitoring of movements is important because it allows diseased stock that has been the subject of recent movements to be traced to other locations that may be infected. This enables swift action to be taken to contain the spread of the disease by imposing quarantine or slaughter requirements upon infected stocks. Officials will clearly need to be empowered to take whatever action is appropriate in a particular case. However, without information from fish movement licensing, there may be formidable difficulties in preventing or reducing the major hazard that disease represents to the aquaculture industry.

Fish movement licensing also has an important role in relation to the problem of escapes from fish farms. In Scotland, a working group from the Scottish Executive was set up to review the situation. It was concluded that there would always be circumstances, such as extreme weather conditions, that would result in escapes, but that improvements could be made in containment. In particular, the adequacy of site-specific containment measures should be incorporated into the environmental assessment process, and contingency plans for escapes should be included in development applications. Assessments of the suitability of containment measures should include a review of likely weather extremes and proposed net maintenance regimes in the light of operations to be conducted at a site. Pursuant to this, the industry has formulated a Code of Practice on the Containment of Farmed Fish.[27] The Scottish Executive is also considering whether regulation is necessary to achieve minimum standards of cage design, equipment and maintenance, and to allow for prosecution for negligent acts or omissions resulting in escapes.[28]

Although the problem of escapes from fish farms has been recognized for some time, the extent of the problem has previously been rather speculative, insofar as there has been no formal mechanism to record such incidents. In respect of Scotland, this position has recently been addressed by the enactment of regulations to require fish farmers to provide information about escapes. The *Registration of Fish Farming and Shellfish Farming Businesses Amendment (Scotland) Order 2002*[29] amends existing requirements for the registration of fish farms to require the Scottish ministers to be notified of the circumstances of any escape of fish, the number of fish involved and the number recovered. For the future, this allows the extent of the problem of fish escapes to be quantified, and for information to be gathered about the *reasons* for escapes. Nevertheless, the assessment of the *impact* of escapes remains a somewhat speculative exercise that may not be entirely dependent upon the numbers involved. The need, formally, to declare escapes is, at least, a beneficial first step in evaluating the ecological problem.

Licensing of genetically modified organisms may seem a little futuristic to some, but the technical capacity for modification of fish species, perhaps to enhance growth rates or to improve disease resistance, is already on the horizon. Clearly, this is a matter of intense international controversy, raising profound environmental, ecological and developmental concerns, and the use of stock of this kind must be subject to the most careful scrutiny. Although fish movement licensing might be seen as an adequate mechanism for enabling thorough prior assessment of projects involving genetically modified fish, it may be thought more appropriate to deal with the issues under a separate system of controls that allows for more extensive and detailed scrutiny. Legislation that fully reflects the breadth of environmental and public health concerns would seem to be more appropriate than regulation within the sphere of aquaculture in isolation.

It may also be relevant to note that, for the future, the regulation of genetically modified organisms will need to be considered in the context of the *Cartagena Protocol on Biosafety to the Convention on Biological Diversity* (2000).[30] The Biosafety Protocol aims at protecting biodiversity and consumers from any adverse impacts that could arise from transboundary movements of living modified organism and products derived therefrom. The protocol has important implications for internationally traded products in requiring the exchange of information between exporting and importing countries. Again, the gravity of the international issues that arise may be a sound reason for adopting a specialized national regime for genetically modified organisms, rather than attempting to deal with them in sector-specific legislation.

Product quality licensing is also a regulatory essential insofar as fish products need to be regulated under public health legislation to avoid any potential hazard to the health of consumers. In times gone by, there may have been a tendency for food safety issues to be addressed at the point of consumer sale; increasingly, the trend is to follow the food chain back into the distribution, harvesting and production processes. The vital importance of ensuring safety of aquaculture products throughout the production chain hardly needs to be re-emphasized. A small number of contaminated products from a single producer are capable of having a devastating impact upon the industry as a whole.

To some extent, measures such as those directed at the misuse of chemicals in aquaculture may have important public health implications in preventing the presence of unacceptable residue levels in the final product. For the most part, however, public health is a technical and specialized area of regulation, one that extends well beyond fish products. Food safety issues need to be addressed within a code of national law that extends to food products of all kinds, so that common principles can be formulated, and consistent implementation and enforcement mechanisms applied. Not least important in this respect is the need for enforcement to be undertaken by an inspectorate with the specialized technical expertise needed to identify food safety issues in practice.

It needs also to be noted that food safety may give rise to specific concerns in relation to aquaculture insofar as products are intended for export. In particular, it may be necessary for national legislation to meet the requirements of the European Community Directive on Fish Products (91/493, as amended) "laying down the health conditions for the production and placing on the market of fishery products" or the requirements of the US Hazard Analysis and Critical Control Point (HACCP) system concerning procedures for the safe and sanitary processing and importing of fish and fishery products.[31] Depending upon the intended export market, these measures will be of critical importance in determining the national standards for food safety in relation to aquaculture products. This issue is illustrative of the growing trend towards internationalization of aquaculture legislation.

On licensing generally, the point is properly made that licensing systems are only as effective as the mechanisms that exist for inspection and checking that license conditions are being adhered to, and the capacity of the enforcement authority to take action where they are not. The issue of enforcement will be returned to later in this chapter.

Consolidated regulation and integrated strategy

From the perspective of a prospective fish farmer, regulation can seem an insuperable obstacle, and the range of possible approaches to licensing that has been provided should leave no doubt as to the potential complexities involved. The range of potential impacts of aquaculture leads into a range of distinct fields of regulation, each with its own peculiarities and intricacy. An important issue to be addressed is the extent to which laws governing aquaculture are capable of being consolidated into a single statute, or at least a small number of legal provisions. In an ideal world, it might be thought possible to bring all aquaculture operations within a comprehensive regime that allowed for a single license or authorization governing all kinds of impacts. Such an approach would avoid the need for multiple applications to different licensing authorities for slightly different purposes, with corresponding multiplication of the bureaucracy involved.

Although consolidated or codified legislative provision for aquaculture seems a generally desirable policy goal, the reality is that it would be tremendously difficult to achieve. Environmental quality issues, public health concerns and activities such as the cultivation of genetically modified species are not matters that are unique to aquaculture, and they need to be addressed consistently across different sectors of activity. For this reason, an "all-purpose aquaculture license" would be formidably difficult to achieve for anything but the most low-intensity small-scale project. Nonetheless, there should be scope for the administrative coordination of activities of regulatory authorities. This might be achieved by "one-stop" licensing, whereby the key authority to which an application is submitted has the responsibility of liaising with other regulatory bodies that are involved, to ensure that the full range of authorizations are secured. From a fish farmer's point of view, this approach has obvious attractions in minimizing administrative complexity and requiring regulators to sort out different kinds of regulatory requirements between themselves.

Whether or not regulatory or administrative consolidation can be achieved, it is essential that governments provide an overall strategy for aquaculture, which plots out the development of the industry and the mechanisms by which development is to be accomplished. Clearly, national strategies will be largely determined by national circumstances and perceptions of priority. However, an illustrative example of a national strategy is to be seen in the Scottish Executive's recently produced draft *Strategic Framework for Scottish Aquaculture*.[32] This document seeks to formulate the respective roles of government, the industry and other stakeholders in aquaculture development across a wide range of issues. The shared vision is that

> Scotland will have a sustainable, diverse competitive and economically viable aquaculture industry, of which its people can be justifiably proud. It will deliver high quality, healthy food to consumers at home and

abroad, and social and economic benefits to communities, particularly in rural and remote areas. It will operate responsibly, working within the carrying capacity of the environment, both locally and nationally and throughout its supply chain.

Revealingly, the draft strategy reports the contention of the industry that costs are higher than those of competitors in other countries because of inefficiencies of scale resulting from regulatory constraints.[33] The Scottish Executive is considering undertaking comparative research to assess the extent to which regulation actually does place Scottish aquaculture at a competitive disadvantage.

It is also notable in the UK context that strategy is a matter that will increasingly need to be addressed at a regional international level, and a strategic position on aquaculture is being developed by the European Community. The European Commission has recently published a communication document, *A Strategy for the Sustainable Development of European Aquaculture*,[34] which builds upon previous Community initiatives that have provided support for the sector. The document affirms the potential for aquaculture development but also notes problems in relation to health protection requirements, environmental impact and market instability. The fundamental issue is perceived to be the maintenance of competitiveness, productivity and durability for the sector, recognizing that further development must take a more integrated approach to farming technologies, socio-economics, natural resource use and governance. Specific legislation relating to aquaculture does not exist at the Community level, though many measures at Community level have significant implications for aquaculture. However, reliance upon national legislation has been alleged to lead to competition distortions, suggesting that this is an area in which the Community might seek to legislate comprehensively at some stage in the future.

Implementation and enforcement

Despite the general coincidence between sustainability and aquaculture that has been previously referred to, there have been regrettable examples of bad practice where aquaculture has damaged itself through pollution, inadequate disease control and practices that have harmed the food safety reputation of aquaculture products generally. Regulation is needed to outlaw the activities of the small number of participants who might wish to secure short-term gains at the expense of the reputation of the industry as a whole, both to protect the environment that it depends on and to secure the quality of its products.

In respect of those activities that are sufficiently serious to be subject to prohibition, the most vital consideration is that of enforceability. There is little point in enacting laws that cannot be enforced in practice, or enacting laws without careful prior consideration of the allocation of responsibilities for enforcement. On this, the relatively specialized nature of aquaculture legislation should be recognized. Aquaculture law is different from laws governing agriculture or fishery activities, or laws governing water quality generally. The counterpart of this is that enacting legislation implies a commitment to the establishment and support of specialized inspection and enforcement agencies that are necessary to ensure proper implementation.

The problem, all too frequently, is that legislators are inclined to make laws without giving sufficient thought to the resources and costs that are necessary for their implementation and enforcement. Arguably, the thinking process should be put in reverse, since there is little point in enacting regulation if, in practice, the means to ensure that it will be properly applied are lacking. Moreover, the cost of implementing and enforcing legislation should be put into perspective by comparison with the potential costs of failing to enforce legal requirements effectively, especially where this has the result of inflicting massive economic damage upon the industry itself through the failure to control disease or to achieve product control requirements because of contamination.

The highly specialized nature of aquaculture legislation should be emphasized. Controls upon fish movements, for example, are generally thought necessary to prevent the transmission of disease between fish farms and to prevent the introduction of non-native species that are likely to give rise to ecologically damaging habitat competition or loss of genetic integrity of native stocks. However, the practicalities involved in securing actual control of individual fish movements are not to be underestimated. Unavoidably, a fairly complex administrative structure is required to secure the registration of origins and destination of fish stocks in relation to all relevant fish movements. The enactment of regulation presupposes the existence, or provision, of fairly intricate administrative structures to ensure that the legal requirements operate effectively. Without such controls, legislation has little prospect of fulfilling its objectives.

Implementation of regulation needs to be considered *before* laws are enacted, rather than afterwards. The process of prior regulatory appraisal has rapidly gained ground in many jurisdictions, involving an explicit attempt by legislators to assess the continuing administrative and enforcement costs to which regulation will give rise and to confirm that these costs will actually be met. It is suggested that the prior regulatory appraisal process has much to commend it in the context of aquaculture legislation, given the particularly specialized nature of administration and enforcement requirements that are entailed.

Beyond regulatory impact appraisal, reflection is needed upon the status and powers of the agency or authority that is responsible for implementation. The importance of providing an appropriate level of resourcing of a regulatory agency has been noted, but another important consideration is the degree of independence with which the agency operates. Clearly. conflicts of interest between regulators and regulated bodies must be avoided, but also it might be seen as desirable to secure some degree of independence between regulatory authorities and government. The role of government is to formulate policies for the development and promotion of the aquaculture sector and to provide the resources necessary to realize those policy objectives. The role of a regulatory authority is to enforce laws effectively and impartially. For this reason, some degree of separation of powers between a government and an aquaculture regulatory authority might be seen as a desirable feature.

In the UK context, there are various mechanisms that do secure a degree of insulation between government and the practical regulation of fish farms. It is fair to note that some regulatory functions are exercised through government departments. In particular, the responsibilities for ensuring that works in tidal waters do not constitute a hazard to navigation[35] are allocated to transport ministries.[36] Similarly, responsibilities for measures concerning fish health[37] are allocated at the central government level.[38] These provisions concern registration of fish farms and provide for actions to be taken in relation to notified diseases.[39]

However, perhaps the most important regulatory powers in relation to fish farms in Scotland are allocated to the Scottish Environment Protection Agency. The Agency has the key power to determine applications for discharge consents for fish farms and may impose conditions to minimize adverse environmental impacts. Such consents are determined against a general duty to promote the cleanliness of waters and to have regard to the need to conserve flora and fauna. The significant point here is that the Agency has its powers and duties determined by statute, but actually has a fair degree of independence from government in determining any particular discharge consent application.

Similar observations could be made about the role of the nature conservation authority, Scottish Natural Heritage, which, again, has a reasonable degree of independence from government in making representations about the appropriateness of fish farm development applications in respect of nature conservation and landscape impacts. Also, the transfer of land-use planning powers to local authorities, which is imminent in Scotland in relation to fish farm applications, will give local authorities quite a degree of autonomy from central government in making individual determinations. The transfer of these powers to bodies that operate with some degree of independence from government has the consequence that any pro-development tendencies of government are capable of being moderated by bodies with specialized sectoral remits.[40]

Concluding observations

Successful aquaculture regulation involves reconciling the interests of aquaculture participants, the communities in which they are based, the consumers of their products and those who advocate vigilance over the environment in which aquaculture takes place. These interests are not necessarily coincident, but should be capable of reconciliation. The consensus is that aquaculture is an activity with such tremendous potential for human benefit that mechanisms have to be found to facilitate its development in a way that respects the competing concerns that have been referred to.

The function of regulation in the development of aquaculture should be neither over- nor understated. Various examples have been provided of information or education-based management mechanisms, which fulfill an important role in aquaculture development by tapping into the self-interest of the industry in ensuring the soundness of its practices and products. Where self-regulation serves as a mechanism for improvement, it should be fully utilized, with the full participation of all involved: producers, administrators and communities. Where self-regulation is not sufficient, or needs to be bolstered, regulatory approaches need to be adopted. However, appropriateness and proportionality are the key imperatives, and, from among the array of legislative mechanisms that have been outlined, flexibility and workability must be paramount concerns. Integration of regulation, precaution and responsiveness to community interests needs also to be carefully weighed in the balance in adopting legislation. The overall goal of sustainability necessitates the maintenance of incentives for the development of aquaculture, but also the drawing of a national and local balance to prevent unacceptable environmental consequences.

Notes

The author is pleased to acknowledge help with the preparation and revision of this paper from Annick Van Houtte of the Development Legal Service of the Food and Agriculture Organization of the United Nations and Professor David VanderZwaag of Dalhousie Law School.

1 Garrett Hardin, "The Tragedy of the Commons," (1968) 162 *Science* 1243–1248.
2 Generally, see Food and Agriculture Organization of the United Nations Legal Office, *Law and Sustainable Development since Rio* (Rome: FAO, 2002), particularly Chapter 7 on fisheries.
3 Generally, see R. Kirk, *A History of Marine Fish Culture in Europe and North America* (Farnham, UK: Fishing News Books, 1987).
4 World Commission on Environment and Development, *Our Common Future* (Oxford: Oxford University Press, 1987) at 43.
5 Food and Agriculture Organization of the United Nations, *Code of Conduct for Responsible Fisheries* (Rome: FAO, 1995).
6 Food and Agriculture Organization of the United Nations, *FAO Technical Guidelines for Responsible Fisheries: 5 Aquaculture Development* (Rome: FAO, 1997).
7 See N. Svennevig, H. Reinersten and M. New (eds), *Sustainable Aquaculture: Food for the Future?* (Rotterdam: A. A. Balkema, 1997), particularly at 343 onwards.
8 Network of Aquaculture Centres in Asia-Pacific and Food and Agriculture Organization of the United Nations, *Aquaculture development Beyond 2000: The Bangkok Declaration and Strategy* (Rome: NACA/FAO, 2000).

9 W. Howarth, R. Hernandez and A. Van Houtte, *Legislation Governing Shrimp Aquaculture: Legal Issues, National Experience and Options* (Rome: FAO, 2001). FAO Legal Paper Online No. 18. Online. http://www.fao.org/Legal/prs-ol/lpo18.pdf (accessed 25 March 2004).

10 Generally, see T. V. R. Pillay, *Aquaculture and the Environment* (Oxford: Fishing News Books, 1992).

11 Generally, see C. Hough, "Codes of Practice and Aquaculture," in R. Leroy Creswell and R. Flos, eds., *Perspectives on Responsible Aquaculture for the New Millennium* (Baton Rouge, LA: World Aquaculture Society, and Ostend: European Aquaculture Society, 2002) at 79.

12 Scottish Executive, Environment and Rural Development Department, *A Strategic Framework for Scottish Aquaculture (Draft)* (Edinburgh: Scottish Executive, 2002), para. 3.83. Generally, see European Aquaculture Society, *Certification, Labelling and Quality of European Aquaculture Produce*, Publication 163 (Brussels: EAS, 2002), and L. M. Laird, "Quality Approaches as Guarantees for Responsible Aquaculture," in Creswell and Flos, *supra* note 11. at 49.

13 *Strategic Framework, supra* note 12 at para. 3.93.

14 Scottish Executive, Environment and Rural Development Department, *Advice Note: Marine Fish Farming and the Environment* (Edinburgh: Scottish Executive, 2003) at para. 67 onwards.

15 Generally, see B. Wildsmith, *Aquaculture: The Legal Framework* (Toronto: Edmond-Montgomery, 1982); and W. Howarth, *The Law of Aquaculture* (Oxford: Fishing News Books, 1990).

16 Generally, see T. V. R. Pillay, *Planning of Aquaculture Development: An Introductory Guide* (Farnham, UK: Fishing News Books/FAO, 1977).

17 Under the *Crown Estate Act 1961*.

18 *Advice Note, supra* note 14 at paras. 2 and 11.

19 Scottish Executive, Environment and Rural Affairs Department, *Locational Guidelines for the Authorisation of Marine Fish Farms in Scottish Waters* (Edinburgh: Scottish Executive, 2003).

20 *Advice Note, supra* note 14 at para. 43 onwards.

21 SI 2002 no. 193, and see Scottish Executive, Environment and Rural Development Department, *What to Do in the Event of an Escape of Fish from a Fish Farm* (Edinburgh: Scottish Executive, 2002).

22 European Council Directive on the assessment of the effects of certain public and private projects on the environment (85/337/EEC) as now amended by Directive 97/11/EC.

23 *Strategic Framework, supra* note 12 at para. 3.71.

24 *Advice Note, supra* note 14 at paras. 39 and 40.

25 *Ibid.*

26 *Ibid.*

27 *Ibid.* at para. 61 onwards.

28 *Strategic Framework, supra* note 12 at para. 3.67.

29 SI 2002 No. 193.

30 *Cartagena Protocol on Biosafety to the Convention on Biological Diversity*, 29 January 2000, Online. Available http://www.biodiv.org/biosafety/protocol.asp (accessed 25 March 2004).

31 Generally, see E. Ruckes, "Aquaculture Products Quality and Marketing," in Creswell and Flos, *supra* note 11 at 63.

32 *Strategic Framework, supra note* 12. The final version of the strategy was published in March 2003. Online. Available http://www.scotland.gov.uk/library5/environment/sfsa-00.asp (accessed 24 March 2004).

33 *Ibid.* at para. 3.4.

34 European Communities, Commission, *A Strategy for the Sustainable Development of*

European Aquaculture, COM (2002) 511 Final (Brussels: European Commission, 2002).
35 Under the *Coast Protection Act 1949*.
36 In Scotland, to the Scottish Executive Development Department Transport Division.
37 Under the *Diseases of Fish Act 1937* and *1983*, and European Community fish health legislation.
38 In Scotland, to the Scottish Executive Environment and Rural Affairs Department.
39 *Advice Note, supra* note 14 at para. 18 onwards.
40 *Ibid.* at para. 25 onwards.

2 Global trends in aquaculture development

Nathanael Hishamunda

Introduction

Aquaculture is the fastest-growing sector of the animal food producing sectors. Between 1970 and 2000, the sector grew at an average annual rate of about 9 percent, compared to 1.4 percent for capture fisheries and 2.8 percent for terrestrial farmed meat production systems.[1] The United Nations Food and Agriculture Organization (FAO) estimates that aquaculture contributions of fish, crustaceans and mollusks have grown from 3.9 percent of total world production by wieght in 1970 to 29.9 percent in 2002.[2]

In many parts of the world, this development was rendered possible by a number of factors including, but not limited to:

- the increasing demand for aquaculture products as a result of the decrease or status quo of the supply from capture fisheries;
- the scientific breakthroughs in production technologies; and
- governments' will and determination to establish enabling policies and legal and regulatory frameworks.

Policy and legal coherence is an essential ingredient in aquaculture, as it acts as a compass to its development. When contemplating this enormously large topic, it is useful to address two main questions: "What is the global trend in the use of law and policy in aquaculture today?" and "What is the way forward?"

Global trends in aquaculture law and policy

The first question is, "What is the global trend in the use of law and policy in aquaculture today, and what are the main elements covered?" It is difficult to provide an exhaustive answer to this question. In an attempt to answer the question, we can, perhaps arbitrarily, classify aquaculture laws and policies into two categories:

- those dealing with the "supply side" of aquaculture; and
- those related to the "demand side" of the industry.

Supply-side legal aspects of aquaculture

Globally, aspects of law looking at the supply side of the industry tend to contain or examine the following three main elements: the secure right to land, the secure right to clean water, and permit and licensing systems. Each will be examined in turn.

First, let us examine the secure right to land. Most nations have come to understand the importance of the aquaculture farmer being able to secure a legal right to the lands on which the farm is located, be it through ownership, a lease or similar legal arrangements. What has been found important is to obtain a right to the necessary lands that is sufficiently secure to allow the farm to be financed, to flourish over an extended period and to enable other people to be excluded from the property.

Unfortunately, such is not always the case. In many countries, especially in the developing world, land rights are not clearly defined, which often leads to serious disputes. In other cases, the land acquisition process is usually long and fraudulent, which deters investors, thereby hampering the development of aquaculture.

The second element is the secure right to clean water. It is generally understood, especially in developed countries, that an aquaculture farmer must also obtain the right to an adequate supply of good-quality water and be able to protect it from the claims of other users. However, most economically disadvantaged countries lack modern legislation for the allocation of water resources, especially during dry seasons or other times of water shortage, and for the control of water pollution that can be enforced at the instance of the farmer.

The third element commonly encountered in the legislative texts of the supply side of aquaculture is the permit and licensing systems. A common feature of aquaculture regulation is the obligation to acquire permits or licenses to establish a farm. On the one hand, the governments' argument for requiring prospective farmers to have permits is that permits enable them to assess the environmental sustainability of aquaculture proposals and to impose conditions that require the farms to be operated in a sustainable manner. On the other hand, permits provide farmers with clear rights to run aquaculture facilities as long as they comply with the terms of the permits, the relevant environmental laws and any applicable codes of aquaculture practices. While permit requirements are often a rule in developed countries, it is only recently that governments in developing countries have introduced policies to require aquaculture farmers to have permits before establishing aquaculture farms. The reason seems to be the emergence of large-scale/commercial farms in these regions. Traditionally, aquaculture was limited to backyard, small-scale fish culture in ponds.

Where permits are required, the permit system generally deals with siting and environmental impact assessment (EIA), control of water quality and control of exotic aquatic species intended for aquaculture. However, most fail to regulate genetically modified organisms and the codes of practice in aquaculture. Codes of practice, often referred to as technical guidelines, are not normally legally enforceable. However, in order to ensure that proper standards for aquaculture operations are adhered to by farmers, the FAO supports inclusion of such codes of practice or technical guidelines in the legislation governing aquaculture so that applicable rules are available to all.

Demand-side legal aspects of aquaculture

On the "demand side," existing laws in many countries deal with aquaculture on one count: the exclusive right to the fish under cultivation. The fisheries laws of most countries consider fish not to belong to anybody until they are captured. In many countries, legislation establishes that operators of fish culture facilities have the exclusive right to the fish under cultivation. In many others, however, aquaculture legislation fails to be specific on this issue.

Supply-side policy aspects of aquaculture

Macro-level policy trends

In addition to legal arrangements, which are established to ensure an orderly development of aquaculture, there are policies whose role is to provide directions for this development. On the supply side, policies tend to follow two paths. The first path concerns the formulation of policies at the macro level. At this level, we can identify a number of common features.

First is the setting of aquaculture as a priority area for economic development. Faced with ever-increasing pressure on fisheries resources, rising demand for fish and fishery products, as well as the need for employment and income-generating activities for both rural and urban populations, many countries with aquaculture potential have made it a priority on their economic agenda. They have defined their goals, set targets and established guiding strategies to achieve them. As a result of this and other policy and non-policy factors, aquaculture development recovered from stagnation or took off in many parts of the world.

The second common policy is the "promotion of sustainable aquaculture development." To the FAO, sustainability of aquaculture has economic, social, environmental and even legal aspects. Economically, aquaculture operations must offer the prospects of competitive and stable profits over the long term. Socially, the farmed species and the farming methods must be acceptable and meet general, cultural, gender and social norms. In addition, the sector's benefits should accrue to a wide socio-economic spectrum as opposed to being retained exclusively by a small elite. Environmentally, sustainability requires intergenerational equity, which necessitates that the environment-derived potential well-being of future generations should at least be as high as that of the present generation. This implies that environmental assets need to be at least maintained over time. Finally, from the viewpoint of law and legal institutions, the conditions that enable aquaculture to occur must be continued over time if the industry is to flourish in the long run.

The third common policy is to establish a good administrative framework for aquaculture management. Where aquaculture has expanded or taken off in recent years, most countries have understood the importance of having a good administrative framework for aquaculture management that provides for the orderly development of the sector. What is apparent, however, is that, in many instances, aquaculture administration falls under the jurisdiction of more than one department. This situation has led to duplication, rivalry and waste.

Compounding the problem of overlapping administrative jurisdiction is the frequent lack of a legislative framework specific to aquaculture. This might be because aquaculture is still in its infancy in many countries and plays only a minor role in the countries' economies.

The fourth commonly used policy is the establishment of an aquaculture investment-friendly climate. This refers to various activities affecting the establishment of a successful aquaculture business. In many instances, government policies have proven to have a profound impact on investment in aquaculture. These policies occur in terms of:

• removal of institutional constraints;
• facilitating reasonable access to credit and, sometimes, providing loans; and
• providing financial incentives, extension and advisory services, and hatchery-produced stocking material.

These and other services are often provided by the government until such time as they are made available by the private sector.

The fifth commonly encountered policy is the supply of good supporting infrastructure and an emphasis on research, technological development and information dissemination. Aquaculture development in the past two decades has been strongly backed by research, technological development, education, training and extension in many countries. Recognizing science and technology as one of the most important drivers of productivity, several governments, including some in developing countries, have placed a high value on aquaculture technological development. Genetics and ecological pools of freshwater and marine species were identified and cataloged, new breeding and culture techniques were developed, production units of artificially formulated quality feed were encouraged, and fish health management was better understood and improved. Research aimed at improving the efficiency of production by reducing production costs made the difference between marginal and profitable investments. However, progress in these areas remains limited to certain countries and regions.

The sixth common policy consists of continuous adjustments in the structure of the aquaculture sector. Government policies and other interventions have greatly contributed to the rapid development of aquaculture. However, as the industry developed, it soon became apparent that there were serious inherent structural problems that needed to be addressed if the industry were to continue growing. In some areas, purely as a result of management practices, fish diseases broke out and spread rapidly, for example in the shrimp industry, which negatively affected the efficiency of the sector and depressed producers' incomes. In other cases, the use of antibiotics and other production inputs was abusive, leading to negative environmental externalities and threatening the image of, and markets for, farmed fish.

In these cases, the industry was self-defeating. Some governments reacted by creating leading agencies with responsibility for the implementation of public policies that were established with the aim of guiding a long-term development of the sector. Producers' organizations were also encouraged to look after farmers' interests and to ensure implementation of self-regulating management codes. The rationale was that because unsound management practices by a farmer could affect that farmer's and others' output, each farmer needed to have an incentive to produce responsibly.

Micro-level policy trends

The second path to discussing global trends in aquaculture policies on the supply side of the industry is the micro level. We have already touched on some of these trends when discussing the macro-level policies used at the farm level, and it is often difficult to draw a clear distinction between the two categories of policies. Nonetheless, government interventions at the farm level tend to fall into three policy directions: start-up, expansion and export promotion.

START-UP POLICIES

In many countries, governments may provide financial assistance to jump-start the industry. There are two arguments to support the provision of start-up funding in the early stages of aquaculture development. First, there is the inability of potential entrepreneurs in infant industries to afford initial investment through their own equity or to obtain private funding. Financial institutions are naturally prudent. Faced with biological, market and other uncertainties, they are generally unwilling to provide credit until risks are known. Without their own equity or bank loans, either the country's aquaculture potential will not be realized at all, or it will be realized partially and the development will be slow. The government must intervene with start-up funding.

The second argument is that, as an infant industry, aquaculture may need support until it reaches a stage at which costs are competitive. If industries learn by doing, costs will decline with experience, and, so the argument goes, such industries need government assistance in their early years. Start-up assistance may be in the form of costs to cover research. Assistance can also be provided in the form of tariff and non-tariff protection, or cash grants to farmers. An alternative is to issue bonds.

In many developing countries, access to initial investment capital is also often complicated by the lack of collateral accessible to farmers and by excessively high interest rates on loans. The lack of knowledge by farmers of the modalities of applying for loans and, on the part of lenders, the lack of, or limited information on, successful aquaculture enterprises in the country or in the region also limit access to capital. In these countries, government policies have encouraged group lending, village banks and solidarity groups that do not require collateral. The use of titled land and moveable property, which in many cases has called for legal and regulatory reforms, is encouraged. When affordable, government loan guarantees and subsidized interest rates are used.

EXPANSIONARY POLICIES

Once aquaculture has taken off, farmers often find it difficult to expand. In most cases, constraints to expansion consist of unavailability and/or high costs of essential inputs, namely feed, seeds and capital. The absence or limited availability of necessary inputs may prove a serious impediment to aquaculture development. Often it is more critical than cost. Government policies that are aimed at assisting farmers to overcome these issues target the increase in supply of the limited or the lacking input.

A policy that is frequently used to develop the feed industry is to encourage the establishment of large farms. By their size, large farms guarantee enough demand for profitable feed production and thus feed companies have an incentive to develop. An alternative policy is to encourage livestock companies to diversify into aquaculture and/or into fish feed production. Where feed can be imported, but at the cost of foreign exchange, import substitution policies are sometimes used.

As with feed production, the availability of stocking material (seed) generally faces the issues of quantity and quality. To ensure enough supply and to maintain high-quality seeds, governments induce a number of farmers to specialize in seed production and train them in modern hatchery techniques. Initially, several government stations are used for this purpose, but, increasingly, private operations replace government stations as a source of seeds.

The problem of capital input has been discussed earlier. To increase the accessibility of farmers to bank loans, one policy is to demonstrate to bankers that aquaculture can be financially viable. By demonstrating the actual profitability of aquaculture, the intention is to create awareness of the sector and encourage lending. The banks targeted are those with high loan portfolios in agriculture and those that are responsive to alternative agriculture.

A complementary policy is to have business plans evaluated for their technical merits by government officials. This reassures bankers who lack aquaculture expertise. If financial institutions remain reluctant to lend, government loan guarantees are used, where possible. If the high cost rather than availability of inputs is a principal constraint on aquaculture development, policies that lead to increased supply may also lessen this problem.

Another policy that has been commonly used or advocated is the provision of subsidies. The argument is that with the high costs of inputs, the industry may lack absolute and competitive advantage, and therefore fail to flourish. However, the high fiscal cost of subsidies, the investment disincentive they create for the private sector and the resulting distortions are forcing their curtailment worldwide.

EXPORT PROMOTION POLICIES

Increasingly, aquaculture is becoming an international business, targeting export markets. This trend may cause a dynamic evolution of market shares internationally. Though globally insignificant, policy assistance is selectively provided to producers to help them compete internationally. These policies consist of tax holidays and tax exemptions on imported production inputs, including equipment, machinery, broodstock, feed and fertilizers, exemptions from permit fees, as well as holidays and/or exemptions from sales and other local taxes.

Demand-side policy aspects of aquaculture

At the micro level, governments intervene on the demand side of aquaculture by kick-starting aquaculture through marketing policies. Some opt to establish market structures that permit hygienic handling and selling of farmed fish. Others hold taste tests at government functions, produce recipe booklets and organize cooking demonstrations on radio and television. Others provide transport and ice for the big buyers. The guiding rationale behind these initiatives is to increase demand by developing new markets and/or expanding existing ones through market promotion and development of new, value added products, thereby triggering more supply. However, as the industry develops, these government facilitation activities decline, with marketing and provision of transport infrastructures increasingly falling to the private sector.

Governments also intervene on the demand side of the industry through regulation of aquatic food safety and the regulation of aquaculture drugs and feed used in the production of fish. Governments may issue special regulations for the processing and handling of aquaculture products, specifically on detailed requirements for food safety and prevention of human health hazards in the processing and packaging of aquatic products. The aim of these measures is to control the business environment of the industry in the interest of both consumers and producers.

The way forward

The second question to consider when discussing global trends and challenges facing aquaculture law and policy is: "What is the way forward?" First, let me underline that in the FAO, we believe that capture fisheries will expand only slowly, if at all, during the coming decades. At the same time, we expect demand for fish to continue to grow globally as a result of population increase, economic growth and shifts in eating habits, among other factors. Because demand creates supply, we are convinced that, globally, aquaculture is set to grow in order to meet the expected demand for fish. In fact, the FAO's recent forecasts indicate that aquaculture output could reach about ninety million tons in 2030 compared to forty-five million tons in 2000.[3] This increase in output is expected to come from the introduction of aquaculture in countries with aquaculture potential that has not yet taken off, expansion of farmed areas where aquaculture already exists, and from the use of intensive technologies and other technological breakthroughs.

A lesson that could be learned from this is that unless an adequate legal and regulatory structure is put in place, there could be unregulated, uncontrolled aquaculture development in many parts of the world, which, in turn, could cause immeasurable harm to the environment and lead to the industry's self-destruction. Thus, the challenge is for each country with aquaculture potential that it wishes to develop to have such a framework. The challenge is also to have this structure for the benefit of the industry and society.

Of the many constraints to aquaculture development in most parts of the world, the legal and environmental restrictions to starting an aquaculture venture are among the most significant. In many instances, requirements to obtain permits are very rigid: public regulations are many and varied, and may demand a considerable investment in time and money before the proposal actually starts, which often deters investors. The challenge for policymakers is to establish simplified and coordinated permit procedures that expedite the processing of permit applications. The further challenge is to eliminate overlapping administrative jurisdictions.

In a world of globalization, economies must open doors to foreign markets. There are also movements of goods, including live aquatic animals, which are sometimes uncontrolled. Where they occur, uncontrolled movements of aquatic animals often produce unintended effects on the industry. In some places, the industry has nearly collapsed. The challenge is for each country with the desire to develop aquaculture to have a national integrated aquatic animal health program that, *inter alia*, provides routine fish health services to the industry, monitors the movement of aquatic animals, and monitors and responds to health emergencies in the aquaculture sector.

One of the most important questions that a farmer has to answer when making a decision to farm is what species to grow. Often the choice has to be made between indigenous and alien species. Today, technology offers farmers a third choice: the use of genetically modified aquatic organisms (GMAOs). While most countries have regulations covering the importation of live aquatic organisms and the introduction of exotic species, legal, regulatory and policy frameworks to guide the use of GMAOs by both producers and consumers seem limited and incomplete, as most focus on the economic interests of producers. Information available to the FAO reveals that, so far, no GMAOs are farmed, at least commercially. But there is no reason to believe that this trend will continue. Like introduced species, GMAOs may become a pest, damaging the environment and even the farming of other species.

In addition, consumers' response to genetically modified organisms, in general, has been unfriendly. Policy-makers would perhaps find it to be in their best interest and that of the industry and the public to join forces with all stakeholders, including industry representatives and consumer organizations, and agree on a protocol on the cultivation and consumption of GMAOs. Once agreed upon, the protocol should become an integral part of national aquaculture legislation worldwide.

Once legal and policy frameworks are in place, countries, at least developing countries, often face the issue of implementing the start or expansion of aquaculture activities. Faced with this issue, governments have turned to the use of economic and other incentives that can be seen, wrongly or rightly, as subsidies. The use of financial transfers and other instruments to enhance the competitive position of the fishing industry is a very controversial issue among World Trade Organization members. Because of the globally increasing presence of aquaculture products in international markets, strong arguments against the use of subsidies and other economic incentives in aquaculture can be expected in the near future. The salmon aquaculture industry has already been accused of dumping, which opponents relate to alleged subsidies.[4] The challenge for policy-makers, at this stage, will be to anticipate the growing international opposition to direct government transfers to producers and the use of other forms of instruments aimed at shielding them against foreign competition. Rather, they must propose policies and other arrangements that will regulate the use of these instruments internationally, while allowing aquaculture to grow harmoniously.

Policy-makers and regulators will also need to anticipate the increase in the use of tariff and non-tariff barriers by countries as a weapon to protect their domestic aquaculture industries against foreign competition. Policy-makers will face the challenge of suggesting countervailing measures, such as the development of internationally agreed-on guidelines for the elaboration of transparent and non-discriminatory certification procedures and the harmonization of aquaculture quality standards.

The implementation of good legal, regulatory and policy instruments requires a solid and sufficient pool of qualified personnel for all aspects of aquaculture. Unfortunately, this requirement remains one of the most serious stumbling blocks to aquaculture development in developing countries. Several policies, which include plans for vocational training and formal educational curricula in harmony with the expanding needs of the industry, have been suggested. In most cases, however, these policies remain difficult to implement in the context of countries' financial austerity. The challenge for the industry as a whole, including policy-makers and regulators, is to come up with a strategy for overcoming this pivotal issue. Perhaps emphasis could be given to further South–South cooperation and networking at subregional, regional and bilateral levels for information exchange and technology transfer in the aquaculture sector.

Conclusion

Aquaculture is poised to grow globally, and therefore needs adequate legal and policy frameworks to guide its development. But it is only when the appropriate legal, regulatory and policy instruments are implemented that the industry will achieve its full potential. We can then look forward to improved food security, more employment opportunities, higher incomes, better nutrition, and enhanced and diversified economies all over this planet.

Notes

1 Food and Agriculture Organization (FAO), *FAOSTAT Data, 2004. Nutritional Data*. Updated 24 August 2004 (Rome: FAO, 2004).
2 FAO Fisheries Department, *The State of World Fisheries and Aquaculture*, (Rome: FAO, 2004) at 14.
3 FAO, *The State of World Fisheries and Aquaculture (SOFIA) 2002* (Rome: FAO, 2004); FAO, FishStat Plus, *Aquaculture Production: Quantities 1950–2002*. Updated 30 April 2004 (Rome: FAO, 2004); and FAO, FishStat Plus, *Capture Production: Quantities 1950–2002*. Updated 30 April 2004 (Rome: FAO, 2004).
4 G. Gonzales, "Trade Conflict with the United States to Open 1998," Inter Press Service, New York, 5 January 1998 at 1; and J. Buxton and C. Southey, "Dumping Claim May Prompt Duty on Norwegian Salmon," *Financial Times*, London, 19 March 1997 at 28.

Part II

Canadian experiences and challenges in aquaculture law and policy

3 Canadian aquaculture and the principles of sustainable development

Gauging the law and policy tides and charting a course

David L. VanderZwaag, Gloria Chao and Mark Covan

Introduction

This chapter examines Canadian law and policy that addresses marine aquaculture proposals and operations, in light of four key principles:[1] integration, precaution, environmental impact assessment, and public participation.[2] The first part assesses the extent to which the federal government, the province of British Columbia and the Atlantic provinces (New Brunswick, Nova Scotia, Prince Edward Island, and Newfoundland and Labrador) have applied the four principles. The second part provides recommendations for charting a clearer course for sustainable aquaculture and raises key questions that need to be considered, such as whether adequate legal foundations exist for integrated coastal/ocean planning.

Gauging the tides: sustainable development principles in Canadian aquaculture law and policy

Introduction to Canadian aquaculture law and policy

Getting a firm grip on the Canadian approach to incorporating sustainability principles into domestic law is especially difficult in light of the complex constitutional division of powers allowing both federal and provincial regulatory powers.[3] The federal government has the power to legislate in relation to sea coast and inland fisheries, allowing it to protect and preserve both marine and freshwater fisheries.[4] Other relevant federal heads of power include shipping and navigation;[5] trade and commerce;[6] interprovincial and international matters;[7] Indians and lands reserved for Indians;[8] and federal works and undertakings and matters declared to be within federal jurisdiction.[9] Provincial governments also play an important role in aquaculture regulation pursuant to two main constitutional heads of power. The farming of aquatic organisms may be subject to the provincial powers over property and civil rights within the province[10] and matters of a merely local or private nature in the province.[11]

A key issue, partly unresolved, is the question of how far offshore provinces may control aquaculture developments. Provincial offshore jurisdiction has been recognized through case law as extending to marine waters *inter fauces terrae* (between the jaws of land), such as some inlets, bays and estuaries,[12] and to marine areas considered to be the part of a province at the time of Confederation.[13] The exact boundaries of provincial jurisdiction thus remain quite uncertain, especially on the east coast of Canada, where provinces may rely on various historical exertions of jurisdiction.[14]

It is important to distinguish between "legislative" jurisdiction and "territorial" jurisdiction. As will be noted below, the federal and provincial governments have chosen to address jurisdictional questions in the development of aquaculture in a manner that does not decisively determine territorial boundaries. Therefore, the following analysis focuses on the agreements that have been reached and on each government's legislative jurisdiction, rather than territorial jurisdiction.

Aboriginal rights, especially title claims and the right to be consulted,[15] add further complexity to the jurisdictional questions raised by aquaculture development. As issues that warrant detailed discussion in their own right, they are beyond the scope of the present chapter.

Integration

Although the integrative principle can be confusing given its various meanings, this chapter will focus on how Canada has implemented the principle of integration in three of its main manifestations:[16] vertical (cooperation among levels of government – national, provincial, local); horizontal (coordinating sectoral government departments/agencies, such as agriculture, fisheries, tourism, transport, energy, environment, defence and communications); and integrated coastal area management (managing multiple land and sea uses in the coastal zone).

Integration: federal

VERTICAL INTEGRATION

Governments have attempted to address issues of vertical integration in two major ways: the delegation of administrative power; and integration by non-legislative agreement.

Federal administrative power has been delegated to provincial levels of government by the transfer of federal control over freshwater fisheries to the provincial governments,[17] and by transferring a number of fish hatcheries to the provinces.[18] The federal government has also signed several memoranda of understanding (MOUs) on aquaculture with provinces and territories, including British Columbia, Nova Scotia, New Brunswick, Prince Edward Island, and Newfoundland and Labrador.[19] In general, these MOUs set out the areas of exclusive jurisdiction and the areas of cooperation between the two levels of government.[20] The MOUs also include provisions creating management committees (or coordinating committees, in the case of Prince Edward Island).

The Canadian Council of Fisheries and Aquaculture Ministers (CCFAM), composed of federal, provincial and territorial ministers, was formed as another vertical integration initiative, to discuss national and global issues affecting the fisheries and aquaculture sectors and to identify shared policy objectives and principles. In 1999, the members of CCFAM signed the *Agreement on Interjurisdictional Cooperation with Respect to Fisheries and Aquaculture*. Under the agreement, both levels of government commit to working together to contribute effectively, with sector stakeholders, to the development of ecologically sustainable and economically viable fisheries and aquaculture resources, habitats, and industries.[21]

Although the delegation of administrative powers and the signing of agreements has led to some degree of regulatory coordination between the two levels of government, much constitutional uncertainty still surrounds the regulation of aquaculture. First, the *ad hoc* delegation of administrative powers has led to inconsistencies in administration between one part of the country and the other.[22] Legislative provisions governing aquaculture have been enacted by both federal and provincial levels of government over an extended period of time, often on a species – or geography – specific basis, and typically as part of fisheries regulation.[23] This fragmented approach was identified by the Office of the Commissioner for Aquaculture Development as a critical area for reform if Canada is to ensure a sustainable aquaculture industry.[24]

HORIZONTAL INTEGRATION

With at least seventeen federal departments and agencies delivering programs and services to the aquaculture industry, horizontal integration remains a considerable challenge in Canada.[25] Some of the principal departments/agencies include the Canadian Environment Assessment Agency (responsible for administering the federal environmental assessment process applicable to most marine finfish aquaculture proposals), the Canadian Food Inspection Agency (responsible for monitoring molluskan shellfish for marine biotoxins and testing aquaculture products for drug residues), Environment Canada (lead department for conducting sanitary surveys of shellfish-growing areas and regulating toxic substances and ocean dumping) and Health Canada (responsible through the Veterinary Drugs Directorate for approving the use of veterinary drugs on aquaculture and through the Pest Management Regulatory Agency for regulating pest-control products such as sea lice therapeutants).[26] Fisheries and Oceans Canada (DFO), through its Aquaculture Management Directorate, is the lead federal department of aquaculture and acts as both a regulator and an enabler of the aquaculture sector.[27] With the Canadian Coast Guard becoming as a Special

Operating Agency within DFO effective 1 April 2005,[28] Transport Canada has become the responsible authority for ensuring that navigational impacts of aquaculture projects are assessed and addressed.[29]

Besides a broad attempt to coordinate federal roles in ocean resource development and management through an Interdepartmental Committee on Oceans,[30] horizontal integration in relation to aquaculture has primarily occurred through two avenues. The federal environmental assessment review process, discussed in the section "Environmental impact assessment" (p. 68) encourages sharing of departmental viewpoints regarding proposed aquaculture prospects. DFO has established the Interdepartmental Committee on Aquaculture (ICA) to conduct federal interagency meetings where coordination and governance issues can be discussed.[31]

INTEGRATED COASTAL AREA MANAGEMENT

The *Oceans Act* is the most important federal statute that includes potential integration requirements for the aquaculture industry.[32] Sections 31–33 of the Act require that the Minister of Fisheries and Oceans facilitate the development and implementation of plans for the integrated management of all activities affecting Canadian estuaries, coastal waters and marine waters. Integrated management is intended to bring together interested parties, stakeholders and regulators to reach general agreement on the best mix of conservation, sustainable use and economic development of coastal and marine areas for the benefit of all Canadians.[33]

Canada is still largely at the "pilot project" stage in developing integrated management plans. *A Policy and Operational Framework for Integrated Management of Estuarine, Coastal and Marine Environments in Canada*[34] sets out a goal of eventually establishing integrated management plans for all of Canada's offshore waters,[35] and proposes two main categories of planning from large scale to small scale – specifically, large ocean management areas (LOMAs) to coastal management areas (CMAs).[36] *Canada's Oceans Action Plan*, released in May 2005, confirms that five priority LOMAs will be focused upon initially: Placentia Bay/Grand Banks, the Scotian Shelf, the Gulf of St. Lawrence, the Beaufort Sea and the Pacific North Coast.[37]

Given the early stages of integrated planning initiatives, it is difficult to assess how adequately aquaculture will be addressed through planning processes. DFO has reported initiating twenty-one CMA efforts, but no specific implementation details were provided in a departmental progress report.[38] The most advanced large marine area planning process, the Eastern Scotian Shelf Integrated Management (ESSIM) Initiative, has chosen to focus initially on addressing ocean use and management issues beyond the twelve-nautical mile territorial sea.[39] The draft management plan, setting out overarching human use and ecosystem objectives and institutional components for further planning purposes, simply recognizes aquaculture as an existing coastal use and places priority on developing strategies to address commer-

cial fishing, oil and gas exploration/development and ship-source pollution.[40] The plan notes the eventual need to develop complementary coastal
management plans.[41]

Integrated planning for the Pacific coast of Canada promises to be especially complex in light of the Wild Pacific Salmon Policy, released in June
2005.[42] The policy proposes to establish a new integrated strategic planning
process for wild salmon conservation but leaves the details to be further
worked out through consultations with First Nations, provincial and territorial governments, communities, and stakeholders.[43] Local planning committees for subregions are suggested,[44] but how they would relate to marine use
planning limitations is left unclear.[45]

Integration: British Columbia

British Columbia ranked as the fourth largest producer of farmed salmon in
the world, with some 131 salmon aquaculture sites producing 72,700
tonnes in 2003, behind Chile, Norway and Scotland, and produced 8,600
tonnes of shellfish (oysters, clams, scallops and mussels) in 2003.[46]

VERTICAL INTEGRATION

The key document setting out a framework for vertical integration between
Canada and British Columbia is the *Canada/British Columbia Memorandum of
Understanding on Aquaculture Development*,[47] signed in 1988. The MOU confirms the division of powers as set out in the *Constitution Act, 1867*, and
identifies a number of coordination duties.[48] Section 10 and Schedule A
of the MOU provide for the formation of a management committee to coordinate such efforts and duties, and to resolve disputes. Since September
2000, British Columbia and the federal DFO have worked together to
prepare a harmonized guidebook for finfish aquaculture management applications.[49] In 2003 the Canada–British Columbia Committee on Regulatory
Reform started conducting several pilot projects in support of the
Canada–British Columbia Agreement on Fish Habitat Management, signed in
2000.[50]

Vertical integration has been attempted through a number of
federal–provincial fora. They include the Salmon Aquaculture Implementation Advisory Committee,[51] the Deputy Minister's Harmonization Committee,[52] the Project Review Team,[53] the Fish Health Working Group[54] and the
Technical Advisory Group.[55]

Furthermore, the federal government has participated in a number of
integrative management initiatives with the British Columbia government.
These include the Georgia Basin Ecosystem Initiatives,[56] the Pacific Marine
Heritage Legacy,[57] DFO's local "sustainability" initiatives,[58] and the Burrard
Inlet Environmental Action Program and Fraser River Estuary Management
Program (FREMP).[59]

More coordination attempts are in the works. For example, the British Columbia Ministry of Agriculture and Lands (BCMAL) is developing a provincial code of practice with the shellfish farming industry in British Columbia. The code will outline standards for shellfish aquaculture that are consistent with provincial and federal acts and regulations. It is expected that the code will be incorporated into mandatory operational standards that will become a condition of holding a shellfish aquaculture license.[60]

In September 2004, Canada and British Columbia signed a *Memorandum of Understanding Respecting Implementation of Canada's Oceans Strategy on the Pacific Coast*.[61] The MOU pledges the development of further subagreements on implementation measures for, among other things, a marine protected areas framework, integrated coastal and oceans management planning, sharing of information related to offshore oil and gas resources, and stream-lining and harmonizing regulatory decision-making for aquaculture.[62]

Despite the various initiatives to formalize vertical integration, the coordination of aquaculture continues to be a challenge, partly as a result of differing policy viewpoints. For example, the Auditor-General of British Columbia has identified federal–provincial disagreement on salmon farm siting criteria as a problem.[63] Among the key issues are how to determine "sensitive habitats" in need of protection and what are adequate distances for buffer zones, such as the distance farms should be from wild salmon streams.[64]

HORIZONTAL INTEGRATION

Getting a firm grip on horizontal integration in the aquaculture approval system in British Columbia has been difficult for at least two reasons. First, the two statutes governing the issuance of property rights, the *Land Act*[65] and *Fisheries Act*,[66] require multiple approvals,[67] with little administrative guidance provided in the past. This opens the door to considerable discretion in application review processes.[68] Second, following a change in government in June 2001, the public service underwent major reorganization, and a new Ministry of Sustainable Resource Management was created. However, following another government reorganization in June 2005, the Ministry of Sustainable Resource Management has been reintegrated within the Ministry of Environment.[69] As a result, governance approaches and mechanisms continued to evolve.

The control of aquaculture activities occurs primarily through four routes having different statutory authorities: Integrated Land Management Bureau (ILMB) (assuming the Crown land tenure responsibilities of the previous Land and Water BC Inc. (LWBC);[70] the Ministry of Agriculture and Lands;[71] the Resource Management Division of the Ministry of Environment;[72] and the Water, Land and Air Protection Programs of the Ministry of Environment.[73]

Integration efforts have occurred in various ways. In 2001, the BC Assets and Land Corporation signed an MOU with selected provincial agencies[74] to clarify the principles and process for *Land Act* application referrals and to establish dispute resolution steps.[75] ILMB, having subsequently assumed the land tenure responsibilities, helps ensure that institutional perspectives are considered in tenure approvals.[76] Aquaculture management plans, required for both commercial finfish[77] and shellfish,[78] provide a common information base for both provincial and federal agencies to review site-specific plans in accord with regulatory mandates. In 2002, provincial agencies entered into an agreement for coordinating compliance and enforcement activities relating to aquaculture. The Ministry of Agriculture and Lands has a lead role in monitoring compliance, and the Ministry of Environment is the leading agency for investigation and enforcement efforts.[79]

INTEGRATED COASTAL AREA MANAGEMENT

Integrated coastal area management is a relatively new principle in British Columbia law and policy. The BC Coastal Resources Strategy Study Steering Committee's 1993 "Towards a Coastal Resource Strategy" was an early provincial initiative that outlined the possible contents of a provincial integrated coastal management strategy. The subsequent 1998 BC *Coastal Zone Position Paper*[80] set out the province's position with respect to the federal *Oceans Act* initiatives.

The Land Use Coordination Office (LUCO) was created in 1994 to serve as the central agency for land-use planning in British Columbia. LUCO no longer exists. Its activities were largely passed to the Resource Management Division of the Ministry of Environment, and then moved recently to the ILMB.[81]

British Columbia has focused its coastal planning on the "nearshore" and "intertidal" (foreshore) areas under provincial jurisdiction.[82] Plans "are intended to address tenuring and conservation/protection opportunities in theses areas, rather than marine resource management, which is primarily a federal responsibility."[83] The British Columbian government has identified two levels of coastal zone planning: strategic-level coastal plans (e.g. at 1:250,000 scale) designed to identify broad goals, objectives and strategies for coastal and marine resources; and local-level coastal plans (e.g. at 1:50,000 to 1:5,000 scale).[84]

One of the largest plan areas in the province is the Central Coast Land and Resource Management Plan, covering about 4.8 million hectares of marine, foreshore and upland area on the west coast of British Columbia.[85] The plan includes forestry and commercial fishing activities; some sixty-eight of British Columbia's coastal salmon farm tenures have been located in this area, and are accordingly affected by the planning process.[86] The planning process began in 1996[87] and the process continues to evolve as the Ministry of Agriculture and Lands has announced its intent to integrate the planning of the Central and North Coasts into a single Land and Resource Management Plan.

Local plans fall into three general types:[88]

1   integrated coastal plans – identify land tenure opportunities including aquaculture, conservation, commercial recreation (e.g. the Nootka Sound Plan, Chatham Sound, Malaspina Complex, North Island Straits);[89]
2   issue resolution plans – attempt to resolve a particular issue or coast land use/activity conflicts (e.g. the Baynes Sound Coastal Plan for Shellfish Aquaculture, Cortes Island Coastal Plan for Shellfish Aquaculture);[90] and
3   special management plans – which aim to "provide more detailed direction for management of specific uses or distinct areas."[91]

Generally, these initiatives are developed on a flexible and "as needed" basis. The ministry does not advocate one use over another but attempts to balance the interests of each group to formulate a sustainable planning process.[92]

In February 2002, DFO, the province of British Columbia, the Nuu-chah-nulth Tribal Council and local governments announced the launch of the West Coast Vancouver Island (WCVI) Aquatic Management Board.[93] The pilot project involves a board of eight government and eight non-government members. The board is to develop recommendations for managing aquatic resources and is specifically asked to address issues in aquaculture and community economic development issues.[94]

Integration: the Atlantic provinces

Aquaculture has become important in all the Atlantic provinces of Canada. In 2003, New Brunswick led the Atlantic provinces in aquaculture production with a total of 36,453 tonnes – 33,650 from finfish and 2,803 tonnes from shellfish.[95] Nova Scotia produced 5,590 tonnes of finfish and 1,852 tonnes of shellfish for a total production of 7,893 tonnes.[96] Prince Edward Island's aquaculture sector produced twenty-four tonnes of finfish and 19,862 tonnes of shellfish.[97] Newfoundland and Labrador's production was lowest, at 3,900 tonnes, with 2,600 tonnes from finfish and 1,300 tonnes from shellfish.[98]

VERTICAL INTEGRATION

In the Atlantic provinces, vertical integration has occurred through delegation of authority and through coordinated assessment of aquaculture applications.

Since 1928, pursuant to an agreement between Prince Edward Island (PEI) and the Dominion of Canada, aquaculture leasing has been the responsibility of the federal government. The 1987 Canada/PEI Memorandum of Understanding (Canada/PEI MOU)[99] subsequently confirmed that the Aquaculture Division of the federal DFO is the lead agency for the administration of aquaculture leasing in PEI. However, the PEI Department of Fisheries, Aquaculture and Environment exercises certain rights and obligations pursuant to both the 1928 agreement and the Canada/PEI MOU.

In Nova Scotia, New Brunswick, and Newfoundland and Labrador, the federal government has largely delegated the administration of aquaculture to the provincial governments through MOUs.[100] For example, the Canada/Nova Scotia MOU provides that Nova Scotia (NS) is responsible for issuing and administering provincial licenses and leases, inspecting aquaculture facilities, and coordinating the approval process for licenses and leases.

The federal government has also largely delegated to the provincial governments the responsibility for ensuring compliance with federal legislation. For example, the Canada/NS MOU provides that "Nova Scotia will conduct periodic on site inspections of aquaculture facilities . . . and will advise the appropriate federal authority of any breach of applicable federal legislative or regulatory requirements."[101]

All four MOUs address the responsibility for monitoring the health of caged stock. In Nova Scotia, the province is primarily responsible for site inspection, including inspection under the federal *Fisheries Act*.[102] The federal government is left to administer the *Fish Health Protection Regulations*[103] and to continue "its national and regional role in the surveillance, detection, prevention, control and eradication of fish diseases in Nova Scotia."[104] In New Brunswick, the MOU provides that both the province and the federal government are to conduct periodic inspections of sites to determine compliance with their "respective Acts."[105] The Canada/PEI MOU does not identify an inspection obligation, although it does provide that Canada "shall continue its role in the prevention, control and eradication of fish disease" in PEI.[106] The Canada/Newfoundland MOU (Canada/NF MOU) also identifies this function as primarily a provincial responsibility, stating that Newfoundland "shall" conduct periodic on-site inspections of aquaculture facilities and that Canada "may" do so as well.[107]

While the four MOUs require that sites be inspected and comply with federal and provincial law, the adequacy of inspection is questionable. For example, under the Canada/NS MOU, Nova Scotia is responsible for monitoring the health of aquaculture stocks.[108] However, while Nova Scotia provides veterinary services to respond to requests for assistance from aquaculturalists,[109] it appears that formal inspection is undertaken only in relation to fish hatcheries and prior to the transfer of aquaculture stock from an aquaculture site, both within the province and interprovincially.[110]

In each of the Atlantic provinces, vertical integration occurs through the coordinated assessment of aquaculture applications. Both federal and provincial government departments have input at the initial application stage of the process. In PEI, a steering committee was formed in the early 1980s to facilitate the leasing application process by developing an aquaculture zoning system. The committee comprised representatives from DFO, the Coast Guard, Public Works Canada, Environment Canada, Parks Canada, the PEI Department of Fisheries, the PEI Department of Environment, and Tourism PEI. This effort led to the PEI Aquaculture Zoning System, which identifies potential conflicts between recreational and commercial uses, wildlife sanctuaries, environmental legislation and aboriginal rights.[111]

The Aquaculture Leasing Management Board (ALMB) is another example of vertical integration in PEI. The Board is made up of federal and provincial representatives, as well as members of industry.[112] ALMB is responsible for the overall management of aquaculture in PEI by approving priorities and developing business and financial plans "associated with the implementation and ongoing management of the PEI Aquaculture Leasing Program."[113] The ALMB also provides management and policy advice to yet another panel, the Leasing Referral Committee (LRC), the members of which are drawn from both levels of government.[114] The LRC assesses lease applications for compliance with ALMB policy (such as habitat, navigation, water quality, public fishery, use conflict and zoning policy), helps resolve contentious issues, develops recommendations and advises on policy amendments.[115]

The other three Atlantic provinces similarly involve federal departments in evaluating aquaculture applications.[116] In New Brunswick, DFO co-chairs the federal–provincial Aquaculture Site Evaluation Committee (ASEC), which makes recommendations to the provincial minister with respect to site applications.[117] However, the province has, on some occasions, ignored recommendations made by DFO staff "against allowing certain sites based on fish habitat considerations."[118] This highlights the difficulty of obtaining effective vertical integration where roles and responsibilities are not clearly delineated.

In Nova Scotia, the Canada/NS MOU requires the province to refer aquaculture license and lease applications to DFO and "other relevant federal and provincial departments and agencies in conformity with their respective jurisdiction relating to such aquaculture matters."[119] For instance, the approval of Transport Canada is required for any aquaculture project that may substantially interfere with navigation through water. Most suspension-type aquaculture operations in Nova Scotia require such approval. The provincial government also consults with the Canadian Food Inspection Agency when considering aquaculture license and lease applications.

In Newfoundland and Labrador, DFO also plays a role in the review of aquaculture applications, as set out under the Canada/NF MOU. Once the Newfoundland and Labrador Department of Fisheries and Aquaculture (NLDFA) receives an application for an aquaculture license, it distributes the application to a number of referral departments at both levels of government, typically including DFO and Environment Canada.[120]

HORIZONTAL INTEGRATION

In the Atlantic provinces, the lead departments charged with licensing aquaculture sites are required to consult with various other departments within the province's government. However, in each province this process is limited only to consultation, since the non-lead departments cannot veto any prospective decision made by the lead department.

Nova Scotia provides the most glaring example. Before making a decision on any application, the Minister of Agriculture and Fisheries must consult with the departments of Environment (now the Department of Environment and Labour), Housing and Municipal Affairs (now Service Nova Scotia and Municipal Relations), and Natural Resources, as well as any additional boards, agencies and commissions prescribed by regulation.[121] However, in at least two reviewed files the concerns of Natural Resources officials were largely ignored, suggesting that absent a veto power, horizontal integration will not be achieved merely through consultation. For example, in an aquaculture expansion proposal for Northwest Cove, NS, an official with the provincial Department of Natural Resources suggested rejection based on the project's planned use of bird nets in close proximity to an active double-crested cormorant colony:

> After reviewing this proposal I have a wildlife concern that must be addressed. This project is planned to operate within approximately 50 meters of Horse Island in Northwest Cove. Horse Island has an active double-crested cormorant colony and possibly Great Blue Herons nesting there as well. Raising any type of finfish in close proximity to such a congregation of fish-eating seabirds would be unwise. The proponent has stated that "bird nets" will be used. I cannot comment on the effectiveness of such devices. . . . Based on the aforementioned factors I cannot support this proposal because of its present location and recommend that it be denied.[122]

The concerns of the official were apparently not taken seriously. The Northwest Cove expansion was approved,[123] subject to fifty-three operating conditions.[124] This also occurred in a project review involving a mussel cultivation proposal for St. Ann's Harbour, Cape Breton. A report by Natural Resources staff recommended that since the project monopolized such a large portion of a shared resource, it should be rejected unless supported by the community at large. Notwithstanding the lack of community support, the minister approved the project on 5 April 2002. The operating conditions of the project's lease and license only superficially accounted for many of the concerns underlying the community's lack of support.[125]

In New Brunswick, the ASEC assesses applications for aquaculture sites, and comprises federal and provincial agencies "with regulatory authority relative to the marine environment."[126] The ASEC consults with the Aquaculture Environmental Coordinating Committee (AECC), which considers aquaculture–environment issues and makes recommendations on a variety of topics, including siting decisions and research priorities. However, as in Nova Scotia, final decisions rest with the minister.[127]

In Newfoundland and Labrador, the Department of Fisheries and

Aquaculture is the key regulator in the aquaculture industry and administers the *Aquaculture Act*[128] and the *Aquaculture Regulations*.[129] Although an
aquaculture application must be examined by a number of provincial referral
departments and agencies, this examination process is reasonably well integrated. The Aquaculture Registrar's Office of the Department of Fisheries and
Aquaculture acts as the "one-stop aquaculture licensing office," working
with referral departments and agencies, as well as industry, in order to
streamline the licensing process by reducing duplication.[130] The various
referral departments and agencies include the Department of Environment,[131] the Department of Municipal and Provincial Affairs,[132] the Department of Tourism, Culture and Recreation[133] and the Department of
Government Services and Lands.[134]

Each referral department or agency has thirty days to review an application and provide comments,[135] and if no response is received within that
period, NLDFA will assume there are no objections.[136] When all of the relevant information is received, the Aquaculture Registrar's Office forwards
this information to the Aquaculture Director for review and recommendation by the Aquaculture Licensing Committee.[137] The committee in turn
reviews the application and makes a recommendation to the Minister of
Fisheries and Aquaculture, who then makes the final decision.[138]

In PEI, horizontal integration is achieved by managing fisheries and the
environment within one government department. However, as already
noted, the federal DFO is the lead agency for the administration of aquaculture leasing in PEI.

INTEGRATED COASTAL AREA MANAGEMENT

Although various integrated coastal management efforts have been initiated
by the Atlantic provinces, most if not all of these initiatives either have been
terminated short of their goals or have yet to be implemented. Few of the
integrated coastal management initiatives have specifically addressed aquaculture.

In 1997, Nova Scotia had close to twenty statutes relating to the oceans
sector, administered by eight provincial departments.[139] The Coastal 2000
program attempted to coordinate these efforts under one umbrella. The
program was based on four key principles: sustainable development, partnerships between community groups and government, integrated and efficient
delivery of government services, and community-based decision-making.[140]
However, the status of Coastal 2000 is unclear. There is nothing to indicate
that these principles are nearing the implementation stage, nor does the
program make specific mention of aquaculture. While an integrated resource
management process has been developed for provincially managed Crown
lands in Nova Scotia, this process has not been used for aquatic lands.[141]

The Guysborough County Sustainable Aquaculture Initiative (GCSAI)
represents a smaller-scale attempt to create an integrated coastal area man-

agement program in Guysborough County, NS. The Guysborough County Regional Development Authority is tasked with the mandate of economic development of the county, and has completed what it calls the "Guysborough County Sustainable Aquaculture Initiative." Finalized in August 2004, the goal of this initiative is to create a "comprehensive information tool" to help decision-makers balance environmental concerns (site characteristics, carrying capacity, etc.), user conflicts, product safety and other factors in designating sites for future development.[142]

While New Brunswick has identified the need for coastal zone management planning,[143] it has yet to formalize integrated coastal planning. The province has developed *A Coastal Areas Protection Policy for New Brunswick*[144] that divides the coastal area into three sensitivity zones (core, buffer and transition). A coastal designation order under the *Clean Environment Act* is proposed, which will provide a framework for managing activities in the coastal area, including designation of prohibited and allowed activities.[145] Given the aim of the policy to primarily address land-based coastal uses, proposals and operations, its effect on aquaculture is likely to be minimal.

Integrated coastal area management is a relatively new concept in Newfoundland and Labrador. Some integrated coastal interests have been mapped on an *ad hoc* basis, but such mapping is difficult, since many coastal interests (including tourism and lobster fishing) are seasonal and do not necessarily use all of the land that they occupy.[146]

Despite the lack of formalized integrated coastal area management plans, it is important to note that Newfoundland and Labrador law and policy require that aquaculture is developed in a manner that does not preclude the development of other coastal activities. In particular, one of the stated purposes of the *Aquaculture Act* is to "minimize conflicts with competing interests and uses."[147] Furthermore, the *Aquaculture Licensing Policy and Procedures Manual* acknowledges that the coastline must be shared with other coastal interests, such as traditional fisheries, other marine resource user groups, and environmental, social, and public health and safety concerns.[148]

In 1998, the federal and Newfoundland and Labrador governments together commissioned a study to examine the underlying sources of resource-based conflict in the aquaculture industry and the strategies that should be used to resolve them.[149] The study examined Green Bay South and Bay d'Espoir, two areas in Newfoundland and Labrador in which the aquaculture industry has expanded rapidly, creating considerable resource-based conflict. In fact, the conflict in Green Bay South reached the point where the Minister of Fisheries and Aquaculture placed a moratorium on new license applications for the area until the conflict could be examined.

The study yielded forty-two recommendations. With respect to integrated planning, the study recommended that NLDFA should coordinate workshops for fishers in areas where aquaculture is or will be occurring; that proposed sites should be discussed in stakeholder meetings with one or two representatives of each interest group, rather than in large public meetings;

and that the Community-Based Coastal Resource Inventory should be continued in order to identify and quantify coastal resource uses and avoid, resolve or mitigate conflicts between competing resource users.[150]

Apart from the PEI Aquaculture Zoning System referred to earlier, there have been no recent integrated coastal management initiatives that directly address aquaculture in PEI.

The federal government has attempted to promote integrated coastal area management in the Atlantic provinces through the Atlantic Coastal Action Program (ACAP), which was initiated by Environment Canada in 1991. The program is based on the principle that coastal zone management plans must be community driven in order to succeed. There are currently fourteen sites across Atlantic Canada, including five in Nova Scotia and five in New Brunswick.[151]

Precautionary approach

Precautionary approach: federal

While most federal statutes do not expressly adopt the precautionary approach, it is emerging in federal legislation and policy. For example, the approach is one of three principles guiding the Minister of Fisheries and Oceans Canada in the implementation of the *Oceans Act* and is defined in the preamble as "erring on the side of caution." In addition, s. 4 of the *Canadian Environmental Assessment Act* was amended in 2003 to state that one of the purposes of the act is to "ensure that projects are considered in a careful and precautionary manner before federal authorities take action in connection with them, in order to ensure such projects do not cause significant adverse environmental effects."[152] The *Canadian Environmental Protection Act, 1999* (CEPA)[153] also adopts precaution as a guiding principle through various provisions, including the preamble, which recites the Rio Declaration version,[154] and s. 2, which imposes an administrative duty to follow the precautionary principle in implementation of the Act. The *Pest Control Products Act*[155] somewhat marginalizes the precautionary principle by only requiring the principle to be taken into account in the course of reevaluation or special review of a pest-control product.[156]

Several federal or national environmental policy documents cite the precautionary approach as an important guiding principle.[157] For example, *Canada's Oceans Strategy* reaffirms the federal government's commitment to wide application of the precautionary approach, pledges priority to maintaining ecosystem health and integrity in cases of uncertainty, and notes that the strategy will be governed by the government's ongoing policy work.[158] *DFO's Aquaculture Policy Framework* adds a specific commitment to develop aquaculture in accordance with the precautionary approach.[159] The *National Code on Introduction and Transfers of Aquatic Organisms*[160] also urges a precautionary approach to the introduction of exotic organisms to natural habitats.

The Government of Canada's *Framework for the Application of Precaution in Science-Based Decision Making about Risk*,[161] adopted in 2003 through an interdepartmental process in order to provide overall guidance on application of precaution through ten principles, supports a rather weak version of precaution. The framework suggests restricting application of the precautionary approach to situations where there is a "sound" scientific basis,[162] fails to take a firm stance on the burden of proof in decision-making[163] and highlights the need for precautionary measures to be cost-effective and minimally trade-restrictive.[164]

Likely to be an especially contentious issue area for application of the precautionary approach is whether transgenic fish should be permitted for commercial use and, if so, under what conditions. CEPA and its regulations have been criticized for not being very precise about risk assessment requirements in relation to the potential effects of modified organisms on biodiversity.[165] Through a May 2004 MOU with Environment Canada and Health Canada, the Department of Fisheries and Oceans has agreed to be responsible for risk assessment under CEPA should there be any applications for aquatic organisms with novel traits.[166] DFO is in the process of drafting regulations for aquatic products of biotechnology[167] but the regulations have yet to be released for public comment. Section 109 of CEPA grants the Minister of Environment considerable discretion to approve animal biotechnology products following information assessments,[168] and such discretion raises concerns whether a strong precautionary approach will be followed.[169]

The *Expert Panel Report on the Future of Food Biotechnology*[170] was prepared by the Royal Society of Canada at the request of Health Canada, the Canadian Food Inspection Agency and Environment Canada. Although the report is not a policy document *per se*, it will likely be considered by the three agencies that commissioned it, should the federal government allow the commercial production of transgenic fish. The Expert Panel concluded that there were significant scientific uncertainties about the potential effects of genetic and ecological interactions between transgenic and wild fish, and about the mitigative utility of rendering genetically modified fish sterile in aquatic facilities. The Expert Panel made a number of recommendations in order to implement the precautionary approach, including a moratorium on the rearing of genetically modified fish in aquatic netpens.[171]

Precautionary approach: British Columbia

The approach is not explicitly identified as a guiding principle in existing environmental and aquaculture legislation. Sustainability principles, designed to guide provincial resource management and approved by cabinet in May 2002, did not include precaution.[172] General principles governing the exercise of authority to issue a salmon aquaculture license include fairness, transparency, efficiency and accountability but not precaution.[173]

However, a number of aquaculture decisions and initiatives in British Columbia are moving towards a more precautionary approach. Between 1995 and 2002, the province decided to impose a moratorium on new salmon farms, and capped coastal salmon farm tenures at 121.[174] The province has approved experimental technology farm sites where closed containment systems with waste recovery, alternative feeds and other "greener" technologies will be tested.[175] The provincial Escape Prevention Initiative, led by the Ministry of Agriculture, Food and Fisheries (as it was then), has established a long-term goal to achieve zero escapes across the aquaculture industry.[176] Pursuant to the *Aquaculture Regulation*,[177] various precautionary measures have been imposed to prevent escapes. These measures establish minimum breaking strengths for net-cage mesh and require jump nets where net-cages do not have a permanently attached mesh top or similar barrier. In addition, a major amendment to the British Columbia *Fisheries Act* was introduced by a private member's bill that would limit licenses for finfish aquaculture to closed containment operations;[178] however, that amendment was not passed.

The decision to lift the marine salmon moratorium fueled further precautionary debates in British Columbia. In 2002, the British Columbia government announced that it would accept applications for new salmon aquaculture sites given improved provisions for siting and relocations, research and development, and fish escapes, health and waste.[179]

Various voices lamented the lack of precautionary approach to aquaculture regulation in British Columbia. Stuart Leggatt, a former judge of the British Columbia Supreme Court, undertook an inquiry into salmon farming with funding from a non-governmental organization, the David Suzuki Foundation. He noted that, despite the government moratorium on new salmon farms, production levels expanded substantially through intensified production allowed at existing farms.[180] The Leggatt Inquiry, among other things, recommended application of the precautionary principle to salmon farming regulation,[181] and urged removal of all net-cage salmon farms from the marine environment by 1 January 2005.[182] West Coast Environmental Law, in a first-year review of the BC Liberal government's environmental performance since the Cabinet's swearing in on 5 June 2001, was critical of the low standards established for aquaculture, including the proposed *Finfish Aquaculture Waste Control Regulation* (now in force),[183] which would focus on only one pollutant – sulphide – and "set standards at levels that are more than double those found to be harmful by scientists."[184]

Precautionary approach: Atlantic provinces

None of the Atlantic provinces specifically adopts the precautionary approach in legislation or government policies directly relating to aquaculture. Furthermore, with the exception of Nova Scotia's *Environment Act*,[185] none of those provinces specifically adopts precaution in environmental legislation that may indirectly apply to aquaculture operations. Each Atlantic province has largely failed to accept precaution in setting standards, preventing escapes and controlling food consumption.

SETTING STANDARDS

Precaution appears to mandate adherence to standards suggested by scientific evidence. However, there are few such standards in the aquaculture legislation and policies of the Atlantic provinces,[186] and where they do exist, they are based upon a reactive or adaptive approach to environmental concerns, rather than a proactive one.[187]

The outbreak of infectious salmon anemia (ISA) virus in New Brunswick's Bay of Fundy region provides an excellent example of a reactive approach to the environmental concerns of aquaculture. In 1998, New Brunswick aquaculturalists followed in the footsteps of the Scottish and Norwegian salmon aquaculture industry[188] by slaughtering thousands of Atlantic salmon in an effort to stop the spread of ISA.[189] Despite these measures, twenty-one New Brunswick fish farms were infected in 1999, and a total of fifty-five farms were eventually found to be infected.[190] In 2002, ISA was detected at Cobscook Bay, near the Maine border, where most of that state's salmon pens are located. Farmers were forced to destroy over 2.5 million fish.[191]

The New Brunswick government responded to the outbreak of ISA, and the resulting closure or fallowing of many infected sites, with the Bay of Fundy Aquaculture Site Allocation Policy (Fundy SAP).[192] However, this policy does not embrace a strong precautionary approach, but leaves applicable standards largely to departmental decision-makers.[193]

Nova Scotia has also failed to adopt firm standards against which to assess both initial applications and ongoing aquaculture operations.[194] For example, the Minister of Agriculture and Fisheries requires that aquaculture license applicants complete a Marine Finfish Development Plan (MFDP). The contents of the MFDP are not prescribed by legislation, but rather are determined by Department of Agriculture and Fisheries (NSDAF) staff.[195] The NSDAF has adopted a highly generalized, subjective self-assessment approach to the provision of information in the MFDP. For example, the MFDP does not address chemical treatments directly, but asks the proponent to "describe your fish health management plans."[196] Furthermore, none of the "recommended standards"[197] found in the MFDP is binding on the applicant. Failure to meet any one of the recommended standards will not necessarily result in rejection of the application, which depends on the personal views of the reviewer. No minimum standards are prescribed by the NSDAF in reviewing an MFDP.[198]

Excessive reliance on consultants in the Nova Scotia license application process further indicates a lack of precaution. Proponents often use consultants to prepare the MFDP and to provide any other information required by the various government departments. Since there is no set form or formula for such reports, the NSDAF may also place considerable reliance on such consultants in addressing the environmental impacts and other aspects of the proposed aquaculture project.[199] Although past consultants have looked to the Nova Scotia Environmental Codes of Practice for Finfish Aquaculture Operations in preparing environmental assessments, the codified provisions themselves lack clear standards.[200]

Nova Scotia also shows a lack of precaution in its regulation of ongoing aquaculture operations. The Nova Scotia *Aquaculture License and Lease Regulations*[201] require that license holders keep records on a number of activities relating to the aquaculture site, which must be made available to an inspector or the Registrar of Aquaculture upon request.[202] This information includes the origin, transport, transfer and introduction of live aquacultural produce, the presence of diseases, the type of medication, dosage, treatment date, and duration of veterinary treatments, and the type and amount of food used in relation to aquacultural produce. However, there are no prescribed limits or standards for those activities that are to be recorded. For example, there are no limits placed on the type or amount of medication used or the frequency of such treatment. Nor is there any requirement to report the presence of diseases. The licensee must merely make a record.

In PEI, the PEI Leasing Referral Committee evaluates potential aquaculture sites on the basis of criteria that include overall site size and dimensions, water depth, existence of natural shellfish populations, proximity to local commercial fishing, navigation concerns, etc.[203] However, there are virtually no regulatory and policy standards for these various factors.[204] For ongoing operations, aquaculturalists need only provide the information contained in their annual lease report, such as harvest information, the number of shellfish remaining on site after harvest, the quantity of seed placed onsite, the time spent operating the lease, any "problems encountered" and any enhancements to bottom culture sites.

Although the Newfoundland and Labrador *Aquaculture Act* and its regulations include a number of measures that could be seen as precautionary, industry-wide standards have not been formally specified. For example, the Act states that the minister may set a number of specific conditions or standards for aquaculture licenses and operations.[205]

Although the permissive nature of this power diminishes its overall precautionary scope, there are specific examples of standards set by departmental policies that emphasize precaution. For instance, the *Shellfish Culture License Policy* states that a regular commercial license will only be granted after a one-year "developmental license" has been issued.[206] This period is used to determine whether the required biological and environmental conditions are present for commercial operations. During this time, the proposed site must undergo regular water testing as required by the Canadian Shellfish Sanitation Program.[207] The NLDFA may also require up to six months to assess the carrying capacity and the needs of other water resource users in a specified geographic area.[208]

ESCAPES

The federal DFO has concluded that "the potential for negative interactions warrant application of the Precautionary Approach for management of Maritime salmon stocks and their interaction with escaped farmed salmon."[209] However, neither the NSDAF nor the New Brunswick Department of Agriculture, Fisheries and Aquaculture assesses the likelihood of fish escapes from particular aquaculture sites, nor do these departments assess the ecological consequences of an escape in light of a site's proximity to wild stocks or potential competition for habitat.

However, license/lease operating conditions may closely regulate the type of enclosure used by fish farms. For instance, Nova Scotia License No. 1169 stipulated that "[t]he cage design and anchoring system will be suitable for the wave action climate expected at the site and certified by a professional engineer." Condition 33 of the license stated that the containment nets must be constructed of heavy twine (210/120), and that the mesh size cannot exceed $1\frac{3}{8}$ inches.[210]

In PEI, aquaculture licenses dictate what enclosure may be used in light of the species being reared. For instance, Arctic char can only be held in land-based systems with specific effluent screening requirements. Saint John River strain Atlantic salmon may be held either in lake cages with proper screening requirements on the outflow to the lake, or in land-based facilities with proper effluent screening.[211]

In Newfoundland and Labrador, the DFA mandates the inspection of aquaculture facilities when a new site is established, an aquaculture facility changes owners, there is a change in status (from developmental to commercial), fish health is a concern, gear and marker buoys are placed on-site (initially), and at any other time deemed necessary by the department.[212] In addition, the Newfoundland and Labrador *Aquaculture Act* allows inspectors to inspect a site where escapes may be a threat.[213] Inspectors may inspect aquaculture operations if reasonably necessary having regard to the measures taken to prevent the escape of cultured aquatic plants or animals; the presence or likelihood of disease; and compliance with the terms, conditions and provisions of the aquaculture license, the *Aquaculture Act* and its regulations. The Act also permits an aquaculture inspector to direct a licensee to take measures to prevent escapes and the development or spread of disease or parasites. To this end, an inspector may even order the destruction of stock and the disinfection of gear and facilities. The *Aquaculture Act* also requires that a person who wishes to transfer live aquatic plants or animals from one body of water or aquaculture facility to another must obtain prior written approval of the minister.

FOOD CONSUMPTION CONTROL

None of the Atlantic provinces specifically regulates the quantity or type of feed provided to aquaculture stock. However, the operating conditions of a lease or license may establish some control in this respect. For example, Nova Scotia License No. 1169 required that a feed camera must be installed and used to minimize feed waste. A further condition prohibited the use of mechanical feeding from 8 p.m. to 8 a.m.[214] The economic implications of good husbandry practices may also entice aquaculturalists to feed their stock appropriately. Poor husbandry practices, such as low-quality or excessive feedings, may involve higher financial costs to deal with sick and moribund fish.[215]

Environmental impact assessment

Environmental impact assessment: federal

The federal statute setting out the environmental impact assessment process for aquaculture is the *Canadian Environmental Assessment Act* (CEAA).[216] Under the CEAA, certain acts of designated federal agencies or departments, also known as "federal authorities," with respect to a particular "project," trigger an assessment of the potential environmental effects of the project.[217] An environmental assessment may also be triggered if there are risks that the project may cause "significant transboundary effects."[218]

Most aquaculture operations involving the construction of aquatic facilities of a permanent or semi-permanent nature will be considered as "projects" pursuant to the CEAA.[219] The most likely trigger of the CEAA assessment process is the issuance of a subsection 5(1) *Navigable Waters Protection Act* (NWPA) approval by Transport Canada. NWPA approval is required when a "work" may substantially interfere with navigable waters. Other possible triggers include the issuance of an authorization for the harmful alteration, disruption or destruction of fish habitat pursuant to subsection 35(2) of the *Fisheries Act* by the Habitat Management Division of Fisheries and Oceans Canada.

Once it is determined that the CEAA process applies to a given aquaculture project, the process follows the procedure outlined in the *Regulations Respecting the Coordination by Federal Authorities of Environmental Assessment Procedures and Requirements*.[220] These regulations require that the federal EIA process is timely and predictable, and that only one federal EIA is conducted for a project. Once the process is initiated, the federal authorities involved in the process are referred to as the "responsible authorities" (RAs).[221]

There are four types of environmental assessment that may be undertaken: screening, comprehensive study, mediation or panel review. There have been several hundred aquaculture records listed on the CEAA environmental assessments registry as being initiated and/or completed.[222] The registry listing of aquaculture records is not intended to be a comprehensive archival collection; therefore, many more projects have undergone or are in the process of undergoing an environmental assessment.[223]

The level of assessment performed with respect to aquaculture projects has consistently been screening.[224] Screenings are self-directed processes in which RAs have the discretion to decide how the assessment is to be conducted, including the extent to which public participation, if any, will be required.[225] Generally, RAs have frequently decided to allow the proponent's environmental impact statement to serve as the screening document to be evaluated; however, screenings for aquaculture have almost always required significantly more detail beyond the environmental impact statement.[226] Where the RAs are of the opinion that the information provided is not adequate to enable them to take a course of action, they shall ensure that any necessary studies and information are undertaken or collected.[227]

In addition to the screenings of individual projects, s. 19 of the act allows for the possibility of class screenings for projects that are repetitive in nature and whose environmental effects are well understood. Following the five-year review of the act, the Minister of the Environment identified the general increased use of well-defined class screenings as a priority.[228] The Office of the Commissioner for Aquaculture Development has also identified class screenings as a potential means of simplifying the CEAA procedure.[229]

Fisheries and Oceans Canada is developing a streamlined approach to CEAA assessments of aquaculture based on the scoping of multiple projects into fewer screening reports in order to better address cumulative effects of multiple aquaculture operations.[230] Fisheries and Oceans Canada is in the process of drafting a CEAA screening reference manual that includes information on how to scope projects and assess cumulative effects based on tables of valued ecosystem components (VECs).[231] VECs include fish habitat, marine water quality and wild fish habitat, etc.[232]

Following a screening, the RAs may exercise any power or perform any duty or function that would permit the project to be carried out, or, where circumstances warrant, they should refer the project to the minister for a referral to a mediator or a review panel in accordance with se. 29. Pursuant to s. 38(1), all forms of environmental assessment may also be subject to monitoring and a follow-up program where appropriate.

The *Species at Risk Act* (SARA)[233] could make special environmental assessment requirements applicable to aquaculture proposals. Pursuant to s. 79 of the act, if a project is likely to affect a listed wildlife species[234] or its critical habitat, the proponent must identify any adverse effects. If the project is carried out, the proponent must monitor effects and take measures to avoid or lessen the adverse impact.[235]

The federal strategic environmental assessment (SEA) process, set out in a Cabinet directive,[236] is meant to apply to policy, plan and program initiatives bound for Cabinet or to ministerial approval, and thus could have importance to aquaculture. However, a 2004 audit by the Commissioner of the Environment and Sustainable Development[237] found that Fisheries and Oceans Canada did not have most of the management system elements in place for SEAs.[238] The audit was also very critical of the department's failure to conduct a strategic environmental assessment for the development of the *Aquaculture Policy Framework* even though the policy document contained a foreword signed by the department's minister.[239]

Environmental impact assessment: British Columbia

Currently, aquaculture activities are not captured as reviewable projects under the provincial *Environmental Assessment Act*.[240] In most instances, neither shellfish nor finfish farms are required to obtain a project approval certificate under the Act,[241] and therefore they are not subject to provincial environmental impact assessments.[242] There are some possible exceptions: for example, groundwater extraction and surface water diversion projects in association with salmon aquaculture may be reviewable under the act.[243] In general, however, the only environmental assessment process for proposed shellfish or finfish sites is that under the *Canadian Environmental Assessment Act*, as detailed earlier. This lack of provincial EIA process is in fact an indication of a degree of vertical integration, as the federal government and the Province of British Columbia entered into an agreement in 1997 and are currently renegotiating an agreement to harmonize their environmental processes.[244]

Environmental impact assessment: Atlantic provinces

None of the Atlantic provinces clearly requires environmental assessments of proposed marine aquaculture developments. Rather, they rely on the federal CEAA process. However, certain activities may trigger such a review.

In Nova Scotia, an aquaculture operation is subject to the environmental assessment process (EAP) of the *Nova Scotia Environment Act* (NSEA) only at the discretion of the Minister of Environment (MOE). The Nova Scotia EAP applies to "undertakings" as determined by the MOE or as prescribed in the regulations. Since marine aquaculture is not likely to be caught under the regulations as a "wetland" undertaking, the only possible trigger for an aquaculture EAP would be qualification as "such other undertaking as the Minister may from time to time determine."[245] This may occur if sufficient public concern warrants.

Assuming the Nova Scotia MOE has the will to activate the EAP, the assessment process itself would involve extensive ministerial discretion. Once the minister orders an environmental assessment for an aquaculture operation, the *Environmental Assessment Regulations* give him or her discretion to determine whether the undertaking is a Class I or Class II undertaking.[246] The only mandatory step in the EAP for Class I undertakings is registration pursuant to the NSEA.[247] After that, the minister has wide discretion in deciding whether to require additional information, a focus report and/or an environmental assessment report. The minister also has discretion to determine whether all or part of the undertaking will be referred to alternative dispute resolution, whether the undertaking will proceed absent further information, or whether it will be rejected because of the likelihood that it will cause unmitigated adverse environmental effects. On the other hand, Class II undertakings require registration, terms of reference, an environmental assessment process and a referral to the Nova Scotia Environmental Assessment Board.[248]

In New Brunswick, aquaculture operations are prima facie exempt from the environmental assessment provisions of the *Clean Environment Act* (CEA)[249] and its associated regulations. Although the CEA broadly defines "environment" and "environmental impact,"[250] an aquaculture project must meet two thresholds to trigger an environmental assessment. First, the project must fall within the definition of "undertaking"; and second, the Lieutenant-Governor in Council must designate the project as the kind of undertaking that "may result in a significant environmental impact." This class of undertaking is defined by an inclusive list in the *Environmental Impact Assessment Regulation*.[251] Subsection (u) of this list defines "undertaking" as including "all enterprises, activities, projects, structures, works or programs affecting any unique, rare or endangered feature of the environment." Since aquaculture has the potential to affect already fragile wild populations of Atlantic salmon, this definition could trigger environmental assessment for aquaculture projects. Indeed, the United States Fish and Wildlife Service and the National Marine Fisheries Service have listed wild Atlantic salmon as endangered in eight Maine rivers,[252] and inner Bay of Fundy salmon have been listed as endangered under Canada's *Species at Risk Act*.[253]

However, as in Nova Scotia, even if an environmental assessment is triggered under the New Brunswick CEA, the Minister of Environment and Local Government retains discretion at all stages. For example, even if an aquaculture project is determined to be an "undertaking" within the meaning of the CEA, the minister retains discretion to obviate the need for environmental assessment.[254] Furthermore, even where an environmental assessment is required, the minister retains broad discretion as to the substance, scope and conduct of the assessment.[255] No environmental assessments have been performed under provincial legislation for marine aquaculture sites in either Nova Scotia or New Brunswick.[256]

In PEI, the *Environmental Protection Act* (EPA)[257] casts a wide net for the types of activities subject to environmental impact assessment. The Act states that "[n]o person shall initiate any undertaking unless that person files a written proposal with the Department and obtains written approval from the minister to proceed with the proposed undertaking." The EPA defines "undertaking" broadly to include any industry, operation or project that will or may (i) discharge any contaminant into the environment; (ii) affect any unique, rare or endangered feature of the environment; (iii) significantly affect the environment; or (iv) cause public concern because of its perceived effect on the environment.

While aquaculture operations may constitute "undertakings" within the EPA's broad definition, as in Nova Scotia and New Brunswick, the PEI EPA has not yet been used to assess the environmental impacts of marine aquaculture operations.[258] This may be due to the fact that the Minister of Fisheries, Aquaculture and Environment has broad discretion to decide whether persons initiating "undertakings" need to carry out an environmental impact assessment, submit an environmental impact statement or provide additional information.[259]

Newfoundland and Labrador also does not generally subject marine aquaculture proposals to environmental impact assessment. The province's *Environmental Protection Act* gives the Minister of Environment discretion to decide whether to subject proposed undertakings to environmental assessment review.[260] Furthermore, the *Environmental Assessment Regulations, 2000* only guarantee environmental assessment registration for fish farms or shellfish production undertakings that have shore-based facilities other than wharves and storage buildings.[261] Thus, marine aquaculture proposals have generally not been subject to provincial environmental impact assessment.[262]

Public participation

Public participation: federal

The principle of public participation appears in federal law and policy on both macro and micro levels. On the macro level, the federal government has, on a number of occasions, attempted to canvas public opinion on relevant federal strategies[263] and legislation. On a micro level, with respect to the assessment and development of individual aquaculture projects, the public has been limited to the CEAA assessment process as a means for participating in the decision-making process. However, consultation with aboriginal communities is mandatory for aquaculture projects in British Columbia.[264] Subsection 55(1) of the CEAA requires that a public registry be established in a manner that will ensure public access to information on every project that has undergone an environmental assessment.

Public participation at the screening stage of a CEAA assessment is generally limited. The responsible authority is given discretion to determine whether public participation in the screening of a project is appropriate under the circumstances.[265] Guaranteed public participation is only found in the other three levels of assessment: comprehensive studies, mediation and panel reviews. For comprehensive studies, members of the public have the right to make written comments on the study report; for review panels, public hearings are required.

Public participation: British Columbia

On a macro level, the finfish industry has been subject to comprehensive public review on a number of occasions from the mid-1980s to the late 1990s. These include the 1986 Inquiry into Finfish Aquaculture in British Columbia; the 1988 *Ombudsman Public Report No. 15*; *Aquaculture and the Administration of Coastal Resources in British Columbia*; the 1993 *Wild and Farmed Salmon and Interactions: Review of Potential Impacts and Recommended*

Action; the 1996 *Environmental Effects of Salmon Net Cage Aquaculture in British Columbia*; and the 1995 *Salmon Aquaculture Review*.

One of the most recent public reviews, the *Salmon Aquaculture Review*, was initiated in July 1995 to comprehensively review all salmon aquaculture legislation and policy. Conducted by the British Columbia Environmental Assessment Office, this review made several recommendations that were approved by the provincial government in 1999 and incorporated in a new salmon aquaculture policy.[266] The policy was devised through various forms of public participation, including oral and written submissions, open houses, and online disclosure of working session briefs. The Salmon Aquaculture Implementation Advisory Committee was established "to involve First Nations, coastal communities, environmental organizations, industry and the federal and provincial governments in the implementation of regulations, policy development, and the strategic development of the salmon farming industry."[267]

In enacting new regulations, the British Columbia government has made an effort to consult with members of the public most affected by possible changes in regulations. For example, before passing the 2002 *Aquaculture Regulation*, the BC government "consulted extensively with salmon farmers, equipment manufacturers, stakeholders and ministry staff."[268] Also, the new *Finfish Aquaculture Waste Control Regulation* was crafted with significant stakeholder and public participation through the Salmon Aquaculture Implementation Advisory Committee.[269]

The developing Shellfish Code of Practice is another provincial initiative that promoted substantial public participation. Approximately 500 letters of invitation to the community stakeholder meetings were sent to shellfish farmers, First Nations, local government, and community and environmental stakeholders.[270] Although there was a relatively low turnout from First Nations, most meetings were attended by those actively interested in the process.[271]

On a micro level, public participation may be available through the application processes. An applicant for the necessary lease or license of occupation under the *Land Act*[272] may be required to publish a notice of application in a local newspaper or *Gazette*.[273] The *Aquaculture Land Use Policy*, revised in October 2005,[274] notes that at the time of application, provincial staff will notify applicants if advertising is required. The Policy states that all new finfish applications will require public consultation which will most often be conducted through an open house session in a local community near the area under application.[275] The Policy provides that where an application for shellfish tenure covers an area which may affect an existing shellfish tenure's ability to expand, the applicant will have to notify neighboring tenure holders by registered letter. The Policy indicates that provincial staff will advise applicants if an upland owner's consent is needed in case of riparian right interferences. The Policy pledges that consideration will be given to integrated resource planning processes and other advisory processes, where they exist, in tenure allocation decisions.[276] Section 63 of the *Land Act* enables any individual to make a formal objection on a land tenure application. If an objection is filed, the minister has the absolute discretion to appoint an individual to hold a hearing concerning the issue(s) raised. At the conclusion of the hearing, the person appointed will submit a recommendation to the minister. This recommendation will be considered by the minister in making a final order.

Finfish Aquaculture Licensing Policies and Procedures for Applications, revised in November 2005,[277] remain quite general regarding public participation. Section 7 of the document states that reasonable efforts are to be made to notify affected parties and to provide them with an opportunity to comment on the application. The section also provides that the Minister of Agriculture and Lands may require the applicant to provide public notice of the proposed application in a manner that is acceptable.

The Ministry of Agriculture and Lands has adopted tenure siting criteria for finfish aquaculture.[278] Tenure will not be granted where it "would pre-empt important Aboriginal, commercial or recreational fisheries as determined by the province in consultation with First Nations and DFO."[279] A proposed site must be one kilometer in all directions from a First Nation reserve (unless consent is received from the First Nation).[280] New sites must be set at least three kilometers from any existing finfish aquaculture site, or in accordance with a local area plan or coastal zone management plan.[281] Other provisions provide for the protection of various public interests.

The *Farm Practices Protection (Right to Farm) Act*[282] is another process that allows for public input. Under s. 1(h) of this Act, "farm operation" includes "aquaculture as defined in the *Fisheries Act* if carried on by a person licensed, under Part 3 of that Act, to carry on the business of aquaculture."[283] Although the Act sets out a number of "right to farm" provisions, it also allows a person who is "aggrieved by any odor, noise, dust or other disturbance resulting from a farm operation" to make a complaint to the Farm Practices Board.[284] If the chair of the board is satisfied that a settlement of the complaint is unlikely, a panel of the board will be established to hear the complaint.[285]

These policy and legislative changes are direct responses to the Salmon Aquacultural Review Committee's criticisms of the past methods employed to garner meaningful public participation as newspaper advertisement and referral to a few selected groups were ineffectual and inconsistently applied, and resulted in an increase of local opposition.[286] The committee was also critical of the past consultation with First Nations, as those interests were not equitably represented, resulting in negative impacts and conflicts.[287]

The most recent inquiry into salmon farming, the Leggatt Inquiry, also recommended increased "involvement of Communities, especially First Nations, in consultation, partnership and ownership of salmon farming operations."[288] Both the federal and the British Columbia governments have been making efforts to enhance aboriginal participation in aquaculture decision-making. As mentioned earlier, on p. 72, aboriginal consultation is mandatory for all environmental assessments of aquaculture in the province.[289]

Public participation: Atlantic provinces

Among the Atlantic provinces, Nova Scotia has the most comprehensive legislative provisions for facilitating public participation in the decision-making process. In reviewing aquaculture license/lease applications, Nova Scotia's Minister of Agriculture and Fisheries is required to consult with a number of government departments, boards and agencies.[290] However, the minister may also refer the matter to a private-sector Regional Aquaculture Development Advisory Committee (RADAC) for comment and recommendation.[291] RADACs are formed in areas that have a potential for aquaculture development and are made up of individuals "who may be affected by the installation of an aquaculture site," including waterfront landowners and business operators.[292] There are ten such committees in Nova Scotia.[293]

After consulting with the various government departments, boards and agencies, the minister may refer the matter to public hearing.[294] There is no set formula used to determine which applications are referred to public hearings. Rather, the matter is one of ministerial discretion based upon the circumstances of the application and any perceived opposition within the community. Such opposition may be generated as a result of the public advertisements required by the Nova Scotia *Fisheries and Coastal Resources Act 1996* (FCRA).[295] NSDAF files indicate that many aquaculture applicants are unhappy with the public meeting component of the application process.

The minister may also refer contested issues to alternative dispute resolution (including conciliation, negotiation, mediation or arbitration) at which virtually any person would have standing.[296] This process may also be used as a substitute for a hearing, for conflict resolution generally, or for resolving disputes over the issuance, cancellation or suspension of a license or lease.[297] The alternative dispute resolution process has yet to be used.[298]

Subject to the Nova Scotia *Freedom of Information and Protection of Privacy Act*,[299] all information held by NSDAF must be available to the public and maintained in a public registry.[300] Furthermore, decisions and orders of NSDAF employees may be appealed to the minister, whose decisions may be appealed to the Supreme Court of Nova Scotia.[301] Appeals to the minister and to the court may only be brought by a "person aggrieved by" the decision.

In New Brunswick, the opportunity for public participation is considerably more limited. Although the minister is required to give public notice of an application for an aquaculture occupation permit[302] in any two local newspapers,[303] there is no provision for public meetings or other fora. Generally, a thirty-day period for the filing of written comments is allowed, and any written submissions must be acknowledged by the Registrar of Aquaculture.[304] The registrar must then communicate the final decision,[305] and residents within 100 meters of the proposed site receive personal notice.[306] Although the applicant must file a site development plan that can be viewed by the public, information about financing, fish health reports and records, and production levels are confidential.[307] Furthermore, only applicants are permitted to appeal the minister's decision,[308] and there is no provision for alternative dispute resolution.

Under the PEI Aquaculture Zoning System, an aquaculture applicant can only apply for a lease in a favorably zoned area. The presumption that zoning decisions have already accounted for the public interest has effectively eliminated most forms of public participation from the aquaculture leasing process. Although the PEI aquaculture division may revise the zoning policy from time to time after consultations with the Leasing Management Board and other stakeholders, there is no legislative or policy mechanism in place to require public involvement in this process.

In PEI, only those members of the public who appear to be directly affected by an aquaculture site (in other words, landowners with riparian rights) receive notice of a proposed aquaculture site.[309] Even so, there is no formal process for receiving comments from such affected persons.[310] Following the site evaluation process, the application is then forwarded to the Leasing Referral Committee. The committee has the discretion to seek input from other parties potentially affected by the proposed site. However, no person other than the applicant is permitted to appeal this decision,[311] and there is no provision for alternative dispute resolution.

Since aquaculture is a relatively new and small industry in Newfoundland and Labrador, there has not been much of an opportunity to obtain public input in aquaculture regulation on a macro level. However, public input was sought and deemed essential in passing the new *Environmental Protection Act*[312] and the *Water Resources Act*.[313]

On a micro level, when an application for an aquaculture facility has been received, an advertisement must be placed in one local and three regional newspapers, notifying the general public and special interest groups.[314] This notice must further advise that information related to the project is available on request, and that written responses regarding any concerns will be considered within twenty working days.[315] Newfoundland and Labrador's *Aquaculture Act* requires that the registrar keep records of aquaculture licenses, leases and environmental preview reports and impact statements, for inspection by the public. However, information deemed "confidential" is not available to the public.[316]

When the Aquaculture Licensing Committee (ALC) refers a proposal to different government departments and agencies for review, it also forwards a copy of the application and development/research plan to other non-governmental groups who have an interest in the proposal.[317] The ALC then forwards written submissions identifying any potential adverse effects to the

Aquaculture Development Officer, who considers and attempts to resolve realistic concerns.[318]

In cases where the aquaculture application concerns a "site congested area," the NLDFA may require up to six months to assess the carrying capacity and/or the needs of other water resource users in the area.[319] In these cases, the NLDFA consults with the Aquaculture Industry Association, provides written notice to the operators in the area of the assessment's details and possible licensing restrictions, and gives public notice of the assessment in one local and three regional newspapers.[320]

Other Newfoundland and Labrador legislation provides a number of additional opportunities for public participation. The *Environmental Protection Act* states that during an environmental assessment, interested persons may have the opportunity to submit written comments and that the project's proponent must provide an opportunity for a public meeting at a place adjacent to or in the geographical area of the undertaking, or as the minister may determine.[321] Members of the public may also play a limited role in the water license application under the *Water Resources Act*.[322] This Act allows the Minister of Environment to post public notices of a license application, presumably to allow the public to comment on the application. Once the license is granted, the Act allows a person to make a written complaint to the minister if the license holder fails to keep works in a proper and safe condition, or to comply with the terms of a license, or with the Act.[323]

Charting a course

Getting a clear picture of future Canadian law and policy reforms likely to be adopted in relation to sustainable aquaculture is difficult, for various reasons. First and foremost, federal government legal reviews and policy/discussion documents to date have been thin on suggesting specific legal and institutional changes. The 2001 *Legislative and Regulatory Review of Aquaculture in Canada*[324] highlights numerous law and policy issues, but deferred proposals for new legislation and amendments to a second phase of legal review. *Canada's Oceans Strategy*[325] re-emphasizes commitment to the principles of sustainable development, integrated management and the precautionary approach, but leaves legislative and regulatory discussion at a general and aspirational level. The strategy pledges to "improve existing legislation and guidelines on marine environmental protection,"[326] to "support new legislation, regulations and policies and programs aimed at protecting marine species at risk,"[327] and to "examine regulatory regimes to ensure effective environmental protection and streamline regulations."[328] *DFO's Aquaculture Policy Framework*[329] gives only broad commitments to reviewing and modernizing laws and regulations and to developing operational policies for existing regulatory responsibilities.[330]

The 2004 aquaculture governance recommendations by the Commissioner for Aquaculture Development, while including a few specific regulatory reform suggestions,[331] focused largely on non-legal initiatives[332] and possible future institutional changes. Three institutional scenarios were suggested for consideration: giving responsibility for aquaculture development to Agriculture and Agri-Food Canada; establishing an aquaculture agency to provide program and research support to the aquaculture industry; and continuing DFO lead agency responsibility for aquaculture but opening up relevant Agriculture and Agri-Food Canada programs to aquaculture.[333]

A second difficulty in predicting future law/policy reforms is the complexity and state of flux of provincial aquaculture approaches. For example, the British Columbia Environmental Assessment Offices' *Salmon Aquaculture Review* made forty-nine recommendations relating to aquaculture governance reform.[334] Provinces are at various stages in developing aquaculture policies and strategies where legal and institutional directions could be clarified.[335]

A third difficulty in predicting future law and policy developments on, for example, how strict and perhaps prohibitory regulatory measures should be is the still limited scientific understanding of the seriousness of risks posed by marine aquaculture operations and the lack of science-based criteria for siting decisions and operational standards. A 2004 report of the Commissioner of the Environment and Sustainable Development to the House of Commons[336] highlighted the significant gaps in scientific knowledge of the risks and potential effects of salmon aquaculture on wild salmon, particularly in the areas of diseases, sea lice and escapes of farmed salmon from aquaculture sites.[337] All three recent audits relating to salmon aquaculture emphasized the continuing absence of science-based criteria for site approvals and waste deposit controls.[338]

A fourth factor making future law and policy predictions problematic is the lack of a common societal vision of what sustainable aquaculture means,[339] which is fueled by differing moral and political views.[340] Voices for deregulation and smart regulation,[341] along with advocacies for allowing the aquaculture industry to be substantially subject to self-regulation[342] and market principles,[343] are clashing with calls by NGOs, environmentalists and others for more and stricter controls.[344]

Whether there should be a new federal aquaculture Act is an especially charged political issue. While the House of Commons Standing Committee on Fisheries and Oceans in 2003 recommended enactment of a federal Aquaculture Act,[345] several committee members expressed dissenting views.[346] The Department of Fisheries and Oceans did not respond enthusiastically to the recommendation and instead noted the need to fully explore the advantages and disadvantages, and various issues such as federal–provincial jurisdictional interactions and whether the *Fisheries Act* could be an adequate basis for developing aquaculture regulations.[347]

Although there are difficulties in foreseeing the precise nature of future aquaculture law and policy reforms, the four principles of sustainable development discussed in this chapter – integration, precaution, environmental impact assessment and public participation – do suggest general directions for strengthening the legal framework for Canadian aquaculture. While the details of law and policy reforms will depend on future review initiatives and political debates, some directions are clear and a number of critical questions need to be addressed.

Integration

The most likely avenue for enhancing vertical integration between the federal government and provinces/territories will be the development of a new framework agreement on aquaculture cooperation, followed by updated and modernized bilateral aquaculture agreements with each jurisdiction. In 2004, the Commissioner for Aquaculture Development recommended such an approach.[348] Discussions on a national Framework Agreement on Aquaculture (FAA) are continuing under the auspices of the Canadian Council of Fisheries and Aquaculture Ministers and their Task Group on Aquaculture. A target completion date has been set for 2007.[349]

Integrated coastal planning activities, presently quite fragmented, *ad hoc* and particularly limited in Atlantic Canada,[350] could be enhanced by further federal-provincial agreements to cooperate in implementing *Canada's Oceans Strategy* and *Action Plan*, with one of the four key pillars being integrated ocean management. Fisheries and Oceans Canada has indicated a desire to enter into further such agreements, as has already occurred for British Columbia with the September 2004 signing of a *Memorandum of Understanding Respecting Implementation of Canada's Ocean Strategy on the Pacific Coast*.[351]

Regardless of where integration in its various dimensions progresses in Canada, two fundamental limitations on provincial jurisdiction to regulate offshore marine aquaculture developments must be faced. Provinces may be substantially restricted in their offshore jurisdiction to be able to validly license aquaculture operations only within waters *inter fauces terrae* and marine waters historically considered to be part of the province.[352] As will be discussed in more detail in Chapter 4 by Phillip Saunders and Richard Finn, previous case law places a further potential hurdle in the way of provincial aquaculture licensing. In the case of *A.G. British Columbia* v. *A.G. Canada* (*Re B.C. Fisheries*),[353] the Privy Council decided that the federal government has the exclusive right to legislate in relation to the public right of fishing in the sea, and no public right to fish could be taken away except pursuant to federal legislation. While provinces can grant property rights over the beds of tidal waters falling within the province, any such grants that interfere with public rights of fishing would have to be authorized by federal legislation.[354]

In light of such provincial jurisdictional restrictions, if Canada wishes to continue supporting offshore marine aquaculture development, it should proceed to adopt federal aquaculture legislation for both preparatory and peremptory reasons.[355] With offshore aquaculture technologies quickly progressing[356] and provincial interests beginning to be expressed over seeing aquaculture move further offshore,[357] Canada should be legislatively prepared to control aquaculture operations beyond provincial boundaries. Federal legislation authorizing aquaculture operations to interfere with existing public fishing rights would also likely serve to pre-empt legal challenges from private parties displaced or hindered in their fishing activities.

Numerous practical questions would need to be addressed in developing new federal legislation. Whether a stand-alone federal Aquaculture Act[358] or a more comprehensive reform effort – for example, a modernized Aquaculture and Fisheries Act or an expanded *Oceans Act* – should be adopted is one key issue.[359] How new legislation should be "nested" within the existing complex array of federal and provincial laws needs to be considered. How exactly federal–provincial jurisdiction should be addressed is perhaps the most difficult question, with various options being possible. Those options include recognizing provincial jurisdiction over aquaculture leasing/ licensing out to 3 nautical miles (historical territorial sea), or 12 nautical miles (present territorial sea) with federal permits required for significant interferences with navigation and fishing; federal aquaculture licensing for areas beyond the territorial sea;[360] developing joint aquaculture licensing and approval arrangements, perhaps modeled on the existing joint federal–provincial offshore petroleum boards in Atlantic Canada, which were created by joint accord legislation;[361] and leaving jurisdictional details to be negotiated through federal–provincial MOUs.

Some key questions must also be faced regarding integrated planning. Does the *Oceans Act* provide an adequate federal foundation for promoting integrated management planning? The Act is not clear on how integrated plans might be given legal force and does not provide incentives or guarantee funding for provincial and local involvement.[362] Is new coastal zone management legislation needed at the provincial level? Such legislation might ensure clear planning mandates, adequate funding and enforceable decisions.[363]

Precaution

Three realities presently stand out in Canada in relation to precaution. First, the broad statement of federal governmental interpretation of the precautionary approach supports a weak version of the precautionary approach by not calling for a reversal in the onus of proof in decision-making processes and by requiring scientific information and evaluation in order to trigger the approach.[364] Second, the federal government has not strenuously sought to further clarify and discuss implications of the precautionary approach to the aquaculture sector but has largely given lip service to the principle and been content to rely on the general federal governmental framework.[365] Third, the provinces have been less than enthusiastic about addressing precaution in the aquaculture context, with no aquaculture legislation explicitly adopting the approach, and the principle largely ignored in discussion documents.[366]

Perhaps one of the greatest challenges, therefore, is how to open up bureaucratic and public discussions and debates over some of the numerous questions raised by the precautionary approach. Those questions include:

1   What is the appropriate role for risk assessment in aquaculture-related decision-making? The federal Commissioner for Aquaculture Development suggested that risk assessment is fundamental to an effective precautionary approach.[367] However, given the limitations of risk assessment created by scientific uncertainties,[368] the assessment of alternatives arguably should become an aquaculture "mantra." The focus might shift towards assessing technological alternatives, to better ensure ecological, economic and social sustainability.[369]

2   Who should decide final approval for aquaculture proposals, and who should ensure that precautionary measures are taken seriously? At both the federal and provincial levels in Canada, aquaculture approvals are largely left in the hands of departments that have the conflicting roles of promoting economic development and ensuring environmental protection.[370] There are a number of possible routes to avoid "capture by industry" and perceptions of conflict of interest, including the establishment of independent aquaculture licensing boards,[371] and ensuring multiple "checkpoints" by giving veto powers to environmental and health agencies.

3   What are the appropriate roles for scientific, social and ethical perspectives in precautionary decision-making?[372] A number of voices in Canada are calling for "science-based" decision-making.[373] Yet such calls may be viewed suspiciously by those who are not committed to a technocratic worldview.[374]

4   To what extent should government officials rely on learning by doing (or "adaptive management"), as opposed to prohibiting developments until applicants can prove that a proposed project will cause no significant environmental or social harm?

Future discussion on law and policy implications of precaution should be considered through various fora. For example, DFO might convene workshops on the precautionary approach and aquaculture in line with the many workshops that have already been held in the fisheries management field,[375] although greater involvement by NGOs, industry, the public and others should be encouraged.[376] Policy discussions could also be enhanced through application of strategic environmental assessment to proposed aquaculture strategies at the provincial level and eventual revision of aquaculture policy at the federal level.[377] Academic institutions in Canada might also play a greater role in aquaculture law and policy reform discussions, with perhaps the Ocean Management Research Network (OMRN) facilitating future research and workshops in the topic area.[378]

Meanwhile, even though the precautionary approach remains subject to ongoing criticisms[379] and controversies,[380] there are a number of law and policy directions that should be considered under the principle of precaution. Precaution should be expressly articulated as a guiding principle in all federal and provincial legislation directly relating to aquaculture,[381] so that administrators, and perhaps even the courts,[382] are responsible for working out the operational details. Pollution prevention in aquaculture should be encouraged by requiring plans for waste minimization and fish escape prevention as preconditions for licensing.[383] The application of existing provisions in Canadian environmental laws to aquaculture should be further clarified, for example the deleterious deposit and harmful habitat alteration/disruption offenses under ss. 35 and 36 of the *Fisheries Act*.[384] Monitoring, surveillance and enforcement should be promoted by requiring that aquaculture operators periodically report fish food, drug and chemical use; mandating "precautionary inspections" of licensed facilities to ensure compliance with regulatory and licensing standards; and requiring the prompt reporting of disease outbreaks. Specific precautionary standards should be established, such as requiring that there be no measurable adverse effects outside aquaculture tenure areas;[385] setting strict limits on the densities of cultured stocks and the distances between aquaculture operations; restricting the use of fish foods to those that are "environmentally friendly";[386] limiting the quantity and timing of drug and chemical use;[387] prohibiting salmonid grow-sites near migratory routes and rivers that support wild salmon stocks;[388] and requiring the use of closed-containment fish-rearing technologies.[389]

Environmental impact assessment

Since environmental assessment of marine aquaculture proposals has been almost completely exempted from provincial processes and subject merely to screening-level assessments under the *Canadian Environmental Assessment Act*, the primary target for EIA law reform logically should be on strengthening the screening category of assessment. Environmental assessment at the screening level has already been the subject of considerable criticisms,[390] and some of the reform coordinates that should be considered include mandating public participation rather than leaving it to the discretion of responsible federal authorities, setting clear timelines for public comments; requiring written governmental responses to public comments; and providing for alternative dispute resolution options such as mediation, fact-finding and arbitration. Funding for NGO and community participation in screening-level assessments also should be considered.[391] An appeal process to allow "checking" of screening decisions also needs to be seriously considered to ensure decision-makers have considered all critical questions, including cumulative effects and potential impacts on endangered or threatened species.[392]

Broader strengthening of the *Canadian Environmental Assessment Act* should also be considered. Government decisions should be required to

include justification that the various sustainability principles have been followed, including pollution prevention, the ecosystem approach, intergenerational equity, social equity, and integration, among others.[393] A transition
from a self-assessment approach by federal departments and agencies, at
times facing conflicts of interest, also needs to occur, with an arm's-length
independent assessment agency being one key option.[394] Providing for the
issuance of EIA permits that are enforceable regarding any mitigation and
monitoring conditions would be a further strengthening.[395]

A further critical area for law reform in Canada is strengthening SEA at
both the federal and provincial levels. Given the fundamental importance of
government policies, programs and plans to achieving sustainable development, and the rather poor record of SEA to date,[396] it seems critical that all
Canadian jurisdictions legally require SEAs with full public participation
ensured.[397] It also seems critical that existing policies and plans be subject to
required periodic review that would also be subject to SEA.[398]

The Report of the Commissioner for Aquaculture Development to the
Minister of Fisheries and Oceans Canada in 2004 recommended development of class-type screenings for aquaculture under EIA for both shellfish
and finfish operations,[399] but such a recommendation raises a key question:
to what extent might class screening of aquaculture proposals promote
administrative efficiency over critical environmental and social assessment?

Public participation

While public participation is important both for the development of regulatory standards and for citizen-monitoring of aquaculture sites, the most contentious public participation issues relate to public acceptance of proposed
aquaculture sites and facility operations.[400] To strengthen public participation in siting approval and in setting operational conditions, at least five
main avenues should be pursued. (1) Those members of the public concerned
with use conflicts and environmental issues should be guaranteed an adequate opportunity to comment on leasing/licensing applications. (2)
Regional and local aquaculture review committees should be established
where they do not presently exist to advise on aquaculture siting and management questions.[401] (3) Alternative dispute resolution (ADR) procedures
should be provided, such as mediation, conciliation and arbitration. Individuals opposed to aquaculture approvals should have the right to request
submission to ADR.[402] (4) The right of appeal should be provided to opponents of leasing/licensing decisions. (5) Priorities for site access by traditional
marine users, riparian landowners, adjacent communities and others should
be clarified, as part of Canada's commitment to ensure social equity.[403]

Full implementation of the other three sustainability principles focused
on in this chapter should also enhance public participation. Integrated
coastal/ocean planning promises to involve communities and members of the
public in long-term visioning of what human uses are appropriate and

should have priority. The precautionary approach encourages discursive processes that should engage the public in debates over what are acceptable societal risks.[404] EIA opens up proposed projects to public scrutiny and should be one of the primary avenues for expressing concerns about potential environmental and social impacts.[405]

A key question relating to public participation still remains for most Canadian jurisdictions. Should specific legislation be developed to address public participation in a broad way, such as an environmental bill of rights?[406]

Conclusion

The principles of integration, precaution, environmental impact assessment and public participation have emerged from international environmental law with important implications for law and policy reform in Canadian aquaculture. However, the task of putting these principles into practice is still largely unfinished. For example, none of the five provinces studied has incorporated the precautionary principle into aquaculture legislation. While the Department of Fisheries and Oceans has pledged allegiance to the precautionary approach in various documents, in relation to aquaculture DFO has yet to go very far beyond general and aspirational commitments.[407]

For better or worse, environmental crises may drive Canadian aquaculture law and practice to further implement the principles of environmental sustainability. One such crisis has emerged in British Columbia, where sea lice infestations from salmon farms near the Broughton archipelago have allegedly caused the reduction of wild pink salmon spawners. This crisis has focused public attention and debate on at least two of the principles highlighted in this chapter: precaution and integration. For instance, the Pacific Fisheries Resource Conservation Council (PFRCC), responsible for providing management advice for protecting wild fish stocks, called for application of the precautionary approach[408] and recommended a number of precautionary options including imposition of fallowing periods for some farms and development of a sea lice control plan.[409] In light of emerging environmental and socio-economic issues, PFRCC has suggested the need for relying on integrated planning.[410]

Moving from the general principles of sustainable development to better governance practices will likely be a long and rocky voyage. Swirling currents must be faced: First Nation claims to offshore title and rights; industry desires for secure access and regulatory efficiency; non-governmental organizations and public environmental concerns; coastal community calls for priority in licensing decisions and for empowerment through integrated coastal planning; bureaucratic calls for scientific rationality;[411] and globalization and foreign investment forces. Canada has just left port in the voyage towards sustainable aquaculture.[412]

Notes

The assistance of past researchers with Dalhousie's Marine and Environmental Law Institute (MELI), Josh Bryson, Melanie MacLellan, Sarah Dyck, Emma Butt, Tricia Warrender, and Mark Tinmouth, is gratefully acknowledged. This chapter represents an updated and substantially revised version of a two-part article under the same title by the authors in (2002) 28(1) *Queen's Law Journal* 279, (2003) 28(2) *Queen's Law Journal* 529.

The views of Mark Covan are his own and are not to be attributed to the Department of Justice. The views of Gloria Chao are her own and are not to be attributed to Blake, Cassels and Graydon LLP.

1 For overviews of the importance of sustainability principles to law and policy reforms and the broad range of principles, see Philippe Sands, "International Law in the Field of Sustainable Development: Emerging Legal Principles," in Winfried Lang, ed., Sustainable Development and International Law (London: Graham & Trotman/Martinus Nijhoff, 1995).

2 Under the heading of public participation, this chapter also briefly discusses how the principle of social equity or justice has been largely ignored in Canadian aquaculture laws in favor of a free-market mentality. For a review of the sources and implications of the social justice principle, see Maarten Bavinck and Ratana Chuenpadgee, "Current Principles," in Jan Kooiman, Maarten Bavinck, Svein Jentoft and Roger Pullin, eds., *Fish for Life: Interactive Governance for Fisheries* (Amsterdam: Amsterdam University Press, 2005) at 250–253. For a review of how British Columbia's salmon aquaculture industry has become increasingly concentrated in the hands of fewer corporations (five multinationals controlling 109 of 131 fish farm licenses as of 2003), see Dale Marshall, *Fishy Business: The Economics of Salmon Farming in BC* (Vancouver: Canadian Centre for Policy Alternatives – BC Office, 2003) at 9.

3 Bruce H. Wildsmith, *Aquaculture: The Legal Framework* (Toronto: Emond-Montgomery, 1982) at 33–91 [Wildsmith, "Aquaculture"].

4 *Constitution Act, 1867* (UK), 30 and 31 Vict., c. 3, s. 91(12), reprinted in R.S.C. 1985, App. II, No. 5.

5 *Ibid.* at s. 91(10).

6 *Ibid.* at s. 91(2).

7 See *R. v. Crown Zellerbach*, [1988] 1 S.C.R. 401 at 436.

8 *Constitution, supra* note 4 at s. 91(24).

9 *Ibid.* at ss. 91(29) and 92(10).

10 *Ibid.* at s. 92(13).

11 *Ibid.* at s. 92(16).

12 Bruce H. Wildsmith, *Federal Aquaculture Regulation*, Canadian Technical Report of Fisheries and Aquatic Sciences No. 1252 (Ottawa: Department of Fisheries and Oceans, 1984) at 3 [Wildsmith, "Federal"]. For a more in-depth look at the difficulties in establishing which waters are considered *inter fauces terrae*, see Wilson's dissent in *Reference Re Bed of the Strait of Georgia and Related Areas* [1984] 1 S.C.R. 388 at 440–446 and 461–470 [*Georgia Strait Reference*].

13 *Georgia Strait Reference, supra* note 12.

14 For further discussions of the uncertainties surrounding federal and provincial maritime boundaries, see Ted L. McDorman, "Canada's Ocean Limits and Boundaries: An Overview," in Lorne K. Kriwoken, Marcus Haward, David VanderZwaag and Bruce Davis, eds., *Oceans Law and Policy in the Post-UNCED Era: Australian and Canadian Perspectives* (London: Kluwer Law International, 1996) at 113–139; Phillip M. Saunders and Richard Finn, Chapter 4 in this volume; and *Georgia Strait Reference, supra* note 12 at 440–446 and 461–470.

15 See, for example, the cases of *Heiltsuk Tribal Council* v. *British Columbia Minister*

of Sustainable Resource Management (2003), 19 B.C.L.R. [*Heiltsuk Tribal Council*], *Haida Nation* v. *British Columbia (Minister of Forests)*, [2004] 3 S.C.R. 511 [*Haida Nation*] and *Homalco Indian Band* v. *British Columbia (Minster of Agriculture, Food and Fisheries)* (2005), 39 B.C.L.R. (4th) 263 (S.C.) [*Homalco Band*]. See also Chapters 8 by Diana Ginn and Chapter 9 by Ronalda Murphy, Richard Devlin and Tamara Lorincz in this volume.

16 At least nine different meanings of integration exist, including the integration of international agreements into national law and the integration of social, economic and environmental considerations in decision-making. See David VanderZwaag, Gloria Chao and Mark Covan, "Canadian Aquaculture and the Principles of Sustainable Development: Gauging the Law and Policy Tides and Charting a Course," (2002) 28 *Queen's Law Journal* 279 at 286–287.

17 For a discussion of federal delegation of power, see Richard W. Parisien, *The Fisheries Act: Origins of Federal Delegation of Administrative Jurisdiction to the Provinces* (Ottawa: Department of Environment, 1972).

18 See Wildsmith, "Aquaculture," *supra* note 3 at 5.

19 Other provinces and territories that have signed a memorandum of understanding with the federal government include Québec, the Yukon and Northwest Territories.

20 For example, most MOUs typically emphasize the province's power to regulate licenses and set out shared research and development initiatives.

21 See Canadian Intergovernmental Conference Secretariat, *Preamble to the Agreement on Interjurisdictional Cooperation with Respect to Fisheries and Aquaculture.* Online. Available http://www.scics.gc.ca/pdf/830662a1_e.pdf (accessed 23 September 2005).

22 Wildsmith, "Federal," *supra* note 12 at 4–5, notes that the 1912 oyster lease agreement the federal government signed with British Columbia granted the province the power to grant and administer oyster licenses, whereas the oyster lease agreements the federal government signed with Prince Edward Island and Nova Scotia, in 1928 and 1936 respectively, transferred the power to lease subaquatic lands for shellfish culturing to the federal government.

23 Wildsmith, "Federal," *supra* note 12 at 1.

24 Fisheries and Oceans Canada (DFO), *Federal Aquaculture Development Strategy* (Ottawa: DFO, 1995). Online. Available http://www.dfo-mpo.gc.ca/aquaculture/Library/index_e.htm (accessed 23 September 2005) at 10 (Principle 10): "Harmonization of federal and provincial policies and regulations is essential to aquaculture development" [Federal Aquaculture Strategy].

25 See Office of the Commissioner for Aquaculture Development (OCAD), "Regulation and Support for Aquaculture in Canada." Online. Available http://www.dfo-mpo.gc.ca/aquaculture/Interjuris dictional/regulationsupport _e.html (webpage no longer available). Depending on how aquaculture involvements are defined, as many as thirty-four federal departments and agencies may be directly or indirectly involved in aquaculture. See Fisheries and Oceans Canada, *Recommendations for Change: Report of the Commissioner for Aquaculture Development to the Minister of Fisheries and Oceans Canada* (Ottawa: Office of the Commissioner for Aquaculture Development, 2004) at 34 [Recommendations for Change].

26 Recommendations for Change, *supra* note 25. For reviews of Canadian regulatory approaches to therapeutants and pesticides used in salmon aquaculture, see Coral Leigh Cargill, "Regulation and Approval of Drugs and Pesticides Used in Canadian Salmon Aquaculture," (2004) 32 *Coastal Management* 331; and Association of Aquaculture Veterinarians of British Columbia, "SLICE (Emamectin Benzoate)." Online. Available http://www.agf.gov.bc.ca/fisheries/health/sealice.htm (accessed 23 September 2005).

27 See Fisheries and Oceans Canada, "Sustainable Aquaculture: The Role."

Online. Available http://www.dfo-mpo.gc.ca/aquaculture/Role/index_e.htm (accessed 23 September 2005).

28 Fisheries and Oceans Canada, "Canadian Coast Guard (CCG) Becomes a Special Operating Agency." DFO Media News Release NR-HQ-05-12E, 1 April 2005. Online. Available http://www.dfo-mpo.gc.ca/media/newsrel/ 2005/hq-ac12_e.htm? (accessed 23 September 2005).

29 Following the announcement on 12 December 2003 by Prime Minister Martin of the structural change, the Government of Canada transferred Canadian Coast Guard policy responsibilities and some operational responsibilities, including protection of navigable waters, from DFO to Transport Canada, with Transport Canada assuming *Navigable Waters Protection Act* (NWPA) responsibilities on 29 March 2004. Pursuant to a protocol between Transport Canada and DFO, oversight of screening report preparation under the *Canadian Environmental Assessment Act* for an NWPA-triggered assessment were delegated to DFO in a transition period starting 29 March 2004 and ending 1 October 2004. As described in Transport Canada, Decision following a Screening conducted by Fisheries and Oceans Canada pursuant to Section 20(1) of the *Canadian Environmental Assessment Act*, Duck-Cove-Finfish Aquaculture Site, New Brunswick (19 May 2004) (copy on file with the authors).

30 A prime focus of the committee has become implementation of *Canada's Oceans Action Plan*, released in May 2005. The Plan sets out four main activity areas: promoting Canada's international leadership, sovereignty and security in the oceans; supporting integrated management planning for the priority large marine areas; enhancing marine ecosystem health; and advancing ocean science and technology. See Fisheries and Oceans Canada, *Canada's Oceans Action Plan: For Present and Future Generations* (Ottawa: DFO, 2005). Online. Available http://www.dfo-mpo.gc.ca/canwaters-eauxcan/oap-pao/foreword_e.asp (accessed 23 September 2005).

31 See Fisheries and Oceans Canada, "Sustainable Aquaculture: Interjurisdictional Cooperation." Online. Available http://www.dfo-mpo.gc.ca/aquaculture/Inter-juris dictional/index_e.htm (webpage no longer available).

32 *Oceans Act*, S.C. 1996, c. 31.

33 See Fisheries and Oceans Canada, "Integrated Management." Online. Available http://www.dfo-mpo.gc.ca/canwaters-eauxcan/oceans/im-gi/index_e.asp (accessed 23 September 2005).

34 Government of Canada, *Policy and Operational Framework for Integrated Management of Estuarine, Coastal and Marine Environments in Canada* (Ottawa: Fisheries and Oceans Canada, 2002).

35 *Ibid.* at iii.

36 *Ibid.* at 15.

37 Fisheries and Oceans Canada, *supra* note 30.

38 Fisheries and Oceans Canada, *Sustainable Development Strategy: Progress Report on 2001–2003 Strategy* (Ottawa, Fisheries and Oceans Canada Communications Branch, 2004) at 14.

39 Fisheries and Oceans Canada, *Eastern Scotian Shelf Integrated Ocean Management Plan (2006–2011) Draft for Discussion*. Oceans and Coastal Management Report 2005-02 (Dartmouth: DFO, ESSIM Planning Office, 2005) at 9–10. Online. Available http://www.dfo-mpo.gc.ca/Library/286215.pdf (accessed 23 September 2005).

40 *Ibid.* at 43 and 49–59.

41 *Ibid.* at 12.

42 Fisheries and Oceans Canada, *Canada's Policy for Conservation of Wild Pacific Salmon* (Vancouver: Fisheries and Oceans Canada, 2005).

43 *Ibid.* at 27.

44 Ibid.
45 The policy notes that marine use planning is proceeding on a pilot basis under
    Canada's Oceans Strategy and Action Plan and recognizes that the results of other
    planning processes will have to be respected. Ibid. at 25 and 28.
46 British Columbia Ministry of Agriculture and Lands (BCMAL), "Salmon
    Aquaculture in British Columbia." Online. Available http://www.
    agf.gov.bc.ca/fish_stats/aqua-salmon.htm (accessed 23 September 2005); and
    BCMAL, "Shellfish Aquaculture in British Columbia." Online. Available
    http://www.agf.gov.bc.ca/fish_stats/aqua-shellfish.htm (accessed 23 September
    2005). British Columbia's aquaculture industry was worth CDN$264,900,000
    in 2003. Finfish aquaculture was responsible for CDN$249,000,000 of the
    total, with shellfish adding CDN$15,900,000.
47 Canada/British Columbia Memorandum of Understanding on Aquaculture Develop-
    ment (6 September 1988).
48 Ibid. at paragraphs 5.3.1–5.3.7. A number of coordination duties are identified
    in the MOU, including the following: (1) Canada shall advise holders of exist-
    ing licenses, issued by Canada, to apply for a license issued by British Colum-
    bia; (2) Canada and British Columbia will develop mutually acceptable
    aquaculture referral processes that consider fish health, fish habitat and
    fish harvesting concerns; (3) Canada and British Columbia will consult in
    exercising existing and in establishing new regulations and policies for
    the aquaculture industry; and (4) Canada agrees to provide, where possible,
    both general and site-specific information necessary to identify critical fish
    habitats, stocks and related matters to enable British Columbia to develop
    policies and programs that minimize the possibility of adverse impacts from
    aquaculture.
49 Office of the Auditor General of Canada, 2000 Report of the Auditor General,
    Chapter 30: Fisheries and Oceans: The Effects of Salmon Farming in British
    Columbia on the Management of Wild Salmon Stocks. Online. Available
    http://www.oag-bvg.gc.ca/domino/reports.nsf/html/0030ce.html/$file/0030ce.
    pdf (accessed 23 September 2005) at s. 30.73.
50 See also, "Federal Requirements" on the marine Finfish Aquaculture Applica-
    tion Guide website. Online. Avaliable http://www.agf.gov.bc.ca/fisheries/
    siting_reloc/marineaff_applic_guide_main.htm. Office of the Auditor General
    of Canada, 2004 Report of the Commissioner of the Environment and Sustainable
    Development, Chapter 5, "Fisheries and Oceans Canada: Salmon Stocks, Habitat,
    and Aquaculture" (Ottawa: Office of the Auditor General of Canada, 2004) at
    s. 5.66. Online. Available http://www.oag-bvg.gc.ca/domino/reports.nsf/html/
    c20041000ce.html/$file/c20041000ce.pdf (accessed 23 September 2005)
    [Report of the Commissioner of the Environment].
51 See Ministry of Fisheries and Ministry of Environment, Lands and Parks (as
    they were then), "Salmon Aquaculture Committee Established," News Release
    No. 00–16 (18 February 2000). The committee was formed to monitor and
    provide advice on Salmon Aquaculture Review [SAR] implementation, review
    data, and serve as a forum for dialogue. It consists of two co-chairs: the Assis-
    tant Deputy Minister of Fisheries and the Assistant Deputy Minister of
    Environment, Lands and Parks (as they were then), and fifteen members from
    various groups, including First Nations, environment, government and indus-
    try. Personal communication, Claire Townsend, Aquaculture Research and
    Development Officer, Ministry of Agriculture, Food and Fisheries (MAFF) (as
    it was then) (26 July 2002). The Salmon Aquaculture Review, initiated in July
    1995, was conducted by the British Columbia Environmental Assessment
    Office (EAO). In October 1999, the provincial government accepted many of
    the recommendations of EAO with the creation of a new salmon aquaculture

policy. MAFF, "Backgrounder: BC Salmon Aquaculture Policy," MAFF No. 02–01 (31 January 2002). Online. Available www.agf.gov.bc.ca/fisheries/reports/aqua_backgrounder.pdf (23 September 2005) [BC Salmon Policy Backgrounder].

52 This committee meets quarterly to provide broad policy direction and to increase efficiency and harmonization in a number of areas, including aquaculture. Townsend, *supra* note 51; and personal communication, Andrew Morgan, Assistant Regional Aquaculture Coordinator, Fisheries and Oceans Canada (26 August 2002).

53 Townsend, *supra* note 51; Morgan, *supra* note 52. The Project Review Team includes one representative from each responsible agency to review aquaculture applications to ensure harmonized tracking.

54 See British Columbia Environmental Assessment Office, *Salmon Aquaculture Review*, Vol. 1, at "Consolidated List of Recommendations," Recommendation 15. Online. Available http://www.eao.gov.bc.ca/epic/output/documents/p20/1036108250753_8f7f46b7f5d34 a4880f886282a95d52f.pdf (webpage no longer available). The group is made up of federal and provincial representatives with appropriate expertise to promote integrated and corporate fish health policy development.

55 See *2000 Auditor General Report*, *supra* note 49 at section 30.75. This committee was formed to provide technical and scientific information and advice in order to finalize the new *Finfish Aquaculture Waste Control Regulation*, BC Reg. 256/2002. Townsend, *supra* note 51.

56 Environment Canada, "Backgrounder: Why the Georgia Basin?" Online. Available http://www.pyr.ec. gc.ca/GeorgiaBasin/reports/whyGB/summary_e. htm (webpage no longer available) at 1.

57 See Fisheries and Oceans Canada, "Pacific Marine Heritage Legacy." Online. Available http://www.pac.dfo-mpo.gc.ca/oceans/fco/oceans_e. html (webpage no longer available).

58 Personal communication, Chris Hillian, Area Coordinator, Habitat Conservation and Stewardship Program, Fisheries and Oceans Canada (31 July 2002).

59 See Burrard Inlet Environmental Action Program (BIEAP) and Fraser River Estuary Management Program (FREMP), "Partners." Online. Available http://www.bieapfremp.org/fremp/partners/index.html (accessed 23 September 2005).

60 See BC Shellfish Growers Association website, http://www.bcsga.ca/industry_cop.html (accessed 23 September 2005).

61 Online. Available http://www.dfo-mpo.gc.ca/canwaters-eauxcan/infocentre/media/bc/index_e.asp (accessed 23 September 2005).

62 *Ibid.* See also Fisheries and Oceans Canada, "Canada and British Columbia Join Forces to Implement Canada's Oceans Strategy," News Release NR-PR-04-053e (18 September 2004).

63 Office of the Auditor General of British Columbia, *Salmon Forever: An Assessment of the Provincial Role in Sustaining Wild Salmon*, 2004/2005, Report 5 (Victoria: Office of the Auditor General of British Columbia, October 2004). Online. Available http://www.bcauditor.com/AuditorGeneral.htm (accessed 23 September 2005).

64 *Ibid.* at 65. Provincial siting criteria for salmon farms are set out in Appendix E of the Report.

65 *Land Act*, R.S.B.C. 1996, c. 245.

66 *Fisheries Act*, R.S.B.C. 1996, c. 149.

67 Section 38 of the *Land Act*, *supra* note 65, provides for the leasing of Crown land, while s. 39 provides for issuance of a license of occupation. Section 13(5)

of the *Fisheries Act* requires a license for carrying on aquaculture at any location in British Columbia or its coastal waters.

68 Section 97(1) of the *Land Act, supra* note 65, allows delegation of leasing/licensing responsibilities. Such delegation occurred in 1998 when the Minister of Environment, Lands and Parks (as it was then) delegated Crown land allocation authority to the British Columbia Assets and Land Corporation. In 2002, that Corporation was renamed Land and Water British Columbia Inc. (LWBC), which was given a broader mandate to include responsibilities for water licensing and allocation. See Land and Water British Columbia Inc., *2001/02 Service Plan Report* (2002); and Government of British Columbia, *News Release*, "Integrated Agency Will Improve Access to Crown Resources" (28 February 2002). In 2005, the programs under LWBC were integrated into several different agencies, with Crown land tenure responsibilities transferred to the Integrated Land Management Bureau. The LWBC no longer exists. See online www.lwbc.bc.ca (date accessed: 6 April 2006)

69 For a review of various reorganization efforts, see Government of British Columbia, *British Columbia Government Annual Report 2001–02: The New Era Update (2002)*; and Office of the Premier, Backgrounder Members of the Executive Council, 16 June 2005.

70 See online at ILMB website, http://ilmbwww.gov.bc.ca/ilmb/index.html (accessed 23 September 2005).

71 The ministry issues and administers aquaculture licenses under the *BC Fisheries Act*, and ensures compliance with the terms and conditions of individual aquaculture licenses. Some of these activities may be undertaken jointly with the Ministry of Environment and ILMB. Townsend, *supra* note 51.

72 The Division leads integrated and issue-specific land use and coastal planning, reviews aquaculture applications and provides referral responses, and is also responsible for archaeological preservation under the *Heritage Conservation Act*, tourism development planning, and coastal resource inventory mapping and information management. Townsend, *ibid.* See also Ministry of Sustainable Resource Management (as it was then), *2001/02 Annual Report: A New Era Update* at Appendix C (Legislation Administered by the Ministry), Online. Available http://www. gov.bc.ca/prem/down/annual_rpts/20ssrmweb.pdf [New Era Update].

73 This ministry has regulatory authority over waste and management matters under the *Finfish Aquaculture Waste Control Regulation*, the *Environmental Management Act* and the *Wildlife Act*. Its conservation officers oversee compliance with the *Environmental Management Act* and *Wildlife Act* and have delegated authority under the *Fisheries Act*. See *Fisheries Act, supra* note 66; the *Environmental Management Act* S.B.C. 2003 c. 53; the *Wildlife Act* R.S.B.C. 1996, 488; and Townsend, *supra* note 51. Also see New Era Update, *supra* note 72 at Appendix B (Legislation Administered by the Ministry).

74 Memorandum of Understanding (MOU) between the BCALC and Provincial Referral Agencies. Agency representatives signing the agreement were from Water, Land and Air Protection; Forests; Energy and Mines; Health Services; Agriculture, Food and Fisheries; Transportation; Land Resource Commission; and Treaty Negotiation Office (as they were then).

75 Referral agencies agreed to develop best management practices and guidelines so that pre-screening might be facilitated to eliminate unnecessary referrals. The MOU set out various dispute resolution steps with executive members responsible for the administration of the MOU being the final level of resolution. *Ibid.*

76 The Bureau reports to six deputy ministers who serve as an advisory board (Agriculture and Lands; Forests and Range; Environment; Tourism, Sport and

Arts; Aboriginal Relations and Reconciliation; and Energy, Mines and Petroleum Resources. See British Columbia Minister of Agriculture and Lands, *Integrated Land Management Bureau Service Plan Summary* 2006/2007–2008/2009 (2006).

77 See *Marine Finfish Aquaculture Application Guide, supra* note 50.

78 See LWBC and BC Ministry of Agriculture, Food and Fisheries (as they were then), *Guide to Completing Shellfish Management Plans* (revised February 2002). Online. Available http://www.lwbc. bc.ca/02land/tenuring/aquaculture/shellfish/mp_guide.pdf (accessed 23 September 2005).

79 Service Agreement on Coordination of Compliance and Enforcement Programs between the Ministry of Agriculture, Food and Fisheries, the Ministry of Water, Land and Air Protection, the Ministry of Sustainable Resource Management and LWBC (as they were then) (2002), Online. Available http://www.agf.gov.bc.ca/fisheries/Manuals/Inspect/10.17-Service_Agreement.pdf (accessed 23 September 2005).

80 British Columbia Land Use Coordination Office (as it was then), *Coastal Zone Position Paper* (Victoria: Land Use Coordination Office, June 1998).

81 Personal communication, Joe Truscott, Senior Coastal Planner, Ministry of Sustainable Resource Management (as it was then) (26 June 2002). See also Resource Management Division section of the Ministry of Sustainable Resource Development (as it was then) website, http://srmwww.gov.bc.ca/rmd/info.htm (accessed 23 September 2005), for further details on the division.

82 See Ministry of Sustainable Resource Management (as it was then), Resource Management Section, Coastal Planning, Projects and Initiatives Section website, http://srmwww.gov.bc.ca/rmd/coastal/ (accessed 23 September 2005).

83 *Ibid.*

84 *Ibid.*

85 Ministry of Sustainable Resource Management (as it was then), *B.C. Central Coast Resource Management Plan, Backgrounder* (4 April 2001). Online. Available http://srmwww.gov.bc.ca/cr/resource_mgmt/lrmp/cencoast/news/bkgrnd 111501.htm (accessed 23 September 2005).

86 Ministry of Sustainable Resource Management (as it was then), *Central Coast Land and Coastal Resource Management Plan.* Online. Available http://srmwww.gov.bc.ca/cr/resource_mgmt/lrmp/cencoast/docs/AIP%20Coastal%20Zone%20Plan.pdf (accessed 23 September 2005).

87 *Ibid.* Progress in implementing the plan is slow, however. See Office of the Auditor General of Canada, Report of the Commissioner of the Environment and Sustainable Development to the House of Commons, Chapter 1, Fisheries and Oceans Canada: Canada's Oceans Management Strategy (Ottawa: Office of the Auditor General of Canada, 2005) at 18.

88 Personal communication, Truscott, *supra* note 81.

89 Ministry of Sustainable Management, Resource Management Division (as it was then), *Nootka Sound Coastal Land Use Plan.* Online. Available http://srmwww.gov.bc.ca/rmd/coastal/north_island/nootka/Nootka.pdf; Ministry of Sustainable Resource Management (as it was then), *Chatham Sound Integrated Coastal Plan Terms of Reference and Map of the Study Area.* Online. Available http://srmwww.gov.bc.ca/rmd/coastal/north_coast/chatham/docs/ChathamTOR.pdf (accessed 23 September 2005); and Ministry for Sustainable Resource Management, *North Island Straits Coastal Plan.* Online. Available http://srmwww.gov.bc.ca/cr/resource_mgmt/lrmp/cencoast/docs/NIS_Final.pdf (accessed 23 September 2005).

90 Ministry of Sustainable Resource Management (as it was then), *Baynes Sound Coastal Plan for Shellfish Aquaculture.* Online. Available http://srmwww.

gov.bc.ca/rmd/coastal/south_island/baynes/docs/Baynes_Plan_Dec19_2002.pdf (accessed 23 September 2005); Ministry for Sustainable Resource Management, *Cortes Island Draft Terms of Reference*. Online. Available http://srmwww. gov.bc.ca/rmd/coastal/north_island/cortes/docs/TOR_Jan_2003.pdf (accessed 23 September 2005).

91   Ministry of Resource Management (as it was then), "Coastal Planning." Online. Available http://srmwww.gov.bc.ca/rmd/coastal/ (accessed 23 September 2005).

92   Trustcott, *supra* note 81.

93   Fisheries and Oceans Canada, "Governments Launch a New Aquatic Management Board on the West Coast of Vancouver Island," *News Release* (21 February 2002). See also: Evelyn Pinkerton, Anita Bedo and Arthur Hanson, *Final Evaluation Report: West Coast Vancouver Island Aquatic Management Board (AMB)* (22 March 2005), which evaluated the Board's progress over its three-year period and recommended continued support for the Board beyond the pilot period.

94   *Ibid*. The Board has established an Aquaculture Working Group and initial discussions have been held with the province about improving siting. See West Coast Vancouver Island Aquatic Management Board "Plans and Activities." Online. Available http://www.westcoastaquatic.ca/strategic_planning_approach.htm#a (accessed 23 September 2005).

95   Fisheries and Oceans Canada, Fisheries and Oceans Statistical Services, "2003 Aquaculture Production Statistics." Online. Available http://www.dfo-mpo.gc.ca/communic/statistics/aqua/aqua03_e.htm (accessed 23 September 2005). New Brunswick's total aquaculture production was worth CDN$188,200,000 in 2003.

96   Nova Scotia Department of Agriculture and Fisheries, "2003 Aquaculture Production Statistics." Online. Available http://www.gov.ns.ca/nsaf/aquaculture/stats/2003.shtml (accessed 23 September 2005). Over this period, the aquaculture sector in Nova Scotia was found to have a value of CDN$40,0001,978.

97   Prince Edward Island Department of Fisheries, Aquaculture and Environment, "Prince Edward Island Fishery Statistics 2003." Online. Available http://www.gov.pe.ca/photos/original/FA_fishstat_03.pdf (accessed 23 September 2005). PEI's aquaculture sector, primarily mussels and oysters, was worth CDN$31,880,000.

98   Newfoundland and Labrador Department of Fisheries and Aquaculture, "Aquaculture Production and Value." Online. Available http://www.fishaq. gov.nl.ca/aqua/prod_value.stm (accessed 23 September 2005). In 2003, the total value of the aquaculture sector's production was deemed to be CDN$15,304,000.

99   *Agreement for Commercial Aquaculture Development between the Government of Canada and the Government of the Province of Prince Edward Island* (1987) [Canada/PEI MOU].

100 *Canada–New Brunswick Memorandum of Understanding on Aquaculture Development* (1989) [Canada/NB MOU]; Canada/Newfoundland Memorandum of Understanding on Aquaculture Development (1988) [Canada/NF MOU]; *Memorandum of Understanding on Aquaculture Development between Minister of Fisheries and Oceans (Canada) and Minister of Agriculture and Fisheries* (Nova Scotia) (2002) [Canada/NS MOU].

101 Canada/NS MOU, *supra* note 100 at para. 8.4.

102 *Fisheries Act*, R.S.C. 1985, c. F-14.

103 *Fish Health Protection Regulations*, C.R.C., c. 812.

104 Canada/NS MOU, *supra* note 100 at para. 9.

105 Canada/NB MOU, *supra* note 100 at 4.

106 Canada/PEI MOU, *supra* note 99 at 2–3.
107 Canada/NF MOU, *supra* note 100 at para. 3(d).
108 Canada/NS MOU, *supra* note 100 at para. 8.4, 9.
109 Nova Scotia Department of Fisheries and Aquaculture, The Aquaculture Division (pamphlet) (Halifax).
110 Personal communication, Toby Balch, Aquaculture and Development Officer, Nova Scotia Department of Agriculture and Fisheries (NSDAF) (19 December 2003). Note that aquaculture sites are inspected yearly to determine "general compliance." The recent Canada/NS MOU provides that "Nova Scotia will be responsible for the implementation of the environmental effects monitoring program and the implementation of a follow-up program." See Canada/NS MOU, *supra* note 100 at para. 7.4. In 2003 and 2004, the NSDAF collected some 800 samples from beneath active and formerly active aquaculture sites in an attempt to conduct quantitative and qualitative assessments of sediments. The results of this study will be released to the public "once they are peer-reviewed and vetted through the regulators." Nova Scotia Agriculture and Fisheries, *A Discussion Document. Growing Our Future: Long Term Planning for Aquatic Farming in Nova Scotia* (February 2005) at 13. Online. Available http://www.gov.ns.ca/nsaf/aquaculture/ (accessed 23 September 2005) [Growing Our Future].
111 Canada and Prince Edward Island, *Prince Edward Island Aquaculture Leasing Policy* (March 2005) at 3 [PEI ALP].
112 *Ibid.* at 3.
113 *Ibid.* at Appendix 1.
114 In particular, the LRC is composed of members of the DFO (Conservation and Protection, Habitat Branch Resource Management), Transport Canada, the PEI Department of Fisheries, Aquaculture and Environment (Fisheries and Aquaculture Division and Water Resources Division) and the Canadian Wildlife Service. *Ibid.* at 43.
115 *Ibid.* at Appendix 2.
116 New Brunswick Department of Agriculture, Fisheries and Aquaculture, *Bay of Fundy Aquaculture Site Allocation Policy* at 11. Online. Available http://www.gnb.ca/cnb/policy/e-fundy.html (accessed 23 September 2005) [Fundy SAP]. Nova Scotia file materials indicate that the NS Department of Agriculture and Fisheries commonly refers aquaculture license applications to DFO, the Canadian Food Inspection Agency and the Canadian Coast Guard. The Canada/NF MOU also sets out similar provisions, including: "2(e) Newfoundland shall refer all applications to the Department of Fisheries and Oceans for comment prior to establishing the conditions of licenses." See Canada/NF MOU, *supra* note 100 at para. 2(e).
117 Department of Fisheries and Oceans, *Interaction between Wild and Farmed Atlantic Salmon in the Maritime Provinces, Habitat Status Report 99/IE* (Ottawa: DFO, 1999) at 20 [DFO Habitat Status Report].
118 Canada, Standing Senate Committee on Fisheries, *Report of the Standing Senate Committee on Fisheries: Aquaculture in Canada's Atlantic and Pacific Regions* (Ottawa: Standing Senate Committee on Fisheries, 2001) at 34 [Standing Senate Committee].
119 Canada/NS MOU, *supra* note 100 at para. 6.2.
120 See Government of Newfoundland and Labrador, *Aquaculture Licensing Policy and Procedures Manual* (DFO, revised 23 June 1998) at 3–5 [NL ALP].
121 *Fisheries and Coastal Resources Act*, S.N.S. 1996, c. 25, s. 47(a) [FCRA]. In Nova Scotia, the FCRA was proclaimed in force on 28 February 1997, repealing and replacing nine provincial statutes, including the *Aquaculture Act*, R.S.N.S. 1989, c. 18, the *Fisheries Act*, R.S.N.S. 1989, c. 173, and the *Nova Scotia Fish*

Inspection Act, R.S.N.S. 1967, c. 209. Aquaculture is regulated by the NS Department of Agriculture and Fisheries, which has subsumed its predecessor, the Department of Fisheries and Aquaculture. References in this chapter to the "DAF" refer to either the former or the current department, depending on the context. For more information on the Department's history, see its website, http://www.gov.ns.ca/nsaf/department/ (accessed 23 September 2005).

122 J. Mills, Wildlife Biologist, Western Region, Memorandum to Richard Hinton, Re: Aquaculture Application No. 1169 – Northwest Cove (13 December 1999) (copy on file with authors).

123 Gerrie Grevatt, "Fish Farm Approval Has Residents Fuming," *The {Halifax} Chronicle-Herald* (27 July 2002).

124 These fifty-three conditions involved a variety of matters, including fish density (at the site, as well as per cage), waste handling, pesticide and therapeutant usage, fish health surveillance monitoring, reporting of escapes and predator control. A number of conditions partly addressed bird issues, including a requirement that bird nets be on the cages at all times; a requirement to provide environmental sensitivity training to site personnel; a direction to site operators to become familiar with the *Migratory Birds Convention Act* and its regulations; and a limitation on scaring birds from the site unless approved by Environment Canada. See License No. 1169 between the Province of Nova Scotia and Aquafish Technology Inc. (18 October 2001) at Sch. B [License No. 1169].

125 For instance, the proponent must "manage resource disputes through good communication, education and negotiation" and "maintain navigational and local access corridors during site operation and provide site schematic identifying these corridors to interested local groups (e.g., tourism operators)." See License Nos. 1186, 1187, 1188, 1189 between the Province of Nova Scotia and Bounty Bay Shellfish Incorporated (5 April 2002) at Sch. B, s. E. Note that in *Sierra Club of Canada* v. *Canada (Attorney General)* (2003), 1 C.E.L.R. (3d) 71 (F.C.T.D.), the Federal Court of Canada set aside a decision of the Minister of Fisheries and Oceans approving the St. Ann's Project, ostensibly for failure to correctly provide for public participation in the approval process.

126 Fundy SAP, *supra* note 116 at 11.

127 *Aquaculture Act*, S.N.B. 1988, c. A-9.2, s. 28 [*Aquaculture Act NB*].

128 *Aquaculture Act*, R.S.N.L. 1990, c. A-13 [*Aquaculture Act NL*].

129 *Aquaculture Regulations*, Nfld. Reg. 1139/96.

130 NL ALP, *supra* note 120 at s. AP.5.

131 The Department of Environment reviews whether an aquaculture activity must be registered for environmental impact assessment and considers whether additional information is required for water-use licensing.

132 The Department of Municipal and Provincial Affairs determines whether a site is within municipal boundaries and whether a development permit may be required from the municipal council or a government service centre.

133 The Department of Tourism, Culture and Recreation may limit or prohibit the approval of an aquaculture application, since the *Historic Resources Act*, R.S.N.L. 1990, c. H-4, s. 18, requires that written consent of the minister be obtained if the application interferes with a provincial historic site, on land or under water.

134 The Department of Government Services and Lands requires that the applicant submit a separate application to Crown Lands for approval of a land or water lot, unless the applicant already has title to the property concerned. See NL ALP, *supra* note 120 at 3.1–3.4.

135 *Ibid.* at 4.1.

136 *Ibid.*

137 *Ibid.*

138 *Ibid.*

139 See Fisheries and Oceans Canada, *Role of the Provincial and Territorial Governments in the Oceans Sector* (Ottawa: Communications Directorate, Fisheries and Oceans Canada, 1997) at 31–35.

140 Nova Scotia Department of the Environment and Nova Scotia Department of Fisheries, *Coastal 2000: A Consultation Paper* (Nova Scotia: Nova Scotia Departments of Environment and Fisheries, 1994). Coastal 2000 evolved into the Coastal Zone Management Committee, reporting to the Nova Scotia Round Table on Environment and Economy. Personal communication, Toby Balch, Manager, Aquaculture Development, Nova Scotia Department of Agriculture and Fisheries (9 August 2005).

141 Personal communication, Hugh Gillis, Planning and Policy Advisor, Nova Scotia Department of Natural Resources (23 August 2002).

142 Guysborough County Regional Development Authority, Guysborough County Sustainable Aquaculture Initiative, Online. Available http://www.gcrda.ns.ca (accessed 23 September 2005). The initiative has been utilized as part of the NSDAF application process in relation to three proposed sites in 2005.

143 New Brunswick Department of the Environment, *Commission on Land Use and the Rural Environment: Final Report* (New Brunswick: Department of the Environment, 1993) at 297.

144 New Brunswick Department of the Environment and Local Government, *A Coastal Areas Protection Policy for New Brunswick*. Online. Available http://www.gnb.ca/elg-egl/0371/0002 (accessed 23 September 2005).

145 New Brunswick Department of the Environment and Local Government, Sustainable Planning Branch, "Coastal Areas Protection Policy: An Update on Implementation" (September 2005) (copy on file with authors).

146 Personal communication, Brian Meaney, Assistant Deputy Minister of Aquaculture, Newfoundland and Labrador Department of Fisheries and Aquaculture (30 May 2002).

147 *Aquaculture Act NL, supra* note 128.

148 NL ALP, *supra* note 120 at s. 1.3.

149 Canada/Newfoundland Agreement on Economic Renewal, *Sharing Coastal Resources: A Study of Conflict Management in the Newfoundland and Labrador Aquaculture Industry* (October 1998). Online. Available http://www. fishaq.gov.nl.ca/publications/aqua_man_study.stm (accessed 23 September 2005). Section 1 of the report states that in cases where significant opposition exists (approximately 8 percent), present conflict management strategies are highly ineffective, resulting in significant use of the NLDFA's time and resources.

150 *Ibid.* at 19, Recommendations 2, 9, 24.

151 Environment Canada, "What Is the Atlantic Coastal Action Program (ACAP)?" Online. Available http://atlantic-web1.ns.ec.gc.ca/community/acap/default.asp?lang=En&n=17F60 AA9-1 (accessed 23 September 2005). For a further review of ACAP and integrated coastal management initiatives in Canada, see Marcus Haward and Larry Hildebrand, "Integrated Coastal Zone Management," in Lorne K. Kriwoken *et al.*, eds., *supra* note 14 at 154–165.

152 *An Act to Amend the Canadian Environmental Assessment Act*, S.C. 2003, c. 9, s. 2(1).

153 *Canadian Environmental Protection Act, 1999*, S.C. 1999, c. 33.

154 The Preamble states:

> Whereas the Government of Canada is committed to implementing the precautionary principle that where there are threats of serious or irreversible damage, lack of full scientific certainty shall not be used as a reason for postponing cost-effective measures to prevent environmental degration.

155 *Pest Control Products Act*, S.C. 2002, c. 28.
156 *Ibid.* at ss. 17(1) and 20(1). For a further critique of how existing legislation implements the precautionary principle/approach, see David VanderZwaag, Susanna Fuller and Ransom Myers, "Canada and the Precautionary Principle/Approach in Ocean and Coastal Management: Wading and Wandering in Tricky Currents," (2002–2003) 34 *Ottawa Law Review* 117.
157 See Canadian Council of Ministers of the Environment, *A Canada Wide Accord on Environmental Harmonization* (Winnipeg: CCME, 29 January 1998). Online. Available http://www.ccme.ca/assets/pdf/accord_harmonization_e.pdf (accessed 23 September 2005), under paragraph 2 of its General Principles; and Federal/Provincial/Territorial Advisory Committee on Canada's National Programme of Action for the Protection of the Marine Environment from Land-based Activities, *Canada's National Programme of Action for the Protection of the Marine Environment from Land-Based Activities (NPA)* (Ottawa: Federal/Provincial/Territorial Advisory Committee on Canada's National Programme of Action for the Protection of the Marine Environment from Land-Based Activities, 2000), s. 1.4B at 16. Online. Available http://dsp-psd. pwgsc.gc.ca/Collection/En21-204-2000-1E.pdf (accessed 23 September 2005).
158 Fisheries and Oceans Canada, *Canada's Oceans Strategy: Our Oceans, Our Future* (Ottawa: Fisheries and Oceans Canada, 2002) at 11 [Canada's Oceans Strategy]. Online. Available http://www.cos-soc.gc.ca/doc/pdf/cos_e.pdf (accessed 23 September 2005).
159 Fisheries and Oceans Canada, *DFO's Aquaculture Policy Framework* (Ottawa: Fisheries and Oceans Canada, 2002) at 24. Online. Available http://www.dfo-mpo.gc.ca/aquaculture/ref/AAP_e.htm (accessed 23 September 2005) [Aquaculture Policy Framework].
160 Canadian Council of Fisheries and Aquaculture Ministers, Task Group on Introductions and Transfers of Aquatic Organisms, *National Code on Introduction and Transfers of Aquatic Organisms* (September 2003).
161 Privy Council Office, *A Framework for the Application of Precaution in Science-Based Decision Making about Risk*, Online. Available http://www. pco-bcp.gc.ca/docs/Publications/precaution/precaution_e.pdf (accessed 23 September 2005).
162 *Ibid.* at Principle 4.3, reads, "Sound scientific information and its evaluation must be the basis for applying the precautionary approach."
163 *Ibid.* Principle 4.3 further restricts the burden of proof treatment to the issue of who should bear the burden of producing scientific information through the words "the scientific information base and responsibility for producing it may shift as knowledge evolves."
164 *Ibid.* at Principles 4.9 and 4.10.
165 Royal Society of Canada, for Health Canada, Canadian Food Inspection Agency and Environment Canada, *Elements of Precaution: Recommendations for the Regulation of Food Biotechnology in Canada, An Expert Panel Report on the Future of Food Biotechnology* (January 2001). Online. Available http://www.rsc.ca//files/ publications/expert_panels/foodbiotechnology/GMreportEN.pdf (accessed 23 September 2005) [Expert Panel Report] at 165. Schedule XIX of the *New Substances Notification Regulations*, SOR/94-260, sets out rather general information requirements for living modified organisms, with section 5(c) of the schedule requiring information on "the potential of the organism to have adverse environmental impacts that could affect the conservation and sustainable use of biodiversity." See also Chapter 13 by Douglas Moodie in this volume.
166 Personal communication, Sarah Cosgrove, Senior Advisor, Biotechnology, Fisheries and Oceans Canada (15 September 2005).
167 *Ibid.*

168 Section 109(1) provides:

> Where the Ministers have assessed any information under section 109 and they suspect that a living organism is toxic or capable of becoming toxic, the Minister may, before the expiry of the period for assessing the information:
>
> (a) permit any person to manufacture or import the living organism, subject to any conditions the Ministers may specify;
>
> (b) prohibit any person from manufacturing or importing the living organism; or
>
> (c) request any person to provide additional information or submit the results of any testing that the Ministers consider necessary.

169 For a further critique of biotechnology legislation, see Canadian Institute for Environmental Law and Policy, *A Citizen's Guide to Biotechnology: Helping Citizens Have a Real Say in the Development of Biotechnology in Canada* (Toronto: CIELP, March 2002). For a recent critique of Canadian regulation of genetically modified foods in light of the precautionary approach, see Canadian Biotechnology Advisory Committee, *Improving the Regulation of Genetically Modified Foods and Other Novel Facts in Canada: Report to the Government of Canada Biotechnology Ministerial Coordinating Committee* (August 2002). Online. Available http://cbac-cccb.ca/epic/internet/incbac-cccb.nsf/en/ah00186e.html (accessed 23 September 2005).

170 Expert Panel Report, *supra* note 165.

171 *Ibid.* recommendations 6.13–6.16. Fisheries and Oceans Canada, as part of the Action Plan of the Government of Canada in response to the Royal Society of Canada Report on Food Biotechnology, has indicated agreement with the need to keep reproductively capable transgenic fish and transgenic aquatic organisms in secure land-based facilities. See Action Plan for the Government of Canada in response to the Royal Society of Canada Expert Panel Report, *Elements of Precaution: Recommendations for the Regulation of Food Biotechnology in Canada* (23 November 2001) at 28. For further review of existing policy, see Office of the Auditor General of Canada, "Response of the Federal Departments and Agencies to the Petition Filed November 21, 2001 by Greenpeace Canada under the Auditor General Act: Concerning the Federal Government's Position and Policy Regarding the Release of Genetically Engineered Fish (Government of Canada, April 2002)." Online. Available http://www.oag-bvg.gc.ca/domino/petitions.nsf/viewe1.0/28719EB6EAE833B885256C56006 89A8F (accessed 23 September 2005).

172 Ministry of Sustainable Resource Management (as it was then), News Release, "Draft Sustainability Principles" (22 May 2002) and News Release, "Sustainability Principles" (22 January 2003).

173 Ministry of Agriculture, Food and Fisheries (as it was then), "Finfish Aquaculture Licensing Policies and Procedures for Applications" (created 31 August 2003; revised 25 August 2003). Online. Available http://www.agf.gov.bc.ca/fisheries/ siting_reloc/MAFF_licensing_policy_update.pdf (accessed 23 September 2005).

174 BC Salmon Policy Backgrounder, *supra* note 51.

175 See Ministry of Agriculture, Food and Fisheries (as it was then), "New Technologies." Online. Available http://www.agf.gov.bc.ca/fisheries/technology/new_tech.htm (accessed 23 September 2005).

176 Ministry of Agriculture, Food and Fisheries (as it was then), "Escape Prevention Initiative," Online. Available http://www.agf.gov.bc.ca/fisheries/escape/escape_prevention.htm (accessed 23 September 2005).

177 B.C. Reg. 78/2002 repealing B.C. Reg. 364/89.

178 For more details, see British Columbia Legislative Assembly, Bill M 201, *Fisheries Amendment Act*, 3rd Sess., 37th Parl., 2002 (1st reading, 27 March 2002).

179 In September 2002, the government of British Columbia announced that it would begin accepting applications for new finfish aquaculture sites; see Minister of Agriculture, Food and Fisheries (as it was then), "Why B.C. Lifted the Moratorium on Fish Farms," Opinion Editorial (News Release) (26 September 2002).

180 See *Clear Choices, Clean Waters: The Leggatt Inquiry into Salmon Farming in British Columbia, Report and Recommendations* (November 2001) Online. Available http://www.davidsuzuki.org/files/Leggatt_reportfinal.pdf (accessed 23 September 2005) [Leggat Inquiry]. Testimony at the inquiry alleged that production levels at existing sites doubled between 1995 and 2000 at 18.

181 *Ibid.* at 26.

182 *Ibid.* at 23–24.

183 B.C. Reg. 256/2002.

184 West Coast Environmental Law, *The BC Government: A One Year Environmental Review* (July 2002) Online. Available http://www.wcel.org/wcelpub/2002/oneyearreview_final.pdf (accessed 23 September 2005) at 22.

185 *Environment Act*, S.N.S. 1994–95, c. 1. Interestingly, while Newfoundland and Labrador's new *Environmental Protection Act*, S.N.L. 2002, c. E-14.2 [NL EPA] does not include a principles section and does not expressly mention the precautionary principle, the government's *Guide to the Environmental Protection Act* states that various principles are reflected in the NL EPA, including the precautionary approach. Government of Newfoundland and Labrador, *Guide to the Environmental Protection Act* (St. John's: Newfoundland and Labrador Minister of Environment, August 2002) at 4.

186 The legislation in each of the Atlantic provinces is "bare bones" legislation, mandating few if any standards against which aquaculture sites are evaluated. For Nova Scotia, see generally the FCRA, *supra* note 121; for New Brunswick, see the *Aquaculture Act NB*, *supra* note 127; for PEI, see PEI ALP, *supra* note 111; for Newfoundland and Labrador, see the *Aquaculture Act NL*, *supra* note 128.

187 In addressing the Standing Senate Committee, *supra* note 118 at 33, Peter Underwood, then Deputy Minister of the Nova Scotia Department of Agriculture and Fisheries, emphasized the adaptive perspective:

> Many scientists and critics of aquaculture would say that sites should not be approved until all the questions have been answered. They want a modeling of the bay and input/output modeling and answers to all questions on impact. I think a more prudent approach is to proceed slowly and monitor the impact as you go. If no significant environmental effects are noted, perhaps operations can increase. There is a combination of upfront assessment and trying to answer as many questions as you can. We think it is as important to monitor the on-farm and off-farm footprint impacts over time in the real world, as opposed to trying to do all of the science in computer models up front.

188 ISA was first detected in Norway in 1984. Outbreaks have since occurred in Scotland, Chile and the Faroe Islands. *Ibid.* at 52.

189 Some 4,500 tonnes of fish, with a market value of nearly CDN$40 million, were slaughtered. The Government of New Brunswick paid out $8.00 per fish, for a total compensation of CDN$18,277,569.00 during 1998 and 1999. New Brunswick Department of Fisheries and Aquaculture, *1998–99 Annual Report* (Fredericton, NB: Department of Fisheries and Aquaculture, October 1999) at 13, 15, 22.

190 Standing Senate Committee, *supra* note 118 at 27. The total cost to New Brunswick taxpayers was approximately CDN$44 million.

191 Bridget M. Kuehn, "Officials Fine-Tune Salmon Virus Response," (1 July 2002) *Journal of the American Veterinary Medical Association.* Online. Available http://www.avma.org/onlnews/javma/jul02/020701e.asp (accessed 23 September 2005).

192 Fundy SAP, *supra* note 116. The Site Allocation Policy divides the Bay of Fundy into zones known as aquaculture bay management areas (ABMAs), the establishment of which is based upon oceanographic, fish health and business considerations. Within each ABMA, New Brunswick "encourages" bay management agreements, with the intention to eventually legislate such agreements. These bay management agreements will be subject to approval by the Department of Agriculture, Fisheries and Aquaculture, and will contain fish management standards and practices concerning, among other things, fish health, waste management, environmental management and a code of practice to minimize escapes. To minimize disease transmission between cages within a farm and between farms, all salmon aquaculture sites are required to adopt "single-year class operating practices," defined as one generation of fish on a site at one particular time. Any fish held over (which cannot exceed 20 percent of the production of the site and must not be held beyond September of the production year) will be subject to annual review by the Fish Health Technical Committee.

193 For examples within the *Aquaculture Act NB, supra* note 127 at s. 11(1), which provides that the registrar may make an aquaculture license subject to additional terms and conditions over and above those prescribed by regulation. However, the only regulation currently under the *Aquaculture Act* is the *General Regulation–Aquaculture Act*, N.B. Reg. 91-158. While the general regulation contains some standards (for example, s. 26(a) mandates a minimum 300 meters between sites), the minister need not refuse the applicant for failure to comply. In March 2004, the New Brunswick Department of Fisheries and Aquaculture released the first draft of a document entitled "Infectious Salmon Anemia (ISA) Management and Control Program." In conjunction with the New Brunswick Fish Health Surveillance Program, this program is designed to provide "a comprehensive and standardized approach to the management and control of ISA, and to minimize the overall economic impact of ISA." In May 2003 and June 2004, the New Brunswick Salmon Growers Association released the "Generic Waste Management Plan and Instructions for Application to Specific Marine Aquaculture Facilities Using a Net-Plan Culture System for Reviewing Atlantic Salmon in the Bay of Fundy" and the "Environmental Policy and Codes of Practice." Online. Available http://www.nbsga.com/science.html (accessed 23 September 2005).

194 The *Aquaculture License and Lease Regulations*, N.S. Reg. 125/2000, are infused with ministerial discretion. For example, under "Location and Marking," the regulations mandate that sites be a minimum 25 meters from the mean low-water level. However, that minimum distance may be overridden if the minister is of the "opinion the area is required for the aquaculture undertaking."

195 The MFDP is one of the first applications an aquaculturalist must complete in applying for an aquaculture license. This application (revised October 1998) requires the provision of the following information: particulars of the applicant (name, address, etc.); description of the marine site; hydrographic coordinates; area of site (diagrams, land ownership of shorefront property and fishery considerations, conflicting uses); site assessment (exposure of site, water depth, water current, bottom type and fish habitat, water temperature, salinity, waste

management); farm design and layout; production and harvest strategies; and business information. The MFDP is currently in the process of being revised, but this revision was not completed at the time of writing. Nova Scotia Department of Agriculture and Fisheries (NSDAF), *Marine Finfish Farm Development Plan (Application form)* (NSDAF Aquaculture, revised October 1998) [MFDP].

196   *Ibid.* Aquaculture license conditions, however, may impose some standards. For example, License No. 1169, *supra* note 124, included conditions regarding mortalities of fish and fish health surveillance. Condition 27 stated:

> Frequent removal of mortalities must be carried out. Abnormal rates of mortalities (exceeding 0.05% of fish in a cage per day) must be reported to and investigated by a veterinarian. Cases of infectious disease must be managed in consultation with a veterinarian in a prudent manner that will protect the well being of the fish stock and prevent infectious disease spread to other farms or wild fish stocks.

> Condition 28 further stipulated that

> A Fish Health Surveillance Program will be implemented and maintained. This program will be carried out at least weekly and will involve examination of mortality, lice counts, measurements of dissolved oxygen, salinity and temperature and plankton sampling and enumeration when warranted. A veterinarian will visit the farm at least every six weeks between April 1 and December 31 and will examine sick and moribund fish.

197   The MFDP, *supra* note 195, contains "recommendations," none of which is a prescribed standard. With respect to water currents, the recommendation is that "average currents at the centre of the proposed site should be equal or greater than 10 cm/sec." The Minister requires information on minimum and maximum water current flow, direction, and the duration of slack tide, if any. The applicant is not required to provide information on duration of minimum and maximum current flows, seasonal variations, how the aquaculture site itself will affect such flows, whether there are any wildlife impacts or effects, and so on. Indeed, all that is required of the applicant is that over the course of a six-hour tidal cycle (on one day), they position floats at the extreme ends and middle of the site, thereafter measuring the change in position each hour. No account is made for seasonal variation. As the sampling is only done on one day, it cannot be seen as representative of the site conditions generally.

Questions 4(B) and 4(C) of the MFDP are also illustrative of the very generalized, subjective self-assessment approach of the MFDP: "4(B) – Compared to the general area where your site is located, how would you describe the strength and nature of the currents at your particular site? Above average, Average, Below average; 4(C) – Are there any important advantages or disadvantages to the water current patterns at the site? Is there a circular or vortex current patterns?"

198   Personal communication, Alan C. Chandler, Aquaculture Licensing Manager, Nova Scotia Department of Agriculture and Fisheries (13 March 2001).

199   *Ibid.*

200   For example, an environmental assessment document prepared by a consultant in support of Lease/License No. 1169, *supra* note 124, a proposal by Aqua Fish Technology Ltd. to raise Atlantic salmon and rainbow trout in Northwest Cove, NS, gave various indications that the company would follow recommendations in the Nova Scotia Codes of Practice. Marci Penney, "Information Regarding Finfish Aquaculture Facility, Lease/License No. 1169," prepared for Aqua Fish Technology Ltd. (Halifax, Nova Scotia: May 2000). The environ-

mental assessment document prepared for Lease/License No. 1169 referred to s. 18 of the Codes of Practice, which provided only general statements of intention regarding blood water containment and treatment:

> To ensure maximum product quality, the accepted, approved method for harvesting farmed fish involves exsanguination or bleeding the fish in an ice-bath. As this resultant blood water may act as a vector for infection to other farmed fish, appropriate biosecurity measures should be followed. . . . As part of the farm's general biosecurity program, industry encourages proper blood water containment, disinfection and disposal during fish harvesting.

The Aquaculture Association of Nova Scotia and the Nova Scotia Department of Agriculture and Fisheries no longer refer to the codes as valid authority. A copy of Nova Scotia's *Aquaculture Industry Environmental Management Guidelines: Finfish Aquaculture Operations* (undated), referred to as Codes of Practice, is on file with the authors.

201 *Aquaculture License and Lease Regulations*, N.S. Reg. 15/2000.
202 *Ibid.* at s. 5.
203 PEI ALP, *supra* note 111 at 16.
204 See, for example, the report by the Prince Edward Island Aquaculture Alliance, "Review of the PEI Aquaculture Leasing Program" (8 January 1999). Online. Available http://www.aquaculturepei.com (webpage no longer available).
205 *Aquaculture Act NL, supra* note 128 at s. 4(5). These include health, safety and environmental matters; resource utilization and sustainable development; use, stocking, investment in or production of the facility; access by contiguous landowners through a site; records and documents to be provided; the source and strain of all stock used; the intensity and concentration of aquaculture activities; escape and disease prevention; protection of other aquaculture facilities; and "whatever other terms may be necessary to carry out the purpose of this Act."
206 NL ALP, *supra* note 120 at s. AP.12 – "Shellfish Culture License Policy."
207 *Ibid.*
208 *Ibid.* at s. AP.21 – "Additional Licensing in Site Congested Areas."
209 DFO Habitat Status Report, *supra* note 117 at 2.
210 License No. 1169, *supra* note 124 at Sch. B, condition 23.
211 Personal communication, Chris Mills, Finfish Biologist, PEI Fisheries, Aquaculture and Environment (27 August 2002).
212 See NL ALP, *supra* note 120 at s. AP. 1 5 – "Inspections." It is anticipated that DFA will visit all licensed aquaculture sites at least annually.
213 *Aquaculture Act NL, supra* note 128 at s. 6.
214 License No. 1169, *supra* note 124 at Sch. B, condition 7.
215 Personal communication, Cyril Boudreau, Aquaculture Development Officer, Nova Scotia Department of Agriculture and Fisheries (28 August 2002).
216 *Canadian Environmental Assessment Act*, S.C. 1992, c.37, as amended [CEAA].
217 There are four "triggers" that will initiate the CEAA assessment process: (1) the project proposal trigger; (2) the financial trigger; (3) the land interest trigger; and (4) the law list trigger. *Ibid.* at ss. 5(1)(a)–(d).
218 Where there is no specific "trigger," the minister has the discretionary power to refer the project to a mediator or review panel where the project may cause significant adverse environmental effects in another province or outside Canada, or on aboriginal lands. *Ibid.* at s. 48(1).
219 See Fisheries and Oceans Canada, *Interim Guide to Information Requirements for Environmental Assessment of Marine Finfish Aquaculture Projects* (15 February 2002). Online. Available http://www-heb.pac.dfo-mpo.gc.ca/publications/

pdf/finfish_ceaa.pdf (accessed 23 September 2005); and Fisheries and Oceans Canada, *Interim Guide to Information Requirements for Environmental Assessment of Marine Shellfish Aquaculture Projects* (15 February 2002). Online. Available http://www-heb.pac.dfo-mpo.gc.ca/publications/pdf/shellfish_ceaa.pdf (accessed 23 September 2005).

220 S.O.R/97–181.

221 Pursuant to the DFO's *Aquaculture Site Application Review Process and Interim Guides* (18 January 2002), the DFO's Regional Aquaculture Coordinator will generally manage the overall review process and coordinate any intersectoral meetings or actions required to complete the review.

222 See the Canadian Environmental Assessment Agency website, http://www. ceaa-acee.gc.ca/050/index_e.cfm (use keyword "aquaculture") (accessed 23 September 2005).

223 Personal communication, William Coulter, Regional Director, Canadian Environmental Assessment Agency (10 August 2001).

224 *Ibid.* It should be noted, of course, that over 99 percent of environmental assessments conducted under CEAA have involved the screening level of assessment. See CEAA, *Review of the Canadian Environmental Assessment Act: A Discussion Paper for Public Consultation* (December 1999). Online. Available http://www.ceaa-acee.gc.ca/013/001/0002/0001/index_e.htm   (accessed   23 September 2005).

225 CEAA, *supra* note 216 at s. 18(3).

226 Personal communication, Allison Webb, A/Director, Sustainable Aquaculture, Fisheries and Oceans Canada (1 November 2004).

227 CEAA, *supra* note 216 at s. 18(2).

228 CEAA, *Review of the Canadian Environmental Assessment Act* (April 2000), Online. Available http://www.ceaa.gc.ca/013/001/index_e.htm (accessed 23 September 2005) at 13.

229 Office of the Commissioner for Aquaculture Development, *Legislative and Regulatory Review of Aquaculture in Canada* (Ottawa: Fisheries and Oceans Canada, March   2001).   Online.   Available   http://www.dfo-mpo.gc.ca/aquaculture/ Library/index_e.htm [OCAD Review].

230 Webb, *supra* note 226.

231 *Ibid.*

232 *Ibid.*

233 *Species at Risk Act*, S.C. 2002, c. 29.

234 *Ibid.* The List of Wildlife Species at Risk set out in Schedule 1 lists Atlantic salmon in the inner Bay of Fundy as endangered. Such a listing in particular could trigger special assessment review, for example where a salmon aquaculture proposal carries the risk of escapes and adverse effects on wild stocks. A "best practice" guide, not meant to be specific to any one piece of legislation, recommends that the onus of proof should be on the proponent to demonstrate to the satisfaction of the decision-maker that adverse effects on wildlife at risk or biological diversity would not be significant in situations where there is a threat of serious or irreversible harm. See Environment Canada, Canadian Wildlife Service, *Environmental Assessment Best Practice Guide for Wildlife at Risk in Canada*, 1st ed. (27 February 2004) at 25. Online. Available http://www.cws-scf.ec.gc.ca/publications/AbstractTemplate.cfm?lang=e &id=1059 (accessed 23 September 2005).

235 For a discussion of challenges and difficulties in implementing SARA in practice, particularly in relation to marine species, see David L. VanderZwaag and Jeffrey A. Hutchings, "Canada's Marine Species at Risk: Science and Law at the Helm, but a Sea of Uncertainties," (2005) 36 *Ocean Development and International Law* 219.

236 The directive was first issued in 1990 and revised in both 1999 and 2004. Canadian Environmental Assessment Agency, *The Cabinet Directive on the Environmental Assessment of Policy, Plan and Program Proposals*. Online. Available http://www.ceaa.gc.ca/016/directive_e.htm (accessed 23 September 2005).

237 Report of the Commissioner of the Environment, *supra* note 50 at Chapter 4, "Assessing the Environmental Impact of Policies, Plans and Programs."

238 *Ibid*. Chapter 4 at 10.

239 *Ibid*. Chapter 4 at 21.

240 *Environmental Assessment Act*, S.B.C. 2002, c. 43. For a critical review of the legislation, see West Coast Environmental Law's brief, "Bill 38: The New Environmental Assessment Act." Online. Available http://www.wcel.org/deregulation/bill38.pdf (accessed 23 September 2005).

241 *Ibid.*

242 See SAR, *supra* note 51, vol. 1, c. 3 – "VIII Environmental Impact Assessment."

243 *Ibid.*

244 See Canada/BC News Release, "Canada and British Columbia Sign Cooperation Agreement on Environmental Assessment" (21 April 1997) and Canada–British Columbia Agreement for Environmental Assessment Cooperation (2004). Online. Available http://www.eao-gov.bc.ca/publicat/canada-bc-agreement/can-bc-agree-mar1104.pdf (webpage no longer available).

245 Section 3 of the *Environmental Assessment Regulations*, N.S. Reg. 26/95 [EAR], provides that undertakings and classes of undertakings listed in Schedule A are subject to the provincial EAP. The Schedule divides undertakings into Class I and Class II, and further divides Class I undertakings into the categories of industrial facilities, mining, transportation, energy, waste management and other, the latter of which includes "an enterprise, activity, project, structure or work which disrupts a total of 2ha or more of any wetland," and "such other undertaking as the Minister may from time to time determine." However, wetland is defined in s. 2(g) as "marshes, swamps, fens, bogs, and shallow water areas."

246 *Ibid.* at s. 11.

247 *Ibid.* at s. 11(1).

248 *Ibid.* at s. 11(2).

249 *Clean Environment Act*, R.S.N.B. 1973, c. C-6 as am. by S.N.B. 1983, c. 17, s.6 [CEA].

250 Section 31.1(1) defines "environment" as "(a) air, water or soil; (b) plant and animal life including human life; and (c) the social, economic, cultural, and aesthetic conditions that influence the life of humans or of a community insofar as they are related to the matters described in paragraph (a) or (b)." "Environmental impact" is defined as "any change to the environment."

251 See the *Environmental Impact Assessment Regulation – Clean Environment Act*, N.B. Reg. 87-83, Sch. A. [EIAR–CER]

252 United States National Oceanic and Atmospheric Administration, Northeast Fisheries Science Center, "Wild Atlantic Salmon in Maine Protected as Endangered Species," Press Release (13 November 2000), Online. Available http://www.fws.gov/northeast/newsrel/asalmon2.html (webpage no longer available).

253 Numbers have been in decline since 1990 and have varied from a peak of 40,000 mature fish in the 1970s to less than 200 wild adult salmon in 2003. Fisheries and Oceans Canada, Canadian Science Advisory Secretariat, *Allowable Harm Assessment for Inner Bay of Fundy Atlantic Salmon*, Stock Status Report 2004/030 at 2.

254 EIAR-CER, *supra* note 251 at ss. 4(a), 6(6).

255 *Ibid.* at s. 9(1)(a).
256 Personal communication, Toby Balch, Aquaculture and Development Officer, Nova Scotia Department of Agriculture and Fisheries (19 December 2003); and personal communication, Darryl Wells, New Brunswick Department of Environment (16 August 2001). The latter indicated that such assessments may be undertaken in the future.
257 *Environmental Protection Act*, R.S.P.E.I. 1988, c. E-9 [PEI EPA].
258 Personal communication, Sean Ledgerwood, Environmental Assessment Officer, PEI Department of Fisheries, Aquaculture and Environment (14 January 2003).
259 PEI EPA, *supra* note 257 at s. 9(2).
260 NL EPA, *supra* note 185 at ss. 47, 2.
261 *Environmental Assessment Regulations 2000*, Nfld. Reg. 48/00, s. 29.
262 Personal communication, Bas Cleary, Director of Environmental Assessment, Newfoundland and Labrador Department of Environment (6 December 2002).
263 See "Background" of Fisheries and Oceans Canada, *Federal Aquaculture Development Strategy* (Ottawa: DFO, 1995). Online. Available http://www.dfo-mpo.gc.ca/aquaculture/ref/FADS_e.pdf (accessed 23 September 2005) at 1, noting the consultation process at the first Canadian Aquaculture Planning Forum in 1992 and the consultation on the *Federal Aquaculture Development Strategy* involving over 350 national and international stakeholders in 1994.
264 See the next subsection under "British Columbia" for more details.
265 Section 18(3) of CEAA, *supra* note 216, broadens public participation to any stage of the screening process but would still leave public participation largely at the discretion of the responsible authority.
266 As of April 2003, of the forty-nine SAR recommendations, almost all were fully or partially adopted, and two were considered not applicable. BC Salmon Policy Backgrounder, *supra* note 51 at 7. For further information see "Status of Salmon Aquaculture Review Recommendations" (April 2003). Online. Available http://www.agf.gov.bc.ca/fisheries/salmonreview_apr03.pdf (accessed 23 September 2005).
267 See BC Salmon Policy Backgrounder, *supra* note 51 at 6.
268 See Ministry of Agriculture, Food and Fisheries (as it was then), "Fisheries and Aquaculture section." Online. Available http://www.agf.gov.bc.ca/fisheries/aquaculture_regs.htm (webpage no longer available).
269 Ministry of Water, Land and Air Protection (as it was then), "Aquaculture Waste Control Regulation Backgrounder" (31 January 2002). Online. Available Aquaculture Waste Control Regulation Backgrounder at 1 (webpage no longer available).
270 See Kingzett Professional Services Report (3 July 2002) at 2. MAFF contracted Kingzett Professional Services to consult with stakeholders regarding a draft Shellfish Code of Practice and submit a report on the consultations and the draft.
271 *Ibid.* at 3.
272 *Land Act, supra* note 65.
273 *Ibid.* at s. 33(1)(3).
274 British Columbia Ministry of Agriculture and Lands, *Crown Land Use Operational Policy: Aquaculture* (2005). Online. Available http://www.lwbc.bc.ca/oilwbc/policies/policy/land/aquaculture.pdf.
275 *Ibid.* at s. 8.1.7.
276 *Ibid.* at s. 8.
277 British Columbia Ministry of Agriculture and Lands, Finfish Aquaculture Licensing Policies and Procedures for Applications (2005). Online. Avaliable http://www.agf.gov.bc.ca/fisheries/siting_rebc/MAl_licensing_policy_update.pdf.

278 British Columbia Ministry of Agriculture and Lands, "Criteria for siting new finfish aquaculture facilities." Online. Available http://www.agf.gov.bc.ca/fisheries/Finfish/Provincial_Siting_Criteri_March_2000.pdf.
279 *Ibid.* at point 12.
280 *Ibid.* at point 1.
281 *Ibid.* at point 15.
282 *The Farm Practices Protection (Right to Farm) Act*, R.S.B.C. 1996, c. 131.
283 *Ibid.*
284 *Ibid.* at s. 3(1).
285 *Ibid.* at s. 5.
286 SAR, *supra* note 51, Vol. 1, "Report of the Environmental Assessment Office," c. 12.
287 *Ibid.* Volume 2, "First Nations Perspectives," c. 4.
288 Leggatt Inquiry, *supra* note 180 at 25, recommendation 3.
289 This policy is strictly adhered to by the federal and BC governments. At the time an application for environmental assessment is made, a joint letter is sent by the federal and BC governments notifying the various aboriginal groups that have potential interests in the area in question of the proposed project. Webb, *supra* note 226.
290 FCRA, *supra* note 121 at s. 47(a)
291 *Ibid.* at s. 47(b).
292 Nova Scotia Department of Agriculture and Fisheries, "Regional Aquaculture Development Advisory Committees: Background Information." Online. Available http://www.gov.ns.ca/nsaf/aquaculture/radac/index.shtml (accessed 23 September 2005).
293 *Ibid.*
294 FCRA, *supra* note 121 at s. 48(c). Note that s. 49 concerns public hearings and sets notice requirements, *inter alia.*
295 *Ibid.* at s. 49(2).
296 *Ibid.* at s. 9(1).
297 *Ibid.* at s. 9(5).
298 Personal communication, Alan Chandler, Licensing Manager, Nova Scotia Department of Agriculture and Fisheries (16 August 2001).
299 *Freedom of Information and Protection of Privacy Act*, S.N.S. 1993, c. 5.
300 FCRA, *supra* note 121 at s. 8(3).
301 *Ibid.* at ss. 118(1), 119.
302 *Aquaculture Act NB, supra* note 127 at ss. 25(3.1), 26(4).
303 *General Regulations – Aquaculture Act*, N.B. Reg. 91–158, s. 24(6) [*General Regulations Aquaculture Act*].
304 New Brunswick Department of Agriculture, Fisheries and Aquaculture, *Bay of Fundy Marine Aquaculture Site Allocation Application Guide* (October 2000) (copy on file with the authors) [Fundy Application Guide]. However, in *Fundy North Fishermen's Assn.* v. *New Brunswick (Minister of Agriculture, Fisheries and Aquaculture)* [2000] N.B.J. No. 493 (N.B.) (QL) [Fundy Fishermen], the court upheld the minister's decision to limit submissions to ten days, finding that the thirty-day period referred to in the application guide is strictly administrative and not binding on the minister.
305 Fundy Application Guide, *supra* note 304 at 15.
306 *General Regulations Aquaculture Act, supra* note 303 at ss. 24(3), 24(5)(b).
307 Fundy SAP, *supra* note 116 at 13.
308 *Aquaculture Act NB, supra* note 127 at s. 28(1). See also Fundy Fishermen, *supra* note 304.
309 PEI ALP, *supra* note 111.
310 The *P.E.I. Fisheries Act*, R.S.P.E.I. 1988, c. F-13.01, does ensure that a

landowner is not deprived of his or her riparian rights. Section 8.1(2) of the
Act provides that landowners are to have reasonable access to the waters adjacent to their land. "Reasonable access" is defined quantitatively by s. 8.1(3).

311 PEI ALP, *supra* note 111 at 37.
312 NL EPA, *supra* note 185.
313 *Water Resources Act*, S.N.L. 2002, c. W-4.01. The Environment Minister
encouraged those who were interested to attend public meetings or to submit
written comments. See Newfoundland and Labrador Department of Environment, *Proposed Environmental Protection Act and Water Resources Act: A Discussion
Paper for Public Consultations* (St. John's: Newfoundland and Labrador Department of Environment, 2001) at 2. See also Newfoundland and Labrador
Department of Environment, "Environment Minister Announces Consultation
on Proposed Environmental Legislation," News Release NLIS 5 (19 September
2001).   Online.   Available   http:www.gov.nf.ca/releases/2001/env/
0919n05.htm (accessed 23 September 2005).
314 NL ALP, *supra* note 120 at s. AP.6.
315 Personal communication, Leonard House, Aquaculture Development Officer,
Newfoundland and Labrador Department of Fisheries and Aquaculture (30
May 2002).
316 *Aquaculture Act NL*, *supra* note 128 at s. 9.
317 NL ALP, *supra* note 120 at ss. 3.4.1–3.4.2.
318 *Ibid.* at ss. 4.4, A-P.6.
319 *Ibid.* at s. A-P.21.
320 *Ibid.*
321 NL EPA, *supra* note 185 at s. 58.
322 *Water Resources Act*, *supra* note 313.
323 *Ibid.* at s. 23(1).
324 OCAD Review, *supra* note 229. A second-phase review has not occurred and
appears unlikely in light of the "amalgamation" of OCAD into Fisheries and
Oceans Canada's Aquaculture Management Directorate.
325 Canada's Oceans Strategy, *supra* note 158.
326 *Ibid.* at 22.
327 *Ibid.* at 23.
328 *Ibid.* at 24.
329 Aquaculture Policy Framework, *supra* note 159.
330 *Ibid.* at 28.
331 Regulatory recommendations included, among others, the need to establish
regulations pursuant to s. 36 of the *Fisheries Act* to authorize the deposit of
deleterious substances from aquaculture operations and the need to amend
provincial and regional regulations under the *Fisheries Act* to exclude aquaculture from application. See Recommendations for Change, *supra* note 25, Recommendations 1 and 5.
332 For example, a special fund to support development of and implementation of
integrated management pilot projects in aquaculture-prevalent areas and new
funding to support aquaculture growth were recommended. Recommendations for Change, *supra* note 25 at Recommendations 8 and 9.
333 *Ibid.* at 46–52.
334 See SAR, *supra* note 51.
335 For example, Nova Scotia has released a discussion document on the future of
aquaculture with plans to incorporate comments into a comprehensive aquaculture development strategy to be released in 2005. See Growing Our Future,
supra note 110. New Brunswick has agreed with the need to develop and
implement a comprehensive aquaculture strategy as recommended by a recent
Auditor General's report. See New Brunswick Office of the Auditor General,

Salmon Stocks, Habitat and Aquaculture, Report of the Auditor General – 2004, Vol. 1. Online. Available http://www.oag-bvg.gc.ca/domino/ media. nsf/html/c200405pr_e.html (accessed 23 September 2005) at paras. 60–63.

336 Report of the Commissioner of the Environment, *supra* note 50 at Chapter 5, "Fisheries and Oceans Canada: Salmon Stocks, Habitat and Aquaculture."

337 *Ibid.* at 15–17.

338 *Ibid.* at ii. The report notes that the British Columbia and New Brunswick audits also reached the same conclusion.

339 This point was emphasized by auditors assessing New Brunswick governance approaches to aquaculture development. New Brunswick Office of the Auditor General, *supra* note 335 at 5.

340 On the "hard choices," which are fundamentally moral and political rather than economic and technical, see Jan Kooiman and Svein Jentoft, "Hard Choices and Values," in Jan Kooiman *et al.*, *supra* note 2.

341 For example, the federal government has been pressing for smart regulation in light of industry perceptions that the current regulatory framework acts as a constraint to innovation, competitiveness, investment and trade. See Privy Council Office, "Smart Regulation: A Regulatory Strategy for Canada: Executive Summary." Online. Available http://dsp-psd.pwgsc.gc.ca/Collection/ CP22-78-2004-1E.pdf (accessed 23 September 2005).

342 For example, in 2004, Fisheries and Oceans Canada responded with a self-regulatory justification to a recommendation from the Commissioner of the Environment and Sustainable Development that regulations to control deleterious deposits from aquaculture operations be developed. DFO was of the view that industry efforts to develop codes of practice for waste management are the best way to achieve environmental protection from potential deleterious substances. Report of the Commissioner of the Environment, *supra* note 50 at 21.

343 For example, a recent multi-agency task force looking at salmon farming in Atlantic Canada called for a "market-based industry" and management based on "market-based principles." See Fisheries and Oceans Canada *et al.*, *Report of the Task Force on Fostering a Sustainable Salmon Farming Industry for Atlantic Canada* (April 2005) Online. Available http://www.mar.dfo-mpo.gc.ca/com-munications/maritimes/taskforce-e.html (accessed 23 September 2005) [Atlantic Canada Task Force].

344 For a review of varying public perspectives of aquaculture and the adequacy of existing regulation, see Fisheries and Oceans Canada, "Overview: Qualitative Research Exploring Canadians' Perceptions, Attitudes and Concerns towards Aquaculture." Online. Available http://www.dfo-mpo.gc.ca/misc/focus-aqua-culture_e.htm (accessed 23 September 2005).

345 Parliament of Canada, Standing Committee on Fisheries and Oceans, *The Federal Role in Aquaculture: Report of the Standing Committee on Fisheries and Oceans* (presented to the House on 8 April 2003). Online. Available http://www.parl.gc.ca/committee/CommitteePublication.aspx?COM=3292&L ang=1&SourceId=37577 (accessed 23 September 2005), at Recommendation 1, at 21–22.

346 The Bloc Québécois viewed the recommendations as "highly interventionist and presumptuous" and opposed the introduction of federal aquaculture legislation since it would be superfluous and duplicative of Québec's regulatory controls over mariculture operations. *Ibid.* at 88. Member of Parliament, John Cummins opposed new federal aquaculture legislation on the basis that new legislation would demand that aquaculture be given priority over wild fish and their habitat, and from a strong conviction that the existing *Fisheries Act* must be enforced and regulations developed thereunder. *Ibid.* at 97, 111.

347 Fisheries and Oceans Canada, *Government Response to the 3rd Report of the Standing Committee on Fisheries and Oceans on the Federal Role in Aquaculture in Canada.* Online. Available http://www.dfo-mpo.gc.ca/communic/reports/ aquaculture2003/let_e.htm (accessed 23 September 2005).

348 Recommendations for Change, *supra* note 25 at Recommendation 7. Various functions of the FAA and individual implementation agreements were foreseen, namely to coordinate policy objectives, delineate respective roles and responsibilities, streamline administrative processes, establish service standards, prescribe consultative mechanisms including procedures for conflict resolution, and identify cost-shared programs and services, where appropriate.

349 See Fisheries and Oceans Canada, *DFO's Sustainable Development Strategy, 2005–2006* (Ottawa: DFO, 2005). Online. Available http://www.dfo-mpo.gc.ca/sds-sdd2005-06/Index_e.htm (accessed 23 September 2005) at 26.

350 For example, the New Brunswick Office of the Auditor General has noted the ongoing need to set up an integrated coastal management system as part of a provincial strategy for aquaculture. New Brunswick Office of the Auditor General, *supra* note 335 at 39, 41.

351 See Fisheries and Oceans Canada, "Canada–British Columbia MOU on Implementation of Canada's Oceans Strategy," DFO Backgrounder BG-PR-04-053e (18 September 2004). Online. Available http://www-comm.pac.dfo-mpo.gc.ca/pages/ release/bckgrnd/2004/bg053_e.htm (accessed 23 September 2005). The *Memorandum of Understanding Respecting the Implementation of Canada's Oceans Strategy on the Pacific Coast of Canada* is available online: http://www.dfo-mpo.gc.ca/canwaters-eauxcan/infocentre/media/bc/ index_e.asp (accessed 23 September 2005).

352 *Georgia Strait Reference, supra* note 12.

353 *A.G. British Columbia* v. *A.G. Canada (Re B.C. Fisheries)* (1913) 15 D.L.R. 308 (P.C.).

354 See Phillip Saunders and Richard Finn, Chapter 4 of this volume, "Summary of Jurisdictional Structure," (p. 115).

355 There are also various *practical* rationales, such as establishing clear objectives and principles for aquaculture, emphasizing federal priority for sustainable aquaculture development, recognizing the roles and interests of First Nations, ensuring long-term institutional and financial support, providing a firm legal foundation for federal provincial agreements and clarifying the federal jurisdictional roles in addressing aquaculture.

356 For example, see Christopher J. Bridger and Barry A. Costa-Pierce, *Open Ocean Aquaculture: From Research to Commercial Reality* (Baton Rouge, LA: World Aquaculture Society, 2003).

357 For example, both Nova Scotia's discussion paper on aquaculture, Growing Our Future, *supra* note 110, and the task force report on sustainable salmon farming in Atlantic Canada, *supra* note 343, mention the future potential for offshore aquaculture operations.

358 The drafting of such legislation was suggested by Bruce Wildsmith in 1985, with an actual draft Act provided as an example. See Bruce H. Wildsmith, *Toward an Appropriate Federal Aquaculture Role and Legislative Base*, Canadian Technical Report on Fisheries and Aquatic Sciences No. 1419 (Ottawa: Department of Fisheries and Oceans, 1985). Wildsmith (p. 34) recommended against simply incorporating new aquaculture provisions into the *Fisheries Act* and regulations thereunder in light of the possible message that aquaculture is simply a subset of the broader field of fisheries.

359 The need to modernize the *Fisheries Act* has been broadly recognized but no clear timeline appears to have been set. Fisheries and Oceans Canada,

2005–2010 Strategic Plan: Our Waters, Our Future. Online. Available http://www.dfo-mpo.gc.ca/dfo-mpo/plan_e.htm (accessed 23 September 2005) at section V, sets a priority over the next five years to "develop a new governance model for fisheries management; including proposals to modernize the Fisheries Act." Addressing aquaculture under a wider legislative umbrella might further complicate the political process.

360 Such a move would be in line with the *National Offshore Aquaculture Act 2005*, presently before Congress in the United States, which would authorize the Secretary of Commerce to issue offshore aquaculture permits in the exclusive economic zone. See NOAA Aquaculture, "NOAA Releases Offshore Aquaculture Bill," Online. Available http://www.nmfs.noaa.gov/mediacenter/aquaculture/ (accessed 23 September 2005). See also Chapter 14 by Jeremy Firestone in this volume.

361 For such a suggestion, see Gloria Chao, "Offshore Aquaculture in Canadian waters: Legal issues and challenges," (December 2003) 103(3) *Bulletin of the Aquaculture Association of Canada* 32.

362 See Aldo Chircop, Hugh Kindred, Phillip Saunders and David VanderZwaag, "Legislating for Integrated Marine Management: Canada's Proposed Oceans Act of 1996," (1995) 33 *Canadian Yearbook of International Law* 305 at 317–321.

363 For an overview of US state efforts regarding aquaculture and coastal planning, see G. R. Nelson, M. Richard De Voe and C. L. Jenseng, "Status, Experiences, and Impacts of State Aquaculture Plans and Coastal Zone Management Plans on Aquaculture in the United States," (1999) 9 *Journal of Applied Aquaculture* 1.

364 Privy Council Office, *supra* note 161.

365 For example, the Aquaculture Policy Framework, *supra* note 159 at 24, is quite general in relation to precaution: "DFO's use of the precautionary approach in the context of aquaculture development will be informed by the *Oceans Act* and federal direction regarding risk management, including application of the precautionary approach."

366 For example, Nova Scotia's aquaculture discussion document issued in February 2005, Growing Our Future, *supra* note 110, does not mention the precautionary approach, and the Atlantic Canada Task Force, *supra* note 343, emphasizes the need for market-based principles.

367 Yves Bastien, Commissioner for Aquaculture Development, "How to Farm the Seas: The Science, Economics and Politics of Aquaculture," Speaking Notes, Rodd Brudenell River Resort, Montague, PEI (30 September 2000) [Speaking Notes]. The Commissioner stated that

> "Sustainability" and the "precautionary approach" are essentially buzz words that will have as many definitions as the number of people sitting around the table. Therefore, these notions are useless in the real life of decision-makers because they do not refer to precise standards, precise objectives or precise deliverables. Moreover, debating such notions between participants may increase misunderstanding, each party interpreting the notion differently.... What is really needed is risk assessment, risk management and risk communication.
>
> (copy of speech on file with the authors)

368 See Mary O'Brien, *Making Better Environmental Decisions: An Alternative to Risk Assessment* (Cambridge, MA: MIT Press, 2000).

369 On the multiple dimensions of sustainability, see Anthony Charles, *Sustainable Fishery Systems* (Oxford: Blackwell Science, 2001) at 188–191.

370 For example, various persons appearing before the Standing Senate Committee

on Fisheries in its review of aquaculture expressed concerns over conflict in the DFO's mandate for both the promotion of aquaculture development and the protection of wild fish stocks and habitat. Standing Senate Committee, *supra* note 118 at 18–19.

371   The possibility for an independent, arm's-length mechanism to make fisheries allocation decisions based on clear policy and criteria was suggested for the fisheries management field. See Fisheries and Oceans Canada, *The Management of Fisheries on Canada's Atlantic Coast: A Discussion Document on Policy Direction and Principles* (Ottawa: Atlantic Fisheries Policy Review, 2001) at 34. Online. Available   http://www.dfo-mpo.gc.ca/afpr-rppa/linksto_discodoc_e.htm (accessed 23 September 2005). However, the Policy Framework subsequently released in 2004 did not adopt the approach. See *A Policy Framework for the Management of Fishereies on Canada's Atlantic Coast* (Ottawa: Fisheries and Oceans Canada, 2004). Online. Available http://www.dfo-mpo.gc.ca/afpr-rppa/link_policy_framework_e.htm (accessed 23 September 2005).

372   See A. Dan Turlock, "Environmental Law: Ethics or Science?" (1996) 7 *Duke Environmental Law and Policy Forum* 193. For the view that the precautionary principle enhances arguments for conservation measures and ecosystem values, see Jaye Ellis, "The Straddling Stocks Agreement and the Precautionary Principle as Interpretive Device and Rule of Law," (2001) 32 *Ocean Development and International Law* 289.

373   For example, the federal Commissioner for Aquaculture Development has called for science to play a central role in decision-making: "The first key message is that science is an essential component of sound decision-making and might be better financed and coordinated." See Speaking Notes, *supra* note 367.

374   On the need to move from centralized management focused on science to new processes of governance that are open to social and community movements, see Michael M'Gonigle, "The Political Economy of Precaution," in Carolyn Raffensperger and Joel Tickner, eds., *Protecting Public Health and the Environment: Implementing the Precautionary Principle* (Washington, DC: Island Press, 1999) at 128–143.

375   See, for example, Fisheries and Oceans Canada, *Proceedings of a Workshop on Implementing the Precautionary Approach in Canada* (5–9 October 1998), Canadian Stock Assessment Proceedings Series 98/18; Fisheries and Oceans Canada, *Report of the Joint Science–Fisheries Management Policy Workshop on the Precautionary Approach (DFO Newfoundland Region)* (27 September 1999), Canadian Stock Assessment Secretariat Proceedings Series 2000/04; Fisheries and Oceans Canada, *Science Strategic Project on the Precautionary Approach on Canada: Proceedings of the Second Workshop* (1–5 November 1999), Canadian Stock Assessment Proceedings Series 99/41; Fisheries and Oceans Canada, *Proceedings of the National Meeting on Applying the Precautionary Approach in Fisheries Management* (10–12 February 2004), Canadian Science Advisory Secretariat Proceedings Series 2004/03.

376   Workshops focusing on precaution in fisheries management have been restricted largely to government scientists and managers.

377   *DFO's Aquaculture Policy Framework* does not provide any clear timeline or process for revision but merely states that DFO will make necessary adjustments of external conditions change to ensure that it remains relevant and effective. See Aquaculture Policy Framework, *supra* note 159 at 32.

378   The OMRN is an interdisciplinary network with an initial focus on social science research and has promoted the holding of regional and national ocean management workshops. It has been funded by DFO and the Social Sciences and Humanities Research Council of Canada (SSHRC). See the OMRN

website, http://www.omrn.ca/eng_home.html (accessed 23 September 2005).

379   Criticisms include the limited direction for decision-makers in light of the high level of generality and the tendency to focus on a narrow subset of risks. See Jaye Ellis and Alison FitzGerald, "The Precautionary Principle in International Law: Lessons from Fuller's Internal Morality," (2004) 49 *McGill Law Journal* 779; and Cass R. Sunstein, "Beyond the Precautionary Principle," (2003) 151 *University of Pennsylvania Law Review* 1003.

380   For example, controversies continue to rage over how weak or strong precaution should be in implementation, with the strongest interpretive versions calling for a shift in the burden of proof. See Richard G. Hildreth, M. Casey Jarman and Margaret Langlas, "Roles for a Precautionary Approach in Marine Resources Management," (2005) 19 *Ocean Yearbook* 32; and David VanderZwaag, "The Precautionary Principle and Marine Environmental Protection: Slippery Shores, Rough Seas, and Rising Normative Tides," (2002) 33 *Ocean Development and International Law* 165. For a case rejecting the petitioners' argument for a strong version of precaution whereby the provincial government and the holder of a fish farm licence would bear the burden of proving that an amendment of a licence allowing aquaculture of Atlantic salmon would pose no risk to wild salmon stocks, see *Homalco Band, supra* note 15.

381   For example, explicit recognition of the precautionary approach is absent in much federal legislation relevant to aquaculture, including the *Fisheries Act*, R.S.C. 1985, c. F-14, the *Navigable Waters Protection Act*, R.S.C. 1985, c. N-22, and the *Food and Drugs Act*, R.S.C. 1985, c. F-27. For a listing of federal legislation relevant to aquaculture, see OCAD Review, *supra* note 229 at Annex IV.

382   The decision by the Supreme Court of Canada in 114957 *Canada Ltée (Spraytech Société d'arrosage)* v. *Hudson (Town)* [2001] 2 S.C.R. 241 gives notice of the potential for opponents of aquaculture approval to challenge discretionary decisions of government administrators, based on a failure to consider the precautionary approach adequately. The court, in upholding a municipal by-law that severely restricted the use of pesticides, recognized the precautionary principle as part of international law and reaffirmed the position that the values and principles of international law may help inform the contextual approach to statutory interpretation and judicial review. In the case of *Brighton* v. *Nova Scotia (Minister of Agriculture and Fisheries)* (2002) 206 N.S.R. (2d) 95 (S.C.), a group of concerned citizens challenged a ministerial decision to allow a fish farm to proceed. One of the arguments was that the minister failed to err on the side of caution as required by the *Oceans Act*, S.C. 1996, c. 31. Justice MacDonald of the Nova Scotia Supreme Court agreed that the minister was under a duty to proceed cautiously, but concluded that the minister had done so in light of the limited term of the license, stringent license conditions and ongoing monitoring provisions.

383   SAR, *supra* note 51, recommended fish escape prevention planning and waste minimization planning as part of Recommendations 12 and 24, respectively.

384   See OCAD Review, *supra* note 229 at 23–24. Since 2002, Fisheries and Oceans Canada has required that aquaculture operators enter into monitoring agreements in order to comply with s. 35 of the *Fisheries Act*.

385   Such a standard was suggested by the British Columbia *Salmon Aquaculture Review*. That review suggested that the standard be imposed in a regulation under British Columbia's *Waste Management Act*, R.S.B.C. 1996, c. 482. SAR, *supra* note 51 at Recommendation 24.

386   The Report of the Standing Senate Committee has emphasized the need to develop environmentally friendly feeds. Standing Senate Committee, *supra* note 118 at Recommendation 7(e).

387 The British Columbia *Salmon Aquaculture Review* has similarly called for enforceable standards on drug use and disease prevention. SAR, *supra* note 51 at Recommendation 17.

388 The Standing Senate Committee on Fisheries has made such a recommendation. Standing Senate Committee, *supra* note 118 at Recommendation 5(b).

389 Such a requirement might be phased in, given the research needs and existing technological investments. The Standing Senate Committee on Fisheries has recommended that the federal government play a leadership role in funding initiatives aimed at developing closed-containment finfish-rearing technologies. *Ibid.* at Recommendation 12.

390 For a substantial critique of screenings particularly in relation to public participation, see A. John Sinclair and Meinhard Doelle, "Using Law as a Tool to Ensure Meaningful Public Participations in Environmental Assessment," (2003) 12 *Journal of Environmental Law and Practice* 27.

391 Section 58(1.1) of the CEAA requires the Minister of the Environment to establish a participant funding program but only in relation to comprehensive studies, mediations and assessments by review panels. CEAA, *supra* note 216.

392 For a previous suggestion that Parliament should consider amending the CEAA to create an expert administrative tribunal to undertake detailed substantive reviews of decisions, see Andrew Green, "Discretion, Judicial Review, and the *Canadian Environmental Assessment Act*," (2002) 27 *Queen's Law Journal* 785.

393 On the need to further develop decision criteria on which decisions would have to be justified, see Robert B. Gibson, "Favouring the Higher Test: Contribution to Sustainability as the Critical Criterion for Review and Decisions under the *Canadian Environmental Assessment Act*," (2000) 10 *Journal of Environmental Law and Practice* 39; and Ted Schrecker, "The *Canadian Environmental Assessment Act*: Tremendous Step Forward, or Retreat into Smoke and Mirrors?" (1991) 5 *Canadian Environmental Law Reporter* (New Series) 192.

394 Regarding the problems of possible conflict of interest and bias concerns associated with the principle of self-assessment, see Alan L. Ross, "The *Canadian Environmental Assessment Act*: An Analysis of Legislative Goals and Administrative Law in Conflict," (1994) 8 *Canadian Journal of Administrative Law and Practice* 21.

395 A system for the issuance of environmental assessment permits has previously been recommended by the House of Commons Standing Committee on Environment and Sustainable Development. See Parliament of Canada, Standing Committee on Environment and Sustainable Development, *Sustainable Development and Environmental Assessment: Beyond Bill C-9* (adopted by Committee on 1 May 2003, presented to the House on 5 June 2003). Online. Available http://www.parl.gc.ca/committee/CommitteePublication.aspx?COM=3293&Lang=1&SourceId=37541 (accessed 23 September 2005) at 19.

396 See Report of the Commissioner of the Environment, *supra* note 50 at Chapter 4.

397 The House of Commons Standing Committee on Environment and Sustainable Development has previously recommended a legislative foundation for federal SEA. See *supra* note 395 at 36.

398 For a discussion of the rather uncertain nature of federal aquaculture policy revision, see discussion on *DFO's Aquaculture Policy Framework, supra* note 377.

399 See Recommendations for Change, *supra* note 25 at Recommendation 4.

400 Regarding common approaches to justify siting decisions and the need to move towards participatory communicative democratic procedures, see Christian Harold and Iris Marion Young, "Justice, Democracy, and Hazardous siting," (1998) 66 *Political Studies* 82.

401 As recommended by British Columbia SAR, *supra* note 51 at Recommendations 1, 7.

402 See Craig Millar and David Aiken, "Conflict Resolution in Aquaculture: A Matter of Trust," in Andrew Boghen, ed., *Cold Water Aquaculture in Atlantic Canada* (Moncton: Canadian Institute for Resources on Regional Development, 1995) at 627, where the authors write:

> In an attempt to address contentious issues, government often organizes public meetings at the community level. Unfortunately, public participation does not ensure the resolution of conflicts. It may, in fact, be inherently adversarial. A community meeting convened by a representative of a government agency may take the form of a court hearing, forcing discussion and raising the stakes. This contrasts with the normal approach in rural communities, where there is an unspoken desire to preserve neighbourliness and avoid confrontation or even aggressive debate that could lead to conflict.

403 The State of Maine provides an example of priority-setting, through a legislative preference given to riparian owners and to traditional commercial fishers if more than one person applies to lease an area for aquaculture. See ME. Rev. Stat. tit. 12, § 6072(8) (1997). Online. Available http://www.maine.gov/dmr/bmp/lawindex/6072.htm (accessed 23 September 2005).

404 See, for example, Andy Stirling, "The Precautionary Principle in Science and Technology," in Timothy O'Riordan, James Cameron and Andrew Jordan, eds., *Reinterpreting the Precautionary Principle* (London: Cameron, 2002) at 61.

405 For recent recommendations on strengthening public participation in environmental impact assessment, see Minister's Environmental Assessment Advisory Panel – Executive Group, *Improving Environmental Assessment in Ontario: Framework for Reform*, Vol. 1 (March 2005) at 69–78, Online. Available http://www.ene.gov.on.ca/envregistry/023747ex.htm (accessed 23 September 2005).

406 In Canada, three jurisdictions (the Yukon, Northwest Territories and Ontario) have adopted environmental bills of rights that provide similar public participation guarantees: allowing residents liberalized legal standing to bring lawsuits against alleged polluters; ensuring public access to environmental information; giving the public a right to demand investigation of alleged environmental offenses; providing members of the public with a right to initiate private prosecutions against alleged offenders of regulatory standards; broadening participation rights in licensing and regulatory development; and granting "whistleblower" protection to employees who report environmental infractions by employers. For an overview of environmental rights legislation in Canada, see Elaine L. Hughes and David Iyalomhe, "Substantive Environmental Rights in Canada," (1999) 30 *Ottawa Law Review* 229.

407 For a detailed discussion, see VanderZwaag *et al.*, *supra* note 156.

408 The Council stated this in relation to precautionary measures:

> While scientific proof is not yet absolute, there is extremely suggestive circumstantial evidence that sea lice are associated with salmon farming. The Council believes that sea lice were associated with the decline observed in Broughton pink salmon. Guidance on how to deal with the level of uncertainty faced in this case comes from the concept of the "Precautionary Approach/Principle." The precautionary approach is a distinctive approach to managing threats of serious or irreversible harm where there is scientific uncertainty. The precautionary approach recognizes that the absence of full scientific certainty shall not be used as a reason to postpone decisions where

there is a risk of serious or irreversible harm. . . . When managing the oceans and its resources, Canada's Oceans Act prescribes that the precautionary approach be applied.

See Pacific Fisheries Resource Conservation Council (PFRCC), *2002 Advisory: The Protection of Broughton Archipelago Pink Salmon Stocks: Report to the Minister of Fisheries and Oceans, Report to the BC Minister of Agriculture, Food and Fisheries* (Ottawa: Fisheries and Oceans Canada, 2002) at 5–6.

409 The sea lice management plan for 2005 included mandatory sea lice monitoring on salmon farms; random auditing visits to 50 percent of the active farm sites; and actions required when three mobile lice levels per fish arise (during the smolt out-migration, either harvest or treatment must occur in addition to increased monitoring). British Columbia Ministry of Agriculture and Lands, "Sea Lice Management 2005." Online. Available http://www.agf.gov. bc.ca/fisheries/health/sealice_MS.htm (accessed 23 September 2005).

410 PFRCC, *supra* note 408 at 8.

411 On the various types of rationalities – scientific/technical, political and ethical – that need to be integrated, see Denis Goulet, "Biological Diversity and Ethical Development," in Lawrence S. Hamilton, ed., *Ethics, Religion and Biodiversity: Relations between Conservation and Cultural Values* (Cambridge: The White Horse Press, 1993) at 17.

412 Regarding the largely yet unmet challenges in moving towards sustainability law, see David R. Boyd, "Sustainability Law: (R)Evolutionary Directions for the Future of Environmental Law," (2004) 14 *Journal of Environmental Law and Practice* 357.

4 Property rights in Canadian aquaculture

A principled approach

Phillip M. Saunders and Richard Finn

Introduction

The 1995 *Federal Aquaculture Development Strategy* summarized some of the difficulties facing aquaculture development in a federal state such as Canada, where the jurisdictional entitlements relevant to this "new" (or at least newly significant) industry are by no means clear:

> Aquaculture is a formidable policy challenge. As a new industry, it straddles the line between fishing and farming, cuts across significant regional differences and is placed in a context involving the participation of municipal, provincial/territorial and federal governments.[1]

Added to the welter of relevant jurisdictions and departmental mandates is the complexity introduced by the application of common law principles to the definition of property rights in aquaculture operations. The fundamental problem is simply stated: aquaculture as a business depends on *some* level of tenure over defined aquatic spaces, but the common law evolved in such a way that it was not fully suited to the effective allocation of property rights in these spaces, or for these uses.[2] In Wildsmith's definitive review of the state of aquaculture law in Canada in 1982, he presented the following assessment of the state of the law with respect to the property rights underlying aquaculture operations:

> The single most important legal issue confronting an aquaculturist concerns the nature and extent of his property rights. Every industry (I can think of no legal exceptions) is premised upon property rights which are on the whole clear and well-defined. Financing is dependent upon the security of these rights. Aquaculture is unique in that it depends almost exclusively on property rights, both real and personal, which are either structured against the aquaculturist or are equivocal as to his position. Only where he maintains his stock in artificial structures located on or in his lands do his rights seem clear.[3]

Wildsmith went on to note that this was "a matter crying out for legislative intervention," and recommended the introduction of legislation that would include provision for aquaculture leases dealing with both the seabed and the water column.[4] This is precisely what has occurred in the years since 1982. The majority of provinces (and all coastal provinces) have aquaculture legislation, and in those most actively engaged in aquaculture, there is provision for some form of lease or analogous entitlement to aquaculture areas, in addition to licensing requirements.[5] The introduction of these arrangements has not, however, answered all of the questions surrounding the nature of property rights involved in aquaculture in Canada. A 2001 review of legislative and regulatory issues conducted by the federal Office of the Commissioner for Aquaculture Development (OCAD) identified a number of outstanding issues relating to the scope, duration and enforceability of property entitlements in aquaculture sites and products, and recommended that improvements be implemented:

> Uncertainty ... exists regarding public rights of access to waters near aquaculture sites [and] prevention of interference with aquacultural activities by other users of aquatic resources. . . . It usually takes several years for aquaculture operations to generate a return on the initial investment. To become established, the businesses require leases that last for a period that is relevant to the commercial activity being carried out and rational, transparent regulatory regimes. Yet, it is unclear what rights and obligations aquaculturists have under the existing legislative and regulatory regime, and how these rights and obligations are upheld and enforced.[6]

What is notable about this assessment is the lack of precision as to the exact nature of the problem. Where and how do the property rights available under the existing lease schemes fail to meet the needs of the industry? What tenure arrangements would satisfy the requirements of the industry, and will this vary with different types of aquaculture operations? Industry representatives have noted problems with duration and security of tenure under existing lease arrangements, but again with little indication of exactly what would be sufficient.[7]

These concerns can all be addressed in the context of modifications to the dominant approach to aquaculture tenure for marine areas in Canada (described in the following sections), which rests on continued Crown ownership of submerged lands, and government issuance of leases or similar instruments granting rights to identified areas. Such measures would seem to respond, at least in part, to the main property-related requirements for successful aquaculture:

> The single most significant question one must ask about any legal framework affecting marine aquaculture is: how secure is the interest that the sea farmer receives from the government? For the interest to

function as a property interest it should have some or all of the following attributes: transferability, duration and renewability, and revocability only for failure to perform specified conditions.[8]

There have also, however, been calls in recent years for the development of full private property entitlements for aquaculture, with rights equivalent to terrestrial freehold property, through the "alienation of the Crown's rights to the foreshore, the water column, and the seabed analogous to the way in which land has been alienated for agriculture."[9] These arguments rest in part on the comparison to agriculture, but also on an ideological conviction that continued government involvement, even to the extent of ultimate control over the issuance of leasehold rights, is bound to be dysfunctional:

> [F]ish farming should be governed by a system of property rights analogous to that which has been so successful in North American agriculture. Like agriculture, aquaculture is *culture*, and should not be governed by rules suited to the hunter/gatherer nature of the wild fishery. Above all, however, property rights would provide the legal framework within which the economic enterprise of aquaculture could achieve *efficiency* – that is, the greatest output for society at the least cost. In the absence of a strongly entrenched, well-defined, rationally constructed set of individual property rights in aquaculture, the assertions of special interests can be given political force through misinformed public opinion or failures in government. The structure of the industry itself then becomes inefficient, inequitable, and dysfunctional in every respect.[10]

There are a number of difficulties with this argument, including the simple fact that the present arrangements for aquaculture in Canada *are* based on the assignment of property rights, in the form of leases or similar instruments, and not on the common property or open access[11] approaches associated with capture fisheries. At a more fundamental level, however, it must be remembered that one is not starting with a blank slate: it is inevitable that aquaculture will often be conducted in an environment shared by a number of other users.[12]

These issues tend to be addressed by way of the regulatory system, which should protect against damage to other resources and uses, and by the development of transparent siting and lease approval processes that allow for other interests to be taken into account.[13] At a more fundamental level, however, it must be remembered that in addition to the existence of other users, there is a complex structure of legal entitlements, protecting some (but not all) of their interests, which has existed for hundreds of years. The displacement of these other interests, whether in full (through privatization) or at least to a greater degree than at present (through enhancement of existing lease rights), raises questions of equity and access that cannot simply be ignored.[14]

The debate over property rights for aquaculture, therefore, cannot be limited to an examination of the functional requirements of aquaculture alone, and whether they are met by the present system. It must also incorporate some understanding of the place, and the legal entitlements, of other interests in the affected marine spaces, and whether the further erosion of those entitlements is both necessary and feasible. Furthermore, as will be seen later in the chapter, the management of marine and other aquatic spaces in Canada engages constitutional doctrines that may affect the validity of current federal–provincial arrangements. These issues are addressed in this chapter through consideration of the following legal elements which combine to create the current structure of marine aquaculture property rights in Canada, and which must be taken into account in any proposals to further alter that structure:

- common law rights relevant to the creation of private property interests in marine and other aquatic areas;
- statutory schemes that have modified the common law position, primarily through the introduction of leasehold arrangements; and
- constitutional doctrines that set limits on the effectiveness of provincial statutory schemes in establishing private rights to marine areas.

The examination of these issues is followed first by a consideration of their impact on current lease arrangements, and by a final section that offers a number of conclusions and recommendations with respect to the policy implications arising from the legal analysis.

Common law property rights and aquaculture operations

Any consideration of the current state of property rights over aquaculture sites[15] in all provinces except Québec[16] must begin with a review of the status of the relevant areas of water and submerged lands at common law.[17] The current statutory framework, which will be dealt with in the "Statutory responses" section (p. 122) was designed in reaction to the pre-existing situation at common law, and can only be fully understood by reference back to the regime it sought to replace or modify. This examination may conveniently be divided into two parts: non-tidal and tidal waters. For the purposes of this section, the constitutional issues related to the respective federal and provincial powers to grant property interests, or to regulate the exercise of those interests, are put aside, and will be addressed in the "Constitutional issues" section (p. 129).

Non-tidal waters

The legal status of lakes and non-tidal rivers in English common law was relatively straightforward: the ownership rights of riparian landowners were presumed to extend to the midpoint of the watercourse (*ad medium filum aquae*),[18] and this ownership of the solum, or soil, extended above and below the bed of the stream or lake in the same way as property rights on land.[19] The ownership rights did not extend to the water itself, or to fish (until captured), but ownership of the soil did bring with it as an "incident of ownership," an exclusive right of fishery.[20] This right to the fishery could, however, be alienated from the ownership of the soil, whether as part of the original grant or by subsequent conveyance.[21]

In English law, it was clear that there was no public right of navigation in non-tidal, as opposed to tidal, waters. In Canada, however, non-tidal waters that were actually navigable came to be treated differently in most provinces (with the Atlantic region as a possible exception),[22] in part because of a recognition of the different physical circumstances in North America, but also because of the assignment of the power over navigation and shipping to the federal government, and the need for one consistent regime.[23] Public rights of navigation in these waters are recognized, and are "dominant" even over validly granted property rights, unless modified or eliminated by legislative action:

> Nothing short of legislation can take away the public right of navigation. The Crown in right of the Dominion or of a province cannot abolish the right in the absence of an authorizing statute. Accordingly, a Crown grant of land does not and cannot give a right to interfere with navigation.[24]

In addition to upholding the public right of navigation, courts have also found that the *ad medium filum aquae* presumption did not apply in non-tidal, navigable waters (again with the possible exception of the Atlantic region). Thus, riparian owners on these water bodies were not presumed to own to the midpoint, and any such submerged lands that were not explicitly included in a grant of land were presumed to be vested in the Crown.[25] There have also been some suggestions that in these waters, where the rule of prima facie Crown ownership was the same as in tidal waters, a "public right" of fishing might also exist, similar to that in tidal waters.[26] It seems clear, however, that the term "public right" is used in this context to mean a mere common right of fishing which exists subject to extinguishment, conveyance or modification by the Crown. In this, it must be distinguished from the public right of fishing in tidal waters, which, as will be discussed in the following section, exists as a "protected" Magna Carta right that cannot be granted or extinguished by the Crown alone.[27]

Tidal waters

The basic framework of common law rights to the seabed of the foreshore and coastal waters dates back to the restraint imposed upon the Crown's exercise of prerogative powers in Magna Carta of 1215.[28] Proprietary rights in the foreshore, and later the seabed of the territorial sea, were normally vested in the Crown, but were subject to the dominant public rights of fishing and navigation. Furthermore, the Crown could make private grants of rights over these submerged areas, but "any private grantees must take title subject to this overriding public right."[29] This position was clearly stated by the Judicial Committee of the Privy Council in the *BC Fisheries Reference*:

> Since the decision of the House of Lords in *Malcomson v. O'Dea*, 10 H.L.C. 493, it has been unquestioned law that since *Magna Charta* no new exclusive fishery could be created by Royal grant in tidal waters, and that no public right of fishing in such waters, then existing, can be taken away without competent legislation.[30]

The major exceptions to the dominance of the public rights are contained within the statement of the law set out above. First, if a grant or prescriptive right of fishery existed pre-Magna Carta, it could be maintained against any public right to fish. Second, and more relevant to the Canadian situation, the public rights of navigation and fishing could be modified or extinguished by an explicit act of the legislature, but not by the Crown. This framework of rights was adopted in Canada in both pre- and post-Confederation cases. In *Meisner* v. *Fanning* in 1842, the Supreme Court of Nova Scotia considered a claim to an exclusive fishery in Deep Cove, arising under a Crown grant. Hill J was prepared to assume for the purposes of argument that a grant to the seabed in the cove could be made by the Crown, but denied the possibility of a grant to the waters[31] and affirmed the general proposition respecting the limitation on the Crown's powers.[32] This view was confirmed in *Donnelly* v. *Vroom et al.* in 1907, in which the defendants owned the foreshore as part of a Crown grant of title to their farm. Their counterclaim against the plaintiffs for the digging and removal of clams from the foreshore was denied, on the basis that the ownership of the land did not remove the public right to fish by digging the clams, notwithstanding the defendants' own activities in this regard.[33] Further, in *Belyea* v. *City of St. John* in 1920, a private lease of the foreshore for purposes of a fish curing operation was used as the basis for a claim to an exclusive fishery. In finding against the lessee, the New Brunswick Supreme Court (Appeal Division) held as follows:

> The settled law of the realm appears to be that ... [w]ithin the territorial waters, subject to the ebb and flow of the tides, the public, being subjects of the realm, are entitled to fish, except where the Crown, or some subject of the Crown has gained a propriety exclusive of the public right, *or Parliament has restricted the common law rights of the public.*[34]

In sum, then, the common law established the following four essential elements which defined the legal status of tidal waters: (1) title was vested in the Crown, which could grant that title (in whole or in part) to others; (2) the Crown rights, and thus the rights of any grantee taking from the Crown, were subject to the public rights of navigation and fishing; (3) new grants of exclusive fisheries required action by the legislature; and (4) the legislature also retained the power to regulate, even to the point of extinguishment, the rights of the public in common law.

The implications of this general structure for the creation of property rights in aquaculture operations are significant, though perhaps not entirely clear in all respects. The basic propositions can be simply stated:

- First, it seems clear that anyone attempting to exert proprietary control over submerged lands in the tidal areas would be a trespasser, against either the Crown or any grantee under the Crown's title, unless they could show their own grant, or that they fell within the exceptions noted above.
- Second, it would be possible, given the validity of Crown grants in the foreshore and other areas, for an aquaculturist to obtain "the right to occupy these subaquatic lands and the water column by grant, lease, or license from the Crown, or from a successor in title to the Crown."[35]
- Third, despite the validity of such Crown grants, no occupier of these lands, including the Crown and its grantees, could in the course of their use and occupation restrict or impede the public rights of fishing and navigation, and to do so would constitute an enjoinable public nuisance.
- Fourth, the interference with the public rights could nonetheless be authorized, but only under the authority of an explicit legislative enactment.

If it is assumed, then, that an aquaculture operation requires protection from interference by others who might otherwise exercise their rights of fishing and navigation, then legislation would be required, either to make the grant or lease explicitly effective in that respect, or to otherwise restrict the public rights in the area by separate regulatory action. It should be remembered, however, that the grant or lease could be effective against *other* uses, not encompassed within the public rights of fishing and navigation. The extent of public rights was limited to the specific categories, as noted by Parker J in *Lord Fitzhardinge* v. *Purcell*, a case in which the court declined to extend similar protection to a claimed right of fowling:

> [T]he public have no rights in the sea itself except rights of fishing and navigation and rights ancillary thereto . . . This beneficial ownership of the Crown, or the Crown's grantee, can only . . . be considered to be limited by well known and clearly defined rights on the part of the public.[36]

While protection against duck hunters may not be of great significance, the principle can be applied to other uses as well.[37] If this reasoning were applied, for example, to the actions of someone fishing inside an enclosed aquaculture pen for species that were the product of that operation, it could readily be argued that this was not an exercise of the public right of fishing, in that it was simply the taking of private property.[38]

Summary

In sum, the common law allowed for Crown grants of private rights in submerged lands in tidal waters. Such grants could conceivably include rights to aquaculture sites, given that this is an activity distinct from fishing, and thus would not constitute a private grant of a *fishery* (which is beyond the scope of the Crown's powers). However, any such *Crown* grant was subject to the dominant public rights of fishing and navigation, so that private rights obtained from the Crown would be ineffective to prevent the continued exercise of those rights. Only grants made under the authority of explicit *legislative* provisions could supersede these public rights. Given that aquaculture sites would generally require this protection, it is assumed that legislative schemes are necessary to provide a sufficiently secure form of tenure. The common law, then, resulted in a requirement for legislative action to create *effective* grants of property rights in marine areas for purposes of aquaculture. The general structure and operation of the legislative lease arrangements that have actually been developed in Canada will be considered in the next section.

Statutory responses

As we have noted, most provinces have legislated to provide for leasehold or similar rights over aquaculture sites, typically in addition to a separate license or permit issued for the conduct of aquaculture operations. This section provides a summary of the main elements of provisions respecting the legislative grants of property rights in five provinces with significant interests in marine aquaculture: Newfoundland and Labrador; Prince Edward Island; Nova Scotia; New Brunswick; and British Columbia. Procedures for the review and processing of applications for tenure, and the involvement of the federal government, will be dealt with separately at the end of the section.

Provincial approaches

Newfoundland and Labrador

The Newfoundland and Labrador *Aquaculture Act*[39] makes no provision for the assignment of leasehold or other property interests in aquaculture sites on Crown lands. However, by s. 4(7)(a), no proponent shall be granted an aquaculture license unless "the proposed licensee owns, leases or otherwise has a right to occupy the parcel of land comprising the site." Therefore, unless the proponent has a private interest in the land at the time of application, they must apply for a grant or lease of Crown land under the provisions of the province's *Lands Act*[40] when applying for an aquaculture license. A number of provisions in the *Lands Act* have a bearing on the grant of land interests for the purpose of aquaculture.

In s. 2(f), "land" is defined as including "land covered by water, both tidal and non-tidal, and the water column superjacent to it," which makes it clear that the entitlements under a lease can encompass both submerged lands and the water. The interests that can be obtained over these areas are of three types: lease,[41] grant[42] or a license to occupy.[43] Given the wording of s. 4(7)(a) of the *Aquaculture Act* (quoted above), it would appear that a prospective licensee could satisfy the requirements by obtaining any of the three forms of entitlement. However, the policy publicized by the provincial government specifies that an applicant for an aquaculture license must have applied for a Crown *lease*.[44]

The *Lands Act* does not specify the nature of the tenure available under a lease with respect to such issues as exclusivity, transferability, divisibility, cancellation and term, although the minister is given wide discretion under s. 3 to specify any terms and conditions that may be required. However, with respect to some of these issues, other provisions and policies should be noted:

- The government has announced a policy of granting aquaculture leases for a period of 50 years.[45]
- Section 3(2) of the *Aquaculture Regulations*[46] states that aquaculture licenses are not transferable, which would render the value of a lease moot, given the lack of a valid license.
- Powers for the minister to suspend or cancel aquaculture licenses for failure to comply with terms and conditions are provided in the *Aquaculture Act*.[47]
- Special provision is made in the *Lands Act* for the 15-metre strips of land adjoining lakes, ponds or the seashore. By s. 7(1), in the absence of an express provision, no lease or grant of Crown lands adjoining the water bodies includes this area. Furthermore, any such grant must be made with the approval of the Lieutenant-Governor in Council, and only for specified purposes, which include aquaculture operations (see s. 7(2)(b)).

In sum, Newfoundland and Labrador relies on the aquaculture licensing provisions for much of the specifics respecting terms and conditions of operations, and the review process for approval.[48] The leasing arrangements are made under a "generic" lands act, without any aquaculture regulations, and as such do not incorporate detailed aquaculture-specific conditions, such as are found in the legislation of other provinces (although these can be inserted in leases, on the decision of the minister).

Nova Scotia

In Nova Scotia, the *Fisheries and Coastal Resources Act* (FCRA) authorizes the responsible minister to issue aquaculture licenses and leases, both of which will be required for aquaculture operations on Crown land.[49] The Act provides for the general content of a lease, as well as terms and conditions to which such leases are subject, including, *inter alia*, the following:

- By s. 52(1), a lease "shall" be "granted for a specific geographic area," "shall" specify the plants and animals to be cultured, and "shall" contain as attachments those permits and approvals that are required (both federal and provincial).
- Leases are for a term of ten years, renewable for five-year terms "at the Minister's option." No provision is made with respect to transferability (s. 52(2)(a)).
- Lessees must submit annual reports to the minister on the lessee's use or the productivity of the leased area (s. s. 52(2)(d)).
- With respect to termination, "the lease may be terminated by the Minister at any time if the lessee breaches any of the terms or conditions of the lease" (s. 52(2)(g)). In addition, leases may be amended upon request of the lessee, with approval of the minister (s. 59(1)).

Exclusivity of access is dealt with in s. 52(3), which provides that, other than where there are restrictions in the lease or legislation, "the holder of an aquaculture lease has, for aquacultural purposes, the exclusive right to use the leased sub-aquatic lands and water column."[50] A similar provision in s. 44(3) states that the grant of a license "carries with it the exclusive right to possession of the water column and sub-aquatic land described in the licence."[51]

New Brunswick

New Brunswick has also adopted a system that provides for both licenses and leases (for Crown lands), issued under the authority of the *Aquaculture Act*.[52] Under s. 4, no person is to carry on aquaculture without a license, licenses being issued under three categories: commercial aquaculture; private aquaculture; and institutional aquaculture.[53] By s. 14(1), the registrar, who is responsible for issuance of licenses and leases,

> shall not issue, renew or amend an aquaculture licence in relation to an aquaculture site on other than designated aquaculture land unless the

applicant is the owner or lessee of the aquaculture site and has a right to occupy the site.

"Designated aquaculture lands" are dealt with in the legislation as follows:

- Section 2 of the *Aquaculture Act* defines the term as meaning "land under the administration and control of the Minister that has been designated by the Minister under section 24 as aquaculture land."
- Section 24(1) simply gives the minister the power to designate lands under his administration and control as "designated aquaculture lands."
- Section 25(1) provides that "the Minister may, in accordance with the regulations, lease designated aquaculture land for the purposes of aquaculture."

In sum, the minister may only lease "designated aquaculture lands," but it is not entirely clear what the significance of that term is, other than that the minister has chosen to so designate particular areas. By s. 25(2), the minister has a general power to make an aquaculture lease subject to "such terms, covenants and conditions as the Minister considers appropriate." In addition to this broad discretion, the Act and regulations specify a number of terms and conditions, including the following:

- Leases are for a term not exceeding 20 years (Act, s. 25(3)(a)).
- Leases may be assigned or transferred, with the consent of the Minister (Act, s. 25(3)(b)).[54]
- The lease "conveys the right to the exclusive use of the land covered by the lease" (Act, s. 25(5)), and "land" is defined in s. 2 to include the water column.[55]
- The minister may cancel a lease for a number of reasons, including failure to abide by any lease or license terms (Act, s. 27(1)).
- Application forms and content are specified (regs., s. 24(3)), and shall include a site development plan.
- Anyone seeking an aquaculture lease must also submit a license application (regs., s. 24(1)).
- Decisions of the registrar respecting leases and licenses may be appealed to the minister, and a process is set out in the regulations (regs., ss. 32, 33).

In addition to leases, the New Brunswick legislation also provides for a second category of entitlement: the Aquaculture Occupation Permit (AOP). The AOP allows the holder to "occupy and use" designated aquaculture lands, for a period of up to three years, and is not assignable or transferable.[56] It is clear that the AOP is intended to create a lower level of property entitlement than the lease, if only because of the shorter term and lack of transferability.[57]

British Columbia

In British Columbia, as in Newfoundland and Labrador, the licensing and leasing authorities derive from different statutes, and different forms of tenure are available, as in New Brunswick. The *Fisheries Act*[58] requires that any person carrying on aquaculture in the province have a permit. For the necessary land tenures, applicants deal with the Integrated Land Management (ILMB) of the Ministry of Agriculture and Lands, which processes applications for leases and other entitlements under the *Land Act*.[59] Three forms of tenure are available to aquaculturists under the Act:[60]

- *Investigative permit.* A permit for a term of up to two years, without exclusive access, "to conduct appraisals, inspections, analyses, inventories, surveys or other investigations of the land or of its natural resources."[61] According to the applicable policy, these are not usually used for aquaculture sites.[62]
- *License of Occupation.* The License of Occupation[63] is the most common form of tenure granted to aquaculturists operating on Crown land. This license does not convey an interest that can be registered or mortgaged, and allows restriction of public access only to the extent necessary to protect the licensee's use.[64] Initial five-year licenses can be followed by the most common form of aquaculture tenure in the province, a twenty-year license of occupation.[65]
- *Lease.* Leases are authorized under s. 38 of the *Land Act*, and offer a higher degree of tenure, including exclusive use, a thirty-year term and the right to make modifications and improvements.[66] Leases are considered the exception for aquaculture operations, and are not typically issued for this purpose.[67]

The primary form of tenure used for aquaculture in British Columbia, the License of Occupation, does not appear to offer the same degree of security of tenure as is available under the legislative schemes in place in the provinces considered above, with respect to both exclusivity and registration of the interest. It is, however, considered to be an assignable interest, with permission.[68]

Prince Edward Island

Prince Edward Island (PEI) is the only province in which the federal government, through the Department of Fisheries and Oceans' (DFO's) Prince Edward Island Aquaculture Division, administers the licensing and leasing of aquaculture operations.[69] There is little information contained in the enabling legislation regarding the procedure to be followed in the issuance of an aquaculture lease, or the nature of the property right obtained by the proponent, but the details of the leasing scheme are found in the *Prince Edward Island Aquaculture Leasing Policy* of the PEI Aquaculture Division.[70]

Under the policy, leases are issued for terms up to twenty-five years, with options to renew. Within the overall term of the lease, the policy distinguishes between a "developmental phase" and a "commercial phase." In the former phase, the lessee "will assess the biological and environmental aspects for a proposed site prior to entering full scale commercial operations."[71] Once the site is fully developed, and obligations under the lease are satisfied, the lease is in the commercial phase, during which the operation is to be periodically assessed to ensure compliance with lease conditions. In addition to the classification of phases, leases are defined with respect to the following types of operation: a "bottom culture lease" covering use of the seabed to cultivate designated mollusk species; and a "water column lease," which is actually a bottom culture lease with a special permission to use the superjacent water column.[72]

Apart from duration, other terms and conditions of these leases are set out in the federal provincial memorandum of understanding (MOU),[73] and expanded upon in the Aquaculture Leasing Policy. These include, *inter alia*, the following:

- Leases are transferable and assignable (including to lending institutions), and may be sublet, but the permission of the Division is required for such transactions.[74]
- Lessees acquire the exclusive rights to species produced within their sites, but with respect to the issue of exclusivity of access, the policy refers to the "use of the sea-bed and water column,"[75] which could indicate a limited approach to exclusion of other uses, as in the British Columbia License of Occupation. However, the MOU states that the coordinating committee established by the agreement can determine conditions, which could include complete exclusivity.[76]

The Prince Edward Island policy also includes an aquaculture zoning system, which divides the province into areas approved or not approved for consideration for bottom or water culture leases. The zoning exercise that resulted in the current structure considered potential conflicts with other uses, as well as the needs of the industry.[77]

Summary

The provincial schemes that have been outlined here all offer some level of property rights in aquaculture sites. They do, however, vary significantly on such key issues as duration, assignability, and exclusivity of access under the property rights obtained. In some cases, notably those of Prince Edward Island and British Columbia, there appears to be more recognition of the possibility of "layering" rights depending on the type of access and use required by a particular operation, so that full exclusivity may not be necessary in all cases.

Review processes and federal involvement

Some of the provincial schemes provide for varying degrees of review and consultation on licensing and leasing decisions. These provisions are not the primary focus of this chapter, which is concerned with the actual proprietary entitlements that result, but it is useful to note some of the provisions currently in place, and to consider the extent of federal agency involvement.

In Nova Scotia, the *Aquaculture Act* provides for a review process, in general terms, which applies to decisions respecting both leases and licenses.[78] The province has established ten Regional Aquaculture Development Advisory Committees (RADACs), as provided for in s. 47(b), and is committed to using these bodies as an integral part of the approval process, including site approval.[79] Recommendations of the RADAC are forwarded to the minister for consideration.

In New Brunswick, the *Aquaculture Act* and regulations make some provision for public review and consultation,[80] including a specific requirement in s. 37(2) that the minister "shall establish advisory committees to advise the Minister in relation to health standards for aquacultural produce and in relation to site selection criteria for designated aquaculture land." More important than the legislative provisions, however, are the policy guidelines that have been developed to deal with the application process in general, which include aspects of relevance to site selection review and lease arrangements. In particular, the Bay of Fundy Marine Aquaculture Site Allocation Policy sets up a system of zoning built around Aquaculture Bay Management Areas (ABMAs).[81] Aquaculturists operating in a given ABMA will collaborate with the government and local management bodies to produce Bay Management Agreements (BMAs), which will define operating standards and practices to be followed in that particular ABMA, including allocation and review processes.[82]

In British Columbia, the aquaculture land use policy incorporates two review processes for applications for tenures. First, the referral process, which feeds into decision-making, provides a means for consultation with interested departments and others.[83] Second, for more complex proposals, the Project Review Team (PRT) process is utilized. This interagency group includes relevant federal agencies, and has a more proactive role to seek out information, consult more generally and make recommendations.[84]

A common element in the provincial processes, even where it is not formalized, is the involvement of federal agencies that are required to give approvals of their own for the aquaculture activity, under their own regulatory mandates.[85] The role of these agencies, however, is largely confined to the *regulation* of aquaculture, and does not address the grant of proprietary rights, except indirectly, in that refusal of a permit may prevent the grant of a lease or other entitlement. Otherwise, it is the provinces (with the exception of PEI) that have taken the lead on the proprietary aspects of aquaculture management, consistent with the MOUs[86] signed by seven provinces and territories (including those considered above) and the federal government.[87] The assumption underlying this approach to the grant of statutory grants of leasehold and other entitlements for aquaculture sites is that the provinces are constitutionally competent to enact legislation that provides for the desired degree of certainty and security of tenure over all potential sites. As will be seen in the following section, this is by no means clear.

Constitutional Issues

The analysis in the previous section demonstrated that, at least to the extent that aquaculture operations require authorization to interfere with public rights of fishing and navigation, or protection from the exercise of those rights, legislative intervention is required to secure the required level of property rights,[88] and those provinces most involved in aquaculture have indeed opted for statutory leasehold arrangements or similar mechanisms. The next obvious question, therefore, is which level of government, federal or provincial, has the constitutional power under the *Constitution Act, 1987*,[89] to make such legislation, and in what circumstances. More particularly, given the approach taken in the majority of coastal provinces, do those provinces have the constitutional authority to legislate for aquaculture property rights in the manner that they have legislated?

It is beyond the scope of this chapter to detail the heads of federal jurisdiction over *regulatory* issues that necessarily impinge on the exercise of provincially granted proprietary rights in aquaculture operations, or to review the provincial regulation of non-proprietary aspects of aquaculture.[90] The concentration here is on the proprietary aspect of aquaculture operations, rather than on the regulatory control that is exerted over it by both federal and provincial governments, and it is assumed throughout that both levels of government have valid jurisdictional interests in other aspects of the regulation of aquaculture.[91] In considering the various relevant heads of jurisdiction as they may affect the subject of this study – the control over proprietary aspects of the industry – a number stand out as potentially relevant. At the provincial level, the following legislative powers, all falling under s. 92, confer extensive control over matters related to property rights on the provinces:

92 (5)   The Management and Sale of the Public Lands belonging to the Province . . . [*in that submerged areas of Crown land may be conveyed for aquaculture*]

(8)   Municipal Institutions in the Province ... [*possibly relevant for municipal zoning and development control under statute*]

(10)   Local Works and Undertakings...

(13)   Property and Civil Rights in the Province...

(16)   Generally all Matters of a merely local or private Nature in the Province.[92]

Of this list, it is "property and civil rights" that has emerged in the case law as the most important to the definition of provincial powers over property rights in submerged areas, both tidal and non-tidal.[93] At the federal level, the *direct* authorization of power over property rights *per se* in s. 91 is more limited, but, as we shall see, non-proprietary powers such as navigation and fisheries have been interpreted as significant limitations on the exercise of proprietary rights:

91   (1A)   The Public Debt and Property...

(10)   Navigation and Shipping...

(12)   Sea Coast and Inland Fisheries...

(24)   Indians, and Lands reserved for the Indians.[94]

It is implicit in the discussion above that jurisdiction over property rights and jurisdiction to legislate respecting an activity are separate concepts, and indeed this distinction is central to understanding the current structure of federal and provincial interests in property rights over aquaculture. Accordingly, before we turn to the question of jurisdiction over property rights as such, it is necessary to consider the significance of this distinction in Canadian constitutional law.

Legislative jurisdiction and proprietary rights

Origin and nature of the distinction

As is indicated above, a fundamental distinction has been drawn in Canadian constitutional law between legislative jurisdiction and proprietary rights. That is, the fact that one level of government has been given legislative jurisdiction over a matter does not imply that it has acquired proprietary rights over the subject of that legislative control. Equally, the existence of proprietary rights in the provincial government does not mean that the assigned legislative powers of the federal level are eliminated.[95]

In the early post-Confederation case of *The Queen* v. *Robertson*,[96] the Supreme Court of Canada considered the relationship between the provincial power over property and civil rights in the province, and the legislative jurisdiction of the federal government over sea coast and inland fisheries. The case involved an attempt by the federal Minister of Marine and Fisheries to issue an exclusive lease of a fishery in the South West Miramichi River in

New Brunswick (an area of non-navigable waters). The lease was successfully contested by the prior holders of a private right in the area in question, and the grantee of the lease sought compensation from the Crown for the loss of the rights and other expenses. Ritchie CJ set out a number of fundamental propositions in his judgment, the first of which concerned the relationship between federal and provincial powers in general:

> [A]s there are many matters involving property and civil rights expressly reserved to the Dominion Parliament, the power of the local legislatures must, to a certain extent, be subject to the general and special powers of the Dominion Parliament. But while the legislative rights of the local legislature are in this sense subordinate to the rights of the Dominion Parliament, I think that such latter rights must be exercised so far as may be consistently with the rights of the local legislatures, and therefore the Dominion Parliament would have only the right to interfere with property and civil rights in so far as such interference may be necessary for the purpose of legislating generally and effectually in relation to matters confided to the Parliament of Canada.[97]

Proceeding from this proposition, which allowed for the coexistence of the two heads of jurisdiction to the extent possible, Ritchie turned to consider the nature of the rights in issue in the case. He noted that there was a public right to float logs on the river, and "a right of passage by canoes &c," but found that such a right was "not in the slightest degree inconsistent with an exclusive right of fishing, or with the rights of the owners of property opposite their respective lands."[98] In sum, he confirmed that in rivers beyond the ebb and flow of the tide, the right to fish was not a *public* right, but a private right connected to ownership of the soil.[99] Building on this common law distinction, Ritchie went on to note the existence in pre-Confederation New Brunswick of private rights of the type in question, and of regulatory legislation dealing with fisheries,[100] and found that, while the previous regulatory jurisdiction had been ousted by s. 91(12) of the *British North America Act* (BNA Act), no such conclusion could be drawn with respect to the control over aspects of the fishery dealing with property and civil rights.[101]

The same position was adopted by the Supreme Court in the 1896 *Provincial Fisheries Reference*. The judgment of Strong CJ confirmed the decision in *Robertson* as it applied to provincial powers over proprietary rights in non-navigable waters,[102] and extended that finding to navigable lakes and rivers within provincial boundaries,[103] including tidal waters.[104] In 1898, the Judicial Committee of the Privy Council, in the *Ontario Fisheries Reference*, followed the same approach and acknowledged the same distinction as did the previous cases.[105]

The significance of these decisions for the structure of common law rights has been touched on earlier, and their impact on the grant of property rights under legislation will be dealt with later, but for the purposes of this section it is clear that *Robertson*, *Provincial Fisheries* and *Ontario Fisheries* all proceeded from the same starting point: that proprietary and regulatory aspects of fisheries could be separated, with federal and provincial levels both having valid constitutional interests under the different spheres.

Implications of the federal regulatory power

The first and most obvious implication for the legal status of aquaculture operations is that, *within* the boundaries of the provinces, the assignment of proprietary interests will fall within the provincial jurisdiction over property and civil rights (see below). The second point arising from these and other cases is that, despite the provincial jurisdiction over proprietary issues, the federal regulatory power over fisheries could be used to restrict, potentially to a very great extent, the exercise of any property rights held or assigned by the province.[106] It might be argued that cases such as *Ontario Fisheries* dealt with fairly direct conflicts between the federal fisheries power and provincial jurisdiction over property and civil rights, and that they may be less applicable to aquaculture, which is on its face a very different activity from the traditional fishery encompassed by s. 92(12) of the *Constitution Act*. These decisions did not, however, limit their effect to provincial measures that were purely fisheries related, but clearly extended to any instances where the valid exercise of the federal power necessarily impinged upon the provincial proprietary interest. In sum, then, a provincial proprietary grant would be effective, *up to the point* that it collided with the valid exercise of a federal regulatory power.

It is assumed, for the purposes of this chapter, that any private exercise of property rights granted under provincial legislation is subject to extensive federal regulation.[107] Given that general context, the next section considers how the provincial power over the proprietary aspect of aquaculture has been structured and limited, apart from the general federal regulatory involvement. In addition, the significance of federal jurisdiction over proprietary aspects of aquaculture in non-provincial waters is examined.

Jurisdiction to legislate property rights

Delineation of constitutional jurisdiction over property rights in aquaculture sites requires consideration of three separate legal regimes, defined with reference to the following categories of waters: non-tidal waters in a province; tidal waters within a province; and waters outside any province. This division is necessary in part because of the common law principles related to public rights of fishing and navigation, as discussed earlier, and in part because of the structure of territorial jurisdiction under the constitution.

Non-tidal waters in a province

The early cases of *Robertson*, *Provincial Fisheries* and *Ontario Fisheries*, referred to earlier, confirmed that provinces generally have jurisdiction over proprietary rights in non-tidal areas within the province. As the Supreme Court found in *Provincial Fisheries*, all waters (tidal and non-tidal) within the provinces at the time of Confederation were vested in the Crown in right of the provinces, except to the extent that they were subject to other existing grants or specific exceptions within the constitution itself.[108]

There are at least three central points that emerge from this general proposition, and from the other cases discussed above, respecting the provincial entitlements. First, areas that were "ungranted" (and not within some category of federal lands) would be held by the Crown in right of the province. Second, where there were pre-existing private rights over the submerged areas in question, whether by operation of riparian entitlements or by explicit Crown grant, those rights survived as private entitlements. Third, and critically important for the development of aquaculture rights, the Crown in right of the province could, by virtue of its power over property and civil rights, make new grants over these areas, whether by lease or by other form of grant. Similarly, the Crown could modify or remove rights gained under existing grants. There are, however, a number of exceptions and limitations to these powers that must be considered.

Perhaps the most significant restriction on the provincial power concerns lands within the provinces that fell under federal jurisdiction by virtue of the *Constitution Act* itself. To begin, the general power under s. 91(1A) to legislate in respect of public property has been interpreted quite broadly, and represents a significant potential source of federal power over proprietary aspects of aquaculture within the provinces.[109]

This position was clearly stated in the *BC Fisheries* case, in which the Judicial Committee of the Privy Council found, *inter alia*, that certain lands of the "Railway Belt" (including river waters) had been conveyed from the Province to the Dominion, as part of the agreement for the building of the Canadian Pacific Railway.[110] For the non-tidal waters of the Fraser River and other bodies of water within the Provinces, the conveyance of the property right meant that the federal government stood in the position that would otherwise have been occupied by the Province, insofar as the control over proprietary rights such as fisheries was concerned.[111]

Further issues arise in dealing with non-tidal waters that are *actually* navigable. As was noted earlier, in at least some, if not all, provinces, the public right of navigation has been extended to non-tidal navigable waters (unlike the approach taken in England). It has also been suggested, as noted above, that there may be a "public right" of fishing in such waters. As was argued above, however, the reference to a "public right" of fishing was not a "protected" Magna Carta right as found in tidal waters, but merely a common right to fish that could be modified or extinguished by a grant of private right by the Crown, with or without legislation. The impact of such a "right" is simply to shift ownership rights from the riparian to the provincial Crown, not to the federal Crown, and the provincial Crown can freely grant both the lands and any fishing rights, unimpeded by Magna Carta rights.

With respect to navigation rights, the position is different. Unlike fishing, which may involve both proprietary elements and use rights, the interest in navigation is of the nature of a right of passage or an easement, maintainable even against the owner of the bed.[112] The resulting irrelevance of the proprietary aspect removes the claim to jurisdiction that supports provincial involvement in the property aspects of fisheries, and the position of navigation rights in non-tidal waters is thus analogous to the status of both navigation and fisheries in tidal waters: a "dominant" right that cannot be interfered with by Crown grant, but only by legislative intervention. Once it is accepted that legislative action is required, it seems clear, as stated by La Forest, that such legislation must be federal, given the power over navigation and shipping.[113]

In sum, any provincial grants in areas of navigable waters would certainly be subject to the federal regulatory power over navigation and shipping, though not to the extent of a federal interest in the proprietary aspect of the submerged lands, which remains provincial. This coexistence of interests means that any private actor wishing to develop a work or undertaking that interferes with navigation in non-tidal waters would require a grant of property rights from the provincial government, and a statutory authorization for the interference with navigation from the federal government.[114] Indeed, this was the basic scheme of rights and requirements identified as early as the *Provincial Fisheries* case, in which it was made clear that a grantee under the province could build a structure in navigable waters, operating under their property rights granted by the provincial Crown, but such activities would of course be subject to federal statutory authorization.[115]

Tidal waters in a province

The status of tidal waters within the boundaries of a province is similar to that of other provincial waters insofar as the granting of property rights is concerned: the river or seabed in these areas may be privately owned, but in the absence of other owners (including the federal Crown), title is in the provincial Crown, and issues relating to property and civil rights are within the jurisdiction of the provincial government.[116] The general position, which is to treat tidal waters on the same footing as non-tidal waters for purposes of determining proprietary interests, was stated in *Provincial Fisheries*,[117] in which it was held that ungranted submerged lands in "all lakes, rivers, public harbours and *other waters* within the territorial limits" of the provinces were vested in the provincial Crowns.[118] Ritchie CJC went on to make it absolutely clear that "other waters" included tidal waters.[119]

Apart from the impact of aboriginal and treaty rights,[120] the most significant limitation on property rights within these waters arises from the existence of the public rights of navigation and fishing and the subordination of any private property interest to the dominant public rights, save for cases in which the legislature has acted either to make the private grant, or to limit the exercise of the public right. From the constitutional perspective, the jurisdiction to grant property rights remains in the province, but the critical question is which legislature has the jurisdiction to authorize any resulting interference with the identified public rights. For navigation, the answer for tidal waters is the same as for non-tidal navigable waters: only the federal Parliament has the power to remove or limit these rights. In the case of public fishing rights, one might expect a different answer, given that the grant of rights in the fishery has a proprietary aspect that does not apply to navigation, and that that proprietary element must be provincial. That is, it could be argued that it is within the competence of the provincial legislatures to modify or eliminate a public right of fishing within their territory, subject of course to regulation of the fishery by the federal government.[121] This approach was, however, rejected by the Judicial Committee of the Privy Council in the *BC Fisheries* case in 1913, which found that the control of public fishing in tidal waters, including grants of new exclusive rights in fisheries, was *ultra vires* the provincial legislature.[122] In light of the centrality of these waters to aquaculture operations, and the significance of the proprietary issues to the relevant lease schemes, it is important to consider the rationale for this decision in some detail.

Viscount Haldane proceeded from the fundamental proposition that the right of fishing in tidal waters was distinguished from that in non-tidal waters by the fact that it was not a matter of property at all, in that the right of fishing in these waters was a public right and was not an incident of ownership of the land.[123] Thus, the Privy Council's previous decision in *Ontario Fisheries*, which recognized the provincial jurisdiction over proprietary aspects of fishery rights, was distinguished as irrelevant, and a different analysis put forth for tidal waters:

> The decision ... does not, in their Lordships' opinion, affect the decision in the present case. Neither in 1867, nor at the date when British Columbia became a member of the Federation, was fishing in tidal waters a matter of property. It was open equally to all the public, and, therefore, when, by sec. 91, sea coast and inland fisheries were placed under the exclusive legislative authority of the Dominion Parliament, there was in the case of fishing in tidal waters nothing left in the domain of the provincial legislature. The right being a public one, all that could be done was to regulate its exercise, and the exclusive power of regulation was placed in the Dominion Parliament.[124]

This reasoning was extended to cover any and all interference by the provincial legislature, whether by regulatory action or by the grant of exclusive rights to individuals:

> Interference with it [the right], whether in the form of direct regulation, or by the grant of exclusive or partially exclusive rights to individuals or classes of individuals, cannot be within the power of the province, which is excluded from general legislation with respect to sea coast and inland fisheries.[125]

While the prohibition from direct regulation of fisheries is clear enough, there are a number of queries that might be raised about the logic of the reasoning related to the grant of exclusive rights. First, while fishing in tidal waters may indeed have been something generally open to the public, it was, even under the Magna Carta restriction, possible to create a private or exclusive right of fishery by explicit act of the legislature. If that power to affect property rights (i.e. to create them) rested in the provincial legislatures at Confederation, why would it not have survived, as did the same power in non-tidal waters (despite the presence of the federal regulatory power in those waters as well)? Second, and related to this point, the decision seems to assume that *only* if the right of fishing is seen as a matter of "property" could it come within the jurisdiction of the provincial legislature. However, if the public right of fishing itself was considered to be a matter of "civil rights" within the province, would it not have come within the scope of "property and civil rights"? Haldane, however, advanced an additional justification for full federal control, based on the fact that non-residents of the province may have access to the right of fishing:

> The right to fish is ... a public right of the same character as that enjoyed by the public on the open seas. A right of this kind is not an incident of property, and is not confined to the subjects of the Crown who are under the jurisdiction of the province.[126]

One might have thought that the exercise of the right *in the territory* of a province would have placed these subjects "under the jurisdiction" of that province, at least for these purposes. Elsewhere in the decision, however, the same point was made in slightly different words: "It was most natural that this should be done [i.e. assigning full jurisdiction to the federal level], seeing that these rights are the rights of the public in general and in no way special to the inhabitants of the province."[127] This argument is also less than convincing: it is not clear why the pre-Confederation jurisdiction of the provincial legislature to grant exclusive fisheries would not have extended, within the territory of the Province, to individuals from outside who have come into the province. Similarly, no clear argument is presented as to why this territorial jurisdiction could not have been continued.

Whatever qualms one may have about the quality of the reasoning in this case and the subsequent *Québec Fisheries* case, which took the same approach, the argument is now moot. It is, however, important to understand the limits of these decisions insofar as they affect the ability of the provinces to make private grants in tidal areas within their boundaries. There is nothing in either decision to suggest that provinces cannot make grants of private rights in these areas, so long as they avoid the two intrusions on federal jurisdiction identified in both cases: an actual grant of an exclusive fishery; or any other grant that has the effect of interfering with the public right of fishing or the regulation of that right.

Thus, in *BC Fisheries* Haldane noted in *obiter* that the situation would be quite different for provincial grants of "fishing" rights based on "kiddles, weirs or other engines fixed to the soil" that would "involve a use of the *solum* which, according to English law, cannot be vested in the public, but must belong either to the Crown or some private owner."[128] In *Québec Fisheries*, this issue arose again, and Haldane confirmed the ability of the province to make such a grant, even for an activity so closely tied to fishing, so long as it was not strictly within the definition of the true public right of fishing. In such cases, however, the limits related to provincial *interference* with the public right of fishing still obtained:

> In so far as the soil is vested in the Crown in right of the Province, the Government of the Province has the exclusive power to grant the right to fix engines to the *solum*, so far as such engines and the affixing of them do not interfere with the right of the public to fish, or prevent the regulation of the right of fishing by private persons without the aid of such engines.[129]

At the same time, of course, the federal Parliament was still restricted from granting proprietary rights, and from exercising its regulatory power "to deprive the Crown in right of the Province or private persons of proprietary rights where they possess them."[130] The result, as was frankly acknowledged in these two decisions, was to create a bifurcated power where there had been a unity, so that, as in non-tidal waters, most grants of private rights would require action by both levels of government to be effective.[131]

Waters outside any province

The fundamental basis of provincial jurisdiction under the Constitution is territorial in nature, and the enumerated powers, including property and civil rights, are explicitly limited in effect to the territory of the several provinces.[132] Thus, as Wildsmith concluded in 1982, in marine areas of federal jurisdiction outside the provinces, the legal status is clear: "The conduct of aquaculture in those areas is clearly a matter for federal control in its entirety."[133] This applies equally to regulatory and proprietary aspects of aquaculture and is subject, of course, to the limitations relating to public rights of fishing and navigation, and aboriginal and treaty rights. The impact of this legal status on the validity of current approaches to leasing arrangements will be dealt with later, but it is useful here to consider the preliminary question of where the boundaries between federal and provincial waters can be found.

The territorial extent of a province includes all of those areas that it brought into Confederation; that is, the extent of the former colony defines the geographical scope of the province.[134] With respect to marine areas, including submerged lands, the general position in British law at the time of Confederation was that the realm, and thus any colony, ended at the low-water mark, in the absence of a legislative enactment to the contrary, and subject to certain exceptions.[135] The exceptions included "waters *inter fauces terrae* (i.e. "within the jaws of the land"), which the common law considered to be . . . within the realm of England."[136] These waters would include bays and estuaries, and possibly straits, but the term is by no means precise and is subject to examination on a case-by-case basis. In sum, coastal waters and submerged lands subject to potential claims by a province could be brought within the province either as part of the general exception, depending on the criteria applied, or by identification of a positive act of the legislature under British rule, as the question was put by the majority in the *Georgia Strait Reference*:

> In order to succeed . . . British Columbia must demonstrate that prior to Confederation either the lands and waters in question were "within the realm" as that term is used in *R. v. Keyn* or else that by some overt act Britain incorporated them into the territory of the Colony of British Columbia.[137]

In that case, British Columbia was able to identify an "overt act," but the central point arising from this and the other cases[138] is that the status of a particular area of water and submerged lands will be dependent on an analysis of its geographical configuration and legal history, all aimed at determining whether it was part of the previous colonial territory prior to Confederation. The result is that the determination of the precise status of many areas of coastal waters is ill defined, and could require close examination and possible litigation to determine,[139] in the absence of some more general settlement of the issues with the provinces.[140]

Other limitations on legislative jurisdiction over property rights

Before we turn to the application of the common law and constitutional principles to existing lease arrangements, it is necessary to briefly note two other general limitations on the jurisdiction of the provincial and federal governments to make grants of property rights in non-tidal and marine areas: aboriginal and treaty rights, and international law.

First, aboriginal and treaty rights, given their constitutional status,[141] can limit or exclude the exercise of federal and provincial legislative powers over fisheries and other relevant natural resources, and it is the exercise of those powers (or of the Crown prerogative) that enables the governments to create new property rights in aquaculture operations. Thus, the presence of aboriginal entitlements, whether through treaties or through aboriginal rights, must stand as a limitation on the ability of federal or provincial governments to issue leasehold rights in any affected areas, if only by virtue of a duty to consult in advance.[142] This complex area of law, which is still evolving, is the subject of Chapter 8 of this volume, and thus will not be addressed here.

Canada is also subject to various obligations at international law which may be of relevance to aquaculture sites located outside internal waters, whether in the territorial sea or in the exclusive economic zone (EEZ), should operations eventually be sited further offshore.[143] Relevant obligations could include the requirement to respect rights of innocent passage in the territorial sea, and the broader navigational rights of other states in the EEZ.[144] In addition, there are more vaguely stated obligations with respect to preservation and protection of the marine environment that may come into play, particularly in the EEZ.[145] For the purposes of this chapter, however, the primary relevance of these obligations is the additional support they give to the assertion of federal jurisdiction, certainly in marine waters, and to a lesser degree in non-tidal waters, to the extent that they engage Canada's international obligations.[146]

Impact on aquaculture lease arrangements

Summary of jurisdictional structure

Allowing for the areas of doubt that have been addressed in the preceding sections, it is possible to summarize federal and provincial powers to create property entitlements over aquaculture sites, and to identify the main potential problem areas that arise when those powers are compared to the leasing and other arrangements discussed in the subsection "Tidal waters" (p. 134). The examination to this point suggests the following general jurisdictional structure with respect to the issuance of property rights in aquaculture sites in Canada.

Non-tidal waters in a province

- The provinces have the power under common law to grant property entitlements over the beds of non-tidal waters within their territory (including the zone above the beds). Such grants could be made by the Crown with or without legislation, although removal of pre-existing rights (including riparian rights) may require legislation.

- Any such grants would be subject to the government's constitutional obligations with respect to aboriginal and treaty rights.
- The exercise of property rights is subject to federal regulatory jurisdiction, particularly over fisheries and navigable waters.
- Any interference with navigation rights can only be justified by legislative authority, and that authority must be federal.

Tidal waters in a province

- A provincial Crown, with or without legislation, may make grants of property rights over the bed of tidal waters that fall within the province.
- Such grants are, as with non-tidal waters, subject to aboriginal entitlements and the paramount federal regulatory power over matters within federal jurisdiction.
- Any grant that interferes with the public rights of fishing or navigation must be authorized by legislation, and that legislation must be federal.

Waters Outside Any Province

- The federal Crown may make grants of property entitlements to the bed of waters outside the provinces, and such grants could be made with or without legislative authority.
- However, where such grants interfere with the exercise of the public rights of fishing and navigation, the federal Crown cannot act in its prerogative, but must be acting pursuant to legislative authority.
- The federal ability to make such grants is also subject to any aboriginal entitlements, and to the international law obligations to which Canada is subject.

Application to the present leasing system

When the general propositions set out above are considered in the light of the existing statutory approach to aquaculture leases in the provinces considered earlier (with the exception of Prince Edward Island), a number of potential problem areas of varying degrees of significance become apparent.[147]

Waters within provinces

The first general area concerns the extent to which the grant of leases or other tenures inside the provinces may interfere with matters within federal jurisdiction. For sites within provincial, non-tidal waters, this issue appears to have been adequately addressed by the requirements for compliance with federal regulations, including the referral of sites for federal review and the necessity of acquiring relevant federal permits.[148] That is, consistent with the various cases that have addressed the issue, the current scheme allows the (provincial) proprietary interest to coexist with the additional (federal) regulatory power, particularly with respect to fisheries and navigation.

Tidal waters and the problem of public rights

In tidal waters within the provinces, however, the situation is more complex. The provincial ability to control proprietary interests remains unquestioned, and the federal regulatory powers are fully acknowledged, as reflected in the requirements for permits under the *Navigable Waters Protection Act* (NWPA)[149] and the *Fisheries Act*.[150] Nonetheless, the existence of the public right of fishing in the tidal areas fundamentally changes the legal situation in a manner that is not fully addressed in the current law. Respect for the public rights of fishing and navigation requires something more than mere provincial avoidance of conflict or incompatibility with federal regulatory powers; to the extent that a provincial grant in tidal waters interferes with either of the public rights, it requires positive authorization under a legislative enactment, and that enactment must be federal. That is, the rights involved reside with the public, and raise questions beyond disputes over constitutional authority.

For navigation, the power under s. 5 of the NWPA is explicit in granting the federal government authority to permit interference with rights of navigation, and so long as permits under this section are issued where relevant, there seems no question that any authorized works will be legally justified with respect to interferences with navigation.[151] For fisheries, however, the issue is more complicated. As a starting point, it seems clear that if a grantee holds no *federal* authorization, under either a lease or another form of permit (as with the NWPA for navigation), then any resultant interference with the public right of fishing is simply a public nuisance, for which an action could be brought by members of the public.[152] The question, then, is whether or not the current system provides for adequate federal authorization to prevent the activity being considered an enjoinable nuisance.[153] The primary federal aquaculture approval of relevance to fisheries is a permit under s. 35 of the federal *Fisheries Act*, dealing with the alteration of fish habitat:

35. (1) No person shall carry on any work or undertaking that results in the harmful alteration, disruption or destruction of fish habitat.

(2) No person contravenes subsection (1) by causing the alteration, disruption or destruction of fish habitat by any means or under any conditions authorized by the Minister or under regulations made by the Governor in Council under this Act.[154]

This section does not seem to constitute an authorization for interference with the public right of fishing in any but the most indirect way, but is rather directed to the regulatory aspect of the federal government's control over fisheries in general, through protection of fish habitat. By contrast, the entire purpose of s. 5 of the NWPA is to permit works that would otherwise constitute an interference with navigation.

The federal aquaculture policy does require consideration of "utilization by other groups," including the traditional fishery and aboriginal fisheries, during the review process for aquaculture siting,[155] but it seems clear that a policy requirement cannot substitute for an explicit legislative authorization for an abrogation of the public right of fishing. In any event, the federal position is equivocal on whether or not its approval is even mandated:

> As a result of the existing regulatory regime, in most provinces, the Minister of Fisheries and Oceans has legal authority by virtue of the *Fisheries Act*, for fisheries management reasons, to be consulted on and to provide recommendations regarding the issuance or expansion of leases issued by Provinces. These recommendations and/or advice will be taken into account by the provincial leasing authority. Considering the wording in the relevant regulations, **DFO's *approval* of the provincial lease, based on fisheries management considerations, in most cases, is *not* required.**[156]

This assessment may be correct from the perspective of DFO's *regulatory* mandate, but it raises problems for the proprietary aspect of the leasing schemes. It would appear that for many leases in provincial tidal waters, presumably including some that interfere with public rights of fishing, the federal government acknowledges no regulatory requirement to *authorize* the private interference with the public right, despite the fact that it may be consulted and regulate "for fisheries management reasons." However, the exclusivity provisions of some provincial lease schemes, which are supported by the MOUs referred to earlier, are intended to provide for an exclusion of the public right of fishing. The federal input to this legislative grant of proprietary rights (leaving aside regulatory involvement) consists of some participation in the siting approval process[157] and possible, but not mandatory, permit issuance. The difficulty with this approach, again from the narrow perspective of the property rights involved, is that it seems to rely on a delegation of federal powers that may be constitutionally invalid.

The delegation "solution"

It has long been accepted in Canada that one level of government can delegate *administrative* powers to another, and this is frequently done by way of government-to-government MOUs, or similar documents. While such agreements have been found to be valid, it is clear that delegation of *legislative* powers is not permitted, in that the parties would otherwise essentially be amending the constitutional distribution of legislative powers by way of a non-constitutional agreement.[158] Thus, any such delegation from the federal level should be based on a clear legislative enactment, the administration of which is then delegated to the provincial authority. In cases where management of freshwater (and some coastal) fisheries has been delegated to provinces, the legislative provisions are still federal.[159]

While there may be some scope for allowing discretion or latitude in the hands of the provincial authority to which the power is delegated,[160] there must still be a legislative basis for the delegation. With respect to interference with the public right of fishing, it seems clear that any authorization of such interference *must* have an explicit legislative origin, and cannot be a purely administrative matter within some larger scheme. This would be true of the delegation of leasing authority from the federal level to the provinces (for extra-provincial waters), and similarly to any delegation of leasing authority to the federal government in provincial waters (as in the case of Prince Edward Island). The courts' treatment of Magna Carta rights in tidal waters, which prohibits the Crown from acting without legislative authority, would make any other characterization impossible.

If the federal Parliament had enacted a lease scheme, the administration of which was then delegated to the provinces via MOUs, that would clearly be an acceptable structure.[161] The question here is whether there is anything in the *Fisheries Act*, the most directly relevant federal legislation, that could be seen as accomplishing this same purpose. There are at least four sections that might be relevant.

- Section 7 of the *Fisheries Act* provides for the issuance of "licenses and leases for fisheries and fishing," wherever no exclusive right of fishing exists. This section, however, restricts the purposes of such leases and licenses to "fisheries and fishing," which does not encompass aquaculture,[162] and s. 3(1) specifically excludes any grant of exclusive fisheries "in property belonging to a province."
- Section 57 provides that the minister "may authorize any river or other water to be set apart for the natural or artificial propagation of fish." It is unclear whether this is meant to be restricted to sanctuary and hatchery operations; that is, for the "propagation" of fish in the natural environment. In any event, it does not appear to have been used for the purpose of permitting aquaculture, and is not explicit with respect to the impact on public rights.
- Section 58 authorizes the minister to issue leases and licenses to "plant" or "form" oyster beds, and provides that "the holder of any such license or lease has the exclusive right to the oysters produced or found on the beds within the limits of the licence or lease." The reference to exclusivity, however, is only to the use of the oysters, and the tenure granted under the lease or license might be limited to the stated purpose of oyster culture, which need not exclude public fishing.
- Section 59(1) does provide for the delegation of leasing powers to

provinces,[163] but only for the culture of oysters, and only for waters within the province in question. As with s. 58, the rights accorded do not explicitly remove the public right of fishing in the lease area (except to the extent of according exclusive rights over all oysters, cultured or not). Similarly, the rights do not appear to extend to exclusive use of the site.

None of these provisions provides for the extent of exclusive use and occupation set out in some provincial schemes, nor for the range of operations that must be accommodated, and they are thus unlikely to be effective as a legislative delegation allowing for the resultant intrusion on the public right of fishing. In any event, the legislative authority for the existing MOUs, which purport to carry out the delegation of powers to the provinces, is not based on these provisions. The most recent MOU between Nova Scotia and the federal government is made (by Canada) on the authority of an Order in Council[164] under the *Department of Fisheries and Oceans Act*, which allows the minister, with the approval of the Governor in Council, "to enter into agreements with the government of any province or any agency thereof respecting the carrying out of programs for which the Minister is responsible."[165] It does not seem likely that the "carrying out of programs" would be a sufficiently explicit legislative authority for removal of the public rights in question.[166]

The situation in Prince Edward Island is, of course, quite different. Here the question is not the extent of provincial authority, but simply whether the federal government has sufficient legislative authority to act so as to restrict or exclude the public right of fishing. As was noted earlier, in "Statutory Responses," the 1928 and 1987 MOUs purport to authorize the federal government to issue leases within the province, but an intergovernmental agreement cannot be legislative authority for anything, let alone removal of the public right of fishing. The federal legislative basis for issuing leases is clear for oysters under s. 58 of the *Fisheries Act* (see earlier), but what of other forms of aquaculture? It may be argued, as already noted, that the power under s. 57 is sufficient, but this is by no means clear.

General regulation under the Fisheries Act

If it is accepted that the provincial leases and other tenures may be ineffective in ensuring exclusivity as against anyone exercising a public right of fishing, an alternative solution is to rely upon the federal power over fisheries, and the licensing provisions of the *Fisheries Act*, to impose the necessary removal of public rights in provincially leased aquaculture areas, whether by specific regulation or by inclusion of license terms. Two issues arise: could the federal government validly legislate to provide the necessary protection to the leasehold areas; and second, does the *Fisheries Act* currently provide for the removal of public rights on this basis?

On the first question, the answer would seem to be that the requisite authority does exist. In the early cases, as noted earlier, it was established that the scope of federal legislation could be extremely broad insofar as it affected provincial jurisdiction,[167] even though it was, in tidal waters in the provinces, always limited to the regulation of the public, and not the proprietary rights.[168] The general approach that was applied up to the 1970s and 1980s was focused on a concept of "protection and preservation" as the basis of management, which could be taken as narrowing the federal power as applied within the provinces.[169] As Meany points out, however, this test arose primarily in the context of potential conflicts with provincial powers, and has been found to be less useful where the issue is the breadth of the federal power in the absence of such conflict. Thus, the courts have found that the federal power over fisheries can extend to the establishment of "close times for catching fish not only for the purpose of conservation, but also for socioeconomic purposes, such as allocation."[170] In a similar vein, sector management rules restricting the operations of vessel types to certain areas were found to be valid, even though they were "directed at the socioeconomic conditions of fishermen."[171] In sum, the federal power, where it does not run up against valid provincial jurisdiction, must be seen as quite extensive.[172] If this reasoning is applied to the situation of aquaculture in tidal waters within the province, the first and most obvious point is that the regulation of the public right of fishing has been found to be entirely within the jurisdiction of the federal government, as has already been discussed. This removes the only *constitutional* impediment, and leaves it to the Dominion Parliament to regulate the matter as it sees fit.

The second question, whether the *Fisheries Act* currently provides for the desired removal of the public right, is more problematic. There is no question that the Minister of Fisheries and Oceans has a very high degree of discretion in the issuance or refusal of fishing licenses, and is statutorily authorized to impose a wide range of conditions by regulation, which could almost certainly extend to prohibitions on fishing within specified areas or types of areas (such as leased aquaculture sites). The minister's discretion may be nearly absolute, but some limited challenges may be possible.

In *Alford* v. *Canada*, Brenner J of the British Columbia Supreme Court considered a challenge by commercial fishers to the federal issuance of communal aboriginal fishing licenses, based, *inter alia*, on a claim that the licenses constituted a grant of exclusive fishery without parliamentary authorization, in that neither the Constitution nor statute had given the minister this power.[173] At the heart of this argument was the contention that the public right to fish had been regulated, but never extinguished, by federal legislation (the *Fisheries Act*):

> The plaintiffs say that the public right to fish has not been extinguished in Canada, it has only been regulated, just as the aboriginal right to fish was found to be merely regulated and not extinguished in *R. v.*

Sparrow.... The plaintiffs contend that the Minister, by granting an exclusive right to the fishery to aboriginal fishers, has violated the public right to fish and that, assuming the latter is a common law and not a constitutional right, so "privatizing" the public fishery would take nothing less than a statute specifically doing so. In particular, regulation and exercise of ministerial discretion are not enough.[174]

The defendant Attorney General applied to have the action dismissed as disclosing no cause of action, but the court accepted the existence of a public right to fish which "cannot be interfered with except by statute,"[175] and further found that it was "not plain and obvious that the public right to fish had been extinguished by competent federal legislation."[176] The low standard of proof required of the plaintiff on an application for dismissal obviously limits the usefulness of this decision, but the basic approach of requiring some proof of extinguishment of the right in question may be of assistance, particularly when the nature of the current problem is considered. In *Alford*, the plaintiffs were in the position of having to challenge the validity of a minister's positive grant or licenses to others, a difficult task given the breadth of discretion afforded the minister under the *Fisheries Act*. If, however, we assume that a challenge to a *private* aquaculture site arises from a claim by a fisher or fishers against the leaseholder, the following line of argument could be put forward:

- Any private interference with the public rights of fishing or navigation in tidal waters, even by a grantee from the Crown, is an actionable public nuisance *unless* it can be shown that it was explicitly authorized by legislation.[177]
- The provincial lease cannot serve as the requisite authorization, as it is merely a provincial grant of proprietary rights, and action by the Dominion Parliament is required.
- It is for the grantee to show that the interference with the public right was justified by an explicit legislative authorization.[178]

As was argued earlier, the authorization for interference with navigation can be satisfied by the permit under the NWPA, but it is not immediately clear where the grantee could turn to show a federal legislative authorization for exclusive use and occupation that extended to a bar on public fishing rights. That is, the leaseholder needs to be able to show their own positive authorization, not merely a permit restriction separately imposed on the plaintiff (although this may be relevant in a procedural sense – see later in the chapter). Section 23 of the *Fisheries Act* does prohibit the taking of any fish "within any fishery described in any lease or licence."[179] This might be argued to be a restriction on the public right of fishing in *any* leases, federal or provincial, but there are two problems with this approach. First, the section refers to taking fish "within any fishery" contained in such leases, which would seem to eliminate aquaculture sites as such.[180] Second, in the statutory context it would be difficult to make the case that "leases" did not refer to leases issued pursuant to s. 7 of the Act (discussed earlier).[181]

In sum, while the law is by no means clear at this point, it is entirely conceivable that a private action in public nuisance could be brought against a leaseholder of an aquaculture site in provincial tidal waters where an interference with the right of fishing could be shown, on the grounds that no valid federal legislative authorization supports that interference. The procedural difficulties that might arise from this action are dealt with below.

Waters outside the provinces

As was shown earlier, outside provincial waters federal authority over both the regulatory and the proprietary aspects of aquaculture is unquestioned. This does not, however, entirely dispose of the matter. There remains the possibility, perhaps unlikely, that a challenge analogous to that in *Alford* could be brought against the minister, claiming that the *Fisheries Act* does not provide the necessary explicit statutory authorization for removal or modification of the public right of fishing. Alternatively, as was earlier suggested with respect to tidal waters, a claim in public nuisance might be brought directly against the leaseholder, putting the proof of legislative authorization on them.

These possibilities may lie in the future, but at present the greater difficulty arises from the imprecision attached to the notion of provincial tidal waters, as discussed, for while the legal *status* of the extra-provincial waters is clear, their *location* is not. The various provincial legislative instruments that authorize aquaculture leases do not purport to extend their effect to areas outside the various provinces, and indeed if they did so they would be *ultra vires* the provinces in any event. There has been no delegation of responsibility from the federal level, for the MOUs only refer to authority over leasing and other activities within the provinces.[182] At time of writing, the only apparent method for application of provincial aquaculture laws outside provincial waters is by delegation as specified in s. 9(1) of the *Oceans Act*, which has not been utilized for this purpose:

> 9.   (1)   Subject to this section and to any other Act of Parliament, the laws of a province apply in any area of the sea
>   (a)   that forms part of the internal waters of Canada or the territorial sea of Canada;
>   (b)   that is not within any province; and
>   (c)   that is prescribed by the regulations.[183]
>
>   . . .
>
>   (3)   For the purposes of this section, the laws of a province shall be applied as if the area of the sea in which those laws apply under this section were within the territory of that province.

The problem of territoriality was briefly addressed in the *Legislative and Regulatory Review* conducted by OCAD, in which it was noted that the provinces "administer the leasing process for all 'near-shore' activities (except in PEI),"[184] a somewhat restrained characterization of a process in which most coastal provinces both *legislate* and *administer* the schemes. Given the acceptance of two distinct geographic areas of responsibility, the Review noted the following problem in a footnote:

> Near-shore is an indefinite term. The federal and provincial governments do not agree on the jurisdictional authority, that is, where provincial authority ends and federal authority starts, with respect to the seabed beyond the tidal mark.[185]

The implications of these statements are clear. First, given that most of the leases in areas of potential doubt are being issued under provincial legislation, it is highly likely that the provinces are issuing leases in areas that are not within provincial jurisdiction. Second, any such leases that are found to be outside provincial waters on the particular facts of a case must be of no force and effect. One response to the apparent untenability of this situation is to rely on the continued cooperation of the two levels of government, neither of which has an interest in pressing the matter. However, this ignores the fact that an action could, as already noted, be brought by private individuals alleging interference with the public right to fish, and the question of location would obviously be relevant to whether *any* authorization (let alone a provincial one) could be raised as a defense.

Consequences and implications

If stability and certainty of tenure is one of the primary objectives of legislative schemes assigning property rights in aquaculture operations, this examination of constitutional issues suggests that the current approach may present a number of difficulties, which can be summarized as follow.

First, for tidal waters inside a province, the provincial power to grant property rights, including leases, over submerged lands does not extend to authorization of any resulting interference with the public right to fish. It is possible that the federal legislative power to restrict fishing for a wide variety of reasons may be sufficient to effectively remove that right for leased aquaculture sites. However, it is also conceivable that in a private action against a leaseholder, the lease itself could be found to be invalid, as an interference with a public right of fishing unsupported by any explicit legislative provision. That is, regardless of the validity of the federal restrictions on the exercise of fishing rights, the grant of the leasehold rights must be legally sound in its own right.

Second, for any waters that are not within the boundaries of a province, the leasehold arrangements as currently structured are ineffective to provide the desired property entitlements. Provincial legislation cannot have extraterritorial effect in this case, and the limited opportunities for assertion of provincial laws outside the province (as set out in the *Oceans Act*) have not been applied. Accordingly, under current provincial legislation the grant of leasehold rights in any area of marine waters that is outside a province would be invalid, in that such a grant would be *ultra vires* the province. Furthermore, the invalidity of the lease would mean that the leaseholder would have no clear defense against a private claim for interference with the public right of fishing.

The first of these difficulties, dealing with tidal waters in the province, may be more theoretical than practical. The private action in public nuisance can be problematic from a procedural standpoint, in that the plaintiffs must show "special damage" above and beyond that suffered by the public at large, in order to maintain the action without the consent of the Attorney-General. While the law is by no means settled, this has been interpreted in the fisheries context to mean that commercial fishers had no special interest, in that they were asserting infringement of the same general right as other members of the public.[186] In any event, the ability of the federal government to regulate fishing in a broad fashion may be seen as effectively removing the truly public nature of the right in such cases. The territorial issue is, however, more problematic. Given that the actual extent of provincial and federal waters is unclear, a site-by-site resolution could be required in order to determine whether a valid provincial lease was even possible. Furthermore, it seems clear that at least *some* aquaculture leasehold areas will be in federal waters, and for those sites there is little doubt that the leases under which they operate could be successfully challenged, leaving a situation in which some operators would have valid leases and others would not, depending on a finding with respect to the status of particular waters. The legal regime should be able to provide for both certainty and general application to all locations and operators, and, given the problems set out here, it may fail on both counts.

Conclusions: the way forward

The analysis in the preceding sections leads to a number of conclusions and recommendations as to how the development of property rights for aquaculture might be pursued in the future. These fall into two general categories: jurisdictional gaps and the need for federal legislation; and the value of a principled approach to any potential expansion of property rights beyond their current status.

Jurisdictional gaps and the need for federal legislation

While the focus of many observers is on improving the scope and intensity of property rights available to aquaculturists, the review conducted in this chapter suggests that the more immediate problem is ensuring the legal validity and effectiveness of those rights that already exist. The potential jurisdictional gaps and other problems with the assignment of property rights in aquaculture, as set out in the preceding sections, point to the necessity for federal aquaculture legislation, for a number of reasons. First, the present reliance on provincial legislative schemes leaves leasehold arrangements subject to a degree of insecurity and vulnerability to legal challenge. Within provincial boundaries, leases in tidal waters could be subject to private claims for interference with the public right of fishing, in the absence of federal legislation explicitly authorizing such interference. As was noted earlier, it is by no means certain that such claims would succeed, but in waters beyond provincial boundaries the situation is clearer, and more difficult.

The provincial acts under which most of the current leases are issued do not purport to extend, nor could they validly extend, to marine areas outside the boundaries of the provinces. This would not present an insuperable problem were it not for the fact that the definition of federal and provincial marine areas, while well set out in some areas, is less than clear in many others. If the validity of a lease is to depend upon whether the waters in question are ultimately defined as federal or provincial, the following choices emerge. A province could decide to issue leases only in areas definitively settled as falling within provincial boundaries, and leave out of consideration any waters whose status is in doubt, letting the federal government act in those areas if it wished. It could also, as seems to be the current unstated practice, simply issue the leases in all requested areas, in the hope that they are not challenged. Neither option is particularly sound: one leads to the elimination of potentially valuable sites, while the other exposes the lease-holders to the risk of losing their investment as a result of the *ultra vires* character of their provincial lease.

It might be argued that this problem could be addressed by federal aquaculture legislation dealing only with federal waters. That is, provinces would continue to legislate for leases within their boundaries, and a federal leasehold system would be in place in adjacent waters beyond the provincial boundaries. This is a less than optimal solution, however, in that it would lead to a patchwork approach to the granting of rights in the same region, and would still not address the problem of identifying all sites as either federal or provincial waters. Nor is the enactment of a federal scheme that would apply to all marine waters a viable approach. While it may provide the prospect of uniformity, the federal government does not have jurisdiction to make grants of private property rights in provincial marine areas (even though it can control the *public* right of fishing in those waters). As was noted earlier, action by both levels of government is needed to make such grants effective.

There are, however, legislative options that could satisfy the requirements of uniform (and constitutional) application to all waters and operators within a region, while at the same time respecting the provincial jurisdiction and recognizing that the provinces have in fact been the lead players in the promotion and management of aquaculture. First, the federal government could act under the *Oceans Act*, ss. 9 and 26, to provide for the application of provincial laws in marine areas outside the provinces. Regulations would be required for the areas adjacent to each province, and would in essence designate the relevant provincial aquaculture and leasing schemes as applicable to those waters. This would have the advantage of ensuring a uniform system across a geographical area, with no uncertainties as to whether waters were federal or provincial.[187] There are, however, potential disadvantages to this approach. As discussed earlier, it could be subject to challenge as an impermissible delegation of legislative authority to the provinces. Alternatively, it might be argued that a regulatory power to allow for the application of provincial laws is not a sufficiently explicit legislative limitation on the public right to fish, opening up the potential for private challenges, as discussed earlier.

There are also broader policy reasons to reject this option. This chapter has not been concerned with the regulation of aquaculture (see Chapter 3), but rather with the assignment of property rights. Given that the proprietary aspect also depends on legislation, however, there would seem to be benefits in integrating the regulatory requirements in one coherent scheme with the critical property elements. This can best be accomplished by the development of a federal aquaculture act, a step suggested by Wildsmith in 1985.[188] In the draft federal Act which he proposed at that time, the focus was on the regulatory aspects, but he included the following proposed provision for empowering the federal government to issue leases:

The Governor-in-Council may make regulations...

(g) respecting the terms of occupation, including leases, of marine or tidal areas outside the boundary of any province for the conduct of aquaculture, including exclusive rights of the occupier, lease fees, and performance standards.[189]

This draft section highlights one of the limitations of federal legislation, with nothing more; for purposes of the proprietary grants (as opposed to regulation) it can only apply outside the provinces, leaving in place the problems of uncertainty (as to the status of waters) and lack of uniformity within the same region. Nonetheless, federal legislation could, at a minimum, put in place a general authorization for the grant of leasehold entitlements that interfere with the public right of fishing, in all marine waters. Within provincial boundaries, this could be implemented by the simple grant of a federal permit to leaseholders, in the same way that a permit under the NWPA deals with the question of interference with navigation issue in provincial waters. That is, the federal legislation would not be effecting the grant, but rather would be authorizing the interference with the public right of fishing, consequent on the provincial grant.

For federal waters, the legislation could provide for the issuance of leases but delegate the operation of the scheme to the provinces. In order to maintain the diversity of provincial approaches, and to ensure regional uniformity, it could be desirable to establish regulations for each province, adopting by reference the provincial lease arrangements in place within that province. The administration of the leases, which would be consistent with those applicable in adjacent provincial waters, could be delegated to the province. This approach would go one step beyond that suggested under the *Oceans Act*, in that the relevant provincial provisions would actually be incorporated into federal regulations by reference, making it clear that only the administration of the scheme (in federal waters) was being delegated to the provinces.

Potential modification of property rights

The arguments presented in this chapter suggest that there may be more pressing issues related to property entitlements in aquaculture than the immediate expansion of those interests: there is sufficient reason to believe that the current system of entitlements is founded on doubtful ground with respect to both constitutional jurisdiction and common law doctrines. Nonetheless, there is a legitimate debate as to whether the existing forms of property entitlements for aquaculture operations are adequate to the functional requirements of the industry, in particular because of the divergence in current provincial approaches. The general examination in the section "Statutory responses" (p. 122) illustrates that, while the provinces have all moved to grant *some* degree of property entitlement in aquaculture sites, these arrangements vary with respect to some of the key descriptors of a property entitlement, noted in the introduction to this chapter: exclusivity, duration, transferability and assignability, and enforceability.[190] Complete uniformity in the provincial approaches to property rights (and even regulation) is not necessary, but a higher level of consistency across jurisdictions might avoid the imposition of comparative disadvantages on aquaculture operators, or alternatively on other resource users, in some provinces.

It must be remembered, however, that this debate cannot be limited to the functional requirements of the aquaculture industry, as was noted at the outset of this chapter. The existence of other users, with long-standing and potentially enforceable rights of their own over the areas in question, means that the debate should be focused on the appropriate balance to be sought between the private and public rights, and simple recourse to full privatization is unlikely to achieve this outcome. While the introduction of more extensive private rights need not lead inevitably to privatization of the resource base, it is important to remember that non-aquaculture users of these areas are likely to view moves in this direction as a threat to their legal rights, and to continued equitable access to important resources.[191] The solution will lie in part in ensuring that open, transparent and *negotiated* processes are used where public rights are, inevitably, reduced or compromised by the introduction of new private rights.[192]

Even before we consider the issues of equity and process, however, there is a need to develop the independent case for enhancement of private property interests in a common resource; in the absence of a demonstrated need to increase the private role, the existing interests of other users would prevail by default. The term "property" in this context denotes a highly variable mix of the characteristics listed above, especially exclusivity, assignability and duration. General assertions that more property rights are needed will not be sufficient; it will be necessary to break down the functional requirements of different types of aquaculture operations with reference to the multiple characteristics of "property" rights. This has not, to date, been a feature of the public debate, but a systematic application of such functional criteria, based on the actual needs of the industry, will at least provide a principled basis for consideration of the industry's needs, and a starting point for the comparison of those requirements with the interests of the wider community of resource users.

Notes

1 Department of Fisheries and Oceans, *Federal Aquaculture Development Strategy* (Ottawa: Minister of Supply and Services, 1995) at 1 [Federal Strategy].

2 See the section "Common law property rights and aquaculture operations" (p. 118).

3 B. H. Wildsmith, *Aquaculture: The Legal Framework* (Toronto: Emond-Montgomery, 1982) at 93. Wildsmith did note the existence of limited exceptions relating to some oyster beds.

4 *Ibid.* at 95, 233–235.

5 See the section "Statutory responses" (p. 122). For a summary of the initial stages of legislative development in the provinces, see S. Coffen and A. Smillie, "The Legal Framework for Canadian Aquaculture: Issues in Integrated Ocean Management," in D. VanderZwaag, ed., *Canadian Ocean Law and Policy* (Toronto: Butterworths, 1992) at 51–63.

6 Office of the Commissioner for Aquaculture Development (OCAD), *Legislative and Regulatory Review of Aquaculture in Canada* (Ottawa: Department of Fisheries and Oceans, 2001) at 18 [OCAD Legislative Review]. The same passage goes on to highlight the need for sufficient duration of tenure, and for the development of a federal approach to leasing (citation omitted):

> Part of the uncertainty relates to the lack of long-term security for various forms of authorization and licensing of aquaculture activities. This is a deterrent to private investment in aquaculture. The lack of a clear federal leasing policy and regulations (or delegation of administrative responsibilities to the provinces) impedes development of the aquaculture sector, particularly as interest increases in developing areas further offshore.

7 *Aquaculture in Canada's Atlantic and Pacific Regions*, Report of the Standing Senate Committee on Fisheries (Ottawa: Senate of Canada, 2001) at 28 [Senate Report 2001]:

Generally speaking, aquaculturists and their representatives told us they needed to have access to new sites, more secure tenures, and longer term approaches to leasing. Lease terms were deemed to be too short; the lack of security of tenure was said to make financing difficult to obtain.

See also *The Federal Role in Aquaculture in Canada*, Report of the Standing Committee on Fisheries and Oceans (Ottawa: House of Commons, 2003) at 28 [Commons Report 2003].

8 A. Reiser, "Defining the Federal Role in Offshore Aquaculture: Should It Feature Delegation to the States?" (1996–1997) 2 *Ocean and Coastal Law Journal* 209 at 212. See also A. Scott, "Property Rights and Property Wrongs," (1983) 16(4) *Canadian Journal of Economics* 555, where the characteristics of property rights are defined as exclusivity, divisibility, enforceability and transferability.

9 OCAD Legislative Review, *supra* note 6 at 22.

10 R. Neill, *Fencing the Last Frontier: The Case for Property Rights in Canadian Aquaculture* (Halifax, NS: Atlantic Institute for Market Studies, 2003) at 2. The debate over the appropriate use of property rights is not confined to aquaculture, but has also been a feature in the management of capture fisheries. For a review of this issue, see R. Shotton (ed.), *Use of Property Rights in Fisheries Management: Proceedings of the FishRights99 Conference* (Rome: FAO Fisheries Technical Paper 404, 2000).

11 Property rights regimes in the natural resource context have been categorized as falling within four main types: private property, with individual ownership; common property, with collective ownership; true open access, with no property entitlements; and state ownership. See the discussion at P. Knight, "Oceans Policy and Property Rights: The Case for Common Property Regimes," (2002) *New Zealand Surveyor* 19 at 21. As will be seen in the following sections, the property regimes in place for sites of interest to aquaculture range from private property (as with some inland freshwater areas) to state ownership (for unallocated submerged lands in tidal waters), and include combinations of the two, where private property rights are assigned by lease but the ownership remains with the Crown.

12 Commons Report 2003, *supra* note 7 at 28. See also Senate Report 2001, *supra* note 7 at (iii): "One of the challenges faced by government now and in the years ahead will be to achieve a delicate balance between various competing users of the marine environment (a common property resource)." For discussions of the conflicts between and among users, see also Reiser, *supra* note 8 at 213–214; and B. Vestal, "Dueling with Boat Oars, Dragging through Mooring Lines: Time for More Formal Resolution of Use Conflicts in States' Coastal Waters," (1999) 4 *Ocean and Coastal Law Journal* 1 at 2–3, 6–10.

13 The challenge is described by Reiser, *supra* note 8 at 213, as follows:

It is crucial, therefore, that the government's process for issuing the lease or license itself protects the sea farmer from conflicts with other marine uses. The statute authorizing the conveyance of a lease of public waters or submerged lands for aquaculture should identify other public and private uses of the marine environment that are potentially affected by aquaculture activities. It should then provide a fair but efficient process for information to be brought forward about those uses in the area proposed for use as a sea farm, allowing the leasing agency to make a balanced and informed decision in which other users believe they have been fairly considered.

See also the discussion at Commons Report 2003, *supra* note 7 at 28–31.

14 See the examination of social impacts and equity of access in the introduction

of commercial aquaculture in the Bay of Fundy, in J. Marshall, "Landlords, Leaseholders and Sweat Equity: Changing Property Regimes in Aquaculture," (2001) 25 *Marine Policy* 335. See also J. Phyne, "Capitalist Aquaculture and the Quest for Marine Tenure in Scotland and Ireland," (1997) 52 *Studies in Political Economy* 73.

15 Prior to the development of the provincial legislative regimes, there was ambiguity respecting the property rights of aquaculturists in the plants and animals contained within their facilities. This issue has been addressed in most provinces by statutory provisions similar in effect to that found in s. 60 of the Nova Scotia *Fisheries and Coastal Resources Act*, S.N.S. 1996, as amended, and will not be addressed further in this chapter:

> 60. All aquatic plants and animals of the species specified in an aquaculture license or aquaculture lease in or on the licensed or leased area, except free-swimming or drifting flora or fauna not enclosed by a net, pen, cage or enclosure, are the exclusive property of the holder of the license or lease.

16 Although the situation under civil law is not addressed in this chapter, the current position in Québec, with respect to the issuance of leases and the rights of aquaculturists is similar to that which has developed in the common law provinces: see Commons Report 2003, *supra* note 7, dissenting report of the Bloc Québécois at 88.

17 It is assumed that aquaculture operations conducted on land, in tanks or similar facilities, will be subject to the normal regime for property in the jurisdiction, and raise no special issues of relevance to this study. Such operations would, of course, still require an aquaculture license under provincial legislation (see later in this section).

18 See, for example, the judgment of Ritchie CJC in *The Queen* v. *Robertson* (1882) 6 S.C.R. 52 at 117. This presumption could, however, be rebutted, as was held in *Wishart* v. *Wyllie*, 1 Macq. H.L. Cas. 389, in a passage cited with approval by Ritchie *ibid.*:

> It may be rebutted, but, generally speaking, an imaginary line running through the middle of the stream is the boundary; just as if a road separates two properties, the ownership of the road belongs half-way to one and half-way to the other. It may be rebutted by circumstances, but if not rebutted, that is the legal presumption.

See also *R.* v. *Lewis* [1996] 1 S.C.R. 921 at para. 56.

19 See the discussion at G. V. La Forest and Associates, *Water Law in Canada: The Atlantic Provinces* (Ottawa: Information Canada, 1973) at 234. As La Forest notes (at 200–201), the ownership rights of the riparian landowner are distinct from their "riparian rights" (including rights of access, drainage and accretion, and some rights relating to flow, quality and use). The most important distinction is that the riparian owner "has certain rights respecting the water therein whether or not he owns the bed" (citation omitted).

20 M. Walters, "Aboriginal Rights, Magna Carta and Exclusive Rights to Fisheries in the Waters of Upper Canada," (1998) 23 *Queen's Law Journal* 301 at 313.

21 *The Queen* v. *Robertson*, *supra* note 18 at 133, per Strong J. See also *R.* v. *Nikal* [1996] 1 S.C.R. 1013 at para. 1053.

22 *Friends of the Oldman River Society* v. *Canada (Minister of Transport)* [1992] 1 S.C.R. 3 at para. 76.

23 La Forest, *supra* note 19 at 179.

24 *Ibid.* at 190.

25 *R.* v. *Nikal*, *supra* note 21 at paras. 67–72.

26 See the discussion in Walters, *supra* note 20 at 325–326, where this view is advanced.

27 This is precisely the distinction noted by the Supreme Court of Canada in the *Provincial Fisheries* case (*Re Provincial Fisheries* (1895) 26 S.C.R. 444 at 527):

> That the Crown in right of the provinces could grant either the beds of such non-tidal navigable waters or an exclusive right of fishing is, I think, clear. Before Magna Charta the Crown could grant to a private individual the soil in tidal waters with the fishery as an incident to it, or the exclusive right of fishing as distinct from the soil. Then, as the restraint imposed by Magna Charta does not apply to any but tidal waters, there is no reason why the prerogative of the Crown to make such grants in the class of waters now under consideration, large navigable lakes and non-tidal navigable rivers, should not be exercised now as freely as it could have been with reference to tidal waters before Magna Charta.

28 See the description of the pre-Magna Carta position in Coulson and Forbes, *Water Law* (1902), as cited in *Donnelly* v. *Vroom et al.*, (1842) 3 N.S.R. 97 (N.S.S.C.) at 589. For a longer discussion of the issues addressed in this section, see P. Saunders, "Marine Property Rights and the Development of Jurisdictional Regimes: Private Rights, Communal Tenure and State Control," in D. Vickers, ed., *Marine Resources and Human Societies in the North Atlantic* (St. John's: Memorial University, 1997).

29 C. D. Hunt, "The Public Trust Doctrine in Canada," in J. Swaigen, ed., *Environmental Rights in Canada* (Toronto: Butterworths, 1981) at 153.

30 *A. G. British Columbia* v. *A. G. Canada* (1913) 15 D.L.R. 308 (P.C.) at 317, per Viscount Haldane, Lord Chancellor [*Re. B C Fisheries*].

31 (1842) 2 N.S.R. 97 (N.S.S.C.) at 99: "There is no pretence for saying the crown could make any such grant. It might as well grant the air around the cove. These waters, fluctuating and in a constant state of change, are not the subject-matter of a grant."

32 *Ibid.* at 99–100:

> [T]he crown could not grant a general fishery – a grant to support that must be as old as the reign of Henry II, and therefore beyond the time of legal memory, for, by Magna Charta, and the second and third charters of Henry III, the king is expressly precluded from making fresh grants.

33 (1907) 40 N.S.R. 585 (N.S.S.C.) at 588:

> It appears . . . that when the defendants dug clams on the flats they did so merely in exercise of a right which belonged to them in common with all other members of the public . . . the question of ownership of the soil, as well as possession thereof, and user, with, or without, leave, seems to me to be immaterial. This right could also be maintained against the Crown, where it stood as owner.

34 (1920) 51 D.L.R. 495 (N.B.S.C. – A.D.) at 497 (emphasis added). The case was somewhat complicated by the fact that the lessor city acquired its title by virtue of its charter, confirmed by an act of the provincial legislature, and had the power to regulate and control fisheries in the area in question. In this sense, both aspects of the legislative power were engaged: the power to make an explicit grant of an exclusive fishery (which was not done on the facts); and the power to regulate and control the public right to fish (as exercised by the city under statute). *Ibid.* at 497, 500–502.

35 Wildsmith, *supra* note 3 at 106.

36 [1908] 2 Ch. 139 at 165–166. It should be noted that the position in the

United States evolved quite differently. There the doctrine of the public trust imposed limits on the alienability of submerged lands to protect dominant public interests, and a critical distinction from the British position was that the list of public trust interests was capable of expansion. See the discussion at Hunt, *supra* note 29, *passim*.

37 See, for example, *Foster* v. *Warblington Urban Council* [1906] 1 K.B. 648 (C.A.). The plaintiff's actions in placing clams on a section of foreshore belonging to another, but over which the plaintiff had prescriptive rights, was seen by Stirling LJ as "totally different from anything that he could have done simply as a member of the public exercising the right of fishing" (at 671). See the discussion of this case at Wildsmith, *supra* note 3 at 106, 108–109. The view expressed in *Foster* was endorsed at the trial level in *Donnelly* v. *Vroom*, which referred to the oysters in *Foster* as the "plaintiff's property," in contrast to the clams in *Donnelly*, *supra* note 28 at 592.

38 This assumes that the "captured" property inside could be designated as private property. This argument would be separate and distinct from any claim that the pen might be an impediment to fishing for wild species within the parameters of the public right.

39 R.S.N.L. 1990, c. A-13, as amended [*Aquaculture Act*].

40 S.N.L. 1991, c. 36. According to the policy of the Department of Fisheries and Aquaculture, the Crown lands lease application should be presented with the application for an aquaculture license.

41 *Ibid.* at s. 3: "The minister may issue a lease to a person of an area of Crown land for the period and upon those terms and conditions and subject to the payment of those rents, royalties or other charges that the minister may set out in the lease."

42 *Ibid.* at s. 4(1) (less than 20 hectares) and 4(2) (greater than 20 hectares).

43 *Ibid.* at s. 6(1): "The minister may issue a licence for occupancy of an area of Crown land subject to those terms and conditions and subject to the payment of those fees, rentals and other charges that the minister may set out in the licence."

44 See the Government of Newfoundland and Labrador website, http://www. fishaq.gov.nl.ca/aqua/licencing.stm#Application (accessed 8 August 2005).

45 See Government of Newfoundland and Labrador website, http://www.env.gov. nl.ca/env/lands/cla/aquaculture_leases-licences.html (accessed 8 August 2005).

46 *Aquaculture Regulations under the Aquaculture Act*, 1996 (O.C. 96-939).

47 *Aquaculture Act*, *supra* note 39 at s. 6(6)–(8).

48 This even extends to the protection of private property rights in the product of operations found in sites, which is found in s. 5 of the *Aquaculture Act*.

49 *Supra* note 15, s. 44(2). On other lands, by s. 44(1), a license is still required. Under s. 44(3), the grant of a license (without a lease) "carries with it the exclusive right to possession of the water column and sub-aquatic land described in the licence." The purpose of this section would seem to be to ensure that the exclusive possession granted by the *lease* on Crown lands is still available to an operator on private submerged lands, where the title would not otherwise extend to the water column.

50 The restriction to "aquacultural purposes" could raise some difficulties of interpretation. Does it mean that others may be excluded only to the extent necessary to conduct aquaculture, or is it more broadly intended to allow for true exclusivity, so long as the purpose of exercising the exclusion is to conduct aquaculture?

51 The purpose of this section would seem to be to ensure that exclusive possession of the water column and submerged lands granted by the *lease* on Crown lands is still available to an operator on private lands, where the title would not otherwise extend to the water column, and where exclusion from the submerged areas might also be in doubt. It is unclear why the exclusivity provision for licenses does not have the same restriction to "aquacultural purposes" as is found

in the lease provision, *supra*. In fact, it is not clear why a separate provision was required for the lease at all, as any aquaculture operator would require a license, and would therefore obtain the desired exclusivity in any event.

52 S.N.B. 1988, c. A-13, as amended [*Aquaculture Act*].

53 There is no similar requirement for a lease, but given that anyone carrying out such activities on Crown lands would be a trespasser, the practical effect is the same.

54 Further specification of requirements for transfer and subletting of leases are set out for the Bay of Fundy area in the *Bay of Fundy Marine Aquaculture Site Allocation Policy* (Government of New Brunswick, 2000) [Bay of Fundy Policy] at 6–7.

55 *Aquaculture Act, supra* note 52: by s. 25(6) the exclusive right does not include mineral rights, and by s. 25(7), the lease may make provision for access by adjacent landowners.

56 *Ibid.* s. 26(1), (3).

57 According to the Department of Agriculture, Fisheries and Aquaculture (DAFA), AOPs are usually issued "prior to the issuance of a lease," as a temporary measure. See DAFA Licensing Policy. Online. Available http://www.gnb.ca/0177/01770002-e.asp (accessed 8 August 2005).

58 R.S.B.C. 1996, c. 149, s. 13.

59 R.S.B.C. 1996, c. 245 [*Land Act*]. The responsibility for processing tenure applications formerly rested with a Crown Corporation, Land and Water British Columbia, Inc. (LWBC), but in 2005 this role was transferred to the ILMB. See the description of the ILMB mandate at http://ilmbwww.gov.bc.ca/ilmb/index.html. As of time of writing, the various policy documents were, however, maintained on the LWBC website at http://lwbc.bc.ca/.

60 Other forms are available in the act, but those available to aquaculturists are limited in a policy document issued by the Ministry of Agriculture and Lands: *Crown Land Use Operational Policy: Aquaculture* (Government of British Columbia: 2005) at 5–6. Online. Available http://www.lwbc.bc.ca/011wbc/policies/policy/land/aquaculture.pdf.

61 *Land Act, supra* note 59, s. 14(1)

62 *Crown Land Use Operational Policy: Aquaculture, supra* note 60 at 5.

63 *Land Act, supra* note 59, s. 39.

64 *Crown Land Use Operational Policy: Aquaculture, supra* note 60 at 5. See also the following clause from a 1988 License of Occupation (on file with author), which is consistent with the policy:

> (9.05) This license shall not entitle the Licensee to exclusive possession of the Land and the Owner [the Crown] may grant licenses to others to use the land for any purpose other than permitted herein, so long as the grant does not materially affect the exercise of the Licensee's rights hereunder. The question of whether a grant materially affects the exercise of the Licensee's rights hereunder shall be determined by the Owner at his sole discretion.

65 *Crown Land Use Operational Policy: Aquaculture, ibid.* at 5: "The standard form of Crown land tenure for a shellfish or finfish aquaculture operation is a 20-year licence of occupation."

66 *Ibid.* at 6. The *Land Act, supra* note 59, s. 38, contains only the essentials: "The minister may issue a licence to occupy and use Crown land, called a 'licence of occupation', subject to the terms and reservations the minister considers advisable."

67 *Crown Land Use Operational Policy: Aquaculture, ibid.* at 6: "Crown land leases are not typically used for aquaculture tenure and where they are, it is more common for shellfish operations than finfish farms."

68 *Ibid.* at 20–22. Both leases and Licenses of Occupation can be assigned or "sub-tenured," which refers to a grant of an interest in the land by the primary tenant to another party, without a full transfer of the right.

69 The origins of this situation date back to a 1928 agreement between the federal and provincial governments in which it was "agreed that the federal government would be responsible for the control, administration, development and improvement of the oyster or mollusk industry, as well as for conducting surveys and issuing leases." *The Prince Edward Island Report: Report of the Standing Committee on Fisheries and Oceans No. 8* (December 1998). This agreement was subsequently confirmed and updated in the *1987 Agreement for Commercial Aquaculture Development between the Government of Canada and the Government of Prince Edward Island* [1987 PEI MOU] at s. 5.1 A, for mollusks, and salmonids and other finfish were added to the agreement by s. 5.1 C: "Canada shall licence and issue leases for salmonids and other fin fish covered under this Agreement and its Amendments."

70 *Aquaculture Leasing Policy Prince Edward Island* (Ottawa: DFO, 2005 ed.) [PEI Policy 2005]. The leasing policy (at 6) states that leases are issued under "the authority" of the 1928 MOU, but, as noted *ibid.*, the authority for species other than mollusks comes with the 1987 agreement. The policy also notes (at 6) that for "more detail on the rights and obligations of a leaseholder, one can reference the Fisheries Act and Regulations, the Management of Contaminated Fisheries Regulations (MFCR) and the 1928 Agreement." No precise reference is included. For the purposes of this section, the validity of leases issued under the policy is assumed, and the implications of the vague legislative basis will be dealt with in "Constitutional issues" (p. 129).

71 *Ibid.* at 7. Under the previous policy, in effect until 2005, these operational phases were dealt with under two distinct leases, the developmental lease (which offered a range of durations, but typically were for three to five years), and the long-term lease (for up to twenty years). See *Aquaculture Leasing Policy Prince Edward Island* (Ottawa: DFO, 2000–2001 ed.) at 5–6.

72 PEI Policy 2005, *supra* note 70. Licenses are available for the collection of seed, but are not considered to be leases.

73 1987 PEI MOU, *supra* note 69 at s. 5.1 B and D.

74 PEI Policy 2005, *supra* note 70 at 25–26.

75 *Ibid.* at 6.

76 1987 PEI MOU, *supra* note 69 at s. 5.1 B. The focus is stability and security of tenure:

> In recognition of the need to convey property rights and stability to licensing, both Canada and the Province agree that appropriate tenure should be granted subject to conditions to be determined by the Coordinating Committee.

77 PEI Policy 2005, *supra* note 70 at 9–11.

78 The main features of the process are as follow:

1 Prior to making a decision on an application, the minister *shall* consult with a number of specified departments and agencies (s. 47(a)) and "*may* refer the application to a private sector, regional aquaculture development advisory committee for comment and recommendation" (s. 47(b) – emphasis added)

2 After the completion of the required consultation, the minister may: issue the license or lease, as submitted or with conditions; reject the application; or refer the application to a public hearing (s. 48). If a hearing is conducted, the matter then reverts to the minister for decision (s. 50).

See the description of the current policy on licensing and leasing of the Government of Nova Scotia. Online. Available http://www.gov.ns.ca/nsaf/ aquaculture/radac/index.shtml (accessed 8 August 2005).

79 *Ibid.* Recommendations of the RADAC are forwarded to the minister for consideration. Where an RADAC is not in place in a region, the intent is to use the public hearing process as a means of obtaining community input on the decisions.

80 The general duty to consult, found in s. 35(1), is vague and discretionary: "The Minister shall undertake such public consultation in relation to aquaculture as the Minister considers appropriate or as is required by or in accordance with the regulations."

81 Bay of Fundy Policy, *supra* note 54 at 4–5.

82 The potential impact of the ABMA scheme is significant. The policy states that the ABMAs will take the form of written contractual agreements, and that they will address such issues as technical aspects of management and day-to-day operations: *ibid.* at 5.

83 *Land Use Policy: Aquaculture*, *supra* note 60 at 13:

> Referrals are a formal mechanism to solicit written comments on an application from recognized agencies and groups. Referrals are initiated as per legislated responsibilities and formal agreements developed with other provincial and federal government agencies. Referrals may also be used to address the interests of local governments and First Nations.

84 *Ibid.* at 14–15.

85 These responsibilities, which are dealt within the next section, "Constitutional issues," and in Chapter 3 by VanderZwaag, Chao and Covan in this volume, may include the following: approval of alteration or disruption of fish habitat; approval for deposits of substances deleterious to fish or fish habitat; environmental assessment requirements; and authorization of obstructions to navigation.

86 See Memorandum of Understanding on Aquaculture Development, Canada–Nova Scotia (2002) [Canada–NS MOU 2002]; Canada–New Brunswick Memorandum of Understanding on Aquaculture Development (1989) [Canada–NB MOU 1989]; Canada/Newfoundland Memorandum of Understanding on Aquaculture Development (1988) [Canada–NL MOU 1988]; and Canada–British Columbia Memorandum of Understanding on Aquaculture Development (1988) [Canada–BC MOU 1988]. The Prince Edward Island MOU is addressed *supra*, note 69. See also the discussion of the MOUs and their role in delegation of regulatory responsibilities at D. VanderZwaag, G. Chao and M. Covan, "Canadian Aquaculture and the Principles of Sustainable Development: Gauging the Law and Policy Tides and Charting a Course – Part II," (2002–2003) 28 *Queen's Law Journal* 529 at 532–536.

87 Senate Report 2001, *supra* note 7 at 28:

> The approval process for site leases is governed by Memoranda of Understanding (MoUs) between the federal and provincial governments. The Memoranda ... were intended to establish a "one-stop shop" approach for lease applicants.... Except in the case of Prince Edward Island, the provinces were said to be the lead agency, and each has developed its own policies and site-specific approval process.

88 Even in areas where legislation is not required, as, for example, where the provincial Crown has full authority to make the necessary grants and there is no actual conflict with fishing or navigation rights, a clearly defined regime is desirable for the purpose of providing certainty and transparency to the process.

89 *Constitution Act, 1987* (UK), 30 of 31 Vict., C.3, reprinted in R.S.C., 1985, App. II, No. 5.

90 See Chapter 3 by VanderZwaag, Chao and Covan in this volume on the scope of provincial and federal regulatory control; and for a full discussion of the current application of regulatory controls under provincial jurisdiction, see VanderZwaag *et al., supra* note 86. For a review of the extent of federal powers to regulate non-proprietary aspects of aquaculture, see B. Wildsmith, *Federal Aquaculture Regulation*, Canadian Technical Report of Fisheries and Aquatic Sciences No. 1252 (Ottawa: Department of Fisheries and Oceans, 1984), and in particular 3–9, 27–52. With respect to provincial regulatory powers and their interaction with federal powers, see Wildsmith, *supra* note 3 at 34–37, 53–57. See also A. Scott, "Regulation and the Location of Jurisdictional Powers: The Fishery," (1982) 20 *Osgoode Hall Law Journal* 720, for an analysis, in the context of fisheries, of the "optimal" assignment of federal and provincial jurisdiction over a common property resource.

91 In 1982, Wildsmith, *supra* note 3 at 71–81, set out an extensive list of federal powers under s. 91 of the *Constitution Act*, including fisheries, public property, navigation and shipping, taxation, and criminal law. See also OCAD Legislative Review, *supra* note 6 at 15–17. The provincial powers under s. 92 derive primarily from jurisdiction over property and civil rights.

92 *Constitution Act, 1867, supra* note 89 at s. 92.

93 For a discussion of the impact of provincial powers in the practical regulation of aquaculture within the provinces, see Senate Report 2001, *supra* note 7 at 11:

> The scope of permissible provincial regulation includes the following: the management and use of Crown land; the licensing of aquaculture operations; the setting of standards for the business of aquaculture and those who conduct it; local marketing and consumer protection; waste management; and labour relations and employment standards ... at the local level, regional districts and municipalities administer zoning bylaws.

94 *Ibid.* at s. 91. It should also be noted that the "public property" of the federal Crown extends to all areas outside of any province, meaning that in significant marine areas the federal government has complete proprietary jurisdiction.

95 In this approach, the constitutional jurisprudence mirrors the separation of these two aspects of entitlements to marine space in common law, as described earlier, and it is perhaps not surprising that a number of the cases that have helped to establish the general position have arisen in the context of fisheries.

96 *Supra* note 18.

97 *Ibid.* at 111. This approach was affirmed in *Venning* v. *Steadman* (1884) 9 S.C.R. 206 at 214–215, per Strong J. The structure of the analysis in this section has drawn on the more extensive review of the early cases found in Wildsmith, *supra* note 3 at 37–53. It should be noted that in an earlier case, *Robertson* v. *Steadman* (1876) S.C.R. 621 at 633–634, the Supreme Court did uphold a federal grant of a "lease" of a fishery in non-tidal waters in New Brunswick. However, that decision was dealing with a grant that did not profess "to give ... an exclusive right of fishery," and the Court found it unnecessary to determine the nature and extent of the *property* rights arising under the lease as between the plaintiff and defendants, noting that it might be the case that "the only liability incurred by a person for fishing without permission, within the bounds of the lease, would be a prosecution for a penalty ... [under the Act]." As such, it seems to stand as an affirmation of the federal legislative power to regulate fisheries, despite some language in the case that might be taken to refer to the grant of proprietary rights.

98 *Robertson, supra.* note 97 at 115.
99 *Ibid.*:

> There is no connection whatever between a right of passage and a right of fishing. A right of passage is an easement, that is to say, a privilege without profit, as in the common highway. A right to catch fish is a *profit à prendre*, subject no doubt to the free use of the river as a highway and to the private rights of others.

100 *Ibid.* at 119–120.
101 *Ibid.*:

> I cannot discover the slightest trace of an intention on the part of the Imperial Parliament to convey to the Dominion Government any property in the beds and streams or in the fisheries incident to the ownership thereof, whether belonging at the date of the confederation either to the provinces or individuals, or to confer on the Dominion Parliament the right to appropriate or dispose of them.

102 *Provincial Fisheries, supra* note 27 at 519.
103 *Ibid.* at 521–522. The impact of this case on the common law with respect to leasehold grants was considered earlier.
104 *Ibid.* at 514–515.
105 *Attorney-General for Canada* v. *Attorneys-General for Ontario, Quebec and Nova Scotia* [Ontario Fisheries Reference] [1898] A.C. 700 at 712, per Lord Herschell:

> Their Lordships are of the opinion that the 91st section of the British North America Act did not convey to the Dominion of Canada any proprietary rights in relation to fisheries. Their Lordships have already noted the distinction which must be borne in mind between rights of property and legislative jurisdiction.... Whatever proprietary rights in relation to fisheries were previously vested in private individuals or in the provinces respectively remained untouched by that enactment.

106 The nature of this interaction, and the outer limits of the federal power, were set out by Lord Herschell in the Ontario Fisheries Reference, *ibid.* at 712–713:

> At the same time, it must be remembered that the power to legislate in relation to fisheries does necessarily to a certain extent enable the Legislature so empowered to affect proprietary rights.... The suggestion that the power might be abused so as to amount to a practical confiscation of property does not warrant the imposition by the Courts of any limit upon the absolute nature of the power of legislation conferred.... If, however, the Legislature purports to confer upon others proprietary rights where it possesses none itself, that in their Lordships' opinion is not an exercise of the legislative jurisdiction conferred by s. 91.

107 See OCAD Legislative Review, *supra* note 6 at 15–17.
108 *Provincial Fisheries, supra* note 27 at 514–515:

> At the time of confederation the beds of all lakes, rivers, public harbours and other waters within the territorial limits of the several provinces which had not been granted by the Crown were vested in the Crown as representing the provinces respectively.... The ungranted beds of all streams and waters were therefore lands belonging to the several provinces in which the same were situated ... subject only to the exception respecting existing trusts and interests mentioned in that section, and excepting the beds of

public harbours, which by operation of section 108, were vested in the Dominion.

109 La Forest, *supra* note 19 at 8 (citation omitted):

> Section 91(1A), which gives the Dominion exclusive legislative power over its public property, is a most important source of federal power in relation to the development of water resources. In the first place, the Dominion may do whatever it wishes with its property, and accordingly where it owns land it may ... make any legislation concerning the property even if such legislation would ordinarily fall within the provincial ambit. So long as it retains title, the Dominion may lease the land and control its development.

110 *Attorney-General for British Columbia* v. *Attorney-General for Canada (BC Fisheries)* (1913) 15 D.L.R. 308 at 311. These measures were incorporated in the Orders-in-Council effecting the union.

111 *Ibid.* at 314–315:

> [Fishing rights] are, in their Lordships' opinion, the same as in the ordinary case of ownership of a lake or river bed. The general principle is, that fisheries are in their nature mere profits of the soil over which the water flows, and that the title to a fishery arises from the right to the *solum*.

For a discussion of the range of lands that might be defined as federal, whether as harbors, reserve lands or otherwise, see Wildsmith, *supra* note 3 at 73–74; La Forest, *supra* note 19 at 18–27.

112 See, for example, *Robertson, supra* note 18 at 115.

113 La Forest, *supra* note 19 at 190: "The only legislature competent to authorize interferences with navigation is the federal Parliament" (citations omitted). This point is of general application, with no distinction as between tidal and non-tidal navigable waters in Canada (again, with possible exceptions noted in some cases). La Forest does go on to note that there is also an exception for provincial statutes passed prior to Confederation and never repealed by either the federal or provincial legislatures.

114 *Ibid.* at 190 (citations omitted):

> Thus federal statutory permission to build a dam in navigable water is necessary, but such permission cannot interfere with the rights to the bed, which will usually be in the province or a private owner. Conversely, while a province may incorporate log boom companies, such companies are not thereby authorized to unreasonably interfere with the rights of others to navigate a river.

115 *Provincial Fisheries, supra* note 27 at 516:

> [I]n the case of a provincial grant such as the question supposes the grantee would have a right to build upon the land so granted, subject only to his compliance with the requirements of the statute [a federal act respecting works in navigable waters] ... and to his obtaining an order in council authorizing the same, and provided the work did not interfere with the navigation of the lake or river.

116 This is, of course, subject to the important exception, as noted earlier, where the federal Crown holds proprietary rights.

117 *Supra*, text quoted in note 108.

118 *Provincial Fisheries, supra* note 27 at 514 (emphasis added).

119 *Ibid.* This was confirmed in *BC Fisheries, supra* note 30 at 318.

120 These issues are addressed in Chapter 9 by Murphy, Devlin and Lorincz and will not be dealt with in detail here. It should be noted, however, that with

respect to the public right of fishing, it may be limited but not extinguished by the existence of aboriginal rights: *R.* v. *Gladstone* [1996] 2 S.C.R. 723 at para. 67:

> As a common law, not constitutional, right, the right of public access to the fishery must clearly be second in priority to aboriginal rights; however, the recognition of aboriginal rights should not be interpreted as extinguishing the right of public access to the fishery.

121 If this line were followed, the position would be the similar to that for non-tidal waters within the province, with the exception that any new grant of a fishery or other restriction on public fishing rights would require action by the provincial legislature, and not by the provincial Crown alone.

122 *BC Fisheries, supra* note 30 at 320, per Viscount Haldane LC. The decision in this case was endorsed and confirmed by the Judicial Committee in the *Québec Fisheries case* in 1920: *Attorney-General for Canada* v. *Attorney-General for Quebec* [*Re Québec Fisheries*] (1920) 56 D.L.R. 358 at 370–371.

123 In *BC Fisheries, supra* note 30 at 320, in a passage dealing with "the right of fishing in arms of the sea and the estuaries of rivers," the Lord Chancellor held the following: "The right to fish is … a public right of the same character as that enjoyed by the public on the open seas. A right of this kind is not an incident of property." Further (at 317), in a discussion directed to Dominion ownership of the soil in the "railway belt," but which nonetheless states the underlying position, we have the following description:

> In the non-tidal waters they belong to the proprietor of the soil. … In the tidal waters, whether on the foreshore or in the creeks, estuaries, and tidal rivers, the public have the right to fish, and by reason of the provisions of Magna Charta no restriction can be put upon that right of the public by an exercise of the prerogative in the form of a grant.

124 *Ibid.* at 317–318.
125 *Ibid.* at 320.
126 *BC Fisheries, supra* note 30 at 320.
127 *Ibid.* at 318.
128 *Ibid.* at 317.
129 *Québec Fisheries, supra* note 122 at 370.
130 *Ibid.* at 370–371.
131 *Ibid.* Haldane, in considering a pre-Confederation statute that covered both the "disposal of property and the exercise of the power of regulation," noted that neither government could now act alone: "The former of these functions has now fallen to the Province, but the latter to the Dominion; and accordingly the power which existed under s. 3 of the Act no longer exists in its entirety."

132 The preamble to s. 92 of the *Constitution Act, 1867, supra* note 89, begins with the following words: "In each Province, the Legislature may exclusively make Laws in relation to Matters coming within the Classes of Subject [as enumerated in s. 92]." This point is further emphasized in both s. 92(13) and 92(16):

> 13. Property and civil rights in the Province…
> 16. Generally all matters of a merely local or private nature in the Province.

For a full review of the development of the territorial limitation on provincial jurisdiction in the case law, see E. Edinger, "Territorial Limitations on Provincial Powers," (1982) 14 *Ottawa Law Review* 57.

133 Wildsmith, *supra* note 3 at 75.
134 See, for example, *Reference Re Offshore Mineral Rights of British Columbia* [*BC Offshore Minerals*] [1967] S.C.R 792.
135 The position adopted in *BC Offshore Minerals* and the subsequent case, *Reference*

Re Bed of the Strait of Georgia and Related Areas [1984] 1 S.C.R. 388 [*Georgia Strait Reference*], was based on the finding in the British case of *R.* v. *Keyn* (1876) 2 Ex. D. 63, where the majority "held that unless specifically extended by Parliament, the realm of England ended at the low-water mark" (*Georgia Strait Reference* at 400).

136 *Georgia Strait Reference, supra* note 135 at 397.

137 *Ibid.* at 400.

138 In the provincial reference to the Court of Appeal of Newfoundland, which preceded the *Hibernia Reference* at the Supreme Court, the Court found that the then 3-nautical mile territorial sea was a part of Newfoundland at Confederation, and thus that area remained part of the province: *Re: Mineral and Other Natural Resources of the Continental Shelf off Newfoundland* (1983) 145 D.L.R. (3d) 9 (Nfld. C.A.). The territorial sea was not addressed in the Supreme Court reference, presumably leaving this ruling in place, but in a subsequent case the Court of Appeal reversed itself on this point, finding that the Supreme Court in the *Hibernia Reference* had effectively assumed that the province ended at the low-water mark (although this was not in issue in the *Hibernia Reference*): *ACE-Atlantic Container Express Inc.* v. *The Queen* (1992) 92 D.L.R. (4th) 581 at 601.

139 See, for example, the situation in New Brunswick in the Bay of Fundy: In *R.* v. *Burt* (1932) 5 M.P.R. 112 (N.B.S.C. App. Div.), a location more than a mile offshore was found to be in the province: see the discussion of this case, and others (including Conception Bay in Newfoundland) in *BC Offshore Minerals, supra* note 134 at 809.

140 The *Oceans Act*, S.C. 1996, c. 31, does not settle the matter, but rather leaves it for case-by-case determination. Section 7 provides that both the 12-nautical mile territorial sea and internal waters "form part of Canada," but says nothing about their status as federal or provincial waters. Section 8(1) provides for the vesting in the federal Crown of title to the seabed and the subsoil of the territorial sea and internal waters, but only for areas outside of any province (and without prejudice to previously held rights and interests). In sum, then, the *Oceans Act* simply relies, as it must, on the general position in constitutional law, and the status of particular coastal areas remains subject to the case-by-case determination described earlier.

141 *Constitution Act 1982*, being Schedule B to the *Canada Act 1982* (UK), 1982, c.11.

142 See *Haida Nation* v. *British Columbia (Minister of Forests)* [2004] 3 S.C.R. 511, in which the Supreme Court confirmed a duty of prior consultation where a government decision might adversely affect potential aboriginal claims.

143 *United Nations Convention on the Law of the Sea*, 10 December 1982 (entered into force 16 November 1994). Online. Available http://www.un.org/Depts/los/index.htm (accessed 8 August 2005) [LOS 1982]. Canada ratified this convention in November 2003. In the EEZ which extends from the outer limit of the 12-nautical mile territorial sea to a maximum of 200 nautical miles seaward from the coastal baselines, Canada's jurisdiction is limited mainly to "sovereign rights" over economic uses of the area, including its natural resources: see LOS 1982, Part V, Articles 55, 56. While this would certainly give jurisdiction to control and regulate aquaculture, the rights are limited as to the extent of permissible interference with foreign shipping, pipelines and submarine cables, all elements which would need to be taken into account in siting decisions.

144 See, for example, LOS 1982, *ibid.* Article 58(1):

> In the exclusive economic zone, all States . . . enjoy . . . the freedoms . . . of navigation and overflight and of the laying of submarine cables and

pipelines, and other internationally lawful uses of the sea related to these freedoms, such as those associated with the operation of ships, aircraft and submarine cables and pipelines, and compatible with the other provisions of this Convention.

145 *Ibid.* Articles 192, 194.

146 See, for example, *R. v. Crown Zellerbach Canada* [1988] 1 S.C.R. 401, in which the existence of international implications, and treaty obligations, respecting marine pollution influenced a finding for federal jurisdiction over the dumping of waste in marine areas within the province of British Columbia.

147 It is assumed throughout that provision for aboriginal and treaty rights must be made to ensure that grants of rights do not infringe upon those constitutional rights. As was noted earlier, this is the subject of Chapter 9 by Murphy, Devlin and Lorincz in this volume, and for the purposes of this section it will be assumed that the necessary consultations are conducted and accommodations are made.

148 For a review of the processes by which the federal and provincial processes are integrated and coordinated, see the discussion of the Atlantic provinces in VanderZwaag *et al.*, *supra* note 86 at 532–562. See also Commons Report 2003, *supra* note 7 at 19–120 for a summary of federal agency involvement in the process; Senate Report 2001, *supra* note 7 at 28–30, on concerns related to the complexity and delays resulting from the roles of multiple agencies, particularly in site selection and approval.

149 R.S.C. 1985, c. N-22.

150 R.S.C. 1985, c. F-14. Apart from any required regulatory approvals, DFO asserts the more general position that its mandate would enable it to object to creation or expansion of lease areas, assuming a fisheries management concern was engaged, while acknowledging that the regulations may not all be in place to permit direct action, see DFO, *Interim Guide to Fisheries Resource Use Considerations in the Evaluation of Aquaculture Site Applications.* Online. Available http://www.dfo-mpo.gc.ca/aquaculture/fisheries_resource_use/pg001_e.htm (accessed 8 August 2005), section 3.

151 *Supra* note 149 at s. 5, which provides, *inter alia*:

> 5. (1) No work shall be built or placed in, on, over, under, through or across any navigable water unless the work and the site and plans thereof have been approved by the Minister, on such terms and conditions as the Minister deems fit, prior to commencement of construction.

DFO has also prepared interim guidelines on some of the requirements related to these approvals: DFO, *Interim Guide to Application and Site Marking Requirements for Aquaculture Projects in Canada Under the Navigable Waters Protection Act* (Ottawa: DFO, 2002).

152 This action is subject to particular restrictions, which will be discussed later.

153 It is assumed for the purposes of this discussion that the grant of an aquaculture lease or other tenure does not violate the prohibition on Crown grants of a "fishery," as that term is applied to the public right of fishing. As was noted earlier, both the *BC Fisheries* case, *supra* note 30, and the *Québec Fisheries Case*, *supra* note 122, distinguished even fishing operations based on "fixed engines" such as weirs from the exercise of the public right of fishing. This reasoning would apply *a fortiori* to aquaculture facilities. Any potential problem, then, will lie in the second element, which is the non-interference with public right. In the words of the *Belyea* case, *supra* note 34 at 498, the federal government could be seen as having "restricted the common law rights of the public" without the authorization of Parliament.

154 *Supra* note 150 at s. 35. Approvals may also be required under s. 36, which deals with deposits of substances deleterious to fish or fish habitat. On the necessity for approval under s. 35 and s. 36, see OCAD Legislative Review, *supra* note 6 at 16. As is noted in the OCAD review, the application of ss. 35 and 5 of the NWPA can also engage the *Canadian Environmental Assessment Act*, S.C. 1992, c. 37. For a detailed description of departmental expectations for the application of s. 35 in one type of operation, see DFO, *Interim Guide to the Application of Section 35 of the Fisheries Act to Marine Salmonid Cage Aquaculture* (Ottawa: DFO, 2002).

155 OCAD Legislative Review, *supra* note 6 at 16. See also *Interim Guide to Fisheries Resource Use Considerations*, *supra* note 150 at 7–8 and Appendix A, for guidance on groups to be consulted and information to be sought in federal reviews from a fisheries management perspective.

156 *Interim Guide to Fisheries Resource Use Considerations*, *ibid.* at 5 (emphasis in original). An exception is noted for Newfoundland salmon operations, and of course in PEI the federal government has a direct role in leasing.

157 This is also clearly the view of both the Senate and House of Commons committee reviews as to how the current system is intended to operate: Senate Report 2001, *supra* note 7 at 32–33; Commons Report 2003, *supra* note 7 at 28.

158 See the statement of the doctrine at *A.G. Nova Scotia* v. *A.G. Canada* [1951] S.C.R. 3. Any delegation of administrative powers must also be considered as revocable, for otherwise a current government could bind future Parliaments, in violation of the principle of parliamentary supremacy: *Reference Re Canada Assistance Plan (British Columbia)* [1991] 2 S.C.R. 525 at 548, per Sopinka J.

159 On the use of this approach in fisheries management, see Wildsmith (1984), *supra* note 90 at 5: "The practice invariably followed where control of aspects of the fishery are turned over to the province is to designate provincial officials, usually Ministers of provincial governments, to administer province-specific regulations, which are still federally enacted."

160 See, for example, *Peralta et al.* v. *The Queen In Right of Ontario* (1985) 49 O.R. (2d) 705 (C.A.), affirmed by the Supreme Court of Canada in *Peralta* v. *Ontario* [1988] 2 S.C.R 1045. In this case, the delegation of fisheries licensing powers, pursuant to regulations, was found to be valid despite the fact that the provincial authorities set quotas for individual species, which were not specifically provided for in the federal regulations. In essence, the setting of individual quotas was seen as consistent with powers provided for in the regulations.

161 See "Conclusions" (p. 149) for suggestions as to how this might be accomplished under federal legislation.

162 Wildsmith (1984), *supra* note 90 at 14, noted that in the Act at that time, some fisheries regulations contained aquaculture licensing provisions, meaning that aquaculture was being construed as part of the fishery for regulatory purposes. He did not, however, take this to mean that this section authorized aquaculture leases. On the current definitions in s. 2 of the *Fisheries Act* for "fishing" ("fishing for, catching or attempting to catch fish") and "fishery" (which refers to methods of catching fish and the localities where they are used), it seems unlikely that this section could be considered applicable to aquaculture.

163 Section 59(1) sets up the power to delegate a power held by the federal government as well, under s. 58. Section 58(2) excludes any additional intrusion on federal use of the lands if they are located in a public harbor.

164 O.I.C. P.C. 2002–1082, 18 June 2002.

165 *Department of Fisheries and Oceans Act* R.S.C 1985, c. F-15, s. 5. See also the Order in Council authorizing the updated MOU with Nova Scotia: *Authority to*

Enter into a Memorandum of Understanding on Aquaculture Development with the Province of Nova Scotia That Will Allow the Parties to Continue Their Collaboration in the Development of Commercial Aquaculture, Order In Council P.C. 2002-1086, 18 June 2002.

166 Neither the Newfoundland and Labrador nor the BC agreements specify a statutory basis, so they might be presumed to fall under the same general authority.

167 See, for example, *Provincial Fisheries*, *supra* note 27; and *Québec Fisheries*, *supra* note 122.

168 See, for example, the discussion in J. Meaney, "Federal Fisheries Law and Policy: Controls on the Harvesting Sector," in VanderZwaag, *supra* note 5, 27–48 at 29; considering the impact of the *BC Fisheries* case, *supra* note 30: "It can be seen that the scope of the federal power to legislate with respect to marine fisheries is not absolute, but limited to the regulation of the public right to fish."

169 See the following discussion of this issue at Meaney, *ibid.* at 29–30 (citations omitted), referring to the Supreme Court's decision in *Interprovincial Co-operatives Ltd.* v. *R.* [1976] 1 S.C.R. 477 at 495:

> The traditional scope was thought to relate to the protection and preservation of fisheries as a public resource – that federal fisheries laws are only valid if they relate to biological conservation. . . . This view continued to be applied by the Supreme Court in the 1970's and 1980's. In *Interprovincial Co-operatives Ltd. v. R.*, Chief Justice Laskin noted that the federal power in relation to fisheries "is concerned with the protection and preservation of fisheries as a public resource, concerned to monitor or regulate undue or injurious exploitation."

170 Meaney, *supra* note 168 at 31.

171 *Ibid.* at 31–32.

172 *Ibid.* at 32. The breadth of the minister's discretion under s. 7 was confirmed in *Comeau's Sea Foods Ltd.* v. *Canada (Minister of Fisheries and Oceans)* [1997] 1 S.C.R. 12 at para. 37, which emphasized the broad discretion available to the minister in virtually all aspects of licensing, assuming minimal requirements for natural justice were met:

> This interpretation of the breadth of the Minister's discretion is consonant with the overall policy of the *Fisheries Act*. Canada's fisheries are a "common property resource," belonging to all the people of Canada. Under the *Fisheries Act*, it is the Minister's duty to manage, conserve and develop the fishery on behalf of Canadians in the public interest (s. 43). Licensing is a tool in the arsenal of powers available to the Minister under the *Fisheries Act* to manage fisheries.

173 [1997] 31 B.C.L.R. (3d) 228 at para. 18:

> [T]he plaintiffs say that the Minister of Fisheries and Oceans, as a minister of the Crown, has violated the public right to fish by granting an exclusive fishery to aboriginal fishers without parliamentary authorization. They contend that neither s. 91(12) of the *Constitution Act, 1867*, nor any provisions of the *Fisheries Act* have taken away the public right to fish or given to the Minister the authority to take it away.

> The plaintiffs had also claimed that the public right of fishing was a constitutional right, but the Court found at para. 19 that this position had already been rejected by the Supreme Court of Canada in *R.* v. *Gladstone*, *supra* note 120, and that it was purely a common law right at stake.

174 *Alford, supra* note 173 at para. 20
175 *Ibid.* at para. 17.
176 *Ibid.* at para. 21. The decision on the motion was appealed, and was affirmed by the BC Court of Appeal: *Alford* v. *Canada (Attorney General)* [1998] B.C.J. No. 2965 (B.C. C.A. 11 December 1998).
177 See *Esson* v. *Wood* (1884) 9 S.C.R. 239, which involved interference with navigation through construction of a wharf on privately owned submerged lands in Halifax harbor. In that case, the plaintiff was the landowner who had created the obstruction, claiming for trespass against a defendant who had destroyed it. The obstruction was referred to as a public nuisance, and although the defendant might have sought an order for removal, their self-help was justified as abatement of the nuisance (at 243–244, per Strong J), defeating the claim for trespass.
178 In Esson, *ibid.*, it was clear from the decisions of Ritchie CJ and Strong J at 242 and 243 that it was for the landowner to show any justification for the prima facie infringement, and that the justification must involve legislative authority.
179 Section 23 provides as follows:

> 23. No one shall fish for, take, catch or kill fish in any water, along any beach or within any fishery described in any lease or license, or place, use, draw or set therein any fishing gear or apparatus, except by permission of the occupant under the lease or license for the time being, or shall disturb or injure any such fishery.

180 This is consistent with the approach taken in the regulations, which distinguish between aquaculture and fishing. See, for example, s. 3(1)(d)(i) of the *Maritime Provinces Fishery Regulations*, SOR 93-55, P.C.1993-188, 4 February 1993, which provides that the regulations do not generally apply to "cultured or cultivated fish" found in or taken from aquaculture sites leased or licensed by the Nova Scotia and New Brunswick governments. This provision does not purport to restrict the public right of fishing, but would permit such fish to be taken by an owner without need for a fishing permit.
181 As was noted earlier, oysters are a special case. Section 59 of the *Fisheries Act* empowers the federal government to authorize provincial governments to grant leases for oyster production within the provinces, and provides that any grantees have the exclusive rights (subject to the fishery regulations) to the oysters "produced or found within the limits" of the lease areas. It is not entirely clear whether it extends to an authorization for any interference with the public right of fishing, save that it would prevent any other fisher from accessing oysters "found" within the lease area.
182 See, for example, the Canada–NS MOU 2002, s. 2.1: "This MOU applies only to aquaculture activities carried out or operated in Nova Scotia." See also Canada-BC MOU 1988, which, in its preamble, refers to "development of the aquaculture industry in British Columbia." This MOU does not deal in detail with leases, but does confirm the continuing effect of provincially issued tenures, defined as falling within the province: " 'Provincial Tenure' means the right to occupy Provincial Crown lands."
183 The regulations, by s. 26(1)(b) of the *Oceans Act*, must be made on the advice of the Minister of Justice.
184 OCAD Legislative Review, *supra* note 6 at 15.
185 *Ibid.*
186 In *Hickey* v. *Electric Reduction Co.* (1970) 21 D.L.R. (3d) 368 (Nfld. S.C.), the Newfoundland Supreme Court held that losses to commercial fishers from a fishery closure caused by pollution were not "special" or distinct from the

damage suffered by the public at large from damage to a public right of fishing. However, in the later case of *Gagnier* v. *Canadian Forest Products Limited* (1990) 51 B.C.L.R. (2d) 218 (B.C.S.C.), which also resulted from a pollution incident leading to a fishery closure, the court found, *inter alia*, that private claims under public nuisance need only show a "significant difference in degree of damage" (at 230).

187   Other federal regulatory provisions under the *Fisheries Act* and the NWPA would still apply, in that s. 9(5) of the *Oceans Act* provides that the application of s. 9 shall not be interpreted as "limiting the application of any federal laws."

188   B. Wildsmith, *Toward an Appropriate Federal Aquaculture Role and Legislative Base*, Canadian Technical Report of Fisheries and Aquatic Sciences No. 1419 (Ottawa: Department of Fisheries and Oceans, 1985) at 30–31.

189   *Ibid.* at 27–28.

190   With respect to enforceability against others (apart from the Crown), the situation is not as clear, in that enforceability against private users in common law would depend to some extent on the nature of the property right asserted, and the extent to which any outside party *actually* interfered with that particular right. Furthermore, the essential problem with the continued existence of public rights, as discussed earlier, is one that relates to enforceability, and the current situation in that regard is untested.

191   See, for example, the description of these fears in a study (Marshall, *supra* note 14 at 350) of the impact of commercial aquaculture on a community in the Bay of Fundy (citation omitted):

> Increasing privatization of the marine commons is fundamentally a disenfranchisement of all traditional fishers, effectively precluding sustainable livelihoods within the wild fishery. The loss of local control threatens to transform the communities into "competitive, atomized, and dependent" entities.

192   See *ibid. passim* on the difficulty of ensuring the introduction of truly negotiated rights in situations of pressure to develop new industries. Advocates of full privatization, on the other hand, would tend to reject the inherently political dynamic involved in transferring rights in this manner, and prefer market-based approaches such as auctions or other forms of sale: see, for example, Neill, *supra* note 10 at 12–16.

5 Conflict prevention and management

Designing effective dispute resolution strategies for aquaculture siting and operations

Moira L. McConnell

Introduction

Social conflict arising from socio-economic system change is neither a new phenomenon nor peculiar to the emergence of fish farming or aquaculture. The term is used in this chapter to refer loosely to activities associated with the growing and harvesting of marine organisms as a coastally based commercial food production industry.[1] Social conflict is to be expected and is, in fact, a very normal human response to the introduction of a new situation and any associated loss or perceived loss of the status quo. As noted in connection with the experience of organizational and individual change,[2] even when a change in situation has been sought as an improvement and is perceived as desirable, inevitably there will be a period of transition, unease and even resistance to the change – often to the point of what may, from some perspectives, be regarded as irrationality or shortsightedness. Where there is uncertainty about the nature or import of the change, a poor change management process, disagreement as to whether the change is an improvement, or where the change involves differing values, then the problem is exacerbated and can lead to various expressions of conflict, including violence, litigation ("court battles") and other forms of political pressure.

Aquaculture is generally understood as the fastest-growing food production industry in the world.[3] According to the United Nations Food and Agriculture Organization (FAO) report *The State of World Fisheries and Aquaculture 2002*, "[A]quaculture is growing more rapidly than all other animal food producing sectors," with more than half of this production being of finfish and with a continued expansion in the already wide range of aquatic species being farmed.[4] Similarly, a report by the IMO/FAO/UNESCO-IOC/WMO/WHO/IAEA/UN/UNEP Joint Group of Experts on the Scientific Aspects of Marine Environmental Protection (GESAMP) points out that "[a]quaculture has great potential for the production of food [and provision of food security], alleviation of poverty and generation of wealth for people living in coastal areas, many of whom are among the poorest in the world."[5] Despite this growth and the apparent and uncontroverted benefits, particularly when combined with the significant decrease in, and in some case even disappearance, of the traditional "wild stock" fishery, conflict relating to the development and growth of aquaculture has increased in countries such as Canada commensurate with the expansion of the industry.

In Canada, conflict or "opposition,"[6] especially with respect to access to high-quality aquaculture sites for new enterprises, is now regarded as "a major impediment to the growth"[7] of this sector on both the eastern and the western coasts.[8] In 2000, the Canadian Commissioner for Aquaculture argued for, *inter alia*, proactive planning of aquaculture siting, including:

- integrated coastal management (aquaculture zoning);
- conflict resolution mechanism; and
- DFO (federal) guidelines and operational policies for decision-making regarding allocation of aquatic space for aquaculture purposes.[9]

Concern about the impact of conflict on the development of aquaculture as a source of future global food security is also articulated in the GESAMP report, which is predicated on adoption of an integrated planning approach to supporting the development of sustainable aquaculture. It points out that

> [i]f an integrated planning initiative leads to litigation, it has clearly failed, since one of the objectives of more integrated planning is to resolve or pre-empt resource use conflict. Indeed, it is arguable that the whole process of more integrated planning is a form of mediation between the various coastal resource users and government sectoral interests.[10]

At the same time, conflict relating to aquaculture, while having some unique features, must be understood as being embedded within the broader socio-economic and demographic context of coastal/ocean settlement and urbanization. The problem of conflict arising from increased demand for use of coastal resources is of sufficient import that it resulted in discussion at the 1992 United Nations Conference on Environment and Development (UNCED). This was followed with action by the United Nations Educational, Scientific and Cultural Organization (UNESCO) in 1996 when it created the platform for Environment and Development in Coastal Regions and in Small Islands (CSI) to focus on the development of "an intersectoral, interdisciplinary and integrated approach to the prevention and resolution of conflicts over resources and values in coastal regions and small islands."[11]

As noted by presenters at a UNESCO-CSI conference in 2001,

> Competition for limited resources and space makes coastal regions flashpoints for conflict. This means that much is at stake for the great majority of the world's countries, 80% of which are coastal, located either adjacent to an ocean or a sea.[12]

Dealing with conflicts has been called the greatest challenge facing integrated coastal management because of the multiuse setting of coastal systems and because most of these systems are a mosaic of "rights" (property rights, fishing rights, use rights) and usually involve common property resources as well.[13]

It is notable that similar points were made in the early 1990s, prior to the concepts articulated at UNCED, specifically in relation to aquaculture development in Canada. Coffen and Smillie argue in *Canadian Ocean Law and Policy* that there were problems for development of this industry in Canada because of:

- vested interests and competition for coastal space;
- privatization of a common "public" resource (erosion of the "public trust" doctrine); and
- human opposition to change.[14]

At that time, the authors proposed a combination of comprehensive zoning or planning and the use of a variety of consensual dispute resolution processes as a means of addressing conflicts. More than a decade later, it is striking that the problems remain much the same, and the regulatory system design and response, although improved in some respects, still presents difficulties. This is the case despite the clear national policy and the allocation of significant budgetary and human resources to foster the development of the aquaculture industry in Canada.[15]

Why is this so? Is the problem of conflict endemic to the situation and therefore an inescapable part of change – something to be borne stoically and with confidence that the economic forces at play will ultimately prevail? How or why is it that in countries such as Norway, where the social, economic and ecological situation and concerns are similar to those of Canada, aquaculture – particularly salmon farming, one of the more environmentally controversial forms of aquaculture – has been able to thrive, with Norway now a world leader in farmed salmon production? As noted in 2001 by the Norwegian Fish Farmers' Association, "In the course of only 30 years Norwegian aquaculture had developed from a side-line into an industry that has turned us into the world's largest exporter of salmon and trout."[16]

This chapter, based on research carried out within the framework of the Law and Policy Project under the auspices of AquaNet,[17] examines the problem of aquaculture siting conflicts in Canada with a view to proposing a regulatory system approach that may reduce the level or nature of conflicts relating to site access. Specifically, it considers the utility and role of oft-recommended procedures[18] such as negotiation, conciliation, mediation and arbitration, generally falling within the term "ADR" (alternative dispute resolution) or, more correctly, "DR" (dispute resolution)[19] for addressing these conflicts. In order to illuminate the problem and explore alternative approaches, this study specifically considers the Norwegian experience with siting conflicts and any legislative solutions for conflict management procedures that have been adopted in the course of the industrial development of aquaculture. For a variety of reasons, Norway has not focused on developing procedures for resolving disputes about individual development applications. Rather, it has adopted an entirely different approach, largely based on the use of coastal zone (including sea-use) planning as a mechanism for conflict prevention and management. In principle, the effect of this approach, from a conflict resolution or management perspective, is to shift the site of conflict upstream, or earlier in the process. It attempts to resolve concerns as a more holistic spatial management exercise before individual development or spatial use applications are involved. This study explores the implications of the Norwegian experience and any regulatory options it may suggest for the Canadian situation.

Using ideas derived from negotiation and conflict management theorists,[20] this chapter concludes by suggesting that the concept of aquaculture conflict, and siting conflicts in particular, needs to be disaggregated and a conscious conflict management strategy developed to identify and address differing kinds of conflicts that arise, taking into account factors such as the actors in the conflict, the underlying interests, the temporal placement of the conflicts, and so on. It is suggested that the more traditional DR techniques, such as mediation or arbitration, are of limited utility to siting conflicts, although integrated management (IM), if understood as a form of macro or system-level mediation, is an exception. Drawing on the example of Norway, the study proposes that aquaculture siting conflicts must be dealt with holistically and must take into account the predictably increasing use of near coastal waters and the seabed for a range of activities – often regulated by differing sectoral agencies and involving transnational actors. A sea-use/zoning approach, perhaps incorporating the planning methodology advocated by communicative planning theorists,[21] is proposed as a regulatory system approach that is the most likely to ensure prevention or reduction of conflict and sustainable development of coastal, including marine, resources and space. It is further proposed that, despite the legal and jurisdictional challenges posed under the Canadian Constitution, which are detailed elsewhere in this book, sea-use planning should as far as possible be carried out at a municipal or the local level of governance, with accountability to provincial and/or national level authorities to provide assurance that broader environmental, health security, trade obligations and social concerns are addressed.

Before moving to the next section of the study, it is important to comment briefly on the relationship between the investigation discussed in this chapter and that carried out by other researchers engaged in the AquaNet Law and Policy Project. An investigation into conflict and conflict management is akin to pulling the thread that unravels an entire cloth. Potentially all issues are raised, including interagency, governmental and jurisdictional conflicts, competing property and other rights, participatory processes, integrated management processes, and so on. These and other topics are dealt with in other chapters in this book. The intention in the study presented in this chapter was, through a mix of discussion with other researchers, literature review, anecdotal/media commentary and reflection, to identify some of the systemic elements that seem to feature in siting conflicts and to consider what may be a viable approach to alleviating this problem in Canada.

The background research reflected in the discussion in this chapter involved case studies and data drawn from Atlantic Canadian and British Columbian experiences. However, the investigation is focused on identifying commonalities as opposed to elaborating the development of a single conflict. Certainly the regulatory climate and systems differ between provinces in Canada, a factor that may itself have an impact on the nature and reasons for conflicts. Similarly, the Norwegian discussion is drawn primarily from the case studies carried out by Norwegian researchers based in one area (Bergen[22] and environs) of Norway,[23] and from discussion with personnel and researchers in that area. The study can therefore only be taken as a sampling of the Norwegian experience. This specificity of experience and the need to design and implement regulatory systems that achieve goals and meet specific standards but also accommodate local needs and factors is, in fact, the essence of modern integrated management.

With respect to this point, it is also important to understand that the geographic and demographic characteristics of a location are a key factor in siting conflicts. Where coastal land and aquatic resources are in abundance and are not under pressure, the likelihood of conflict is, of course, significantly reduced. However, given the need for access to rapid and inexpensive transport of harvested fish farm products for global markets, it is predictable that desirable or "prime" locations will involve both ecological and economic/transport considerations, and will likely conflict with other users. As noted in a report of Canada's Standing Senate Committee on Fisheries, *Aquaculture in Canada's Atlantic and Pacific Regions*:[24]

> When selecting a site for aquaculture, many factors are taken into account, such as water depth, current flow, salinity, temperature, wind and waves, oxygen content, pollution, ice conditions, proximity of other resource users and tourist sites, patterns of marine traffic and proximity to suppliers and services (e.g., wharves, roads, air transportation, and communications). Thus the industry is generally constrained by the availability of suitable grow-out sites. Although Canada has 244,000 kilometres of coastline on the Pacific, Arctic and Atlantic Oceans, the environment, in particular, is an important limiting factor.[25]

The next section of this study outlines on a "snapshot" basis the experience in Norway specifically in terms of siting decisions and any regulatory mechanisms that either assist to resolve or may instead serve to exacerbate or create conflicts. The third section of the study presented in this chapter discusses the implications of the Norwegian experience and any lesson or options it might suggest as an approach to better address the problem of siting conflicts in Canada.

Aquaculture and siting conflicts in Norway

This section of the study provides an overview of some of the issues that have arisen in relation to siting and the development of aquaculture in Norway. As noted earlier, it is not intended to present a comprehensive examination of the industry or regulatory regime, but rather is intended to focus on the elements relating to conflict prevention and resolution.

Aquaculture has been carried out in Norway for just over thirty years and has rapidly achieved prominence as a significant contributor to the Norwegian economy.[26] Although there is an increasing interest in shellfish aquaculture and the culture of other finfish species,[27] the main exports are Atlantic salmon and rainbow trout. In 2000, Norway was the world's largest producer of Atlantic salmon, which it exports mainly to the European Union.[28] It is useful to note that Norway, like Canada, has an extensive coastline,[29] an increasingly declining wild fish fishery, a well-developed offshore oil industry and related shipping industry activities (boat-building, etc.). In some areas, particularly in the south, there is increasing use of the coastal areas and coastal waters for recreational homes, boating and conservation. As Bennett notes, "[t]hroughout this century [the twentieth], the number of uses of marine resources has proliferated and environmental impacts have become heavier."[30] However, as he also points out,[31] coastal planning concerns are not uniform and differ between regions, a fact that is important in the Canadian context as well. For example, in the south and east of Norway, the pressures relate more to development of the coastal land and nearshore areas, while in the west and north, with less densely populated coastal areas, the concerns relate more to regulation of sea areas and management of marine resources.

An important geophysical feature of Norway, aside from the famous fjords or inlets, is the small islands or skerries that fringe much of its coastline. The Norwegian baseline, which marks the seaward limit of Norway's internal waters and land areas and the point from which the territorial sea[32] is measured, runs along the outermost of the skerries. This is an important factor in Norway's approach to coastal and sea area regulation,[33] and one that distinguishes it from Canada's. This means that much of the near coastal water that would be used for aquaculture is not in the territorial sea but in internal water – marine spatial areas governed by domestic law. Although Norway has various levels of government and authorities, unlike Canada it has a unitary system of government with one (national) government that has ultimate legislative authority and responsibility. The NORCOAST study explains[34] the relationship between the various governmental actors as follows:

> Central government is responsible for the formulation of national policy objectives for all sectors. The government system in Norway is characterized by a division of tasks and responsibilities with a hierarchy of local and county institutions based on local elections on one hand, a hierarchy of decentralised state agencies on the other. Communes [the "lowest" or most local institution] are not subordinate to the county councils, it is more a division of responsibilities and tasks.[35]

Aquaculture in Norway is governed by several different pieces of national legislation administered by different ministries. The first legislation for fish farming, adopted in 1973, was a temporary Act on fish farming, with a permanent law first being enacted in 1981.[36] The current legal framework[37] comprises[38] the following primary legislation and related regulations:

- *Aquaculture Act* (Ministry of Fisheries);
- Act No. 54, 1997, *Act Relating to Measures to Counteract Diseases in Fish and Other Aquatic Animals* (the Fish Diseases Act) (Ministry of Agriculture, Department of Veterinary Services);
- Act No. 77 of 1985, *Planning and Building Act*, as amended (Ministry of the Environment);[39]
- *Pollution Control Act* (Ministry of the Environment);
- *Act Concerning Quality Control of Fish and Fish Products* (Ministry of Fisheries);[40]
- *Working Environment Act* (Labour Inspection Authority)
- *Harbors and Coastal Waters Act*;[41]
- *Animal Protection Act*;[42]
- *Animal Feed Inspection Act*.[43]

In addition to the foregoing, an important regulatory tool in Norway is found in the application of the outcome of the National Evaluation of the Suitability of the Norwegian Coast and River System for Aquaculture (known as LENKA). LENKA was an interdepartmental project initiated in the mid-1980s (ending in 1990) by the Department of the Environment, largely to respond to the administrative demands posed by the boom in fish farming, particularly in the coastal areas. Bennett describes LENKA as follows:

> Its general aims were to ensure "continued positive development and growth of the fish-farming industry without causing huge conflicts with other users and conservation interests" and "contribute to communal and county planning in the coastal and river systems and decision-making concerning the location of aquaculture." . . . In the sea LENKA

was designed to measure suitability for the farming of salmon and rainbow trout in open cages.[44]

LENKA sought to identify areas that were biologically and environmentally suitable for aquaculture and to estimate capacity, which, as Rogers notes, was then estimated at 700,000 tonnes, a figure which using today's technology would be even greater.[45]

However, as Bennett points out, although LENKA generated a huge amount of potentially useful data, it was viewed with skepticism because of its,

> huge cost, top-down organization and weak methodology ... [and its failure to take] account of actual water quality and circulation, and was said frequently to lead to the "wrong" classification of areas that were known to be suited for fish farming.... [However, irrespective of the failures of LENKA] ... it did generate discussion and awareness of issues and methods in coastal zone planning, highlighting the need for other methods of determining recipient capacity and locating fish farming.[46]

Although the *Aquaculture Act*, which sets up a system of licensing for aquaculture, can be seen as the primary regulatory instrument for aquaculture development, from the perspective of aquaculture siting and conflict management the *Planning and Building Act* of 1985, under the Ministry of the Environment, is the key legal instrument affecting this issue and has greatly influenced the approach that has been adopted in Norway. Of course, the operation of this Act also interacts with other regulatory instruments dealing with factors noted earlier, affecting siting. For example, legislation aimed at preventing the spread of fish diseases by regulating distances between farms, stock density and interaction between species can have a significant impact on the location of a farm.[47]

The *Planning and Building Act* requires that communes (municipalities) develop "commune plans" for spatial use. These plans, developed at the local level, are subject to review and approval of the national-level ministries affected by them, with final agreement under the Ministry of Environment. As noted in the NORCOAST study of Hordaland,

> The Ministry of Environment is the main planning authority at national level. Through policy guidelines and by monitoring at county and commune level, the Ministry is responsible for planning within a national policy framework.... The responsibility for spatial planning lies basically with the political branch of government, whereas state administration is responsible for implementing the central government policy within the specific sectors and for monitoring local government performance in relation to national directives and norms. However, the

expectations of cooperation and interaction between state administration and local government institutions at all levels is a key feature of planning legislation and should be emphasized. . . . Responsibility for planning in accordance with the Planning and Building Act is therefore decentralised to the county councils and communes. Plans on local and regional level due to the *Planning and Building Act* are to be revised every four years.[48]

Originally addressed only to land-use planning, the Act now provides for national, county and municipal/commune planning of use of marine spaces under the control of the communes (often including large marine spaces) out to the national territorial baseline.[49] The *Planning and Building Act* provisions are supplemented by a number of national-level coastal zone planning guidelines.[50] It also contains obligations on the part of the various government agencies to cooperate both vertically[51] (national–county–municipal) and horizontally[52] (interstate sectoral, intercounty and intermunicipality planning), and also provides for environmental impact assessments as part of the development application process for some activities.[53]

Bennett, explaining the evolution of coastal zone planning in Norway, points out that although awareness of the need to plan for coastal use had been building since the 1960s with the rise in recreational homes and other coastal uses,

[t]he biggest single development precipitating a need for coastal zone planning in western northern Norway, and to a great extent setting the agenda and pace, was the rapid expansion of fish farming, particularly farming of salmon and rainbow trout, which took place from the mid-1980s onwards.[54]

The 1999 NORCOAST study comments similarly:

Norway had several reasons for establishing coastal zone planning. An increasing number of second homes were being built along the coast, especially in the south. These were often the result of dispensations from the general directive of no building within a 100-meter belt from the shoreline and the terms of local structure plans. In addition to this, comes pollution around many urban areas. Still, it was fish-farming which led to the first initiatives to plan in sea areas. The fish-farming business steadily increased in extent from the mid 1980s and needed larger areas to avoid contamination of infectious fish-diseases. The business was also often in conflict with traditional uses of the coastal areas. This led politicians to understand that something had to be done to avoid further problems. Questions about ownership rights arose in an area where everyone earlier had a right to free use, but where fish-farms now demanded significant areas for their floating constructions. The

Norwegian solution was to enable the communes and the counties to plan in sea areas through the *Planning and Building Act of 1985*. Each commune is expected to prepare a Commune Plan for the onshore areas. The planning of marine areas is not mandatory, but communes are strongly advised to do so.[55]

Aside from the more obvious economic returns from fish farming, the early stages of government promotion of fish farming also served a deeper socio-political purpose in that they countered the trend to urban migration from the coast. It was "realized early in the 1970s that the industry could be used to generate jobs and wealth and help to maintain population in the periphery."[56] This was achieved largely through the licensing system, which provided a control mechanism with respect both to local ownership and to farm size. Unfortunately, this approach also led to a proliferation of small-scale fish farms in inhabited areas without regard to the ecological suitability of the site for farming.[57] This sudden growth was in a period of infancy for the industry when technical and scientific knowledge about aquaculture operations on a commercial scale was only emerging. It resulted in significant problems with fish disease, pollution and conflicts with other users. Changes in the licensing regime to allow owners to hold multiple farm licenses and changes in size restrictions to promote industrialization of operations have largely removed the demographic, if not the economic development, component of fish farming. The industrialization of fish farming to the scale of global marketing necessarily requires proximity to rapid international transport centers. The growth of the industry and the operational and food safety demands of the industry led, in the 1980s, to the understanding that,

> decisions on licensing and location should no longer be made on a piece-meal basis, even though each application was subject to an elaborately democratic process. Conflict potential with other uses of the coastal zone made it imperative to solve questions of resource management and spatial ordering within the framework of communal structure plans for marine areas.[58]

Bennett and others have detailed the evolution of marine space planning in Norway through the 1980s and 1990s, including extensive government-funded research and reports into planning law, spatial ordering, conflict analysis and the development of LENKA, as discussed earlier. The main point of interest for this study is that the primary response to the coastal zone use conflict and the industrial development of marine space was through planning and planning law under the Ministry of the Environment, as opposed to other regulatory mechanisms. It is useful, therefore, to understand how planning was viewed in Norway, and the purpose behind the planning law. In terms of conflict resolution, the fact that the issue is already characterized as one of "no planning" or poor planning is itself of significance. Section 2 of the *Planning and Building Act*, the purpose clause, provides that,

> [p]lanning pursuant to the Act is intended to ensure the right conditions for coordinating national, county and municipal activity and provide a basis for decisions concerning use and protection of resources concerning development.
>
> By means of planning, and through special requirements concerning the individual planning building project, the Act shall promote a situation where the use of land and the building thereon will be of greatest possible benefit to the individual and the community.
>
> When carrying out the planning pursuant to this Act special emphasis shall be placed on securing children a good environment in which to grow up.

The philosophy behind the Norwegian approach to planning has been described as a combination of "project planning" with elements of "strategic planning."[59] The difference between the two is important in that planning can have various meanings in a regulatory system. For example, as pointed out by Bennett,[60] a development plan can be a control mechanism – a blueprint, if you will – that is to be implemented, with success measured in terms of adherence to and implementation of the blueprint. A project planning approach might arise in the context of the particular problem that has to be dealt with and the focus is on ensuring that there is an accurate diagnosis of the problem and appropriate implementation. A development plan can also be intended as a "strategic planning" activity in that it is intended to function as a "store of policy principles intended to guide but not control decisions,"[61] in which case the focus is more on inclusiveness in policy formulation, communication and flexibility with implementation and policy development ongoing processes. Finally, development planning can also be a "system of conflict mediation,"[62] with the planning process itself allowing for a working out of the conflict through an articulation of interests, views, values and concerns, and then the negotiation of solutions in the plan to meet varying interests.

Bennett, a long-time commentator and observer of Norwegian coastal planning, suggests that the *Planning and Building Act* reflects the view that "development is a normative social process rather than a rational technocratic exercise."[63] The emphasis on cooperation and dialogue, with vertical and horizontal integration among institutional actors, has already been referred to. At the same time, once the plans are adopted, they are legally binding, a situation that to some degree counters the flexibility of strategic planning. Despite the four-year review, once interests are recognized in a plan, then they are vested; that is, successor plans must work with the status quo.

From a slightly different perspective and analysis, the NORCOAST study describes the *Planning and Building Act* (PBA) as embodying,

> democratic ideals of openness and public participation, requiring communes to inform and consult all who have an interest in the plan at an early stage of the planning process. The PBA thus builds on concepts of social learning and participatory planning. . . . One practical instrumental objective of this planning act, was to render public planning more effective by making it more legitimate. The prescription was to root planning in local communities, by implanting it in the sphere of local political discourse and social interaction. Decision making on spatial matters was to be brought into a structured institutionalised framework within which county and state authorities also exercised considerable influence. . . . [In addition] Regional government bodies [regional offices of national ministries e.g. Fisheries] also have responsibility for planning and management of their own sector interests through legal acts as law of nature conservation, law of aquaculture, law of salt water fisheries etc. Integration of these interests is taken care of through the PBA. . . . Planning under the Act is intended to form a basis for subsequent decisions concerning the use and conservation of resources and building development.

Despite this ethos of cooperation and communication, coastal zone planning, particularly in relation to aquaculture and marine space planning, has been a "battlefield"[64] and a site of power struggles, largely between the sectoral agencies. Until the advent of sea-use planning under the PBA, the Ministry of Fisheries was for the most part the primary sectoral agency responsible for marine space and activity regulation. It still retains that responsibility and jurisdiction through sectoral control over the activities of fishing and aquaculture.

However, the Ministry of the Environment is in charge of spatial planning. Bennett provides a detailed account of some of the early commune and regional planning exercises, which were supported by the Norwegian government, in part as a means of developing a methodology and identifying issues arising in connection with sea-use planning. In many cases, studies indicate that although the communes were able to develop plans, problems were encountered in getting approval of the plan from the various state authorities that were required to agree, all of which had been involved in the planning exercise: that is, Fisheries and/or Environment. In addition, the Ministry of the Environment itself came out with a national parks coastal conservation plan that impacted on commune planning and apparently constituted a "surprise" to the Ministry of Fisheries and the affected commune. This coastal conservation plan directly affected areas where Fisheries had managed the biological resources. Thus, a battle erupted between the agencies, which was then fought out over the planning exercise.[65]

Further issues arose and continue to arise in connection with interpretation of the PBA. One view holds that communes are obliged to consult with the state agencies, but ultimately the communes determine the extent to which they plan the sea areas. In principle, a commune could exclude aquaculture. However, the Fisheries authority contested this interpretation, arguing, *inter alia*, that national policy favored maximum flexibility for fish farming and that the *Aquaculture Act* allows Fisheries to reserve areas for fish farming. Ministry of Fisheries authorities, therefore, have the view that the communes should plan only in areas where there is conflict, otherwise areas should remain unplanned and be available for aquaculture (assuming ecological suitability). In addition, Fisheries authorities hold the view that aquaculture is an industry that changes rapidly and moves locations: the communal planning structure lacked sufficient flexibility as a planning approach. Bennett, examining the (continuing) history of this struggle, posits several ideas as to why so much trouble was encountered despite enthusiasm on the part of communes for planning.[66] For example, from a planning theory point of view, he points to what is sometimes seen as a weakness of territorial planning compared with sectoral management.

Furthermore, the shift to planning directly affected the historically powerful position of the Ministry of Fisheries. Although the intent of planning was to develop an inventory to help in siting of fish farms, in fact planning became regarded as a mechanism for production and control and therefore resulted in what we may call "turf wars" between sectoral agencies for control over coastal space. Bennett observes that:

> The struggle over management of the coastal zone is at least partly attributable to the lack of a tradition of cooperation between sector authorities. Until the mid 1980s the fishery authorities had a virtually unchallenged monopoly of the management of marine resources and of ideological production in the field. Up to about 1990 the approach to management of aquaculture can be characterised as typical piecemeal engineering based on the rational appraisal of a limited number of variables with little concern for externalities. With little or no experience of the procedures or complexities of public planning, the fishery authorities were not well prepared for the introduction of coastal zone planning by the *Planning and Building Act of 1985*. This was a challenge to their accepted view of the world and their role in it. . . . Established in 1971, the Ministry of the Environment was originally conceived as a ministry over ministries in charge of the environment, resources and spatial planning: but it is common knowledge that it has never been accepted as such by other ministries, which regarded it as just another sector department.[67]

A related problem arose from the lack of clarity in the legislation and, in particular, the relationship between sectoral legislation and the PBA. In addition, the PBA does not require sea-use plans or aquaculture planning, but where they are developed, the Act provides for planning for "areas for traffic, fishing, aquaculture, nature and recreation, either separate or together" (s. 20.4). It remains unclear and debated as to whether this allows for supplementary rules to be developed within these areas. Some new areas of aquaculture conflict that were observed as emerging in relation to this issue are, for example, whether a commune plan can designate a particular kind of aquaculture. Conflicts also arise between forms of aquaculture as they compete for space for either salmon or other finfish or shellfish. This has provoked some frustration in communes which have found that they cannot fully control or plan for resources or space – for example, to avoid fish diseases – under the PBA because it is an issue regulated by the veterinary authorities and not the Act.

Other areas of conflict that have been noted are the following:

> Planning in the sea represents something quite different from planning on the shore. Planning on the shore and offshore are supposed to be integrated and seen in connection with each other. It is meaningless not to see them closely tied to each other. This is also in most cases what is attempted, but it has often been hard to do. . . . The sea represents a 3-d medium which contains a surface the volume between and the bottom. This makes it difficult to translate the planning principles from two-dimensional physical planning onshore. . . . More and better knowledge about the coastal zone and its related problems is relevant to good planning solutions, since decisions made on insufficient knowledge destroy the legitimacy of plans and confidence in planners. To register data and problems in cooperation with other agencies could ease this problem. To build up a GIS-database for the coastal zone would probably also contribute a lot to problem solving.[68]

Although there are formal administrative appeal processes provided for in the Act, and in fact there are some letter-writing campaigns and complaints filed with respect to individual development applications,[69] it appears that the level of litigation and resort to courts seen in Canada is not common in Norway.[70]

Another factor that may have affected the acceptability of aquaculture, despite the above-noted coastal zoning "battles" and its ability to develop as an industry, is simply that of timing and history. Aquaculture developed to a commercial/industrial scale in Norway through the late 1970s and early 1980s. Although there were environmental concerns during this period (the United Nations Environment Programme was created in the mid-1970s), the level of information and immediacy and breadth of communication through media and the Internet simply did not exist. Nor were the plethora of national and international environmental non-governmental organizations (ENGOs) that now impact significantly on public awareness and the level of concern in existence. However, as was pointed out in the NORCOAST study, increasingly conflicts are now experienced between the more urbanized desire for recreational space and the more traditional interests that favor farming and are dependent on fish farming.[71]

To some extent, at least in connection with salmon farming, this tension is said to be easing because salmon cages are being moved further offshore and therefore out of onshore sight lines. This also means that the harvesting, and indeed much of the access, to the farmed stock is occurring by water. This is seen as helping to address the aesthetic concerns posed by farms, although the movement of fish cages further out to sea is also potentially posing greater problems for navigation. There is also a foreseeable impact on shipping and tanker routes to service the oil industry, particularly as information and concern about the impact of invasive species (e.g. red tide) and diseases introduced through discharges of ships' ballast water increase. The movement of farms further offshore, when combined with increasing corporatization and non-resident ownership of the farms, is also seen as leading to new tensions as local residents experience the costs, ecological and otherwise, of the industry but are not necessarily experiencing the benefits in terms of revenue through taxation or employment in the community.

As was noted earlier, another emerging tension or conflict relates to possible competition between forms of farming/species for fish farming space. Other agency-related tensions also exist in addition to those between sectoral agencies, some of which are now actively being addressed through a coordinated or "team" approach to working with communes on planning exercises. One such tension is experienced within the Ministry of Fisheries, where traditional fishing "wild stock" interests, spatial claims and fish health concerns may appear to conflict with the other activities of the department related to the promotion and development of aquaculture.

The foregoing story has briefly outlined some of the issues and solutions that have arisen in the development of aquaculture – and, in particular, salmon farming – in Norway. The regulatory response to problems of conflict in the siting of farms has been dominated by the adoption of a planning approach to marine spatial management. Sea-/land-use planning was the central response to dealing with conflicting uses and interests and the need to provide an adequate number of ecologically appropriate and healthy sites for fish farming. As is often the case with integrated coastal management, the introduction of a new actor or spatial claim on the coastline, in this case the development of aquaculture in Norway, is a catalyst for the broader process of coastal/marine zone planning. The emphasis on local planning and involvement is also a notable part of the overall planning exercise, which appears, at least initially, to support the development of the industry.

Finally, although problems with diseases did generate immense difficulties and some reconsideration of the industry, the fact that aquaculture developed largely from within communities and on a commercial scale about thirty years ago, prior to the current heightened level of environmental awareness, access to information and activism, cannot be ignored. Information about environmental concerns and fish disease, and the level of general access to information and stories of conflicts now available on a global basis, are clearly factors relating to greater community mobilization and possibly greater conflict than would have existed in Norway, or anywhere, thirty years ago. It is evident that conflicts still exist in Norway with respect to some siting decisions, but the industry is now well established as an active participant and contributor to the coastal community. The problems and conflicts that are now emerging appear to be what might be called "second-generation" problems relating to globalization, such as transnational ownership, intra-industry conflict, and competition for space and markets, as well as the greater social awareness of the need for environmental and biosafety protection.

Implications and options for Canadian siting conflicts

A number of salient points emerge from the story[72] of Norwegian aquaculture development and siting concerns outlined in the foregoing section. First, the conflicts relating to aquaculture siting appear to change over time and the development of the industry. Thus, there are what can be called "first-generation conflicts"; that is, where aquaculture is a newcomer or entrant new user of coastal and marine space. Where space is relatively abundant or there is no perceived competition, then initial development can be fairly easy,[73] particularly if local residents carry out the activity. This was largely the case in the very early stages of aquaculture development in Norway. However, this situation is clearly unlikely for contemporary fish farming activities that are intended to operate on a commercial scale with a global market: they will need to occupy premium marine space that is also sought for recreational use, shipping or other activities.

A "second generation" of emerging siting-related conflicts was also noted in Norway. These relate to a number of issues, including competition between differing forms of aquaculture for prime aquaculture areas; concerns about non-resident ownership and benefits, with costs being borne locally; and ecological concerns about the impact of farmed fish on biodiversity and on the ecological balance in an area. The Norwegian response to the first-generation conflicts, many of which were ultimately triggered by the need for larger and better-situated spaces to avoid fish disease, was to invest in a strategy of promoting and developing marine spatial planning to resolve immediate conflicts and operational concerns relating to existing sites and to identify ecologically and operationally appropriate sites for industrial development of aquaculture. At present, the same strategy remains as the main tool to deal with the more complex second-generation conflicts.[74]

Although, as was indicated earlier, there were and are, in fact, jurisdictional conflicts between sectoral regulators in Norway, it was relatively simple as a legislative exercise to add sea-use planning to land-use planning activities, because the marine space involved is, for the most part, internal waters (although the *Planning Act* can apply in the economic zone). Finally, although perhaps not perceived to be so, it is clear that despite some tendency to regard fish farming as in opposition to some "environmental concerns" (biodiversity) but complementary to other such concerns (how to feed the world without destroying all the wild fish stocks to meet market needs), there is a clear common ground. In order to operate high-production, "high-quality fish" farms (irrespective of the form of farming, i.e. shellfish or finfish), a clean and healthy marine ecological system is required. Water that is contaminated by chemicals, other substances, disease, or invasive or harmful organisms is not viable for the production of farmed species. Equally, fish farms must be operated in an environmentally friendly manner to have long-term productivity or they must be prepared to move as sites get fouled. This mix of concerns, largely caused by the spread of fish diseases among farmed species and the need to move farms for the sake of having a clean or healthy location, was an important trigger, in combination with coastal use competition, in generating a more holistic planning response from the Norwegian government.

Does the Norwegian experience have any relevance to the Canadian situation? Does Canada also have to go through the same "teething" process of fish disease, siting problems, and so on as its industry develops? Are there some lessons to be learned from the foregoing "story"? A brief review of recent conflicts relating to siting of aquaculture farms in Canada[75] suggests that the answer is both yes and no.

Certainly it can be observed that conflict and managing conflict is a preoccupation in Canada and is perceived as a major problem by many on both the east and the west coasts, to the extent that it has even generated Senate inquiries into the matter.[76]

As suggested earlier, a very important factor to consider relative to the Norwegian situation is that Canada is seeking to promote aquaculture at a much later period in history.[77] There is a markedly different global social political climate, where interest groups and the media of communication are vastly different and highly influential. In addition, the level of sophistication required for commercial operations, particularly in connection with finfish farming, as well as the competitive and integrated global marketing involved, means that large-scale industrial developments, probably involving transnational and, at a minimum, non-resident ownership to some degree, will be the norm.[78]

Another central factor in the Canadian context is that (unlike in the 1970s and 1980s, when, although the wild fishery was the subject of increasing concern, there had not been the wide-scale experience and conflicts relating to the "death" of a fishery) now there is ongoing, even violent, conflict and resistance to government regulation and restrictions on the wild fishery on both coasts. While aquaculture is not necessarily a direct replacement for, or equivalent industry to, the traditional "wild" fishery, obviously in some communities it may have been seen by some as an alternative indus-try to provide employment for displaced fishers.[79] However, the lifestyles, skills and cultures differ: for various reasons, fishers often resist location of aquaculture sites in coastal communities. Shellfish farming is seen as a nui-sance or as providing problems for nets and access to fishing areas or posing a danger to navigation. Salmon farming is more a matter of concern for its possible impact on wild fish stocks and the introduction of fish disease to the wild stocks.[80]

A further confounding factor is the prevailing climate of distrust of gov-ernmental regulation of fisheries and any science associated with the regula-tion of fisheries, in particular regulation involving the Department of Fisheries and Oceans.[81] In many cases, on both the east and the west coasts it is evident that there is an ongoing refusal to accept that the wild stock fishery is no longer viable and may not ever recover. The issues differ on the two coasts in terms of causal factors, but the refusal to accept change is a common thread. Efforts to introduce a new actor into communities that do not accept that the traditional fishery is no longer possible are, understand-ably, viewed with suspicion and resisted.

The following comments from a residents' committee in Nova Scotia in February 2003 reflect concerns and a general suspicion of the industry and any perceived benefits, including government-related information as to the economic benefit of the industry:

> The "benefits," often quoted by open-cage finfish aquaculture interest groups, cite job creation and economic contribution. *In reality most of these jobs are part-time and the loss of traditional jobs, due to probable* harm to the natural fishery are not taken into account.
>
> Aquaculture operations receive subsidies, soft loans, grants and com-pensation from government and a false picture is created of the eco-nomic viability of aquaculture operation. . . . *In reality, jobs are not provided, and needed revenue is not provided to the government*[82] [emphasis in original].

These are factors that render the climate and context for conflict preven-tion and management with respect to the siting of aquaculture, and indeed the introduction of any new coastal use in Canada, complex, to say the least. A further complication is the fact that legislative and regulatory authority combine a vertical and horizontal mix of federal and provincial jurisdictions – that is, some standards are national and some are provincial – and, in each case, multiple agencies are implicated at both levels, none of which are necessarily in hierarchical relationship to each other.[83] The fact that control over the near offshore has been and remains a somewhat fraught question in federal–provincial relations in Canada adds to the problem. These constitu-tional and jurisdictional questions, as well as the emerging issues such as the nature and extent of First Nations peoples' constitutional authority in decision-making over spatial use and natural resources, the impact of international trade obligations and the effect of the 1992 Rio Declaration principles[84] such as precaution, public participation and subsidiarity, are considered in greater depth in other investigations in the AquaNet Law and Policy Project that are presented in this book.

Finally, it appears that most siting "conflicts" – that is, ones that reach a stage of expression in the courts or media – involve citizens or citizen groups against the government and are usually in the form of a citizen challenge to a ministerial decision to grant a permit or a license to a farm.[85] Thus, although they may appear to be citizen versus citizen conflicts, in fact the form of the complaint or dispute is generally citizen disagreement with a process or a decision of the government. In addition, in some cases the issue, although appearing as a siting problem, is often, in fact, operational in that the industry is already in the location but problems are being encountered with other actors located nearby.[86]

In many cases, the concern is that government did not employ an appropriate process or fully take into account relevant concerns, including accurate scientific data, when making the decision to allow the site. This leads to ongoing operational conflicts for existing sites and creates a barrier to development of new sites. For example, in some cases community groups may have the view that they were given insufficient notice to respond to public hearing notices, even where there have been regional or advisory community processes provided.[87] Community concerns often arise from conflicting data or lack of information, especially with respect to scientific data about the ecological suitability of an area or the environmental impact of the farming, particularly finfish cage farming. For example, the study of conflicts in New-foundland noted that:

> [a] striking feature among all interest groups was the lack of awareness of the positive impact aquaculture has had on the community. Though aware of some employment generated by the farms, people doubted the actual magnitude of this benefit as well as the overall profitability of aquaculture. Also misconceptions abound regarding the negative impacts of aquaculture. Though not a source of conflict, this lack of knowledge certainly makes conflicts more bitter and debilitating for the aquaculturists.[88] ... Some local residents attribute any negative environmental change to fish farming operations. Often residents don't always realize that growers need to maintain clean water to produce healthy fish.[89]

The result is that when conflicts become manifest or erupt, they tend to do so in the context of an individual siting decision or application, often after extended negotiations or investment has occurred on the part of the applicant to evaluate and prepare a proposal that may be acceptable to the various government departments involving approvals in Canada (even with the "one-stop shopping" approach).[90]

Various forms of public participation approaches or transparency-based siting and licensing criteria to avoid conflict and bring "stakeholders" into the process have been tried or proposed to prevent or reduce conflict. For example, in Nova Scotia local Regional Aquaculture Development Advisory Committees (RADACs)[91] can be set up to help with the public consultation process. However, the creation of these has itself become a source of conflict where the government has chosen not to create them or where the RADAC is viewed as "stacked."[92] These do have not appear to have solved the problem of citizen complaints, many of which have ended up in court.

The British Columbia government, through Land and Water British Columbia Inc., a Crown corporation dealing with land-use programs, including agricultural and aquaculture (now Integrated Land Management Bureau), has also developed very detailed aquaculture tenure policies and procedures as part of the government's support for aquaculture development involving Crown land. This policy, effective since the autumn of 2002,[93] articulates strategic principles underlying the government's approach to aquaculture development decisions. The policy is based on the British Columbia government's acknowledgement of aquaculture as a legitimate user of coastal resources and its support for sustainable development of aquaculture that is conducted in an "environmentally, socially and economically suitable manner." It also expresses the government's commitment to reduce red tape and the regulatory burden, expedite decisions, increase access to Crown land to protect and create jobs, protect both private property and resource tenure rights, eliminate government subsidies to business and eliminate delays to Crown land applications. In 2002, the aquaculture policy was said to have been built on ten strategic principles (summarized):

1 certainty (timely and clear resource decisions with a predictable regulatory framework);
2 competitiveness (removing barriers to investment and promote open trade);
3 efficiency (maximizing net benefits arising from the allocation, development and use of natural resources);
4 shared responsibility (encouraging cooperation among departments, First Nations, industry, non-governmental organizations (NGOs), etc. in developing and implementing resource management);
5 innovations (encouraging innovative technologies, skills, etc. to ensure sustainability of natural resources);
6 integration (ensuring that resource decisions integrate economic, environmental and social considerations);
7 accountability (performance-based standards, compliance, reporting, auditing and enforcement mechanisms);
8 continual improvement;
9 transparency (open, understandable decision-making processes, including consulting with key interests prior to making a decision); and

10 science-based decision-making (justifiable decisions informed by science-based information and risk assessment).[94]

However, laudable though these principles are, they do not appear to have resolved the conflicts in British Columbia, at least with respect to salmon aquaculture. Although it is outside the scope of this chapter, one can partially attribute the problem to another kind of conflict or perceived conflict: government regulatory agencies appear to find themselves in what can be seen as a conflict of interest or, at the very least, tasked with meeting mandates that may appear to be in conflict. Although the mantra of "sustainable development" or environmentally sustainable operations is oft invoked as the point of reconciliation or mediating paradigm to achieve myriad, sometimes conflicting, agendas, increasing public skepticism about the term renders it meaningless in this function.[95]

In sum, both the timing of the Canadian desire to promote aquaculture industry and with the general nature of industrial activity in the twenty-first century suggests that the Norwegian experience in responding to and managing siting conflicts, while of interest, is not applicable or especially relevant to Canada and that we are perhaps doomed to battle it out until the passage of time itself changes views and attitudes. And to some extent this is correct. The best regulatory system and conflict dispute resolution system possible will not reverse or alter some of the factors referred to above. Regaining public confidence in regulatory activity in Canada, particularly in this and other sectors such as health, will not be easy or rapid. It is, then, perhaps less than helpful that regulatory leaders and officials, such as the Commissioner for Aquaculture Development located within a department responsible for overall regulation of fisheries and ocean activities in Canada, are on record with views such as the following:

c) Allocation of Aquatic Space
There is considerable competition for the use of public waters among various groups, including recreational boaters, fishers, aquaculturists, shippers, offshore oil and gas developers, etc. *Integrated coastal zone management is one means whereby long-term, balanced decisions could be made on the use of coastal and open sea areas, including use for aquaculture and enhancement purposes. The time frame for implementation of an effective mechanism for integrated coastal zone management, however, is lengthy. To meet the current needs of the aquaculture sector, the Commissioner does not consider it acceptable to wait for the implementation of a system of coastal zone management.* He also considers that the lack of guidelines to assist operational staff with resource allocation decisions is delaying the decision-making process and constraining growth in the sector.[96] [emphasis added]

To be fair, however, this comment is part of a broader position urging the development of other mechanisms to address a wide range of conflicts, including regulatory complexity and uncertainty.

The lack of definition of aquaculture in Canadian law and the lack of any guidance as to the application of the precautionary principle and risk assessment to expedite decision-making are factors that can lead to conflicts that affect the development of the industry. Nonetheless, there is a clear message from some of the highest levels of government: where a process such as integrated management takes "too long," at least in terms of a government agenda to expedite the development of the aquaculture industry, then the processes will be ignored in favor of the government and industrial development agenda. This is the message despite the fact that integrated management is officially promoted as a means of preventing poor decision-making that can negatively affect aquaculture operations and the environment in the long term. Given this stance, citizen suspicion of hearings and processes designed to be "independent" is not surprising. The receiving climate, while welcoming and simplified (no red tape) from a government perspective, is not welcoming from the community perspective, and applicants and government will face citizen resort to courts, media and consumer boycott tactics, all of which counteracts government efforts to expedite the development of sites. The siting application and the applicant become the catalyst and crucible in which these broader tensions are fought out.

These are largely political issues that underlie any proposal to develop conflict management and prevention approaches to siting applications in Canada. Can they be resolved by resort to ADR or DR approaches such as arbitration, mediation, negotiation or conciliation? These four ADR procedures are listed as examples of dispute resolution procedures in the *Fisheries and Coastal Resources Act* of Nova Scotia and are generally endorsed elsewhere.[97] These processes, which are often understood as referring to a spectrum of procedures ranging from consensual (negotiation) to coercive (litigation/legislation), can be useful as tools to help resolve some types of conflict. For example, operational disputes or beach access or nuisance complaints appear particularly well suited for supported negotiation, conciliation, mediation (third party to assist with process but usually not the decision) or even arbitration (third party chosen by parties to make a decision or finding) processes involving citizen-to-citizen matters. However, as suggested earlier, it is questionable whether they are useful to siting conflicts where the complaint is largely with respect to government process and, often, ministerial decisions. Are governments genuinely willing to enter into processes such as mediation to reach consensual decisions that may in fact alter the outcome and require that they relinquish some degree of decision-making authority? Can a minister actually do this and still be within her or his jurisdiction if charged with responsibility for making a decision? Will a government accept negotiated outcomes on siting, or non-siting as the case may be, of aquaculture in any one case? It is not at all clear that this is likely or even desirable.

Nevertheless, this does not mean that the idea of developing DR and conflict management prevention approaches should be abandoned. To the contrary, they are, if anything, clearly essential in the Canadian context. The argument in this chapter is that the notion of dispute resolution and the nature of siting conflicts must be disaggregated and a much broader view of the procedures and processes be taken. There are many different kinds of siting and other conflicts warranting different strategic responses.

It is in this context that although in some respects the Norwegian story is not easily applicable to Canada's situation, except perhaps warning about the danger of *ad hoc* decision-making, there are some features in the Norwegian response and approach that are instructive and are applicable or should be tried in Canada.

The first and most essential lesson – and it may simply be a *post hoc* reification of the events – is that it appears that the Norwegian government specifically recognized marine spatial conflict as problem in and of itself. Resources were devoted to exploring the problem, and a specific strategic approach was adopted to preventing and managing these conflicts, which were obviously going to increase with the development of the industry and its increasing need for marine space allocation. Similarly, what is needed in Canada is a consciously focused strategy for addressing conflicts relating to use of marine space, including aquaculture development. Of course, this can and should be part of a broader integrated management process in which aquaculture is the catalyzing issue or trigger event around which the process is centered. In addition, a more strategic approach can be adopted for specific siting issues or conflicts. There are numerous models and tools that can be adopted.

One useful approach to dealing with the kinds of issues and actors involved in siting conflicts is sometimes known as the "wheel of conflict," a diagnostic and strategic tool propounded by Christopher Moore in the late 1980s for working with group conflicts.[98] This approach focuses on a careful evaluation of the nature of the conflict and key elements in it, which are then tied to a range of strategic interventions to address the particular problem or problems. It posits that sources of conflict can be understood as falling into one or more of the following categories: data conflicts, interest conflicts, structural conflicts, value conflicts and relationship conflicts.

Very often a group conflict may exhibit more than one feature. Each of these categories itself has various causes. For example, a data conflict can be caused by lack of information, misinformation, and differing views on what is relevant, by differing interpretations of data or by differing assessment procedures. The point of interest is that each of these elements or categories or sources of conflicts may need to be, and can be usefully, addressed by some specific interventions aimed at their resolution. Data conflicts, depending on the source of the conflict, can perhaps be resolved by interventions such as agreeing on what data are important, agreeing on a process to collect data, developing common criteria to assess data, or the use of third-party experts to gain an outside opinion or break deadlocks. Such an approach requires that time and expertise be devoted to considering the nature of conflicts as they develop and to strategically resolving them. Positive resolutions will in turn generate a "counter-history" to poor conflict-processing histories and can be a basis for developing greater trust among actors. This kind of approach can be adopted within a more long-term overall marine and coastal integrated management approach.

A second lesson, and one that may appear more difficult in the Canadian context, is adopting a prevention-based comprehensive planning approach. It is suggested that planning, particularly planning that involves community interaction and dialogue (communicative planning), can be understood as a conflict resolution/prevention mechanism. For example, use of joint map-making and other tools can help to facilitate communication among actors and to identify areas of disagreement and areas of mutual concern. Sea-use planning, especially if done at the municipal or local level, may seem more difficult to achieve in the Canadian constitutional system, especially given the geographic layout of the Canadian coastline. Nonetheless, the Norwegian experience demonstrates quite clearly that piecemeal *ad hoc* siting is likely to result in reduced efficiency of operation and fish disease problems.

Perhaps one of the most useful aspects, from an industrial development perspective, of the Norwegian approach is that making some initial zoning/use decisions about marine space allocation pushes major community conflicts "upstream" in the process. Major issues and concerns are brought to the surface and resolved *before* any individual development or licensing application is involved. In principle, this can allow for more objective assessment of elements of the ecological and other utility of the space without the tension of a community concerned that it is facing a "done deal" and that the consultation is simply *pro forma*. This approach can also better ensure a higher level of predictability for an applicant and, in turn, generate a better "climate" for sectoral development. As in the case of land-use planning, where zones are designated for certain kinds of buildings and activities, there may still be hearings and assessments and, in some cases, concerns or conflicts relating to any individual applicant, depending on its nature and the unavoidable NIMBY ("not in my backyard") factor. Nonetheless, the underlying decision as to the nature of the use has been made beforehand, and the issues then may simply be more related to specific placement within a zone or specific features of the operation. The Norwegian emphasis on local planning is also important in that, as noted, the second generation of concerns about transnational ownership, locally experienced consequences and cost are themes in contemporary conflicts in Canada and are emerging as concerns in Norway.

There has been some experience in Canada with zoning. For example, on Prince Edward Island (PEI) an aquaculture zoning system was developed in the 1980s that essentially classified the waters of PEI in terms of their acceptability or appropriateness for aquaculture. In so doing, it took account of other uses in the area at the time.[99] The main difficulties that appear to have emerged with this system are that uses and concerns have changed over time, but the 1980s zoning policy is not easily altered. The zoning system did not formally provide for further public participation in siting decisions unless required under the federal environmental impact assessment process. In fact, most decisions are made at an administrative level by a leasing management board. However, unless zoning of marine space is comprehensive and not focused on a single industry, it does not really achieve the goals of marine spatial planning in resolving conflicting or competing uses. Given the rapidity of coastal use change as well as the need in some cases to move sites from time to time, it is also important that plans be reviewed fairly regularly.

The 1997 study of possible conflict resolution strategies for Newfoundland notes that:

> The application of land use planning methodologies and techniques to the Province's coastal resources would be of considerable benefit in avoiding, resolving or mitigating conflict situations in the provinces and the aquaculture industry. . . . In the case of aquaculture, there is a desire to have a land (water) planning process that will allow designation for exclusive integrated aquaculture development. Such a process would be designed to recognize the value of aquaculture to regional and community development and protect that value from activities that would be detrimental to development of the industry (e.g., waste disposal).[100]

However, the same study also observes:

> Fishers are generally opposed to establishing land (water) use designations. However, they have cooperated with such efforts to date by providing information regarding their traditional fishing areas. They see merits for its use as a tool to keep aquaculture away from traditional fishing areas but are not in favor of using it to divide a resource between two (or more) users.[101]

Similarly, in Nova Scotia there has been some site identification and mapping;[102] however, it has not incorporated sea-use planning or ocean zoning processes as such. There is, nonetheless, increasing interest in ocean zoning issues in Nova Scotia, with workshops relating to zoning for ballast water discharge to avoid introducing invasive or harmful species into sensitive areas (for example, red tide into shellfish aquaculture farm sites) and an ocean zoning workshop held in 2004, organized in part by an NGO, the Ecology Action Centre.[103] The fact that Nova Scotia is the site of several marine uses, including a marine protected area, a strong interest in encouraging offshore oil and gas development, a commercial port and major coastal tourism, points to the need to develop a planned and strategic response if aquaculture is to develop further in this province.

Conclusion

Although not strictly speaking a siting conflict as such, the story of Norway is similar to that of Canada in terms of the impact that inter- and intra-agency jurisdictional conflicts can have on siting and industrial development. The development of memoranda of understanding and general adoption of a "one-stop shopping" approach (as much as possible) to regulatory approvals means that federal–provincial and interdepartmental jurisdictional conflicts, while not resolved totally, have to some degree been worked out.

However, we can expect differing kinds of interagency conflicts to surface relating to control over marine space and the coastline, and at times conflicting mandates. For example, the development of marine protected areas, heritage areas, oil and gas development, the increasing interest in use of marine space for wind farms, communication channels and pipelines, as well as more traditional uses such as shipping, recreation and traditional fisheries, all have the potential for generating conflicts unless a strategic approach is adopted. Again, it is suggested that a strategic sea-use planning approach combined with a well-developed conflict prevention and management strategy could be useful in preventing or managing such conflicts. The exercise of developing plans for all potential users and, with it, processes for resolving conflicts among potential licensees of marine space and coastal users, through municipal authorities – often representing particular socio-economic and ecological systems – could ultimately result in a climate which recognizes that conflicts will occur, but has developed a means of resolving disputes as they arise with strategic and flexible planning process.

This process cannot, of course, occur in isolation from the development of other aspects of siting – namely, issues of fish health or disease control, ecologically appropriate siting, data collection, and protection of fish safety to ensure that biodiversity and ecological systems are protected. In addition, a licensing system/space allocation system should be developed with a view to addressing concerns about the need to ensure that local benefits are explicitly provided for in order to offset any short- and long-term costs that are incurred for lost access to public resources, including the aesthetic impacts of aquaculture.

All of these activities will require a legal framework that articulates and supports this approach rather than exacerbating conflicts. To the fullest extent possible, a federal aquaculture law and policy should focus on development of these approaches in a facilitative role, perhaps much as the role envisaged for the Minister of Fisheries and Oceans under the *Oceans Act* of Canada for integrated management, to date a role perhaps less than fully utilized. Certainly flexibility and responsiveness to local concerns and differences in values and kinds of conflicts on each of Canada's coasts should be a key theme at a federal level.[104] A provincial-level effort should be made to develop conflict management strategies and to develop greater human capacity and awareness of approaches to analysis and intervention in sources of conflict, in order to de-escalate conflicts and to also plan for likely conflicts in a strategic manner.

Conflicts will occur even with a well-planned approach. Indeed, conflict is a normal and even healthy process in a society where interests and values are diverse and changing rapidly. To fail to acknowledge this process does, however, risk the very experience that is increasingly present in discussions about marine space activities conflict leading to violence and court battles. At all levels of governance, it is quite clear that these tensions and concerns will only increase as pressures on the coastline and the environment increase. Although time is perhaps of the essence for some in achieving a prominent role for Canada in the world aquaculture market, the folk wisdom about "haste making waste" also comes to mind. It suggests that time and resources devoted to preventing or resolving conflicts would be well spent and might provide a more solid foundation for the establishment and long-term growth of the industry.

Notes

The research assistance of Josh Bryson, a former student, now an Associate with the law firm of Hennigar Wells Lamey & Baker, and Kalen Brady, LLB candidate, Dalhousie Law School, is gratefully acknowledged. The financial/research network and support provided by AquaNet is also gratefully acknowledged.

1 The specific focus in this chapter is on the problems arising from the siting of enterprises on the coastline, traditionally a "public resource" or commons, rather than from land-based farms or ponds set on private property. The latter can also result in conflict with neighbors; however, the nature of the coastal activity provides unique challenges, aside from the fact that farmed ocean space is the site of most growth in the industry in Canada and elsewhere.
2 See, generally, William Bridges, *Transitions* (Cambridge, MA: Perseus Books Publishing, 1980).
3 *DFO's Aquaculture Policy Framework* (Ottawa: Fisheries and Oceans Canada, 2002) at 11.
4 FAO, *The State of World Fisheries and Aquaculture 2002* (Rome: FAO, 2002) at 26 and 27. Published every two years. Online. Available http://www.fao.org/sof/sofia/index_en.htm (accessed 17 March 2004). The same comment is made in the 2004 report at 13. In 2002, more than 220 different aquatic animal and plant species were reported, with more exploration for farming other species under way (see, *The State of World Fisheries and Aquaculture 2004* (Rome: FAO, 2004) at 17).
5 GESAMP, *Planning and Management for Sustainable Coastal Aquaculture Development*, Reports and Studies GESAMP No. 68 (Rome: FAO, 2001) at vii.
6 The 1998 report *Sharing Coastal Resources: A Study of Conflict Management in the Newfoundland and Labrador Aquaculture Industry*, Canada/Newfoundland Agreement on Economic Renewal. Online. Available http://www.gov.nf.ca/fishaq/publications/aqua_man_study.stm (accessed 14 March 2004) draws a distinction between conflict and opposition from interest groups and argues that opposition should be addressed before it develops into conflict.
7 FAO, 2002, *supra* note 4 at 21.
8 See, for example, *supra* note 6; and D. J. Noakes, L. Fang, K. W. Hipel and

D. M. Kilgour, "An Examination of the Salmon Aquaculture Conflict in British Columbia using the Graph Model for Conflict Resolution," (2003) 10 *Fisheries Management and Ecology* 123.

9 Y. Bastien, "Legal and Policy Challenges for the Canadian Aquaculture Sector," PowerPoint presentation made at Dalhousie University, Halifax, NS, November 2000.

10 GESAMP, *supra* note 5 at 68.

11 UNESCO, *Managing Conflicts over Resources and Values: Continental Coasts. Results of a Workshop on "Wise Practices for Coastal Conflict Prevention and Resolution,"* Maputo, Mozambique, 19–23 November 2001. Coastal Region and Small Island Papers 12 (Paris: UNESCO, 2002). Online. Available http://www.unesco.org/csi/pub/papers2/map.htm (accessed 14 March 2004).

12 *Ibid.*, "Introduction," speaker noted as Dirk Troost.

13 *Ibid.*, "Resolution and Prevention of Conflicts," citing Rijsberman (1999).

14 S. Coffen and A. Smillie, "The Legal Framework for Canadian Aquaculture: Issues in Integrated Ocean Management," in D. VanderZwaag, ed., *Canadian Ocean Law and Policy* (Toronto and Vancouver: Butterworths, 1992) at 49.

15 In 1995, Fisheries and Oceans Canada (DFO) was confirmed as the lead federal agency for aquaculture through approval of a Federal Aquaculture Development Strategy (FADS). In 1999, a Commissioner for Aquaculture was appointed to advise the minister for DFO and a budget request for $36 million was made ($15 million allocated). In 2000, a six-point action plan and Program for Sustainable Aquaculture was launched. See "DFO's Role as Lead Federal Agency for Aquaculture Development. Positioning Canada for the Future," presentation to the Canadian Aquaculture Law and Policy Workshop, 26–28 February 2003, Halifax, NS.

16 *Aquaculture in Norway – 2001* (Trondheim: Norwegian Fish Farmers' Association, 2001) at 3. More recently the Norwegian Minister of Fisheries and Coastal Affairs has focused on the promotion of quality products and sustainability as important features of the aquaculture industry: see, for example, "Norwegian Fisheries Policy – with Main Focus on the Promotion of Sustainability," Lecture by the Minister at the University of Washington, Seattle, 20 October 2004. Online. Available http://odin.dep.no/fkd/engelsk/p100001957/047041-090002/dok-bu.html. In his lecture, the Minister noted that trade in seafood is Norway's second largest export industry after oil and gas. Of that trade, 40 percent by value (US$1.4 billion), or 577,000 tons of 2.7 million tons by weight, is derived from the high-value farmed salmon and trout.

17 AquaNet is an initiative of the federal government under its Networks of Centres of Excellence program. It is aimed at encouraging collaboration between university researchers, industry and government in the development and advancement of aquaculture in Canada.

18 See, for example, *Fisheries and Coastal Resource Act*, S.N.S 1996, c. 25, s. 9; GESAMP, *supra* note 5 at 68, which also includes litigation.

19 The term "ADR" has been criticized as implying that litigation is the norm, while all other methods for settling disputes are "alternative" to the norm. DR tends to be used to imply a panoply of methods, including litigation, all of which are potentially useful, depending on the features of the problem.

20 In particular, the wheel of conflict developed by Christopher W. Moore, *The Mediation Process: Practical Strategies for Resolving Conflict* (San Francisco: Jossey-Bass, 1989), is a useful tool for designing systems to prevent or respond to conflict.

21 Communicative or collaborative planning has been seen as a mid-ground between a "top-down" social-engineering rationalist approach to planning and

social mobilization/activism that is sometimes viewed as difficult for government to implement or adopt. The term "governance" as opposed to "government" serves to capture the idea that the activity of collective decision-making
is a partnership including all actors in a system. There are many different
authors, studies and theoretical perspectives, ranging from the most abstract to
the more applied, regarding the utility of communicative planning techniques
in increasingly pluralistic societies. See, for example, J. Habermas, *The Theory
of Communicative Action*, Vol. 2: *Lifeworld and System: A Critique of Functionalist
Reason* (Boston: Beacon Press, 1987); P. Healey, *Collaborative Planning: Shaping
Places in Fragmented Societies* (London: Macmillan Press, 1997); R. G. Bennett,
"Coastal Planning on the Atlantic Fringe, North Norway: The Power Game,"
(2000) 43 *Ocean and Coastal Management* 879; J. Amdam, "Governance and
Communicative Planning in Practice in a Welfare State: Experience from
County Level in Norway," (2002) Paper presented to Association of European
Schools of Planning Congress, Volos, Greece; T. Sager, *Communicative Planning
Theory* (Aldershot, UK: Avebury, 1994). A useful description and discussion of
the relationship between communicative planning and greater use of GIS
and expert information is found in B. Goldstein, "How Communicative
Planning Can Help Counties to Reduce the Impact of Land Use Change
on Biodiversity, while Other Planning Frameworks Fall Short: A Conceptual
Blueprint for the California Biodiversity Project" (1996). Online. Available
http://gis.esri.com/library/userconf/proc96/TO50/PAP036/P36.HTM (accessed
14 March 2004).

22 Hordaland, the county in which Bergen is located, is "the leading county in
fish farming.... In 1996, 150 fish farms in Hordaland produced 68 793 tons of
the total Norwegian export. 93% of the production is exported." *NORCOAST –
Review of National and Regional Planning Processes and Instruments in the North Sea
Region – Full Study, Phase 1 Report, Hordaland Fylkeskommune, Norway* (County of
North Jutland, Denmark: NORCOAST Project, 1999) at 11.

23 I am particularly indebted to Professor Roger G. Bennett, Department of Geography, University of Bergen, first for providing me with his very interesting
scholarly articles and case studies on coastal planning in Norway (which led
me to Bergen), and second for his generous collegiality in hosting my visit to
the Environmental Law Center, University of Bergen, and organizing and
accompanying me to several key interviews with relevant personnel in the
County of Hordaland. I am also grateful to the kind assistance and information
provided by Ernst Nordtveit, Dean of Law, University of Bergen.

24 *Interim Report*, June 2001. Online. Available http://www.parl.gc.ca/37/1/
parlbus/commbus/senate/com-e/fish-e/rep-e/repintjun01-e.htm. Although the
Report is still at this location on web search engines as of March 2004, it is
not possible to access it directly using this URL. It is also not easily accessible
through the Senate website. The best link to the report is through the Aquatic
Network website. Online. Available http://www.aquanet.com/Resources/
aqua/aq_aqua5.htm (accessed 14 March 2004).

25 *Ibid.* at 8.

26 An estimate of value added in 2001 is 16.200 million Norwegian kroner. See
also *supra* note 16, where the Minister of Fisheries estimated the value of aquaculture in 2004 as US$1.4 billion.

27 Cod, char, halibut and other species.

28 In its 2002 report, the FAO noted that Norway has now identified Asia as the
future growth market and that international trade in farmed salmon has
moved from "virtually zero to about 1 million tones (2001) in less than two
decades." Atlantic salmon accounts for about 88 percent of this trade, with
Norway being the main exporter. FAO, 2002, *supra* note 4 at 36.

29 R. G. Bennett, "Norwegian Coastal Zone Planning," (1996) 50 *Norsk Geografisk Tidsskrift* 201, notes that it has over 55,000 kilometres of coastline (including islands) and a relatively small population. However, he continues, "[s]ettlement is largely concentrated in the coastal zone and many activities are localized and intensive."

30 *Ibid.*

31 *Ibid.* at 202.

32 In January 2003, the Norwegian government issued a White Paper proposing extension of its territorial sea from 4 to 12 nautical miles (nm), largely to respond to marine pollution concerns and the need to be able to regulate a greater area of near coastal activities. See Press Release 9/03, Ministry of Foreign Affairs, 17 January 2003. Online. Available http://www.odin.dep.no/ud/engelsk/aktuelt/pressem/032081-070019/index-dok000-b-n-a.html (accessed 17 March 2004). The *Act relating to Norway's Territorial Sea and Contiguous Zone*, extending the nautical sea limit to 12 nm, was adopted on 27 June 2003 and entered into force on 1 January 2004. See Press Release 212/03, Ministry of Foreign Affairs, 30 December 2003. Online. Available http://www.odin.dep.no/ud/engelsk/aktuelt/pressem/ 032001-070316/index-dok000-b-n-a.html (accessed 17 March 2004).

33 Bennett, *supra* note 29 at 201.

34 In order to better reflect and perhaps even shed further light on the situation in Norway, as much as possible, Norwegian-authored descriptions of the Norwegian "system" and relationship between agencies and the processes have been used, rather than author observations or a review of legislation. This is a critically important aspect of comparative research in that, although a review of legislation or an outsider description of situation has some utility, it can and often does fail to capture the subtle differences, or what we might call a sense of "how things really work"; that is, the view from inside. The relevance of the perception of how things work resides in the fact that responses to problems reflect the comprehension of the problem by those experiencing it and how they frame the issues. This seemingly obvious point is more difficult in practice and is central to ideas about effective negotiation, conflict resolution and prevention.

35 NORCOAST, *supra* note 22 at 5.

36 Bennett, *supra* note 29 at 203.

37 References taken from interviews, Bennett, *ibid.*, copies of legislation provided during a meeting with the Fiskerdirektoratet in Bergen, and from Rogers Consulting, *A Review of Legal and Policy Frameworks Used to Regulate and Legislate Aquaculture in Australia, Japan, New Zealand, Norway, United States*, prepared for the Office of the Commissioner for Aquaculture Development, Fisheries and Oceans Canada, June 2000 at 53–58.

38 This list is derived from the 2001 Report, *supra* note 16 at 7. The name of the legislation varies when translated: for example, the Act of 14 June 1985 No. 68 relating to Aquaculture (Aquaculture Act) (Online. Available http://www.ub.uio.no/ujur/ulovdata/lov-19850614-068-eng.doc (accessed 17 March 2004)) is also referred to in some reports as Act No. 68 of 14 June 1985 Relating to the Breeding of Fish, Shellfish, etc. (see Rogers Consulting, *supra* note 37), or the Act of Fish Farming (see notes for presentation of Dr. Bjarne Aalvik, May 2000, provided to the author by the Ministry of Fisheries, Bergen, August 2002). Some but not all Norwegian legislation has been translated into English. See the University of Oslo, Faculty of Law Library website listing translated legislation. Online. Available http://www.ub.uio.no/ujur/ulov/ (accessed 17 March 2004). As noted by the Minister of Fisheries and Coastal Affairs, *supra* note 16, the *Aquaculture Act* is undergoing revi-

sion to "make it a modern tool for administering the industry." The revised Act is scheduled to come into force in 2006. It will be administered within a more integrated system, to be established in 2005, which will better incorporate environmental concerns and improved technical standards. A new regulation is also under development to address sea ranching for lobster, shellfish and other species, which will broaden the sector. It should be noted that the role of government in developing the regulations remains, as noted by the Minister, one of "wealth creation" and facilitation rather than of being a "restraint" on the industry.

39 Interestingly, a handout (hand-dated May 2000) comprising speaking notes provided to the author by the Ministry of Fisheries allocates responsibility for this Act to the Ministry of Local Government and Labour.
40 Rogers Consulting, *supra* note 37, or, as it is also known, the *Act Relating to the Regulation of Exports of Fish and Fish Products* (1990).
41 This is listed in the report, *supra* note 16. In the "Presentation Notes" referred to above in note 39, it is listed as Ports and Harbours under the jurisdiction of the Ministry of Fisheries; however, it is not listed in the legal database and does not appear to have been translated.
42 Listed in the Report, *supra* note 16, but not elsewhere and does not appear to be available in translation.
43 Listed in the Report, *supra* note 16, but not elsewhere and does not appear to be available in translation.
44 Bennett, *supra* note 29 at 204–205.
45 Rogers Consulting, *supra* note 37 at 54.
46 Bennett, *supra* note 29 at 205. Bennett also provides a useful summary of LENKA's methodology (and problems associated with it) for assessing capacity as a basis for developing regulatory responses. Problems noted included the overly large zones for capacity assessment, which did not correspond to the local governance/planning units and did not, therefore, prove to be much help for local planning and management decisions.
47 For example, Chapter 3 of the *Fish Diseases Act*, No. 54 of 1997, which applies seaward up to and including Norway's economic zone, provides that

> 7. No one may establish, expand or move aquaculture establishments without the approval of the Ministry [of Agriculture]. [in translation]

48 NORCOAST, *supra* note 22 at 5.
49 Section 1 of the Act (as amended to 1990 in translation – which is the most recent English translation). Interestingly, it provides that the King may fix the limit of application of the Act to beyond the baseline in "certain sea areas" (s. 1). Bennett, drawing on case studies, particularly those involving the development of fish farming, has traced the evolution of coastal zone planning in Norway in several articles. Specifically, he examines the impact of different planning processes used by communes and considers the relationship between institutional actors and the varying degrees of public participation in coastal zone use planning under the *Planning and Building Act*. See Bennett, *supra* note 29; Bennett, *supra* note 21; R. G. Bennett, "Challenges in Norwegian Coastal Zone Planning," (1996) 39 *GeoJournal* 153.
50 For example, Note T-1078 National Political Guidelines for Protected Watercourses; Note T-9/1985 About Coastal Zone Planning; Note T-4/1996 About Laws and Guidelines for Planning and Resource Use in the Coastal Zone, *supra* note 22 at 4.
51 For example, sections 9.3, 9.4, 10.2, 12.3 and 29.
52 For example, sections 19.2 and 20.3.
53 Section 33.3ff.

54 Bennett, *supra* note 29 at 202–203.

55 NORCOAST, *supra* note 22 at 3. For example, at the time of the author's research in Bergen, it was believed that the government was about to enter a new round of licensing for salmon farms (highly lucrative and not easily available) and would perhaps give preference to communes where a sea-use plan was in place. Although, in principle, no building is allowed within a 100-meter belt of the shore (s. 17.2 of the *Planning and Building Act*) except for some specified exceptions, local authorities have some authority to allow "dispensations" from this national planning rule. Informal comments indicate that this is a source of tension in that local populations had differing views regarding what areas needed protection. In fact, one project (Hordaland) was under way to experiment with allowing local authorities/population to draw their boundaries. The thought was that in many places the local population might well draw a line that more closely accorded with local understanding of the geography and ecology of areas and places that needed more or less protection than that afforded by a rigid application of the 100-meter rule (a rule that seemed perhaps more often honored in the breach).

56 Bennett, *supra* note 29 at 203.

57 In fact, it is understood by the author that under traditional Norwegian law, people resident on the coast – usually fishers or farmers – had a property right over the marine area adjacent to the property out to the depth at which a horse could stand at low tide (horse's height). In effect, fish farming evolved in many communities in the fjords and along the coast as "out the back door" for many who were also fishers and engaged in this as a supplemental activity: it was not necessarily an incursion into a "public space" *per se* or an activity that might compete with the traditional fishery.

58 Bennett, *supra* note 29 at 204.

59 NORCOAST, *supra* note 22 at 5.

60 "Challenges in Norwegian Coastal Zone Planning," *supra* note 49 at 154, drawing on the work of Healey, *supra* note 21.

61 *Ibid.*

62 *Ibid.*

63 Bennett, *supra* note 29 at 206.

64 Bennett, *supra* note 21.

65 This particular pattern of coastal conflict, and indeed competition for regulation of the coast and marine space, is increasing and will become more common with the proliferation of marine protected areas (MPAs), which also seek to regulate activities within marine ecosystems "zones." See, for example, the 2002 dispute, described as a dispute "between two state government departments," over the siting of a finfish farm in Sceale Bay, on the west coast of South Australia, in an area that had been proposed for a protected marine park that is also a breeding site for Australian sea lions. Concern was expressed that "PIRSA [Primary Industries South Australia] Aquaculture appears to be an active advocate for the proposal [finfish farm], rather than assessing the proposal independently." See Friends of Sceale Bay media release, "Major Conflict Looms over Proposed West Coast Fish Farm Development" 12 October 2002, *GROWfish News* ref: 455/02. Online. Available http://www.growfish.com.au.

66 Bennett, *supra* note 21 at 895ff.

67 *Ibid.* at 896–897. He notes also the contest regarding a "monopoly of knowledge" and the refusal of the Fisheries Department to agree for some time that escaped salmon were contributing to the parasite problem, despite the views of the Department of the Environment on the problem at 899.

68 NORCOAST, *supra* note 22 at 21–23.

69 This appeared to be perceived largely as a NIMBY ("not-in-my-backyard")

problem rather than general opposition to aquaculture, and appeared to relate to recreational and aesthetic concerns. The fact that many areas are already zoned for aquaculture use means that decisions on individual applications relate more to reviewing the ecological/food security suitability of a specific operation and siting within a zone. Although things may be changing, it appears that Norway also has fairly restrictive rules regarding non-resident ownership of countryside property and second homes. Informal comments suggested that those who were not full-time residents in an area were perhaps seen as having less of a "stake" or legitimacy in voicing concerns about local land-use decisions, particularly those relating to economic development decisions.

70 It was suggested in discussion that Norwegians are perhaps more accepting of the idea of a social contract and the accommodation of broader social interests, less litigious and more accepting of regulatory authority. It was noted, for example, in asking about recourse to mediation or DR in other areas of law such as divorce, custody, etc., that whereas the practice is well developed in Canada, it appeared that the norm is for the couple to settle matters and come to an arrangement between themselves, with the agreement then registered with the court: people are expected to be able to sort out these social matters themselves.

71 NORCOAST, *supra* note 22 at 11.

72 The word "story" is used to acknowledge the fact that this account of the evolution of aquaculture in Norway is told from a coastal conflict, planning and regulatory development perspective. The secondary materials, case studies and discussions used are derived from sources with a "stake" in that view: planners, geographers, legal academics and regulators. Other stories or explanations can also be advanced, each presenting equally plausible descriptions for a series of events. For example, this history of developments can also be understood in terms of the development of the science and practice of aquaculture, or perhaps also in terms of industrialization of farming and food production to keep up with population growth and international marketing. Other chapters in this book reveal other versions of the Canadian "story."

73 Informal observation on the part of a regulatory official to the effect that there has not been a major problem of conflict in the community, largely because there is "a lot of coast in Newfoundland," with major problems mainly relating to a lack of clarity regarding the jurisdiction between regulatory agencies. It is, however, notable that a study was carried out in Newfoundland regarding ways to prevent or manage conflict and opposition to aquaculture; see *Sharing Coastal Resources*, *supra* note 6 at 19, which noted:

> Newfoundland and Labrador, in many ways, is at the cutting edge of the aquaculture industry ... however, in some regions, the incidence of conflict has significantly delayed aquaculture development.

74 For example, there has been some discussion about whether the commune can designate for specific forms of aquaculture when designating aquaculture zones. Ecological issues can also be worked out as part of the planning process. Questions of residency and benefits are more likely to be worked out through the licensing system process.

75 As noted earlier, the interest in this study is to identify common themes and issues rather than to trace the course of a particular conflict. In addition, the specifics of a conflict necessarily vary from location to location, a fact that underlies an integrated management approach. In Atlantic Canada, various reports on conflicts, including the discussion in the Senate study, *supra* note 24; J. A. Percy, "Farming Fundy's Fishes: Aquaculture in the Bay of Fundy" (Autumn 1996) 7 *Fundy Issues*. Online. Available http://www.bofep.org/

aquacult.htm (accessed 18 March 2004) (an overview of conflicts in the Bay of Fundy area, primarily New Brunswick, where aquaculture has developed quite rapidly), as well as data relating to recent Nova Scotia siting conflicts (1999–2002: Northwest Cove, St. Margaret's Bay, and St Ann's Bay, Victoria County, Cape Breton, Nova Scotia), were considered. The situation in British Columbia more recently relates to the decision to lift the seven-year moratorium on salmon farming and to concerns about intertidal shellfish farming activity on some of the Gulf islands.

76 See Senate report, *supra* note 24; Newfoundland report, *supra* note 6.

77 Aquaculture began in Canada at about the same time as in Norway. However, the major government support and development of the industry on a large scale appears to have occurred later.

78 For example, as noted by the Georgia Strait Alliance, "Wrong Direction on Fish Farms," (November 2002) 8(6) *Strait Talk* at 14, reporting on the lawsuit filed by a First Nations people against the BC government for its failure to consult before issuing two fish farm licenses thought to affect First Nations' interests:

> As for the "jobs" claim, it is worth noting that world wide fish farms have expanded and that the number of people they employ has decreased. There is no reason to think it will be different in BC, especially since the same foreign-owned multinationals control the industry here (only one of the big "five" companies operating in BC is even Canadian).

79 For example, see comments describing federal and provincial agencies' views in Percy, *supra* note 75.

80 The complaints voiced in February 2003 by the Stop (application) 1169 Horse Island Fish Farm, Residents' Committee from Northwest Cove, Southwest Cove and Tilley Cove, Nova Scotia, areas where there remains a strong lobster fishery as well as mackerel and tuna industry, are typical of concerns expressed regarding siting in traditional fishing communities:

> Conclusion
> Inshore open-cage fin fish farms are not compatible with traditional fishing communities. The present system of aquaculture can do serious harm to established fisheries and the existing economy of the area.
> Featuring largely in the concerns are: the impact of escaped farm fish on wild fish stocks; the spread of sea lice; and possible water quality deterioration from fecal matter.

81 See, for example, R. Barron, "Fish Farm Campaign Kicks Off," 29 October 2002, *Nanaimo Daily News* A1, on the public education campaign about the environmental and potential health risks associated with farmed salmon, mounted on the west coast by the Coastal Alliance for Aquaculture Reform (CAAR) and the quoted views of media spokespersons that: "the government and industry have not been forthright in providing the necessary information to allow the consumer to make an informed decision about farmed salmon."

82 Stop application, *supra* note 80 at 2.

83 Aspects of the structural problem of multiple decision-makers, each with multiple objectives, provide ample fodder for conflict resolution game theory and systems modeling approaches to conflict resolution. See, for example, "Salmon Aquaculture Conflict in British Columbia" and other case studies. Online. Available http://www.eng.uwaterloo.ca/~lhamouda/interest.htm (accessed 18 March 2004), developed by various faculty members working with systems theory and systems design engineering methodologies at the University of

Waterloo, and "the graph model of conflict resolution," co-author Hipel, *supra* note 9. Online. Available http://www.systems.uwaterloo.ca/Faculty/Hipel/conflict_resolution.htm (accessed 18 March 2004).

84 The Declaration, issued with a global systems management document, Agenda 21, was produced by the 1992 United Nations Conference on Environment and Development. For the text of both, see online. Available http://www.unep.org (accessed 18 March 2004).

85 For example, the 2003 complaint referred to in note 80 is with respect to a license that had been granted. In the context, there is, of course, even more concern in British Columbia about the idea that some areas of surface water would be brought into an agricultural reserve by amending "the Right to Farm Act" (the *Farm Practices Protection Act*) to include aquaculture farms. Such an amendment would, to some degree, insulate aquaculture, along with other agricultural operations, from nuisance by-laws and other regulations that may affect operations. See Georgia Strait Alliance, *supra* note 78 at 14.

86 See, for example, the Newfoundland study, *supra* note 6, which details concerns about moorings for mussel farms being unmarked or otherwise creating problems. Of course, apprehension about the possibility of these impediments to the operation can lead to siting concerns at the outset. However, for purposes of this study they are distinguished. One cannot, however, ignore the fact that operational conflicts, although usually manifesting themselves as involving different parties than siting conflicts (i.e. citizen-to-citizen), in turn fuel resistance to the development of new sites, as they can provide evidence of predictable "problems" that will occur if a site is allowed. Thus, the collective experience is shaped by the earlier history of poor processes and decisions or difficult experiences. This historical element or "legacy," as it is sometimes called in negotiation and conflict resolution theory, is a factor that must be explicitly considered in developing a strategic approach to conflict prevention.

87 The Nova Scotia case of the St. Ann's Bay siting conflict is a good example of such concerns.

88 Newfoundland study, *supra* note 6 at 10.

89 *Ibid.* at 12.

90 For fuller discussion of the various departments involved and their roles, see Office of the Commissioner for Aquaculture Development (OCAD), *Legislative and Regulatory Review of Aquaculture in Canada*, DFO/6144 (Ottawa: Fisheries and Oceans Canada, March 2001) at 15ff.

91 Regional Aquaculture Development Advisory Committees (RADACs) can be created to review an aquaculture application. In principle, RADACs are used to "facilitate economic development while simultaneously providing information to the local residents and determining the level of public support." Online. Available http://www.gov.ns.ca/nsaf/aquaculture/radac/ (accessed 18 March 2004). The committee is usually made up of local fishers, aquaculturalists, recreational boaters, waterfront landowners, business operators and local politicians. On occasion, provincial bureaucrats will also be appointed to the committee. The committee provides recommendations to the minister on whether to accept or reject the proposal for aquaculture activity. However, these recommendations do not fetter the minister's discretion in making a final decision. The various provincial agencies that the provincial government must consult with are contained in s. 47 of the *Fisheries and Coastal Resources Act*, SNS 1996, c. 25 as amended.

Before making a decision with respect to the application, the Minister

a) shall consult with
(i) the Department of Agriculture and Marketing, the Department of

the Environment, the Department of Housing and Municipal Affairs and the Department of Natural Resources, and

(ii) any boards, agencies and commissions as may be prescribed; and

b) may refer the application to a private sector, regional aquaculture development advisory committee for comment and recommendation.

92 This was controversial in the case of the St. Ann's application. No RADAC was formed, although comments in the media suggest it should have been. However, the industry has been less enthusiastic and the government less inclined to create them, as there is concern that the RADACs are being stacked with opponents of aquaculture that are not interested in developing workable solutions.

93 LWBC file 12075-00, Volume 3, Chapter 3, Section 3.2.0400. This policy has been updated and revised in 2004 and 2005. Online. Avaliable http://lwbc. bc.ca/0llwbc/policies/policy/land/aquaculture.pdf.

94 *Ibid.* at 2–3.

95 This is rather unfortunate, since in principle many of the ideas behind sustainable development – that is, how to have economic development without destroying the environment – are useful. The problem lies less in the inherent fallacy of the notion that in the failure actually to implement any of the associated concepts such as the precautionary principle. In many cases, from a public perspective there has been a lot of "process" and discussion but decisions and impacts remain largely the same as they would have been without the concept of sustainable development. Ultimately, knowledge of the "product" affects the degree to which participants engage wholeheartedly in processes, hence the resort to the familiar measures of litigation and media and political battles.

96 OCAD, *supra* note 90 at 25.

97 SNS 1996, c. 25, s. 9. Similar procedures are listed in the GESAMP report, *supra* note 5 at 68.

98 Graph entitled "Sources of conflict in groups and types of interventions," in Moore, *supra* note 20 at 27.

99 Information drawn from the PEI "Aquaculture Leasing Policy 2000–2001." Aquaculture is regulated somewhat differently in PEI than in other provinces, in that PEI signed a memorandum of understanding with the federal government in which the federal government has responsibility for aquaculture leasing, much of which relates to shellfish farming.

100 Newfoundland Report, *supra* note 6 at 13.

101 *Ibid.* at 6.

102 See the "Aquaculture Site Finder" developed by the Aquaculture Division of the Department of Agriculture and Fisheries and the Nova Scotia Geomatics Centre of Service Nova Scotia and Municipal Relations. Online. Available http://gov.ns.ca/GeoNova/home/applications.asp (accessed 5 October 2005).

103 See, for example, the policy backgrounder N. Hynes and J. Graham, *Coastal Zone Planning in Nova Scotia*, prepared for the RCIP Rural Policy Forum, 17–19 February 2005 (Halifax: AHPRC, Coastal Communities Network and Dalhousie University, 2005). Online. Available http://www.ecologyaction.ca/coastal_issues/coastal_issues.shtm (accessed 5 October 2005).

104 It is acknowledged that this study has not at all addressed the fact that Canada is tri-coastal and also an island country, with all the differences and concerns the extremes between each raises.

6 Mariculture and Canadian maritime law

An unexplored relationship

Aldo Chircop

Introduction

Much of the burgeoning literature on aquaculture law and policy focuses on industry development, property rights, regulatory standards and environmental impact, but the relationship of mariculture to maritime law remains largely unaddressed. The layperson may not easily see the relevance of this body of law for mariculture activities. In reality, there are actual and potential connections between the two that can result in far-reaching legal and fiscal consequences.

Also known by the older term "admiralty law," maritime law governs shipping and navigation. In ancient times, this law emerged to support commercial marine transportation and maritime trade, but it eventually evolved to address all types, aspects and impacts of navigation at sea and inland waterways, including obstacles to navigation. Whether a vessel is a container ship, fishing vessel, drillship or recreational vessel, while at anchor or navigating (using own propulsion or in tow), there are commercial, safety and environmental aspects governed by maritime law. Obstacles in navigable waterways are a maritime law matter. If harm is caused to persons or property, off or on board a ship, as a result of faulty navigation, or if a ship causes environmental or resource loss, maritime law governs the claims that may be advanced and the liability and compensation that will apply. Ocean uses such as fishing, recreational boating, whale watching, laying of submarine pipelines and cables, and offshore oil and gas activities may not be shipping *qua* marine transportation, but still trigger the application of Canadian maritime law because very often there is a navigation element involved.

Similarly, the reason why mariculturalists should take an interest in maritime law is not that this body of law governs farming; unlike aquaculture, fisheries and environmental law, maritime law does not regulate the farming activity proper. Rather, maritime law is relevant where there is navigation or an impact on safe navigation, with reference both to the location of a farm and to support activities. Although this may appear as fairly confined relevance, in Canadian maritime law "shipping and navigation" matters may engage a broad range of activities and issues beyond what the layperson would normally associate with this area of law. Characterization of a marine activity as a "maritime matter" may invoke uniquely admiralty institutions, such as limitation of liability for maritime claims.

Initially occurring in sheltered inshore coastal areas, mariculture has gradually expanded to include offshore areas. Offshore cage culture has been found to have minimal environmental impact, is less likely to find objection by coastal communities, has access to cleaner and better marine areas, and can be located in areas where there is less likelihood of conflicts with other uses. Huge cages engineered with space-era materials are now constructed and may be located on the surface or submerged at various depths. Also, disused offshore oil and gas installations may be converted for use as platforms for offshore farms.[1] The technology used includes offshore installations, supply ships or boats, a diversity of cages and, occasionally, superstructures to provide workspace (including living space) for workers and anchoring systems, and floating automated feeders. Not all coastal and offshore mariculture activities fall within the ambit of maritime law, hence the need to explore the relationship between the two to determine which activities and under what conditions they do constitute maritime law matters. The question ultimately relates to where the line is to be drawn to determine the applicable body of law.

This chapter addresses this largely unexplored relationship by looking at some of the most important maritime institutions. It commences with a discussion of Canadian maritime law, the jurisdiction it generates and the criteria to apply when considering whether mariculture activities are maritime matters. Its main discussion focuses on the actual and potential relevance of several maritime law institutions. The consequences of applying maritime law to mariculture activities are also discussed. It is submitted that federal and provincial regime-building in support of mariculture should factor in maritime law applications and consequences.

Preliminary maritime considerations

Maritime law as a body of federal law

As the law governing navigation and shipping in Canada, maritime law is a federal constitutional responsibility.[2] Section 2(b) of the *Federal Court Act* defines Canadian maritime law as

> the law that was administered by the Exchequer Court of Canada on its Admiralty side by virtue of the *Admiralty Act*, chapter A-1 of the Revised Statutes of Canada, 1970, or any other statute, or that would have been so administered if that Court had, on its Admiralty side, unlimited jurisdiction in relation to maritime and admiralty matters, as that law has been altered by this or any other Act of the Parliament of Canada.[3]

This complex provision puts a historical dimension on the content of Canadian maritime law. It refers to English maritime law as it was received in Canada and administered by the Exchequer Court, and further developed by Parliament and Canadian courts. The origins of this law are civilian, but over time it received significant inputs from the common law and international maritime law. Thus, the modern sources of Canadian maritime law are: (1) federal statutes (e.g. the *Federal Court Act*, the *Admiralty Act 1891*,[4] the *Admiralty Act 1934*[5] and any other maritime law statute enacted by the Parliament of Canada, such as the *Canada Shipping Act*,[6] the *Canada Shipping Act 2001*[7] and the *Marine Liability Act*;[8] (2) case law, including the jurisprudence of English courts until 1934 and the jurisprudence of Canadian courts before 1934 and decisions since then (federal and provincial); (3) principles of civil law and the common law applied by the Federal Court; and (4) maritime law conventions to which Canada is a party (or which Canada has implemented without becoming a party), and presumably also international maritime customary law.

The exact limits of Canadian maritime law are uncertain, in that commercial, technological and environmental factors continue to influence the development of this law to encompass new issues. As the Supreme Court of Canada has had opportunity to note, Canadian maritime law is not static or frozen in time, but rather continues to evolve in the modern context of navigation and shipping.[9] Maritime law is not part of the provincial law, but rather is subject to the power of Parliament to repeal, abolish and alter. It is uniform throughout Canada irrespective of origins or location of cause of action. There is overlap between maritime law and provincial law, and different outcomes are possible depending on which body of law is applied.[10] Federal law is not "foreign law" to the provinces, but rather is an integral part of the law of each province.[11] This body of law relates to both domestic and international activities; there is thus a policy imperative in support of uniformity, both nationally and internationally.[12]

Admiralty jurisdiction

Subject matter of jurisdiction

Admiralty jurisdiction – that is, the jurisdiction of a court to administer maritime law – may not be exercised by any court, but only by the Federal Court, provincial superior courts and such other courts that are specifically so empowered.[13] The Federal Court was established for better administration of the laws of Canada, and as the successor to the Exchequer Court and earlier Vice-Admiralty Courts, it is the Admiralty Court of Canada.[14] For historical and statutory reasons, its admiralty jurisdiction is original and concurrent with that of provincial courts. Whereas the provincial superior courts possess inherent jurisdiction over maritime matters, the Federal Court does not.[15] Its jurisdiction is statutory and limited only by what may reasonably be interpreted as being maritime, thus pertaining to navigation and shipping, and within its statutory limits. The Federal Court's admiralty jurisdiction is set out in the *Federal Court Act* and several other maritime law statutes nourishing that statutory grant.

Historically, the jurisdiction of admiralty courts in common law systems was based on a combination of maritime subject matter and location of cause outside the jurisdiction of common law courts. Today in Canada, the geographical dimension is a factor, but not a necessary requirement to ground admiralty jurisdiction.[16] The Admiralty Court's jurisdiction and the administration of maritime law by any other court of competent jurisdiction is grounded *ratione materiae*, that is, by reason of subject matter. Hence, characterization of a cause of action as "maritime" will ground this jurisdiction and invoke Canadian maritime law.

In general, admiralty jurisdiction is engaged whenever a claim is advanced under or by virtue of "Canadian maritime law or any other law of Canada relating to any matter coming within the class of subject of navigation and shipping, except to the extent that jurisdiction has been otherwise specially assigned."[17] The presumption is that all navigation and shipping matters fall within the Federal Court's admiralty jurisdiction, unless jurisdiction, over specific matters is assigned to other bodies instead.[18] Without limiting the general grant of jurisdiction, the *Federal Court Act* provides a comprehensive inventory of subject matter giving rise to maritime actions, much of which is directly or potentially relevant for mariculture activities:

- title to and possession of a ship or share;
- disputes between co-owners concerning possession and use of ship;
- mortgages, hypothecs and other securities;
- damage, loss of life and personal injury caused by ship in collision or otherwise;
- damage to or loss of ship, cargo, equipment and any property during loading or unloading;
- damage to goods during carriage by ship and transit;
- loss of life or personal injury during the operation of a ship;
- damage to cargo and passengers' luggage;
- hire and charter parties;
- salvage;
- towage;
- pilotage;
- necessaries;
- construction, repair and equipping of ship;
- wages of seafarers;
- disbursements;
- general average;
- marine insurance;
- port, canal, dock and related charges.[19]

The discussion so far has assumed a scenario where there is no issue on the characterization of causes of action. A cause that falls within any aspect of the subject matter mentioned above or that arises by virtue of any other maritime law statute raises no issue. Thus, because ships are a maritime law matter, the utilization of supply vessels for mariculture is subject to several maritime causes of action, *inter alia* disputes concerning supply ship ownership, wages due to those who sail the ship and claims for damage done by the ship in a collision. However, where the vessel used is not immediately characterized as a ship, the application of maritime law may not be obvious and determination of admiralty jurisdiction will depend on one of two factors, namely the extent to which technology used in mariculture ought to be considered a vessel or ship, and the nature and degree of connection between the mariculture activity in question to navigation and shipping. The latter will concern not only the technology involved, but also the nature of the claim advanced when it touches upon other maritime aspects – for example, relating to the personnel involved. Canadian courts have developed analytical approaches attempting to establish when the threshold is crossed.

When a vessel, installation or structure is considered a ship

In addition to the utilization of vessels, potentially a wide variety of structures are employed in mariculture, including floating platforms, converted offshore installations, floating and submerged cages, and floating and automated structures, among others, and the technology can be expected to develop new structures. As seen earlier, if a structure employed for mariculture is characterized as a ship, it is clear that any claim in connection with it is within the ambit of maritime law. In Canadian maritime law, there is no one definition of ship for all purposes. The *Federal Court Act* defines "ship" as follows:

> "[s]hip" means any vessel or craft designed, used or capable of being used solely or partly for navigation, without regard to method or lack of propulsion, and includes
>
> (*a*) a ship in the process of construction from the time that it is capable of floating, and
> (*b*) a ship that has been stranded, wrecked or sunk and any part of a ship that has broken up.[20]

The legal definition of ship has a broader scope than everyday usage of the term. Thus, an unpropelled craft (an undefined generic term) that is capable of being used in part for navigation is caught by this definition. The key element is a navigation capability of sorts. Other statutes, such as the *Canada Shipping Act*, similarly emphasize navigation capability.[21] The *Canada Shipping Act 2001*, which is not yet in force, does not define "ship" and instead uses a generic definition of vessel.[22] However, even this definition emphasizes navigation capability.

Canadian courts have considered a wide range of vessels and craft to determine what ought to qualify as ship. The leading case defining "ship," in a non-exhaustive manner, is *Canada* v. *Saint John Shipbuilding & Dry Dock Co.*, wherein the criteria that were applied to determine whether a crane barge was a ship were the following:

- construction for use on water;
- capability of being moved from place to place, even if only occasionally (e.g. with towage assistance);
- cargo-carrying capability, even if only occasionally;
- people-carrying capability.[23]

Assessed collectively, the criteria indicated that the vessel, although not having its own means of propulsion, was "built to do something on water, requiring movement from place to place."[24] The crane barge was able to perform a number of maritime functions and was consequently considered a ship.[25] Generally, the courts have considered barges, including dumb barges (i.e. having no propulsion), to be ships.[26] Further, for the purposes of the *Collision Regulations*, "barge" is defined to mean " a non-self-propelled barge, scow, dredge, pile-driver, hopper, pontoon or houseboat," clearly further widening the range of vessels to which these regulations, which are designed for navigation safety, would apply.[27]

A different conclusion was reached in relation to a floating dry dock, which was considered in *R.* v. *Star Luzon (The)* with reference to its characteristics and function.[28] The vessel resembled a rectangular barge with high sidewalls and open ends. It was built in Japan and was towed to North Vancouver, where it was moored in a manner that allowed it to be raised and lowered while remaining centered. It had navigation lights when it was towed from Japan, but these were removed once it was moored. Its function was to raise ships for repair. It could be moved, but only after extensive work to remove it from its moorage. It could not be used for navigation, and when moved was akin to cargo (possibly as machinery) at best. Even so, the owners had it registered as a ship under the *Canada Shipping Act*. However, the court found that the floating dry dock was not used for navigation, despite its towage from Japan. It was probably a vessel (at least during towage), but not a ship for the purposes of the *Oil Pollution Prevention Regulations*.

There are vessels that are incapable of navigation on their own, but which are not necessarily disqualified from consideration as ships. It is important that such vessels must have been designed to be able to undertake some form of movement, even without their own means of propulsion. In *Falconbridge Nickel Mines Ltd.* v. *Chimo Shipping Ltd.*, a contract of carriage was performed in part by barges used to lighter cargo from ship to shore. The lighters were considered ships.[29]

Especially useful by way of analogy are a wide diversity of vessels, including what might be more appropriately termed installations and structures, used for offshore oil and gas drilling and production. These may not specifically be designated as ships.[30] Although not necessarily included in formal definitions of "ship," a "drilling unit" means "drillship, submersible, semi-submersible, barge, jack-up or other vessel that is used in a drilling program and is fitted with a drilling rig, and includes other facilities related to drilling and marine activities that are installed on a vessel or platform."[31] The inference is that they are vessels and have some navigational capability. Some installations (e.g. jack-up rigs, semi-submersibles and tension leg platforms) are towed in and out of a location, and their main function is performed in a stationary capacity. Some may be anchored, whereas others may have some propulsion and dynamic positioning (e.g. semi-submersibles). The principal purpose of these installations is clearly drilling and production, not navigation, but they all need to move from one place to another, and are designed to do so. Offshore installations are required to be classed, have to comply with international standards and are registered like ships. They can be mortgaged and require marine insurance.[32] The courts have considered them ships.[33] The Court of Appeal of Newfoundland in *Bow Valley Husky (Bermuda) Ltd.* v. *Saint John Shipbuilding Ltd.* considered Bow Drill III a drilling platform and a navigable vessel. The vessel was "capable of self-propulsion; even when drilling, is vulnerable to the perils of the sea; is not attached permanently to the ocean floor and, can travel world wide to drill for oil."[34] Significantly for mariculture, the Supreme Court of Canada in the same case held that "even if the rig is not a navigable vessel, the tort claim arising from the fire would still be a maritime matter since the main purpose of the Bow Drill III was activity in navigable waters."[35] The inference to be drawn here is that tort claims in a marine context, even if not directly navigation related, but occurring in navigable waters, may still come within the ambit of Canadian maritime law.[36]

A final consideration here is what is included in the ship. The term "vessel" includes the appurtenances, tackle, apparel, furniture, engines, and boilers,[37] as well as the wreck thereof.[38] The notion of appurtenances is an inclusive concept and continues to evolve to include other items and situations,[39] but does not include bunkers.[40] As we shall see, the range of equipment considered to be part of the ship has far-reaching significance.

In conclusion, it is likely that the broad definitions of ship and vessel for maritime law purposes will encompass vessels used for mariculture purposes. Insofar as characterization of a mariculture vessel or structure as "ship" is concerned, although function may be considered, it is not necessarily determinative. What is determinative is whether the structure concerned has a connection to navigation and/or navigable waters.

Integral connection to maritime matters

When an activity is not immediately apparent as maritime, what are the criteria applicable to determine whether it falls within the class of subject "navigation and shipping"? The question as to when a particular activity in the marine environment constitutes a maritime matter has been considered by the Supreme Court of Canada in a variety of situations, but not in a mariculture setting as at the time of writing. Accordingly, in order to seek guidance on the question as to when and under what conditions a mariculture activity or issue would be subject to maritime law, pertinent Supreme Court case law needs to be assessed with a view to drawing appropriate analogies.

In *Buenos Aires Maru*, which concerned theft from a warehouse, the Supreme Court of Canada considered whether the subject matter of the case at bar was sufficiently "integrally connected to maritime matters as to be legitimate Canadian maritime law within federal legislative competence"[41] and thus avoid what might be in pith and substance a provincial matter. "Integral connection" was identified on the basis of three facts relating to the incidental storage of the goods as part of the contract of carriage:

- proximity of the warehouse terminal operation to the sea in the port of Montreal;
- the connection between the terminal operator's activities in the port area and the contract of carriage for the stolen goods; and
- the temporary nature of the storage pending final delivery to the consignee in accordance with the carriage contract.

The maritime connection consisted of a spatial relationship between the warehousing and the maritime carriage, the function of the warehousing in the maritime contract, and the temporary nature of the warehousing to enable completion of a maritime undertaking.

A similar type of analysis was undertaken by the same court in *Monk Corp.* v. *Island Fertilizers Ltd.* concerning a claim for various expenditures and whether the contract in question was simply a sale of goods governed by provincial law or a carriage by sea contract. The court was divided on the characterization of the contract. The purchase of marine insurance, the chartering of a self-geared vessel and discharge (including demurrage and dispatch provisions) suggested a contract of maritime carriage. The claim before the court concerned the obligation to discharge and not the sale of goods proper. Again significantly for mariculture, Iacobucci J held that litigants

> can assume maritime obligations governed by maritime law even though they may not formally be parties to a charter-party or even a contract of carriage by sea. What is important for purposes of maritime law jurisdiction is that their claim be integrally connected with maritime matters.[42]

In *Shibamoto & Co.* v. *Western Fish Producers*, a case where the use of a fish processing ship, financing and fish processing at sea were concerned, the basic issue underpinning the maritime matter was the provision of funds for the acquisition of a ship and processing of salmon and salmon roe at sea. A key question was whether the supply of monies on board the ship constituted necessaries; that is, "any claim in respect of goods, materials or services wherever supplied to a ship for the *operation* or maintenance of the ship, including, without restricting the generality of the foregoing, claims in respect of stevedoring and lighterage" (emphasis added).[43] Rouleau J held:

> It might well be that the word "operation" in that paragraph does not refer only to the actual navigation of a ship over the water but to its operation generally where it has another function such as receiving delivery of fish on the high seas and processing same, even though the actual processing might well be the same as the operation carried on by a fish processing factory situated ashore.
>
> . . .
>
> The contract between the parties was for the "Nicole N" to proceed on the high seas to acquire fish in a specified fishing area and receive, process and deliver same. If one looks at the reasoning of the Supreme Court of Canada in the I.T.O. case [*Buenos Aires Maru*] . . . I think the analysis referred to at page 657 wherein the Court considered the proximity of a terminal to the operations at sea sufficient to bring it within Maritime law, I am satisfied that I am by no means exceeding the bounds of jurisdiction conferred on this Court and the issues are integrally connected to Maritime law.[44]

The connection between a cause of action arising from a mariculture setting and maritime matters could be similarly analyzed. If the cause of action relates to an activity or event that happens in navigable waters, the spatial dimension is satisfied. If the cause relates to an activity or event happening on land, but in support of mariculture, this *per se* does not render the cause not maritime. It would be necessary to examine the function. The functional criterion is met where the cause of action relates to what is essentially the performance of a function of a ship or provision of a maritime service, such as towage or carriage (including temporary storage on land). If the cause of action does not clearly relate to a maritime function, then a more probing analysis of the circumstances of the claim would be necessary. A court would need to consider whether marginal cases are more appropriately provincial "private and local" or "property and civil rights," or possibly whether the federal context of navigation and shipping has evolved further to encompass new subject matter.

Relationship between federal maritime law and provincial law

Where a mariculture issue is characterized as a maritime matter, it is federal maritime and not provincial law that will be applied to dispose of the matter. As we have seen earlier, the reason is that Canadian maritime law is a body of federal law uniform throughout the country, and consequently does not comprehend provincial statutory law.

Before *Ordon Estate* v. *Grail*, it was possible to consider the application of provincial law in a variety of maritime settings, such as marine insurance, death and personal injury at sea, and occupiers' liability.[45] However, in *Ordon* the Supreme Court of Canada developed a test to determine when a provincial statute may be applied in a maritime context. Since maritime law is by definition federal law that is uniform across the country, maritime law cannot include provincial law, because provincial law is not uniform. Thus, a court will first need to characterize the issue before it as navigation or shipping, or, more appropriately (in this case), property and civil rights.[46] On the facts, the court must be able to discern an integral connection to maritime matters in order to determine the application of maritime law, and if this is the case, it then moves to the second step of applying maritime law.[47] Where the issue is characterized as maritime and yet there is insufficient maritime law to apply, the courts are encouraged to develop non-statutory maritime law as an exercise in judicial reform.[48] This is significant because the absence of substantive maritime law does not leave a vacuum to be filled by provincial law. Finally, and assuming that the first three steps do not find applicable federal law, it is only at this stage that a constitutional analysis may be needed. Such an analysis may result in the reading down of a provincial statute where it trenches upon a core federal power.[49] Commenting on Beetz J's effort at distinguishing between the pith and substance and interjurisdictional immunity doctrines in *Bell Canada* v. *Quebec*, Professor Hogg comments as follows:

> According to this [i.e. Beetz J's] formulation, provincial laws may validly extend to federal subjects unless the laws "bear upon those subjects in what makes them specifically of federal jurisdiction." This formulation seems to involve a judicial judgment as to the severity of the impact of a provincial law on the federal subject to which the law ostensibly extends. If the provincial law would affect the "basic, minimum and unassailable" core of the federal subject, then the interjurisdictional immunity doctrine stipulates that the law must be restricted in its application (read down) to exclude the federal subject. If, on the other hand, the provincial law does not intrude heavily on the federal subject, then the pith and substance doctrine stipulates that the provincial law may validly apply to the federal subject.[50]

In earlier dicta, the Supreme Court considered the possibility that in some cases provincial law may be applied. In *Buenos Aires Maru*, it was stated that "where a case is 'in pith and substance' within the court's statutory jurisdiction, the Federal Court may apply provincial law incidentally necessary to resolve the issues presented by the parties."[51] This may be useful to consider where the provinces have aquaculture legislation and have extended the application of other provincial law (e.g. workers' compensation and occupiers' liability statutes) to this industry. The *Ordon* analysis does not eliminate the application of provincial law where a case involves a mixture of issues clearly invoking parallel federal and provincial laws. Where one body of law ceases to apply and the second begins to apply will depend on the circumstances of the case.

Planning and financing considerations

Registration and licensing

The provincial licensing of a fish farm does not automatically mean that vessels used in association with the farm are thereby registered or licensed. Under the *Canada Shipping Act*, ships over fifteen gross tons require registration.[52] In order to qualify for Canadian registration, there must be Canadian ownership of the ship or shares, and registration in Canada is prima facie evidence of ownership.[53] Ships with a smaller tonnage may still be registered at the owner's option.[54] The registration process involves various steps, including the classification of the ship. Ship classification is conducted by surveyors employed by classification societies in order to ascertain that the ship conforms to international standards for safety. Mariculture vessels of less than fifteen gross tons and equipped with a motor of 7.5 kilowatts of power or more, and not registered as ships, are covered by the licensing regime for small commercial vessels administered by the Department of Transport.[55]

Insofar as mariculture structures other than ships are concerned, there is at this time no requirement for registration or licensing in Canada. This is similarly the case with many other jurisdictions. The practice in relation to offshore oil and gas installations is clearer in this respect, where registration is required, and classification societies have a well-established practice of certifying a wide variety of structures and equipment for marine safety purposes. Mariculture, like offshore wind farming, is a relative newcomer to maritime regulation; however, at least one major ship classification society has now commenced certification of wind farms.[56]

The lack of registration requirement does not mean that marine safety standards are not applied; indeed, conditions may be imposed in the license. For example, aquaculture regulations in Norway address safety issues.[57] Interestingly, the major Norwegian classification society, Det Norske Veritas (DNV), has discontinued the certification of standards for fish farm equipment.[58] Standards seem to be addressed by the Ministry of Fisheries. In Canada, the provincial licensing process includes input from the Coast Guard, an organization responsible for maritime safety and administering much of the *Canada Shipping Act* and *Navigable Waters Protection Act* (NWPA).[59] Particularly relevant for mariculture is the NWPA review process. The Coast Guard's prior approval is needed for any work built or emplaced in, on, over, under, through or across navigable waters that interferes, or may interfere, with navigation.[60] The Coast Guard may require the provision and maintenance of lights and markers for navigation safety purposes.

Whereas shipowners are subject to a duty of seaworthiness in relation to the vessels they put to sea, it is not clear that there is a similar duty for mariculturists, other than for mariculture vessels. Other than statutory duties, it remains to be seen what the standard of care might be for mariculturists that emplace structures in navigable waters.[61]

Mortgages

Depending on whether the mariculture vessel under construction or in use is characterized as a "ship," mariculture operators may be able to benefit from maritime mortgage financing. In addition to the regular mortgage available for ships in operation, the *Canada Shipping Act* also provides the builder's mortgage, which is available for a ship to be or under construction. Both mortgages require registration and will in fact appear on the Register of Ships entry. In addition to these mortgages, mariculturalists may also benefit from mortgages of vessels under the *Bank Act*.[62] However, it appears that aquaculture mortgages may be at a disadvantage if, although registered under the *Bank Act*, they are not similarly registered in the Register of Ships. The disadvantage will arise in a competition for a limited fund, where the registered mortgages will be discharged before unregistered mortgages.[63]

The ship is a valuable asset and may be used to generate financing to support operations. For mortgaging purposes, it is important to consider what components of a ship would be considered as forming part of the vessel to determine the full value of the security. There may be an issue for mariculture equipment suppliers and the security they might hold for payment. As seen earlier, the ship includes not only the hull, but also equipment installed into it. Once installed, such equipment is considered part of the ship. Maritime law may operate to the exclusion of provincial conditional sales legislation with the consequence that title over the equipment passes with the title in the ship.[64]

Operational considerations

Safety of navigation and safety of life at sea

There are three major maritime law safety of navigation concerns for mariculture. The first concerns the emplacement of mariculture installations and structures that may become obstacles to navigation. The second is collision avoidance regulations. The third is safety of human life at sea.

Safety of navigable waters

Coastal and offshore mariculture farms come in various sizes. At their largest, they can occupy significant tracts of ocean space, but large size can be a problem for navigation safety. The controversial 1988 American Norwegian Fish Farm proposal for an offshore aquaculture license twenty-seven miles from Cape Ann, Massachusetts, failed in part for reasons of size and impact on navigation. The proponents planned a forty-seven-square mile facility with ninety floating salmon pens (ninety feet in diameter and ninety feet deep) attached to nine moored barges. In addition to objections from the fishing industry and environmentalists, there were navigational concerns, including concerns regarding the passage of submarines in the area. A major issue was whether it was in the public interest to allocate for exclusive use large areas of public waters and thereby exclude mariners.[65] The subsequent and more successful SeaStead pilot project twelve miles off Martha's Vineyard occupied nine square miles, but had to be moved by five miles to avoid overlap with an active trawl fishing ground.[66]

The *Navigable Waters Protection Act* addresses the problem of potential obstacles to navigation. As noted earlier, the Act prohibits the building or emplacement of any work "in, on, over, under or across any navigable water" without prior ministerial (Minister of Fisheries and Oceans) permission.[67] "Work" includes a variety of infrastructural works, but especially relevant for mariculture is "any structure, device or thing, whether similar in character to anything referred to in this definition or not, that may interfere with navigation."[68] The minister has discretionary power to approve the work and site, and impose construction and maintenance conditions.[69] Alterations must not interfere with navigation and also require permission.[70] The minister may order a mariculturist to refrain from proceeding with the work where, in the minister's opinion, the work interferes or would interfere with navigation. Where the work is emplaced without permission or ministerial conditions are not abided by, the minister may direct the alteration or removal of the work, and may even remove and destroy the work at the expense of the owner.[71]

Norwegian regulatory safety requirements for mariculture installations are more specific than those in Canada. Mariculture floating installations must not be anchored in such a way as to pose danger to ordinary traffic. They must be equipped with lights, and such as not to have a blinding effect on ordinary traffic. The extreme ends of the installation must carry flashing yellow lights at night, and colored buoys or poles (yellow or orange) in daytime. License numbers must be visibly displayed.[72] Although not specific to mariculture, the *Navigable Waters Works Regulations* have generic requirements that may be applied to mariculture. Any work in navigable waters will require the installation and maintenance of lights, buoys and other marks to the satisfaction of the minister.[73]

Where a mariculture vessel, or for that matter any structure, sinks in navigable waters and causes an obstruction or impediment to navigation, the owner has a duty to notify the minister and remove it.[74] Until removal, there is a continuing duty on the owner to maintain warning signals and lights. Failure to remove the wreck may result in its removal by public authorities at the expense of the owner.[75]

Collision avoidance

Mariculture vessels must be navigated with good seamanship and are subject to the *Collision Regulations*, just as is any other ship.[76] A vessel towing a structure such as a cage could be considered a vessel restricted in its ability to maneuver and will therefore enjoy the status of stand-on vessel in relation to most other vessels.[77] It has to carry prescribed signs and lights.[78]

It is less clear to what extent, if at all, the regulations apply to mariculture installations and structures, other than ships, when on site. The rules are potentially important not only for the mariculture operation, but also for all other navigation in its vicinity. Where the legislator intended that the regulations apply to exploration and exploitation vessels (i.e. in relation to non-living resources of the seabed), the regulations clearly state so. Thus, the Additional Canadian Provisions in the *Collision Regulations* provide requirements for signs, lights and sound systems to enable communication with surrounding traffic.[79] They are also required to maintain safety zones of 500 meters in all directions and fifty meters beyond the boundaries of the anchor pattern, and in some cases may have smaller or larger safety zones.[80] No traffic extraneous to the exploration and exploitation activity is allowed within the safety zones. In the realm of marine scientific research, the regulations also apply to ocean data acquisition systems (ODAS), consisting of "any object on or in the water that is designed to collect, store or transmit samples or data relating to the marine environment or the atmosphere or to the uses thereof."[81] Where special vessels or vessels undertaking particular operations or that are in a certain condition are concerned, the regulations likewise take into consideration their requirements for safe navigation.[82]

Despite the obvious safety importance, there appear to be no similar specific requirements for mariculture installations and structures in the *Collision Regulations*. For the purposes of these regulations, one would need to rely on the definition of "vessel" in the Rules, which includes "every description of water craft, including non-displacement craft and seaplanes, used or capable of being used as a means of transportation on water."[83] However, this poses a problem for floating and submerged cages, which cannot be interpreted as craft and which potentially are threatened by surface navigation in their vicinity, while at the same time posing a hazard to navigation.

Instead, various other regulations provide an incomplete patchwork of safety measures. For instance, as seen earlier, the *Navigable Waters Works Regulations* require the demonstration of lights, buoys and other marks over works in navigable waters. Under the *Boating Restriction Regulations*, an aquaculturist may be authorized "to place a sign in an area for the purpose of indicating that a restriction on the operation of vessels established by these Regulations exists in respect of that area."[84] However, these regulations are restricted to boating and designated waters only, which tend to be mostly inland waters and bays.

Should mariculture in Canada move from coastal to offshore waters, the existing collision avoidance regime ought to be reconsidered. In addition to the requirements in the NWPA, it would be advisable to consider regulatory intervention to enhance safety around mariculture operations. One possibility is new regulations under the Additional Canadian Provisions in the *Collision Regulations*. Provision in the *Collision Regulations* is important because mariners are expected to be fully familiar with those rules. A second, and possibly complementary, option is for dedicated regulation similar to offshore installations and structures engaged in the Newfoundland and Nova Scotia offshore.[85]

Safety of life at sea

For shipping generally, safety of life at sea is regulated under international standards established under the *International Convention for the Safety of Life at Sea* (SOLAS), 1974, as amended.[86] The standards concern various aspects of the equipping and operation of ships, such as construction, life-saving appliances and arrangements, and safety of navigation. SOLAS requirements do not apply to all types of ships (e.g. construction requirements for small recreational vessels not used for trading), but at the same time there are particular requirements for specific ships (e.g. fishing vessels and passenger ships). SOLAS safety standards have been applied to the offshore oil and gas industry. In addition, safety of life at sea in this industry is regulated further. For instance, the *Nova Scotia Offshore Petroleum Installations Regulations* require that offshore installations be constructed with reference to a quality assurance protocol and be classified by a designated international classification society.[87] Human safety is paramount, and therefore there are requirements for navigation equipment, communication systems, emergency evacuation procedures and life-saving equipment, among others.

It is not clear what the full safety of life at sea regulatory requirements applicable to mariculture are in situations other than when a ship is utilized. It is not even fully apparent whether the federal *Marine Occupational Safety and Health Regulations* apply to mariculture, although they are applicable to Canadian ships.[88] The regulations apply to persons working on board ships registered in Canada, persons employed in the loading and unloading of ships, and employees on government non-military ships.[89] Again, unless mariculture structures are characterized as ships, these important work safety standards at sea are probably inapplicable. In contrast, the legislator saw fit to regulate safety and health matters in the oil and gas industry through dedicated regulations.[90] This leads to a potential problem in situations of death and personal injury at sea in the mariculture industry.

Death and personal injury at sea

The problem

In *R. v. Jail Island Aquaculture Ltd.*, a judicial review case, the issue arose as to whether New Brunswick's *Occupational Health and Safety Act* was applicable to an accident occurring at sea during an aquaculture operation.[91] The application for review was from a provincial court oral decision ruling that New Brunswick had jurisdiction to try Jail Island on charges under the provincial statute. A worker died as a result of an accident on an aquaculture harvest barge during the unloading of smolt at a salmon cage site at sea. The oral decision found that aquaculture was of a local and private nature under s. 92 of the *Constitution Act 1867*.[92] Counsel for Jail Island argued that the death occurred on a ship registered under the *Canada Shipping Act*, and accordingly death at sea is a federal navigation and shipping matter as per s. 91 of the *Constitution Act 1867*.[93] Moreover, maritime law is a body of federal law, and therefore if the accident in question was a maritime matter, provincial law could not be applied.

The logic of this argument is that the Province of New Brunswick did not have constitutional authority to lay the charges under the Act in question, but would have had to lay charges under federal maritime or other law. In this case, although not considered, it is likely that the *Marine Occupational Safety and Health Regulations* would have been held applicable, as the death occurred on a ship.[94] The reviewing court did not characterize the application as one going to jurisdiction, but rather on error of law, so that a potential jurisdictional conundrum was not addressed. The court decided that the lower court had the right to be wrong until a final determination of the case, at which point there could be an appeal, but until that point it permitted continuation of the case. As a result, the extent to which federal law applies in a mariculture setting to the exclusion of provincial occupational health and safety law has not been tested post-*Ordon*, but can reasonably be expected to arise again.

There are a number of issues that arise. First, in the absence of statutory provision, what is the legal status of mariculture workers? Second, depending on their legal status, what law applies to work safety matters (federal *or* provincial, or federal *and* provincial)? Third, what law governs claims from death and personal injury at the workplace, provincial workers' compensation or federal maritime law?

Legal status of mariculture workers

Mariculture workers are not one homogeneous group of workers with a well-defined legal status. Subject to an exception in relation to those working on ships, they are neither seamen nor fishermen, and most of the marine safety requirements are for those two categories of marine workers and offshore industry workers. And yet mariculture employs a variety of professionals and laborers whose status merits classification. Inclusion within a particular class of marine occupation is important for various reasons, including certification of competence, occupational health and safety, workers' compensation, and remedies available for unpaid wages. There is first the aquaculturist, who may be the entrepreneur and license-holder (i.e. management). A second category of persons is researchers and technicians. A third category is skilled and unskilled laborers, and may include mariners.

There is no one definition of "aquaculturist" in use for all regulatory purposes. There are two federal statutes with such a definition, and the proposed definitions are inconsistent. The *Bank Act* defines this person as "tenant of an aquaculture operation,"[95] whereas the *Bankruptcy and Insolvency Act*'s definition more generously "includes the owner, occupier, landlord and tenant of an aquaculture operation."[96] Both definitions have limited application to the entrepreneur, owner or manager. Neither definition is helpful for determining rights and responsibilities beyond fiscal issues. The Nova Scotia *Fisheries and Coastal Resources Act* defines "aquaculturist" simply as a person who practices aquaculture.[97] In contrast, in shipping, the shipowner is a well-understood concept and, as a consequence, designation as an "owner" carries a complex set of rights and responsibilities in the ownership and operation of ships.

As for technicians and laborers, there is no standard term or definition for "aquaculture workers," although provincial occupational health legislation may include employees in the aquaculture industry as an eligible occupation. But even when this is the case, it needs to be considered to what extent, if at all, such workers are also subject to maritime law and in particular to labor law, standards of safety and responsibility. It must be borne in mind that the working conditions of seamen are federally regulated.[98]

Where aquaculture workers are part of the crew of a ship, maritime law is likely to consider those persons "seamen" with consequent rights and responsibilities for the worker and shipowner alike. Seamen are:

(a) *every person*, except masters, pilots and apprentices duly indentured and registered, *employed or engaged in any capacity on board any ship*, and

(b) for the purposes of the Seamen's Repatriation Convention, every person employed or engaged in any capacity on board any vessel and entered on the ship's articles,

but [do] not include pilots, cadets and pupils on training ships and naval ratings, or other persons in the permanent service of a government except when used in Part IV where [the definition] includes an apprentice to the sea service.[99] [emphasis added]

Similarly, the *Crewing Regulations* include as seafarer a person who "is employed or is to be employed in any capacity on a ship."[100] The crew are not necessarily exclusively those persons that actually sail the ship, but may also include other employees performing functions on board and for which the ship was intended.[101] Thus, Canadian courts have recognized fisherpersons,[102] musicians[103] and seamen acting as watchmen as crew members.[104] Similarly, workers on mariculture vessels may be considered members of the crew if their responsibilities on board the vessel are an integral part of the functions of the ship concerned.

Seamen operate in a dangerous environment and may be the subject of abuse by those who employ them. Over time, there have emerged international and national maritime safety and labor standards for seamen and employers alike. The implications for those working on board mariculture vessels can be far-reaching. Persons who work on board a ship are deemed to be members of the crew and are subject to the *Canada Shipping Act* and regulations. There are international training and certification standards that apply to the master, officers and crew.[105] Although the contract of crew employment may be governed by the *Canada Labour Code*, this fact *per se* does not supersede the application of the *Canada Shipping Act*.[106] Thus, the shipowner or master is always bound by the *Canada Shipping Act*. Moreover, the master and crew enjoy special prioritized privileges for unpaid wages, most especially through a maritime lien on the ship.[107] A maritime lien travels with the ship irrespective of a change of ownership, and lasts until discharge by payment or judicial sale. A seaman's wages may be garnished in limited situations.[108] There are also maritime limitation periods applicable and restrictions on which courts can hear suits for seaman's wages.[109]

If workers are only temporarily on board the vessel (e.g. for maintenance purposes) or are simply using the ship as a platform for the performance of a function not connected to the ship, they do not form part of the crew complement. However, although persons on board a ship may not be part of the crew complement and therefore not subject to the training and certification standards for seamen, they are still subjects of maritime law, whether as passengers or visitors, if they suffer death or injury on board the ship. Thus, persons injured while on board an inflatable dinghy for whale watching purposes have been covered by maritime law.[110] Not all cases of death and injury where a ship is present necessarily engage maritime law. In *Dreifelds* v. *Burton*, a chartered vessel carried scuba divers to a dive site. A diver had a gas embolism and died, but the vessel was not involved in the accident. Both trial and appeal courts held that Canadian maritime law was not engaged, and provincial tort law applied instead.[111]

Workers' or seamen's compensation?

A question arises as to whether mariculture workers are entitled to workers' compensation. Workers' compensation is a provincial field of legislation, but seamen's compensation is an area occupied by federal legislation, notably the *Merchant Seamen Compensation Act*.[112] In general, under the federal Act a seaman is entitled to claim compensation from the employer as a result of injury suffered by reason of an accident arising out of and in the course of employment.[113] The significance of the issue is not whether "mariculture seamen" are entitled to compensation, but rather whether that compensation ought to be provincial workers' or federal seamen's compensation.

Because of the relative novelty of this industry, provincial legislation has not always moved fast enough to address the entitlements of mariculture employers and workers to the benefits normally applicable in such legislation. For instance, whereas New Brunswick's *Workers Compensation Act* is drafted in a general manner such as to apply to any industry in the province,[114] the counterpart statute in Nova Scotia extends workers' compensation only to designated occupations, and aquaculture is not specifically so designated. It is possible that this might give rise to uncertainty in that province.[115]

In a post-*Ordon* scenario, a question may be raised as to the extent to which provincial workers' compensation would still apply to mariculture workers who may come within the ambit of Canadian maritime law. Workers' compensation is in pith and substance a provincial matter, and consequently such provincial legislation may apply to federal undertakings.[116] At the same time, seamen are a central maritime law matter and consequently fall within Parliament's prerogative over navigation and shipping.[117] The *Merchant Seamen Compensation Act* has a provision that might serve to avoid the need to apply the *Ordon* test insofar as the application of provincial law to the compensation of seamen is concerned. The federal Act provides that where a seaman is entitled to compensation under a provincial workers' compensation statute, that seaman is not entitled to compensation under the federal Act.[118] This provision lends support to the view that provincial workers' compensation may also apply to federal undertakings.[119]

Towage

The emplacement of mariculture structures at sea may involve on-site construction and/or towage to the emplacement area. An interesting hypothetical question is whether the contract governing the towage of a cage is a contract of towage proper or a contract of affreightment. The significance is for liability and insurance purposes. In *Burrard Towing Co.* v. *Reed Stenhouse*, Southin JA stated that a "contract to move goods from one place to another by means of a tug and barge, both supplied by the tugowner, is a contract . . . for the carriage of goods." She added that "the fact that a vessel called a tug has a tow does not mean that the undertaking upon which the tug is engaged is a contract of towage."[120] She then raised a question as to whether a contract to tow a log boom constitutes towage rather than carriage of goods.[121] Case law has held that a log boom is not a ship and could thus be argued to be cargo.[122] By analogy, if a cage is a good to be towage-delivered by the supplier to a particular site indicated by the mariculturist, there might be an argument for carriage of goods law to apply. Naturally, the intention of the parties to the contract will be key.

Salvage

Salvage is an ancient maritime institution concerning the saving of life and property at sea by volunteers or persons contracted for this purpose. The key issue to address in determining the relevance of salvage for mariculture is the extent to which mariculture property qualifies as a subject of salvage. It is conceivable that, owing to weather conditions or other misfortune, workers and mariculture property may need to be rescued. For instance, a two-person team who lived on a former oil industry platform in the Gulf of Mexico which was converted temporarily for the cultivation of red drum in cages had to be evacuated by oil personnel for their safety, and cages with fish were lost or damaged.[123] Where salvage applies to mariculture activities, the salvor is entitled to a reward for his or her efforts, but only where there is success (i.e. property has been saved, on the basis of which the reward will be determined). The basic principle is "no cure, no pay."

Historically, in English and Canadian maritime law the subject of salvage was maritime property, and this was generally understood as centering on the saving of the ship and associated property interests, which could be equipment, stores, fuel, cargo on board and the freight to be earned by the ship on completion of the voyage. If the ship suffered calamity, the wreck, flotsam and derelict were also property subject to salvage.[124] There had to be a ship and/or property associated with a ship.[125] In the United Kingdom, there were exceptions to this conception of eligible property, and, most interestingly, royal fish (fish and marine animals deemed to belong to the Crown, such as whales and sturgeon) were subject to salvage.[126] Thus, a beached whale could be salved and returned to the Crown.[127]

US maritime law seems to include a wider scope of salvage subject matter. In addition to vessels and their cargo, the term "maritime property" includes all kinds of objects, such as money on a floating corpse and floating logs.[128] The property may be located onshore or at sea, but does not need to be maritime in nature.[129] According to Schoenbaum,

> [T]he limit as to property that can properly give rise to a salvage award should be properly understood as jurisdictional in nature: whatever of value found in or upon navigable waters that is properly within admiralty jurisdiction is subject to the law of salvage.[130]

The *International Salvage Convention*, 1989, to which Canada is a party and which is implemented through the *Canada Shipping Act*,[131] broadened the subject of salvage by adopting a wider definition of vessel to include "any ship or craft, or any structure capable of navigation,"[132] and including also "any other property in danger in navigable waters or in any other waters whatsoever."[133] Property is defined as "any property not permanently and intentionally attached to the shoreline and includes freight at risk."[134] It seems that although, on the one hand, "any other property in danger" comprehends any other property other than the traditional subjects of salvage, on the other hand, the convention underscores navigation (i.e. movement) capability at the time the salvage service is rendered. Consistently, the convention further provides that it does not apply to platforms and drilling rigs when these are on location engaged in exploration, exploitation and production of seabed resources.[135]

Canadian courts have not considered the extent to which mariculture equipment is a legitimate subject of salvage. Consideration of this depends in part on the actual property that may be the subject of salvage. Clearly, ships and the equipment on board are subjects of salvage. The reference to "craft" in the convention potentially refers to a broad range of structures, but there must be a navigational capability, which at a minimum is the possibility of being towed. It is conceivable that the aquatic product may be considered cargo for salvage purposes when salvage is rendered to a mariculture vessel in distress. However, the saving of fish in a floating cage would be more difficult to encompass as a subject of salvage, unless the cage can somehow be characterized as a craft, or possibly as a platform.[136] Given that, as indicated earlier, one potential use of an offshore installation decommissioned from oil and gas activity is mariculture, it is conceivable that salvage services to it might be eligible subject matter.

Vessel-source pollution

Maritime law has a well-developed regime for compensation of oil spill victims, in both Canadian maritime law and international law. Unlike in most of the other areas discussed so far, there is no argument that mariculturists benefit from this regime. Although the Canadian mariculture industry has been spared loss from oil spill damage, the mariculture industry in many other countries has not, and perhaps it is only a matter of time before a similar loss may be experienced in Canada. There have been numerous incidents of losses resulting in claims in Denmark,[137] France,[138] Greece,[139] Japan,[140] South Korea,[141] Singapore,[142] Spain,[143] the United Kingdom[144] and the United States,[145] among others. In the case of the *Braer* casualty in Scotland, 84,700 tonnes of crude and 1,600 tonnes of bulk fuel oil were spilled and a 400-square mile area was closed to fishing and aquaculture. As many as eighteen farms in the exclusion zone and the salmon intakes at each farm had to be disposed of. Salmon farmers outside the exclusion zone also suffered loss of income as a result of reduced prices for Shetland salmon.[146] Even worse, in the aftermath of the *Erika* casualty, French mariculture and oyster farmers submitted 989 claims for compensation.[147]

The closest that Canada has been to a similar threat to mariculture was in 2000, when the *Keta V*, a Canadian-registered tug, grounded and eventually sank in Liverpool Harbour, Nova Scotia, with approximately 27,000 liters of diesel fuel on board. The wreck was located only a half-mile from a large salmon farm, which had to be boomed. After several attempts, all vessel parts that may have been contaminated with oil were removed in January 2001. Fortunately, the only costs that arose were those incurred by the Canadian Coast Guard, whose claim was settled by the Ship-Source Oil Pollution Fund (SSOPF) in November 2001.[148]

Where oil spills have caused harm, such harm has produced various losses, including mortality of fish and other organisms (e.g. contact toxicity), tainting of flesh, public health risks, loss of access to mariculture sites, fouling of cages and other equipment, loss of market share, impossibility of fulfillment of contracts, and pure economic loss for those farms not suffering physical loss, but suffering economic loss as a result of public association of their product with the incident.[149] For instance, after oil was detected in the oyster beds in the case of the *New Carissa*, the Oregon Department of Agriculture closed the Coos Bay commercial oyster farms. Four of the farms concerned had US$10 million worth of young oysters seeded in the bay, and soon enough 3.5 million oysters died. The tissues of every single oyster sampled tested positive for oil from the *New Carissa*.[150]

The compensation scheme is a multi-tiered system operating under the *Marine Liability Act*, the *Convention on Civil Liability for Oil Pollution Damage* (CLC)[151] and the *International Convention on the Establishment of an International Fund for Compensation for Oil Pollution Damage* (IOPCF), to which Canada is a party.[152] The two conventions have been implemented through the *Marine Liability Act*, and the applicable Fund in Canada is IOPCF 92. The Act and conventions establish three closely related funds that cover mariculture losses. These are explained briefly and are followed by an assessment of the criteria that mariculturalists have to meet in order to be compensated for oil spill-related loss. Insofar as the threat posed by bunkers from other ships is concerned, there is a new convention establishing a compensatory regime similar to the CLC, but this instrument is not discussed, as it has not yet entered into force.[153] A similarly modeled compensation regime for pollution from hazardous and noxious substances is established in another international convention that is also not in force.[154]

CLC and IOPCF

The CLC provides a regime of strict liability for the shipowner. Unlike other forms of strict liability for public welfare offenses, strict liability in the case of vessel-source oil spill pollution does not allow a due diligence defense, and consequently the grounds for avoidance of liability are fewer.[155] At the same time, the right to damages is supplanted by the statutory compensation process. The rationale is that, irrespective of fault, the shipowner is the first line of defense and has to compensate victims of pollution damage subject to limitation of liability based on a tonnage formula. In Canada, the owner must establish a fund in the Federal Court to which claims will attach.[156] Not all shipowners can benefit from limitation: the ships must be registered in CLC states parties, the ship concerned must be a tanker (i.e. a ship designed to carry oil in bulk) and the cargo carried must be persistent oil.[157] Prior to 1992, only laden ships were covered, but since then ships on ballast voyages are also covered as long as they carry residues of an oil cargo. This is important because a ship carries several thousand tons of fuel oil (bunkers).

In those instances where the claims against the shipowner exceed the limit of liability or where the shipowner is incapable of meeting the claims, there is the possibility of supplementary compensation from the IOPCF 92.[158] Unlike the IOPCF, the Fund represents the cargo-owner's share and in fact consists of contributions made by large importers in Fund member states. The CLC rules on the nature of compensable damage apply to and are administered by the IOPCF. It is essential that the oil be from a ship.

The IOPCF was established to provide an administrative process to handle claims, in lieu of judicial recourse, and thus enable an expedited process of compensation. In many instances, the Fund has not been able to provide full compensation to claimants, and from time to time the Fund Council has set levels of compensation depending on the number of claimants and the extent of their claims.[159] For the most part, mariculture claimants have generally opted to accept compensation from the IOPCF.

Occasionally, mariculture-related claimants have had recourse to the courts, either because they saw the possibility of more compensation through litigation or because their claims were not considered admissible by the IOPCF. Central issues have tended to be causation and the extent to which pure economic loss may include relational or secondary economic loss, and the extent, if at all, to which the common law rules on causation and loss are to be read into strict liability provisions in the maritime law statute implementing the IOPCF convention.

In order for mariculturalists to be eligible for CLC and IOPCF compensation, they have to provide evidence of pollution damage. "Pollution damage" is defined as

> [l]oss or damage caused outside the ship carrying oil by contamination resulting from the escape or discharge of oil from the ship, wherever such escape or discharge may occur, and includes the costs of preventive measures and further loss or damage caused by preventive measures.[160]

Thus, eligible losses are property damage, economic loss, cleanup, preventive measures, and reinstatement of the environment where feasible. The most common mariculture claims are those concerning damage to property (cleaning costs and equipment replacement) and business interruption (e.g. additional management and feed costs). Occasionally, where product has to be destroyed and thus there is lost production, management and feed become saved costs and are calculated to determine profit loss. As indicated earlier, the IOPCF has considered many mariculture damage and loss claims, but in general it has not considered these claims differently from other claims, such as those in the fishing and tourism industries.[161] Guidelines for compensation have been developed, and the criteria include the following:

- Any expense/loss must actually have been incurred.
- Any expense must relate to measures that are deemed reasonable and justifiable.
- A claimant's expense/loss or damage is admissible only if, and to the extent that, it can be considered as caused by contamination.
- There must be a link of causation between the expense/loss or damage covered by the claim and the contamination caused by the spill.
- A claimant is entitled to compensation only if he or she has suffered a quantifiable economic loss.
- A claimant has to prove the amount of his or her loss or damage by producing appropriate documents or other evidence.[162]

The compensation provided may also include pure economic loss. The central elements to be satisfied are: (1) causation and proximity; (2) quantification; and (3) documentation. An Inter-Sessional Working Group of the IOPCF meeting in 1994 developed four criteria to determine reasonable proximity:

- close geographical proximity between the claimant's affected activity and the contamination;
- economic dependence of the claimant on the affected resource;
- the claimant's access to alternative supplies; and
- the extent to which the claimant's activity is integrally connected with the area affected by the spill.[163]

Measures incurred to prevent pure economic loss are also eligible, but again the cost of proposed measures must be reasonable, proportionate (to the loss they seek to mitigate), appropriate (promising a measure of success) and related to actual target markets (e.g. where a marketing campaign is launched to mitigate losses).[164]

In the United Kingdom, the issue of proximity was addressed in *Land-catch Ltd.* v. *International Oil Pollution Compensation Fund*, which concerned a supplier of smolt to the Shetland salmon industry that lost its market as a result of the *Braer* casualty in 1993.[165] Landcatch ran its business not in the Shetlands, where the spill occurred, but on the Scottish west coast. The rearing process took some two years, and therefore Landcatch's business was heavily dependent on advance planning and commitments to purchase its product. More than half of its business was with Shetland farmers when the spill occurred. However, because of the government exclusion zone around the Shetland spill area, farmers could not buy the usual smolt for their fish farms. In the past, Landcatch had always been able to rely on its Shetland buyers. Landcatch therefore proceeded against the Fund for compensation, and, having failed to obtain compensation, sought relief in a Scottish court, arguing that its business was integrally connected with the Shetland economy, which had suffered loss. Landcatch was not successful. The Outer House and Inner House (Second Division) of the Scotland Court of Session saw its claim as one of relational economic loss. Landcatch did not claim under the common law, although if it had done, it would in any case not have been successful. It had advanced its claim under the UK statute implementing the CLC and IOPCF conventions, which provided for strict and limited liability of the shipowner if pollution damage could be proven. It was not in the position of a person who had an existing contract disrupted, and it was not bound with the Shetland fish farming industry. The court applied general principles of causation to determine whether the claim advanced was a result of oil pollution damage, did not find a reasonable degree of proximity and instead found the connection between supplying smolt and the activities affected by the exclusion zone remote. Thus, a claim advanced on the basis that the loss would not have occurred "but for" the casualty is insufficient without the establishment of a persuasive consequential link and evidence to that effect.[166]

In assessing claims, the IOPCF is normally assisted by the International Tanker Owners Pollution Federation (ITOPF), an industry organization possessing extensive expertise in oil spill response and damage assessment.[167] ITOPF seeks to provide objective advice to the IOPCF on the claims before it. On the basis of IOPCF guidelines, ITOPF undertakes a detailed analysis and assessment of the technical merits of claims for compensation, reasonable claims for response and cleanup costs, and claims for damage to economic resources. ITOPF is normally present on-site shortly after a spill has occurred. This is important because what ITOPF considers "reasonable" measures in the circumstances may mean what is "compensable" by the IOPCF.[168] That measures may be ordered or taken by a government agency *per se* does not make those measures reasonable. Measures taken in order to be seen to be responding are not necessarily compensable. Because water quality is critical to avoid tainting, mariculturists face further difficulties when attempting to relocate cages with product to other areas. The towage (if at all possible) can cause damage to the product by way of abrasions, and forecasting of spill directions may not be accurate, so that the spill may still reach the relocated farm.[169] In any case, there must be likelihood that the measures will be reasonably successful in minimizing environmental harm and damage to economic resources.[170]

The actual claims must satisfy a number of requirements. It has often happened that in the aftermath of a spill, not all claims are legitimate.[171] Accordingly, the compensation regime that has evolved requires transparency and accountability on the part of claimants, and claimants must fulfill key requirements:

- adhere to published international guidelines on the admissibility of various classes of claims;

- keep good records, and follow international advice on claims presentation and the provision of supporting evidence;
- do not submit claims that are speculative in nature or are inflated beyond their true value; and
- cooperate and share information with those who will pay the compensation (i.e. the shipowner's third-party liability (P&I) insurers and, if relevant, the 1992 Fund), as well as with those who are working on their behalf, especially during the actual spill.[172]

Ship-Source Oil Pollution Fund (SSOPF)

Formerly operating under the *Canada Shipping Act*, the SSOPF now functions within the framework of the *Marine Liability Act*. It was established in 1973 as the Marine Pollution Claims Fund on the basis of a levy imposed at the time on every barrel of imported oil. In 1989, the SSOPF became the successor fund and is today responsible for compensating claims for oil pollution damage or anticipated damage in Canada, Canadian waters and the exclusive economic zone (EEZ).[173] It also operates on the principle of strict liability, when the polluter is known.

Unlike the IOPCF, the SSOPF is a fund of last and first resort. It serves as a third tier of compensation after the shipowner, its insurer and IOPCF, and is, in fact, joined as defendant with the others.[174] Importantly, and unlike the IOPCF, the SSOPF provides compensation for damage from mystery spills.[175] The compensation may include costs incurred for preventive measures in respect of actual or anticipated pollution.[176] The SSOPF enables claimants to proceed against it first, subject to a right of subrogation on compensation of claims. Subrogation entitles the SSOPF to eventually proceed against the shipowner and IOPCF, if necessary. Like the IOPCF, SSOPF operates on an administrative basis, and the administrator is empowered to consider, investigate and assess claims, and make offers of compensation to claimants.[177] Claimants may choose to accept the offer or appeal it to the Federal Court.[178]

Claims from mariculturists for loss of income are compensable claims when those claims are not recoverable otherwise under the *Marine Liability Act*.[179] Mariculture claimants are persons who derive income from "the production, breeding, holding or rearing of fish, or from the culture or harvesting of marine plants."[180] It is not necessary that the loss has actually been suffered, so long as the claimant believes that he or she will suffer loss. However, the claimant must be a national or permanent resident, or be incorporated federally or provincially, and have been carrying out a lawful activity (e.g. licensed).[181] The SSOPF administrator considers the claim and may settle it with the claimant. If the claim is not settled, the administrator communicates the claim to the Minister of Transport, who in turn will appoint an assessor for the claim.

The assessor, who has powers of a commissioner, may receive evidence given by the SSOPF administrator or the claimant, and determine whether it would be admissible before a court. The assessor has to determine whether the following requirements are met:

- The loss alleged by the claimant has been established.
- The loss resulted from the discharge of oil from a ship.
- The loss is not recoverable otherwise under this part.[182]

The assessor then sets the amount of the loss, and the SSOPF administrator is directed by the minister to pay the loss. In essence, the claimant has to meet an evidentiary requirement. However, if the loss resulted from a mystery spill, the claim is still eligible. If a mariculture claimant opts to proceed against the shipowner or insurer directly, he or she is still required to join the SSOPF administrator in the suit.[183]

Procedural matters

Like all other ships, mariculture ships possess a special legal status as maritime property, entailing far-reaching procedural consequences for mariculturists and creditors alike. The ship can be mortgaged and earn liabilities secured by maritime and other liens. A ship at fault in a collision automatically earns a maritime lien on the date of the event, and, unless discharged, this lien travels with the ship irrespective of a change in ownership.[184] Similarly, the salvor's claims for assistance to the ship[185] and the unpaid wages of the master and crew generate maritime liens.[186] In addition to maritime liens, the ship can also earn other liabilities secured by statutory rights *in rem*, some of which also survive a transfer of ownership.[187] Other than discharge of these liabilities by the owner, it is only a judicial sale of the ship that can launder the claims against it, after which the ship may start to earn new liabilities again.[188]

Unless otherwise provided by statute, creditors may not be able to proceed against maritime property in the same way as other property. For instance, in a situation of parallel proceedings in bankruptcy and an admiralty action over the ship, the admiralty action is not necessarily stayed.[189] In Canada, claims against the ship are pursued by means of an action *in rem* (the ship being the *res*) and, with few exceptions, only where there is also the shipowner's liability *in personam*.[190] Hence, the action *in rem* is then accompanied by an action *in personam*.

Also significant is the court within which a claimant institutes proceedings. The Federal Court is normally the court within which the action *in rem* is instituted. The action is commenced with the "arrest" of the ship. It is possible for a claimant to institute an action *in rem* in those provinces whose rules of court allow for such a procedure, as in the case of British Columbia.[191] The action *in personam* against the owner can otherwise be instituted in any court of competent jurisdiction.

One of the most important consequences of proceeding against a ship is that its owner, charterer, manager, operator and insurer are entitled to limitation of liability.[192] Limitation of liability is based on a tonnage formula and might well result in a claimant (whether for property damage or death and personal injury scenario) not being able to recover fully from the shipowner. In order to benefit from limitation, the shipowner will need to constitute a limitation fund in the Federal Court up to the amount of the claims or, if these exceed the limitation amount, up to the limitation the owner is entitled to.[193] All claims will then attach against the ship, and claimants cannot proceed against any other assets of the shipowner.

Conclusion

It is likely that Canada will need to legislate a national framework to complement provincial aquaculture law in order to ensure that there is a functional and appropriate legal regime for activities in federal waters. Mariculture poses jurisdictional, regulatory and other legal challenges similar to those posed by the offshore oil and gas industry. There are lessons to be learned on how an appropriate regime can be developed to facilitate orderly development. In the case of offshore hydrocarbons, Parliament legislated dedicated statutes in cooperation with interested provinces, and there is no reason why this could not serve as a policy precedent for mariculture. After all, the same arguments of proximity and regional economic development are equally valid for mariculture.

In addition, the *Oceans Act* provides the Governor-in-Council with the power to extend the application of any federal and provincial law to Canada's maritime zones.[194] The exercise of this power in relation not only to provincial aquaculture statutes, but also to other pertinent statutes, such as occupational health and safety and workers' compensation, would go a long way to address some of the gaps, ambiguities and concerns raised in this chapter.

However, even in a scenario where Canada opts to develop and adopt a national aquaculture statute as federal law or to extend the application of provincial legislation to offshore mariculture, maritime law will remain relevant, because this body of law will continue to apply to navigation concerns in navigable waters. A discussion of maritime law alone shows that there are legal issues of concern to mariculturists and legislators alike. For example, the legal status of mariculture installations, applicable safety standards and legal status of mariculture workers at sea are significant concerns.

Canadian maritime law, perhaps more than the maritime law of any other Commonwealth jurisdiction, has evolved to respond to a broad range of navigation and shipping concerns. It certainly has not been frozen in time or remained limited to shipping, and has shown a capacity to respond to the needs of new ocean uses and technologies as necessary. Thus, Canadian maritime law has responded to the fishing, cruise ship, and oil and gas industries, as well as a broad range of recreational uses (e.g. boating), so it can be expected to address the concerns raised by newer ocean uses in a Canadian context, such as mariculture and wind farming. What legislative intervention could also do is to better define where one body of law ceases and another commences to apply.

It is thus important for those in the mariculture industry to consider the relevance of maritime law and how it might transform their rights and responsibilities, and its procedural implications. At the same time, and in the interests of an integrated approach to mariculture regime building, federal and provincial bodies responsible for the promotion of this industry need to consider the complex interactions among different ocean uses and the bodies of law that govern them.

Notes

The author is grateful for AquaNet research assistance for this chapter. The research assistance of Carolyn Boyd, LLB, is gratefully acknowledged. Emma Butt, a Dalhousie Law School student, assisted with footnoting. The author is also grateful for the assistance received from David Moulder and Susan Wangechi-Eklöw (World Maritime University Library, Malmö, Sweden), John Liljedahl (WMU, Malmö, Sweden), Joe Nichols (International Oil Pollution Compensation Fund, London, UK), Tor Ystgaard (Det Norske Veritas, Norway) and John Bainbridge (International Transport Workers Federation, London, UK). The support of the Dalhousie Law Library and the Public Archives of Nova Scotia is also gratefully acknowledged.

1 See "The Grace Mariculture Project." Online. Available http://www.lib. noaa.gov/docaqua/presentations/hubbsseaworld_files/TextMostly/Slide10.html (accessed 11 March 2004). See also the discussion of the SeaFish Mariculture project in the Gulf of Mexico, where a disused oil industry platform was used temporarily for mariculture purposes, in Biliana Cicin-Sain, Susan M. Bunsick, Rick DeVoe, Tim Eichenberg, John Ewart, Harlyn Halvorson, Robert W. Knecht and Robert Rheault, *Development of a Policy Framework for Offshore Marine Aquaculture in the 3–200 Mile U.S. Ocean Zone* (Delaware: Center for the Study of Marine Policy, University of Delaware, 2001) at 53.
2 *Constitution Act 1867* (U.K.), 30 & 31 Vic., c. 3, s. 91(10), reprinted in R.S.C. 1985, App. II, No. 5 [*Constitution Act, 1867*].
3 *Federal Court Act*, R.S.C. 1985, F-7, as amended by S.C. 1990, c. 8.
4 *Admiralty Act 1891*, S.C. 1891, c. 29.
5 *Admiralty Act 1934*, S.C. 1934, c. 31.
6 *Canada Shipping Act*, R.S.C. 1985, c. S-9.
7 *Canada Shipping Act*, 2001, S.C. 2001, c. 26.
8 *Marine Liability Act*, S.C. 2001, c. 6.
9 *ITO International Terminal Operators Ltd.* v. *Miida Electronics Inc.* [1986] 1 S.C.R. 752 at 779 [*Buenos Aires Maru*].
10 *Associated Metals and Minerals Corp.* v. *Evie W (The)* [1978] 2 F.C. 710; *Triglav* v. *Terrasses Jewellers Inc.* [1983] 1 S.C.R. 283 at para. 7; *Buenos Aires Maru, supra* note 9; *Q.N.S. Paper Co.* v. *Chartwell Ltd.* [1989] 2 S.C.R. 683 at 725.
11 *Buenos Aires Maru, ibid.* at 753.
12 *Ordon Estate* v. *Grail* (1996) 30 O.R. (3d) 643, aff'd [1998] 3 S.C.R. 437.
13 *Ontario (AG)* v. *Pembina Exploration Canada* [1989] 1 S.C.R. 206.
14 *Constitution Act*, 1867, *supra* note 2, s. 101; *Federal Court Act*, *supra* note 3, s. 3.
15 *Ordon Estate* v. *Grail*, *supra* note 12 at para. 46.

16 *Federal Court Act, supra* note 3, s. 22(3).
17 *Ibid.* at s. 22(1).
18 For example, the *Merchant Seamen Compensation Act*, S.C., c. M-6, ss. 3, 15, 19 establishes a board with exclusive jurisdiction to examine, hear and determine all matters under the act, and with final and conclusive decision-making.
19 *Federal Court Act, supra* note 3, s. 22(2).
20 *Ibid.* at s. 2(1), as amended by *Marine Liability Act*, S.C. 2001, c. 6, s. 115.
21 "'Ship,' except in Parts II, XV and XVI, includes (*a*) any description of vessel used in navigation and not propelled by oars, and (*b*) for the purpose of Part I and sections 574 to 581, any description of lighter, barge or like vessel used in navigation in Canada however propelled"; in turn, "'vessel' includes any ship or boat or any other description of vessel used or designed to be used in navigation." *Canada Shipping Act, supra* note 6, s. 2.
22 "'Vessel' means a boat, ship or craft designed, used or capable of being used solely or partly for navigation in, on, through or immediately above water, without regard to method or lack of propulsion, and includes such a vessel that is under construction. It does not include a floating object of a prescribed class." *Canada Shipping Act*, 2001, *supra* note 7, s. 2.
23 *Canada* v. *St. John Shipbuilding and Dry Dock Co.* (1981) 43 N.R. 15 (F.C.A.). See also *Cyber Sea Technologies, Inc.* v. *Underwater Harvester Remotel* [2003] 1 F.C. 569 at 579.
24 *Canada* v. *St. John Shipbuilding and Dry Dock Co., supra* note 23.
25 In this case, the Federal Court of Appeal seems to have decided differently from similar English case law, e.g. *Merchant's Marine Insurance Ltd.* v. *North of England Protection and Indemnity Association* (1926) 25 Ll. L. R. 446, which concerned a crane placed on a pontoon that could be moved from place to place. The English court held that the movement was the exception, rather than the rule. This decision was upheld on appeal, with the Court of Appeal further emphasizing function.
26 *R.* v. *Gulf of Aladdin* (1977) 34 C.C.C. (2d) 460, (1978) 2 W.W.R. 472; *Clark* v. *Kona Winds Yacht Charters Ltd.* (1990) 34 F.T.R. 211 (T.D.), aff'd 12 June 1995, F.C.A.
27 *Collision Regulations*, C.R.C., c. 1416, s. 2(1).
28 *R.* v. *Star Luzon (The)* [1984] 1 W.W.R. 527 (B.C.S.C.).
29 *R.* v. *Gulf of Aladdin, supra* note 26; *Falconbridge Nickel Mines Ltd.* v. *Chimo Shipping Ltd.* [1969] 2 Ex. C.R. 261, aff'd (1973) 37 D.L.R. (3d) 545 (S.C.C.).
30 *Canada Oil and Gas Installations Regulations* define diving installation, drilling installation, production installation and accommodation installation, their dependent systems (e.g. diving system, personnel accommodation), floating platform, mobile offshore platform, and offshore loading system. None of these is included in any definition of ship, but the inference might be drawn that at least some ought to be considered so. S.O.R./96-118, s. 2(1).
31 *Ibid.*
32 W. W. Spicer, *Canadian Maritime Law and the Offshore* (Calgary: Canadian Institute of Resources Law, 1984) at 16.
33 *Seafarers' International Union of Canada* v. *Crosbie Offshore Services Ltd*, [1982] 2 F.C. 855, (1982) 135 D.L.R. (3d) 485. A contrary view is *Dome Petroleum* v. *Hunt International Petroleum Co.* [1978] 1 F.C. 11, which stands as a unique decision and has been criticized. In this regard, see Spicer, *ibid.* at 17.
34 *Bow Valley Husky (Bermuda) Ltd.* v. *Saint John Shipbuilding Ltd.* (1995) 126 D.L.R. (4th) 1.
35 *Bow Valley Husky (Bermuda) Ltd.* v. *Saint John Shipbuilding Ltd.* [1997] 3 S.C.R. 1210.
36 However, see *Dreifelds* v. *Burton* [1998] O.J. No. 946 (Ont. C.A.) (QL)

[*Dreifelds*], to be discussed later, concerning a scuba diving accident. The Ontario Court of Appeal held that not every tortious activity in inland waters is necessarily a maritime law matter.

37 *Dundee (The)* (1824) 1 Hag. 109, 166 E.R. 39.

38 *Aline (The)* (1840) 1 W. Robb. 111, 166 E.R. 514.

39 See J. D. Buchan, *Mortgages of Ships: Marine Security in Canada* (Toronto: Butterworths, 1986) at 33; S. F. Friedell, ed., *Benedict on Admiralty*, 7th ed. (San Francisco: Lexis Nexus, 2002), v. 1 at § 167 and v. 2 at § 32.

40 *Fraser Shipyard & Industrial Centre Ltd.* v. *The Atlantis Two* (1999) 170 F.T.R., varied in part by No. T-111–98 (F.C.T.D.).

41 *Buenos Aires Maru, supra* note 9. Of more general relevance to the admiralty jurisdiction of the Federal Court, the court also set out a three-part test to determine that jurisdiction, namely: (1) there must be a statutory grant of jurisdiction by the federal Parliament; (2) there must be an existing body of federal law which is essential to the disposition of the case and which nourishes the statutory grant of jurisdiction; and (3) the law must be a "law of Canada" as the phrase is used in s. 101 *Constitution Act 1867, supra* note 2.

42 *Monk Corp.* v. *Island Fertilizers Ltd.* [1991] 1 S.C.R. 779 at 781. Dissenting, L'Heureux-Dubé J saw the contract as a sale of goods matter. At 815, she stated that in distinction "to the generally strict construction of Federal Court jurisdiction, this Court has, in the area of Federal Court jurisdiction over maritime law, pursued an expansive method of interpretation."

43 *Federal Court Act, supra* note 3, s. 22(2)(m).

44 *Shibamoto & Co.* v. *Western Fish Producers*, upheld by 29 F.T.R. 311 (F.C.T.D (1989), 63 D.L.R. (4th) 549 (F.C.A.).

45 *Ordon Estate* v. *Grail, supra* note 12.

46 *Ibid.* at para. 73.

47 *Ibid.* at para. 74.

48 *Ibid.* at para. 76.

49 *Ibid.* at paras. 80ff.

50 Peter W. Hogg, *Constitutional Law of Canada*, 2001 stud. ed. (Toronto: Carswell, 2001) at 386–387.

51 *Buenos Aires Maru, supra* note 9 at 781.

52 *Canada Shipping Act, supra* note 6, s. 16.

53 *R.* v. *S. G. Marshall (The)* (1870) 1 P.E.I. 316. Other cases have held that registration is more than prima facie proof, but not conclusive evidence of ownership. See, for example, *Stone* v. *Rochepoint (The)* (1921) 21 Ex. C.R. 143, which followed a string of cases to this effect.

54 *Canada Shipping Act, supra* note 6, s. 17.

55 *Small Vessels Regulations*, C.R.C., c. 1487, s. 7.

56 Det Norske Veritas. Online. Available http://www.dnv.dk/windturbines (accessed 15 March 2004).

57 *Regulations Relating to Establishment, Operation and Disease-Prevention Measures at Fish Farms (Operation and Diseases Regulations)*, sections 3 and 4. Ministry of Fisheries and Agriculture, 18 December 1998.

58 Communication from Tor Ystgaard, Det Norske Veritas, dated 2 September 2003 (on file with the author).

59 *Navigable Waters Protection Act*, R.S.C. 1985, c. N-22 [NWPA].

60 *Ibid.* at s. 5. See the NWPA application guide prepared by the Canadian Coast Guard. Online. Available http://www.ccg-gcc.gc.ca/nwp-pen/Application/ApplicationGuide_e.htm (accessed 11 March 2004).

61 Older cases tended to apply a standard of reasonableness when an obstruction to navigation was created: *Crandell* v. *Mooney* (1878) 23 U.C.C.P. 212 (C.A.): an obstruction of a river by logs for several days was unreasonable; *McNeil* v.

Jones (1894) 26 N.S.R. 299 (C.A.): a vessel tied to a wharf that prevented the navigation of another vessel was unreasonable.

62 *Bank Act*, S.C. 1991, c. 46.
63 See comments by Hugh Kindred in E. Gold, A. Chircop and H. Kindred, *Maritime Law*, Essentials in Canadian Law Series (Toronto: Irwin Law, 2003), Chapter 3: "Maritime mortgages and liens."
64 *Hoover-Owens Rentschler Co.* v. *Gulf Navigation Co.* [1923] 54 O.L.R. 483. Engine cylinders and accessories became part of the ship once installed.
65 See Cicin-Sain *et al.*, *supra* note 1 at 46–50.
66 *Ibid.* at 50–53.
67 NWPA, *supra* note 59, s. 5.
68 *Ibid.* at s. 3(c).
69 *Ibid.* at s. 5.
70 *Ibid.* at s. 10.
71 *Ibid.* at s. 6.
72 *Operation and Diseases Regulations*, *supra* note 57, s. 4.
73 C.R.C., c. 1232, s. 4.
74 NWPA, *supra* note 59, s. 16.
75 *Ibid.* at ss. 16 and 17.
76 C.R.C., c. 1416. The regulations are based on the *Convention on the International Regulations for Preventing Collisions at Sea, 1972*, 20 October 1972, 1050 U.N.T.S. 16, as amended. Part F of the regulations concerns Additional Canadian Provisions. Several international rules are also subject to Canadian modifications.
77 *Collision Regulations*, *ibid.* Schedule 1, Rules 3(g) and 18.
78 *Ibid.* Schedule 1, Rules 21(d) and 24.
79 *Ibid.* Schedule 1, Rule 42.
80 *Ibid.* Schedule 1, Rule 43.
81 *Ibid.* at ss. 2(1) and 3(1).
82 *Ibid.* Schedule 1, Rules 3(d), (f) and (g), concerning vessels engaged in fishing, a vessel not under command and a vessel restricted in its ability to maneuver (e.g. laying or servicing of navigation marks, cables; dredging, surveying and underwater operations) respectively. See also Annex II to the Additional Canadian Provisions, which concern Additional Signals for Fishing Vessels Fishing in Close Proximity.
83 *Ibid.* Schedule 1, Rule 3(a).
84 C.R.C., c. 1407, s. 8.
85 *Newfoundland Offshore Petroleum Installations Regulations*, SOR/95–104; and *Nova Scotia Offshore Petroleum Installations Regulations*, SOR/95–191 [NSOPI].
86 1 November 1974, 1184 U.N.T.S. 2 (entered into force on 15 May 1980).
87 NSOPI, *supra* note 85 at s. 4.
88 SOR/87–183. These regulations are under the *Canada Labour Code*, R.S., 1985, c. L-2; R.S., 1985, c. 9 (1st Supp.), s. 1, c. 24 (3rd Supp.), s. 3; 1993, c. 42, s. 3; 1998, c. 26, s. 55; 2000, c. 20, s. 2.
89 *Ibid.* at s. 1.3.
90 See *Oil and Gas Occupational Safety and Health Regulations*, SOR/87–612.
91 [2000] N.B.J. No. 338.
92 *Constitution Act 1867*, *supra* note 2, s. 92(16).
93 *Ibid.* at s. 91(10).
94 In *Jail Island*, the ship concerned was a barge, which in maritime law is considered a ship.
95 *Bank Act*, *supra* note 62, s. 2.
96 *Bankruptcy and Insolvency Act*, R.S., 1985, c. B-3, s. 1; 1992, c. 27, s. 81.2(2).
97 S.N.S. 1996, c. 25, s. 2(b). Rather than defining "aquaculturist," the *New*

Brunswick Aquaculture Act, c. A-9.2, s. 1 refers to the lessee (aquaculture lease holder) and licensee (aquaculture license holder).

98 *R.* v. *Pacific Coyle Navigation Co.* [1949] 1 W.W.R. 937, 94 C:C:C: 113, [1949] 3 D.L.R. 157. In this case, provincial statutory holidays legislation was held inapplicable to seamen.

99 *Canada Shipping Act, supra* note 6, s. 2.

100 SOR/97–390 under the *Canada Shipping Act.*

101 The US Court of Appeals Fifth Judicial Circuit Pattern Jury Instructions require a two-part test to determine whether a person is a seaman.

> He performs the work of the vessel if and only if: 1. he was assigned permanently to a vessel or performed a substantial part of his work on a vessel; and 2. the capacity in which he was employed or the duties that he performed contributed to the function of a vessel or to the accomplishment of the vessel's mission or to the operation or maintenance of the vessel during its movement or while at anchor for the vessel's future trips. A person need not aid in the navigation of a vessel in order to qualify as a seaman.

Online. Available http://www.admiraltylawguide.com/documents/5th.html (accessed 11 March 2004).

102 *Morrissette* v. *"Maggie" (The)* (1916) 16 Ex. C.R. 494, 27 D.L.R. 464. Salmon fisherpersons, who doubled-up as sailors and fisherpersons, who spent most of their time fishing and without having sleeping quarters on board, were considered seamen. They were entitled to a maritime lien for their wages.

103 *Metaxas* v. *Galaxias (The)* [1990] 2 F.C. 400 (T.D.).

104 *Balodis* v. *The Prince George* [1985] 1 F.C. 890 (T.D.). However, the courts found otherwise where the watchmen concerned did not have a prior link to the ship. See *Jorgensen* v. *Chasina (The)* [1926] 1 W.W.R. 632; *Nicholson* v. *The Joyland* [1931] Ex. C.R. 70.

105 *International Convention on Standards of Training, Certification and Watchkeeping for Seafarers*, 7 July 1978, as amended (especially by the 1995 major revisions), 1361 U.N.T.S. 2 [STCW]. STCW is implemented in Canada through the *Crewing Regulations*, SOR/97-390, as amended. Under the auspices of STCW, see also the *Seafarers' Training, Certification and Watchkeeping Code*, 7 July 1995, A.T.S. 1997 No. 3 [STCW Code]. There are also similar standards for fishing vessel personnel: *International Convention on Standards of Training, Certification and Watchkeeping for Fishing Vessel Personnel*, 7 July 1995, 43 S.I.D.A. 148 [STCW-F 1995]. Although Canada is not a party to STCW-F, standards for fishing vessel personnel in Canada are set out in the *Crewing Regulations.*

106 *Makar* v. *"Rivtow Lion" (The)* (1982) 43 N.R. 245 (Fed. C.A.). The procedure for wage payment and deductions from the seamen's salary is not necessarily in violation of the *Canada Shipping Act* where the master and seamen enter into a contract that incorporates the terms of a collective agreement.

107 *Federal Court Act, supra* note 3, s. 22(2)(o), master's and crew's wages. *Castlegate (The)* [1893] A. C. 38 at 52; *Fugère* v. *The Duchess of York* [1924] Ex. C.R. 95.

108 *Canada Shipping Act, supra* note 6, s. 203.

109 *Ibid.* at s. 209.

110 *Efford* v. *Bondy* [1996] B.C.J. No. 171 (B.C.S.C.).

111 *Dreifelds, supra* note 36. The Court of Appeal held, "Not every tortious activity engaged in on Canada's waterways is subject to Canadian maritime law. Only if the activity sued about is sufficiently connected with navigation or shipping . . . will it fall to be resolved under Canadian maritime law."

112 R.S., c. M-11.

113 *Ibid.* at s. 8(1). Note that the definition of "seamen" under the *Merchant Seamen*

Compensation Act, ibid. at s. 2(1), is more restricted than that in the *Canada Shipping Act, supra* note 6, s. 2.

114 The Act defines "industry" as "the whole or any part of any industry, opera-tion, undertaking or employment within the scope of this Part; and in the case of any industry, operation, undertaking or employment not as a whole within the scope of this Part means any department or part of such industry, opera-tion, undertaking or employment as would, if carried on by itself, be within the scope of this Part." *Workers' Compensation Act,* R.S.N.B. 1973, c. W-13, s.1. See also ss. 2(1) and 7(1).

115 *Workers' Compensation Act,* S.N.S. 1994–95, c. 10. Whereas employer includes "any person operating a boat, vessel, ship, dredge, tug, scow or other craft usually employed or intended to be employed in an industry to which Part I applies and, with respect to the industry of fishing, the owner or operator of a boat or vessel rented, chartered or otherwise provided to a worker employed in the fishing industry and used in or in connection with an industry carried on by the employer to which Part I applies," aquaculturists, or for that matter those persons working in aquaculture, have not been included in the list of occupations subject to Part I of the act. See Annex A, *Workers' Compensation General Regulations,* N.S. Reg. 22/96, as amended up to O.I.C. 2002-539 (28 November 2002, effective 1 October 2002), N.S. Reg. 146/2002.

116 *Workmen's Compensation Board* v. *CPR* [1920] A.C. 184.

117 Seamen are covered by the *Canada Shipping Act, supra* note 6, Parts II and III and regulations under the act. Seamen's wages are a specific head of jurisdic-tion in the *Federal Court Act, supra* note 3, s. 22(o). See also *Reference re: Indus-trial Relations and Disputes Investigation Act (Canada)* [1955] S.C.R. 529.

118 *Ibid.* at s. 5(a).

119 Hogg, *supra* note 50 at 386.

120 [1996] B.C.J. No. 1000 (B.C.C.A.) [*Burrard*]. See also *British Columbia Mills, Tug & Barge Co.* v. *Kelley* [1923] 1 W.W.R. 597, 1 D.L.R. 1015, where a con-tract to tow a log raft was treated as towage.

121 *Burrard, supra* note 120 at para. 21.

122 *Paterson Timber Co.* v. *British Columbia (The)* (1913) 16 Ex. C.R. 305; *McLeod* v. *Minnesota Pulp & Paper Co.* [1955] Ex. C.R. 344.

123 Reported in Cicin-Sain *et al., supra* note 1 at 53.

124 John Reeder, ed., *Brice on Maritime Law of Salvage,* 4th ed. (London: Sweet & Maxwell, 2003) at 221 [*Brice*]. Francis D. Rose, *Kennedy and Rose: The Law of Salvage,* 6th ed. (London: Sweet & Maxwell, 2002) at 78–79 [*Kennedy and Rose*].

125 The pre-International Salvage Convention position was typified by *Wells* v. *The Owners of the Gas Float Whitton (The Gas Float Whitton No. 2)* [1897] A.C. 337, where both the Court of Appeal and the House of Lords pronounced that the gas float, which was moored in tidal waters as a navigation aid, was not a ship. Although it was an aid to navigation, it could not navigate itself and practic-ally could not be towed. Salvage was not allowed. Discussed in *Kennedy and Rose, ibid.* at 79–81.

126 *Kennedy and Rose* suggests that probably other species were also included, mainly grampuses, porpoises, dolphins, riggs, graspes and generally large and fatty fish. *Ibid.* at 152.

127 *Brice, supra* note 124 at 221. According to *Kennedy and Rose*, the procedure seems to have been as follows: "Unless action was taken by the Admiralty, the captors kept royal fish. Where action was taken, it was by process in the High Court of Admiralty for condemnation of the fish, or the oil, etc., produced from it, as a droit of Admiralty. In that event, the captors could intervene in the proceedings and claim an award of salvage. Where captors claimed, it was

the practice to award salvage, at least where their claim was not contested."
Ibid. at 152–153.

128 Thomas J. Schoenbaum, *Admiralty and Maritime Law*, 3d ed., vol. 2 (St. Paul, MN: West Group, 2001) at 361.
129 *Brice, supra* note 124 at 222.
130 Schoenbaum, *supra* note 128 at 361.
131 *Canada Shipping Act, supra* note 6, ss. 449.1ff. and Schedule V.
132 *International Convention on Salvage*, 28 April 1989, in force 14 July 1996, A.T.S. 1998 No. 2, article 1(b).
133 *Ibid.* at article 1(a).
134 *Ibid.* at article 1(c).
135 *Ibid.* at article 3.
136 An important consideration here is that aquaculture statutes, such as Nova Scotia's *Fisheries and Coastal Resources Act, supra* note 97, ss. 60 and 61, establish property rights in the cultivated aquatic animals, including within a 100-meter radius of the licensed or leased area where the fish escape.
137 For example, the *Baltic Carrier*, 2001.
138 For example, the *Erika*, 1999.
139 For example, the *Iliad*, 1993 and *Kriti Sea*, 1996.
140 For example, the *Taiko Maru*, 1993, *Toyotaka Maru*, 1994 and *Senyo Maru*, 1995.
141 For example, the *Keumdong No. 5*, 1993, *Sea Prince*, 1995, *Yeo Myung*, 1995, *Yuil No. 1*, 1995 and *Honam Sapphire*, 1995.
142 For example, the *Natuna Sea*, 2000. This incident also had an impact in Indonesian and Malaysian waters.
143 For example, the *Portfield*, 1990, *Aegean Sea*, 1992 and *Prestige*, 2002.
144 For example, the *Braer*, 1993.
145 For example, the *New Carissa*, 1999.
146 *International Oil Pollution Compensation Funds: Annual Report 2002* (London: IOPCF, 2002) at 47–49, 162–163. See also earlier IOPCF annual reports dating back to 1995 [*IOPCF Annual Report 2002*].
147 *Ibid.* 95–107 at 98. As of 2002, 807 of these claims were settled, eighty-six rejected, and the remaining claims were still under consideration.
148 *Ship-Source Oil Pollution Fund: Annual Report 2001–2002* (Ottawa: SSOPF, 2002) at 19 [*SSOPF Annual Report 2001–2002*].
149 One expert opinion holds that for "shellfish and caged fish, it is often concentration of oil in the water column which is of more concern, high concentrations present a risk of toxic effects and even low levels may taint or impart an oily flavour to the product." See Hugh D. Parker, "The Role of ITOPF in Major Oil Spill Responses," paper presented at the PAJ International Oil Spill Conference 2000, Response to Major Oil Spills and Implementation of Effective Training, 1–2 March 2000, Tokyo (on file with the author). Also, "Aquaculture operators tend to sell their produce intermittently and so the timing and duration of harvesting bans in relation to the normal farming cycle will largely determine the extent of economic loss." T. H. Moller and B. Dicks, "Fishing and Harvesting Bans in Oil Spill Response." Online. Available http://www.itopf.com/fishban.pdf (accessed 11 March 2004).
150 See *Clausen* v. *M/V New Carissa*, No. 01-35928 D.C. No. CV-00-06078-TC, US Court of Appeals for 9th Circuit, 12 August 2003.
151 *International Convention on Civil Liability for Oil Pollution Damage*, 29 November 1969, 973 U.N.T.S. 3, as amended by Protocols on 19 November 1976, 1225 U.N.T.S. 356, and 27 November 1992, U.K.T.S. 1996 No. 87. Canada is a party to the convention and protocols.
152 *International Convention on the Establishment of the International Fund for Compensation for Oil Pollution Damage*, 18 December 1971, 1110 U.N.T.S. 57, as

amended by protocols on 19 November 1976, 16 *International Legal Materials* 621, 25 May 1984, 23 *International Legal Materials* 195, 27 November 1992, A.T.S. 1996 No. 3, and 16 May 2003, LEG/CONF.14/20, 27 May 2003. The last amending protocol establishes a voluntary third tier of compensation. Canada is party to the convention and the 1992 protocol.

153 *International Convention on Civil Liability for Bunker Oil Pollution Damage*, 23 March 2001 (not yet in force), IMO LEG/CONF.12/19. Modeled on the CLC, the pollution damage covered is

> (a) loss or damage caused outside the ship by contamination resulting from the escape or discharge of bunker oil from the ship, wherever such escape or discharge may occur, provided that compensation for impairment of the environment other than loss of profit from such impairment shall be limited to costs of reasonable measures of reinstatement actually undertaken or to be undertaken; and (b) the costs of preventive measures and further loss or damage caused by preventive measures.

The convention makes insurance compulsory and enables direct action against the insurer.

154 *International Convention on Liability and Compensation for Damage in Connection with the Carriage of Hazardous and Noxious Substances by Sea* (HNS), 3 May 1996 (not yet in force), 35 *International Legal Materials* 1406.

155 *Marine Liability Act, supra* note 8, ss. 52(3), 54(2).

156 *Ibid.* at s. 58.

157 See the definition of "Convention ship," *ibid.* at s. 47.

158 *Ibid.* at s. 75. See also IOPCF Convention, *supra* note 152, article 4.

159 *IOPCF Annual Report 2002, supra* note 146 at 33ff.

160 *Marine Liability Act, supra* note 8, s. 47.

161 Communication from Joe Nichols, Deputy Director/Technical Advisor, IOPCF, 6 August 2003 (on file with author).

162 *IOPCF 1992 Claims Manual* (London: IOPCF, November 2002) at 17–18. Additional guidelines to assist experts assessing fisheries, mariculture and processing sectors are currently in preparation. Nichols, *supra* note 161.

163 Report of the Seventh Inter-Sessional Working Group of the Fund, 1994 at para. 7.2.30 (London: IOPCF, 1994).

164 José Maura, "Future Developments of IOPCF," paper presented at the PAJ Oil Spill International Symposium, Tokyo, March 2002.

165 [1997] Scot. J. No. 323 (Outer House), [1999] Scot. J. No. 303 CA155/9/95 (Inner House, Second Division).

166 See also *R. J. Tilbury & Sons (Devon) Ltd. (t/a East Devon Shellfish)* v. *Alegrete Shipping Co.* [2003] E.W.J. No. 383 (Court of Appeal), a non-mariculture claim, but addressing an issue of remoteness. After the grounding of the *Sea Empress* in 1996, a fishing ban around Wales was declared. The whelk fishery could not proceed and a whelk processor, who could not purchase Welsh whelks, lost its Korean business as a result. Its claim was held to be relational economic loss. In Canada, a claim for relational economic loss was allowed in *Canadian National Railway* v. *Norsk Pacific Steamship Co. (Jervis Crown)* [1992] 1 S.C.R. 1021, but not in *Bow Valley Husky (Bermuda) Ltd.* v. *Saint John Shipbuilding Ltd.* [1997] 3 S.C.R. 1210. Neither case concerned an oil spill. See A. M. Linden, *Canadian Tort Law*, 7th ed. (Markham, ON: Butterworths, 2001) at 440–443.

167 Parker, *supra* note 149.

168 *Ibid.* For instance, an at sea response would normally not be able to recover more than 10 percent of spilled oil. At-sea cleanup may be less costly than shore cleanup, but it still must be commensurate with the likelihood of

success, whether in retrieving spilled oil or preventing the oil from reaching the shore.

169 See Tim Wadsworth, Brian Dicks and Clément Lavigne, "The Adaptation of Mariculture Practices in Response to Spilled Oil." Online. Available http://www.itopf.com/maricult.pdf (accessed 11 March 2004).

170 *Ibid.* A difficulty pointed out by Parker is "the lack of guidelines in terms of the levels of hydrocarbon contamination that can be tolerated." Parker, *supra* note 149. This has implications as to the extent of compensable measures to be taken in relation to a fish farm.

171 In the case of such claims in the aftermath of the *Braer*, see *IOPCF Annual Report 2002, supra* note 146 at 48. See also *Shetland Seafarms Ltd.* v. *Braer Corp.* [1998] Scot. J. No. 239 (Outer House); *Shetland Farms Ltd.* v. *Assuranceforeningen Skuld* [2001] Scot. J. No. 212 (Outer House); *Assuranceforeningen Skuld* v. *International Oil Pollution Compensation Fund* [2003] Scot. J. No. 182 (Outer House).

172 Ian C. White, "Facilitating the Speedy Payment of Oil Spill Compensation Claims under the CLC and Fund Convention." Online. Available http://www.itopf.com/tianjin.PDF (accessed 11 March 2004).

173 *Marine Liability Act, supra* note 8, sections 77 and 84. See also *SSOPF Annual Report 2001–2002, supra* note 148.

174 *Marine Liability Act, supra* note 8, s. 84.

175 *Ibid.* at s. 84(e).

176 *Ibid.* at s. 85(1).

177 *Ibid.* at s. 86.

178 *Ibid.* at s. 87.

179 *Ibid.* at s. 88(3). The claim must be filed within three years of the oil discharge or having come to be known by the claimant, and within six years after the occurrence, unless the period is shortened by the Federal Court, *ibid.* at s. 88(6).

180 *Ibid.* at s. 88(2)(a).

181 *Ibid.* at s. 88(4).

182 *Ibid.* at s. 89(4).

183 *Ibid.* at s. 90.

184 *Federal Court Act, supra* note 3, s. 22(2)(d).

185 *Ibid.* at s. 22(2)(j).

186 *Ibid.* at s. 22(2)(o).

187 *Ibid.* at s. 43(3).

188 *Federal Court Rules*, S.O.R./98–106, rule 490(3).

189 *Holt Cargo Systems Inc.* v. *ABC Container Line N.V.* [2001] SCC 90; *Re: Antwerp Bulkcarriers N.V.* [2001] SCC 91.

190 On this point, see the discussion on maritime liens in Gold *et al., supra* note 63, Chapter 6.

191 B.C. Reg. 221/90, rule 55 (British Columbia Supreme Court Rules).

192 *Marine Liability Act, supra* note 3, Schedule 1, article 1(2) and (6).

193 The *Marine Liability Act* assigns the administration of a constituted limitation fund exclusively to the Federal Court, *ibid.* at s. 32. This means that although eligible persons may claim limitation of liability in a provincial court, they may not establish a fund to which the claims would attach.

194 *Oceans Act*, S.C. 1996, c. 31, ss. 20, 21, 26.

7 The taxation of aquaculture in Canada

A comparison with the taxation of agriculture and its policy implications

Faye Woodman

Introduction

In Canada, at both the federal and the provincial government levels, the tax rules applicable to agricultural producers under the *Income Tax Act*[1] and other taxing statutes often apply with relatively few modifications to the aquaculture sector. The agriculture rules differ in significant aspects from those applied to other taxpayers. They also tend to be more generous. Thus, the aquaculture sector operates under regimes of taxation in Canada that may be characterized as preferential, but may also have been developed with the needs and circumstances of agriculture, not aquaculture, in mind. This chapter will examine the rationales underlying the various special rules and their application to the aquaculture sector. The policy implications of the agriculture model when it is applied to aquaculture will be addressed.

There are, of course, many different taxes levied on aquaculture producers – whether incorporated or unincorporated – by the federal and provincial/municipal levels of Canadian government. The federal government is responsible for the imposition of an income and a capital tax on certain large corporations.[2] In addition, it imposes a value added tax, the goods and services tax (GST).[3] The provinces also levy income and capital taxes on individuals and corporations, although in some provinces the federal government may collect the tax on their behalf.[4] They impose retail sales taxes,[5] except in three Atlantic provinces in which the proceeds of a 15 percent value added tax (harmonized sales tax, HST) are shared between particular provinces and the federal government.[6] Provinces also levy property taxes.[7] These taxes may be imposed in addition to or in lieu of property taxes levied by local units of government. Finally, provincial governments exact license and leasehold fees from the operators of aquaculture concerns. These license and leasehold fees are not, strictly speaking, taxes. Taxes are compulsory levies by government that are not, at least directly, for goods or services. Nonetheless, these levies will be briefly considered in this chapter.[8] Their connection with market values is in many instances tenuous, and they are worth canvassing to derive a more complete picture of the government–aquaculture sector fiscal relationship.

It is not possible to survey the specific rules governing the taxation of aquaculture under all these regimes. The following discussion, while considering the three main categories of taxes – income, property and sales – will concentrate on income taxation. The income tax system is more complex than the others, draws more completely on the agricultural model and ultimately seems to yield more insights concerning the particular position of aquaculture. Nonetheless, the other types of taxes have significant effects on aquaculture operations, and in unprofitable years may be the only taxes to which they are subject. So too, the focus will be on the federal/provincial taxation regimes of British Columbia and the Atlantic provinces. While aquaculture operations can be found in all the provinces of Canada, in 2001 British Columbia and New Brunswick accounted for over 81 percent of the gross value of the Canadian sector.[9]

The emphasis here will be on primary producers in the marine aquaculture sector. The situation of the suppliers, processors and marketers who surround the producers will not be specifically addressed. Like their equivalents in the agricultural sector, these others do not have an unique taxation regime devoted to their special circumstances.

Finally, it must be remembered that primary producers in aquaculture, like primary producers in agriculture, are a diverse lot. They are of different sizes, they operate differently and they are connected to the market in different ways. Most producers, it is true, are incorporated. The corporations include, however, a range of operations. Many of the shellfish farms on both coasts are run through small family corporations. In the west, salmon producers are generally Canadian subsidiaries of large multinational corporations. In New Brunswick, there are locally based but substantial "independent" finfish operations. On both coasts, First Nations may be involved in aquaculture, and for reasons unrelated to the industry, but due to historical entitlements, may be exempt from some taxes. In addition, aquaculturalists may carry on their businesses differently. There are many instances of vertical integration where one entity controls production from hatcheries through to and including value added processing. Other producers may concentrate on fish-raising only but be contractually tied to other concerns. They may own the fish in the operation but be constrained by marketing and supply relationships. In some cases, the corporation may simply offer "management" services of aquaculture sites to some other body. In considering the various tax regimes, it is important to keep these differences in mind. Tax rules that lack an appropriate policy rationale when applied to taxpayers with one type of profile may be quite justified when imposed on other taxpayers with different characteristics. In this regard, the distinction between shellfish operators, which tend to be smaller, closely held corporations, and finfish concerns, which are, in many cases, large multinationals, is particularly important.

Income tax

The federal *Income Tax Act* does not specifically refer to aquaculture. Rather, the courts have held that a "farmer" includes a fish farmer. Thus, the Tax Review Board in *Les Immeubles Dramis Inc.* v. *M. N. R.*[10] held that a trout farmer was a "farmer" for the purposes of the Act. It said:

> The fact that fish are raised in the water rather than on land or in the air has nothing to do with the point at issue. In my view, there is no real distinction for income tax purposes between growing, keeping and catching marine animals – that is, fish-breeding – and performing the same activities with respect to other animals.[11]

Interestingly, in that case the Crown, not the taxpayer, argued for the designation in order to have the "hobby farming" limitation of losses provisions in the Act apply. It is not clear whether at any time the government, through the Department of Finance (the department responsible for tax policy), ever actively considered the position from a public policy perspective.

The provisions relating to the taxation of farmers in the Act, while extensive, do not in themselves constitute an altogether separate system for the taxation of agriculture.[12] The calculation of income for tax purposes is generally the same as that for other taxpayers. So too, the rate of tax, either corporate or individual, is the same as for other corporations or individuals. The main differences are when income is recognized for tax purposes (timing) and the taxation of capital gains.

These differences, or tax preferences, which are really deviations from the normative tax system, are difficult to quantify. In the 2002 tax expenditure budget, it is estimated that the CDN$500,000 capital gains exemption for all farm property will cost CDN$230,000,000 for the year.[13] Many of the other benefits extended to farmers through the tax system, such as the inter-generational rollover and cash-basis accounting, are not assigned a value since data are not available to support a meaningful estimate. Aquaculture, of course, generates only a fraction of the total Canadian agricultural revenue and it could be expected to generate only a corresponding fraction of tax expenditures.

The differences between the taxation of farmers and the taxation of other taxpayers developed over many years and, some would argue, on an *ad hoc* basis with *ad hoc* rationales. One of the earliest concessions was to amend the Act to recognize the prevailing practice of permitting farmers to account for income on a cash, not on an accrual, basis. In its 1966 report, however, the Royal Commission on Taxation (the Carter Commission) recommended that this and other preferences for farmers be abolished. In its words:

> In general, we have found that many of the special tax provisions and practices are no longer appropriate. Because of the changing nature of

the industry, farmers, or at least those with larger incomes, should now be able to report income on a basis similar to that followed by other small businessmen.[14]

As the Commission reasoned, and its remarks apply, in part, to aquaculture operations today:

> The taxation of farming income must take into consideration the special characteristics of this natural resource industry, the vagaries of nature and markets, the prevalence of small individual operators, and the close relationship of personal and business activities. On the other hand, if equity is to be achieved, the importance of these special characteristics must be considered in comparison with those encountered by taxpayers in other lines of endeavor. In making this comparison it is necessary to keep in mind the changes which have been taking place in agriculture and, in particular, the increase in the size of the farm unit, the increased technical assistance from government authorities, improved marketing arrangements, and the increased use of scientific knowledge and business methods.[15]

However, despite the Commission's recommendations, the 1972 reform legislation left many of the preferences intact. Indeed, tax reform, and in particular the introduction of the taxation of capital gains, which many felt had a particularly adverse effect on farmers, precipitated a further spate of special concessions to farmers. So too, the reformed tax system incorporated a number of income recognition and averaging provisions that were either specifically orientated to farmers or, in some aspects, modified for farmers. The policy implications for aquaculture of these provisions can be best addressed under four rationales for the adoption or continuation of special concessions to agriculture.

Administrative expediency

One of the most significant differences between the taxation of farmers, including fish farmers, and other taxpayers is that the former, unlike the latter, are specifically permitted to use the cash method of accounting in computing their (farming) incomes.[16] The cash method of accounting is a considerable advantage to a taxpayer. It permits the concern to recognize income for tax purposes when it is received and to deduct expenses (including the cost of inventory) when they are paid. A taxpayer's inventory is otherwise not recognized until it is sold. In contrast, the accrual method recognizes income when it is earned and expenses (generally) when they are incurred. The difference between the cost of the taxpayer's opening and closing inventories reduces the cost of goods sold in any particular taxation year. Since the cash method permits the calculation of taxable income based on cash flows rather than income earned and because of the treatment of inventory, it is far more susceptible to manipulation by the taxpayer for the purpose of deferring the recognition of income and therefore tax.

Until the introduction of a specific provision in the *Income Tax Act 1948* there was some uncertainty whether, under the Canadian Act, the accrual method of accounting was authorized for any taxpayers. The definition of income in the 1948 Act clarified that the accrual method was permitted (and, in fact, practically required) in most calculations of business income. Farmers and fishermen, the latter by administrative fiat, were permitted to continue using the cash method of accounting. Farmers were allowed to use the cash basis of accounting because it was thought that some, at least, of the many farmers (at that time, many more Canadians lived on farms) would find it too burdensome to deal with the more complicated accounting required under the accrual method.

The use of the cash method of accounting by farmers was subsequently reviewed by the Carter Commission, which recommended its abolition in most circumstances.[17] It said:

> The failure of the cash basis to reflect accounts receivable and payable would not materially affect the income of most farms, but its failure to take inventories into account is serious because of the substantial inventories of livestock or grain which are maintained on many farms. In such cases, the cash basis permits the cost of building up the inventories to be deducted immediately, thereby giving the farmer the advantage of a tax deferment equal to the tax which would have been exigible on an amount equal to the cost of the inventory. It is true that the advantage under the present tax system is only a deferment of tax in that the cost would ultimately be allowed as a deduction; however, the deferment is equivalent, in relative terms, to an interest-free, unsecured loan, which could be of material amount, and is not granted to business generally.[18]

The government failed to adopt the Commission's recommendation for this reform, and, indeed, formally extended the provisions to fishermen. Further, two important adjustments were introduced to address some of the difficulties inherent in the cash basis system. The provisions adopted sought both to enhance and to limit the advantage of cash-basis accounting. They are the optional inventory adjustment[19] and the mandatory inventory adjustment.[20] The former provision is intended to assist farmers who stand to "lose" loss years. Under the cash method of accounting, the costs of inventory are recognized when they are paid, but revenue is not recognized until it is sold. As a result, some farmers might generate a number of loss years and those loss years, might "expire" before they can be set off against profitable years. The optional inventory adjustment eliminates this problem by permitting a farmer to elect to recognize all or part of the value of inventory in a year. In addition, the provision operates to permit the farmer to average income. The mandatory inventory adjustment, on the other hand, was introduced to limit the ability of the farmer to generate losses and gain significant tax deferral and income averaging benefits. It requires a farmer who incurs a loss in a year to recognize the value of purchased inventory up to the amount of the loss.

The obvious complexities of the inventory adjustments undermine the justification of the cash basis method of accounting as a method to assist unsophisticated taxpayers. The cash method of accounting as modified by the inventory adjustment rules in the Act requires considerable expertise to master and to use effectively. But even if we accept that the rules are simpler than the accrual method of accounting, the profile of aquaculture concerns, which may be run by younger and more educated individuals or as part of large multinationals (though admittedly this varies from region to region and type of operation), does not suggest any compelling justification for this tax preference. It is a concession that exists because it has always existed.

It is worth noting that fishermen are permitted to use the cash method of accounting, and always have been, although this was not specifically recognized in the tax Act until 1972. Presumably they also have the requisite lack of sophistication in accounting knowledge. The inventory adjustment rules do not apply to them since, presumably, they do not carry significant inventories over time.

Interestingly, the cash method of accounting and the one-year class rules in New Brunswick have intersected in, perhaps, unexpected ways. The move to the one-year class system has meant that under the cash basis of accounting and with one site, the aquaculturalist will have no income in year 1, and all the income from that crop will typically arise in year 2.[21] This can be accommodated to a certain extent by the optional inventory adjustment.

Preservation of the "family" farm: intergenerational tax-free transfers

Under the Canadian *Income Tax Act*, farmers, including fish farmers, are permitted a more or less tax-free intergenerational transfer of the family farm.[22] In tax parlance, this transaction is called a rollover. In order to qualify for the rollover, the recipient (transferee) of the next generation must be the child or other lineal descendant of the taxpayer.[23] The rollover was initially available only for property used in unincorporated farms. With some tax planning, however, the benefits of the rollover could be obtained for shares of family farm corporations and interests in family farm partnerships. Further, the obvious unfairness to farmers who choose for good family or business reasons to carry on business in corporate or partnership form resulted in the extension of the rollover to property of an individual where the property is either (1) shares of the capital stock of a family farm corporation;[24] or (2) interests in a family farm partnership.[25] The rollovers were introduced as a result of the changes to the Canadian tax system arising from the tax reform in 1972. One of the Carter Commission's most significant recommendations was the full taxation of capital gains.[26] While the 1969 White Paper on Tax Reform eschewed the Carter approach, encapsulated in the aphorism (incorrectly attributed to Carter) that "a buck is a buck is a buck," it did propose that half, not the full amount, of capital gains be taxed.[27]

The reluctance to adopt the Carter position of full taxation reflected a number of concerns. Because capital gains are normally realized only on gift, sale or death, there is the problem of income "bunching" in one year, which in a progressive rate system can push the taxpayer up into a higher tax bracket. The taxation of capital gains also imposes a hardship on taxpayers in times of inflation. Tax will be levied on "nominal" rather than the "real" appreciation in the value of an asset. Finally, the "deemed disposition" and taxation of capital gains on death is a necessary part of the taxation of capital gains. Otherwise, accrued but unrealized capital gains that were not taxed during the taxpayer's lifetime would escape the tax net altogether. Nevertheless, the fact that all the property of the taxpayer is deemed to be disposed of can contribute to the forced sale of the deceased taxpayer's estate assets.

For farmers, even the 50 percent taxation of capital gains eventually adopted was, they asserted, an unacceptable burden. Farmers were concerned that the taxation on death of capital gains accrued on farm assets, especially land, would undermine the institution of the family farm. Instead of the farm being passed on to children and grandchildren, the farm would have to be sold to pay the tax. Hence, the intergenerational rollover was introduced. Of course, this argument applies to a lesser or greater extent to other types of businesses. And, indeed, a provision was introduced for approximately a ten-year period ending after 1987 with the enactment of the CDN$500,000 capital gains exemption, to permit a limited rollover of up to CDN$200,000 of capital gains on shares of "small business corporations" transferred by a taxpayer to a child.[28] Farm businesses, however, may have faced a heavier tax burden because of the accelerating appreciation in the value of land during that period. Furthermore, farms, especially family farms, were an integral part of the rural landscape and rural towns that governments, then and now, were ostensibly committed to preserve.

Fish farmers are an integral part of the coastal rural landscape. Whether they are more like farmers or other rural businesses, which do not enjoy the rollover, is a more difficult question. It is noteworthy that the intergenerational rollover has not been extended to fishermen. Fishermen can take advantage of the cash basis of accounting but they cannot pass on their business to their children tax free. Of course, fishermen do not own large amounts of appreciating land, and maybe that was, at first, the justification for the difference. But fishermen may own substantial depreciable properties (which might benefit from the rollover), and fishermen and family fishing businesses are mainsprings of many coastal communities.[29]

It is significant that the rollover, which is intended to preserve the family farm for the next generation, does not require the next generation to continue farming, nor does it require family farms to be "family-sized." The transferors or their children must farm before the rollover, because the rollover can only be claimed for any amount of farm property that was "used principally in a farming business in which the taxpayer, the taxpayer's spouse or common-law partner or any of the taxpayer's children was actively engaged on a regular and continuous basis."[30] The transferees do not have to be farmers, however, or, if they are, there is no penalty if they discontinue farming. Therefore, the rollover may, in some circumstances, simply be a means to facilitate the tax-free intergenerational transfer of family property that will then be sold off. It may be possible, as well, for the rollover and the CDN$500,000 capital gains exemption to operate together to multiply possible tax advantages for a family who wants to extricate themselves from farming.[31]

Thus, the intergenerational rollover, as we have seen, is not a particularly well-targeted provision. It probably extends tax advantages to the undeserving and possibly withholds them from individuals/businesses who are important contributors to the rural economy. There are a number of alternatives, most of which would include a claw-back of benefits if the operations are sold out of the family. Further, a direct expenditure program in which specific grants are made, in lieu of a tax preference, would have the further advantage of transparency and greater accountability.[32]

Finally, even if the intergenerational rollover is successful in encouraging the retention of the aquaculture operation in the family, it is arguable that more attention should be paid to the assertion that this result makes good social and economic policy. In some cases, it could be suggested, it may lock in family members whose efforts would be better directed elsewhere. Also, one effect is probably to make aquaculture sites scarcer and hence more expensive – at least in some of the more developed regions – so that new and, generally, young entrepreneurs may be priced out of the market. On the other hand, in some systems, such as the one in place in southwestern New Brunswick, the effect of the rollover may be to encourage intrafamily transfers of smaller concerns, which might preserve some of the few "independents" in the region against the continuing pressure for integration with larger concerns.

Providing funds for retirement

The introduction of the capital gains tax at tax reform galvanized farmers to press for relief for family farms. But even with the intergenerational rollover described above, farmers still perceived that they were particularly adversely affected by the taxation of capital gains. Many farmers, after all, experienced low lifetime earnings while sitting on highly appreciating or appreciated assets. Because of their low incomes, they argued, they were not able to save for retirement in an ordinary tax-subsidized savings vehicle.[33] For many of these individuals, the expected reward for a lifetime of marginal income was a retirement secured by the (untaxed) proceeds from the sale of the farm. Even a half-rate of inclusion of 50 percent of capital gains was not enough relief. Farmers lobbied for a further, special tax reduction.

Of course, other taxpayer interest groups could and did rail against the capital gains tax. Being a "new" tax, and a tax that disproportionately affected upper-income taxpayers,[34] it was almost inevitably challenged. In any case, in 1985 the government of Brian Mulroney introduced a lifetime exemption from taxation of CDN$500,000 of capital gains for all individuals.[35] The lifetime exemption was ostensibly to encourage risk investment in Canada.[36]

As originally formulated, the CDN$500,000 capital gains exemption was to be phased in over a six-year period except for dispositions of "qualified farm property." Taxpayers disposing of qualified farm property were to immediately enjoy the whole $500,000 exemption.[37] Moreover, there was no requirement that the exemption be limited to proceeds destined for use as farmers' retirement savings. Qualified farm property was initially defined to be:

- real property used by the individual, his spouse or any of his children, family farm corporation, or family farm partnership in the course of carrying on the business of farming in Canada:

   (a) in the year the property was disposed of by the individual, or
   (b) in at least five years during which the property was owned by the individual, his spouse and his children;[38]

- the share of a capital stock of a family farm corporation[39] and
- an interest in a family farm partnership.[40]

It did not take long for the deficiencies of the new lifetime $500,000 capital gains exemption to be evident. As the provision was originally drafted, taxpayers could claim the exemption against capital gains accruing on assets outside Canada, and in respect of "non-risky" assets such as real estate. In any case, after only three years the 1988 Tax Reform halted the phase-in of the lifetime exemption at $100,000 and increased the inclusion rate to two-thirds.[41] The $100,000 exemption was subsequently eliminated in 1994.[42] Significantly, however, the $500,000 capital gains exemption was preserved for farmers, although rules were introduced to attempt to limit the exemption to "real" farmers. In addition, the $500,000 exemption was extended to taxpayers who held qualified small business corporation shares.[43] These corporations have to qualify as Canadian-controlled private corporations, carry on an active business and comply with certain other conditions. Farming qualifies as an active business, so some shares of farming corporations may qualify both as shares of a family farm corporation and shares of a qualified small business corporation. Any shareholder is, however, limited to one $500,000 exemption.

The amendments in 1988 referred to above also limited the ability of farmers to claim the exemption. The amendments represented another effort, more extensive and complicated than the rollover rules, to target benefits under the tax system only to "real" farmers. Since there is not the same element of recreational or "hobby" use in fish farm operations, these rules are of less concern to aquaculturalists, but they nevertheless must be addressed if the exemption is claimed.

The rules distinguish between the disposition of farm property other than the shares of a family farm corporation and shares of a family farm corporation. Where a taxpayer disposes of farm property other than shares of the family farm corporation, and the property was owned by an eligible user, either of the two tests described below must be met for the taxpayer to claim the exemption. The gross revenue test requires that the gross revenue of the individual or other eligible users (including a spouse or children) from the farming business in which the property was principally used must exceed the income of the eligible user from all other sources for at least two years. Further, the eligible user must have been actively engaged on a regular and continuous basis in the farming business in which the property was principally used. Alternatively, the property must be used by a family farm corporation or a family farm partnership principally in the business of farming throughout a period of at least twenty-four months during which time the individual or other specified persons (including a spouse or children) was actively engaged on a regular and continuous basis in the farming business.[44]

Holders of shares in a family farm corporation are not subjected to a gross revenue test. Generally, in order for the shares to qualify as property eligible for the exemption, the following conditions apply:

- The corporation has to be in existence for at least twenty-four months.
- Throughout any twenty-four-month period ending before the disposition, more than 50 percent of the fair market value of the property owned by the corporation was used by the taxpayer, spouse, child, or parent principally in the business of farming.
- At the time of the disposition, all or substantially all of the fair market value of the property was used principally in the business of farming by the corporation or the taxpayer, spouse, child or parent.[45]

As mentioned previously, some shares may be both shares of the capital stock of a family farm corporation and qualified small business corporation shares.

The policy rationale behind the initial introduction of the lifetime capital gains exemption is, as has been explained, somewhat suspect. Further, its continuation solely for farmers and holders of qualified small business corporation shares has been criticized, most notably in the *Report of the Technical Committee on Business Taxation* (the Mintz Committee).[46] The report was the result of the efforts of a technical committee of the Department of Finance established in 1996 to review taxes paid by Canadian business. Its objectives were to suggest ways to:

- improve the tax system to promote job creation and economic growth;
- simplify the tax system to facilitate compliance; and
- enhance fairness in the system.

It recommended

> elimination of the enhanced lifetime capital gain exemption for farm property and qualifying shares of small business corporations, with transitional relief for all gains accrued to the date of the change (to be obtained by election similar to that used for the repeal of the general lifetime capital gain exemption).

> It also recommended the exemption be replaced by an enhanced RRSP [registered retirement savings plan] contribution system that would allow taxpayers to use taxable capital gains on farm property and qualifying small business shares that are earned in a year to increase their RRSP contribution room for previous years, up to the maximum room that would be available if they had had sufficient earned income.[47]

Three reasons were cited for the recommendation.[48] First, the committee did not favor taxing capital gains differently, depending on the nature of the particular asset. In the committee's view, the differential taxation of capital gains is contrary to the principle of neutrality in the business tax system. By neutrality, the committee meant that "total tax paid on income earned from different business activities is similar so the decisions of business are largely unaffected by the tax system."[49] For the committee, the pursuit of the principle of neutrality (together with internationally competitive taxes) is essential if the goals of job creation and economic growth, simplification, and fairness are to be attained.[50]

Second, the committee found little evidence that the capital gains exemption or its more limited version, the capital gains exemption for farmers and small business, has had any measurable positive impact on encouraging risk-taking and investment – ostensibly the reason for their introduction.

Finally, although the committee found some evidence to support the proposition that farmers and lower-income business owners do not benefit from tax-assisted retirement savings as much as others, it favored adjustments to that system rather than the wholesale exclusion of the capital gains of some assets from taxation. To quote the committee, "A measure such as the lifetime capital gains exemption provides too much benefit to some who do not need it and not enough to those who do."[51]

As an alternative to the CDN$500,000 exemption, the committee proposed that the capital gains arising on the disposition of qualifying farm property or a qualified small business corporation share be transferable on a tax-deductible basis to a registered retirement savings plan. The maximum amount transferable would be the lesser of CDN$375,000 (in 1996, three-quarters of capital gains were taxable) and the maximum annual registered retirement savings plan deduction multiplied by the number of years the property was held.[52]

Fluctuating incomes

Income averaging

For over fifty years, farmers have had the advantage of provisions in the *Income Tax Act* intended to ameliorate the adverse effects of fluctuating incomes under the system of progressive rates. The farm sector, which faces the vagaries of weather, the vicissitudes of natural storms, droughts, and disease and insults effected by unreliable markets both at home and abroad, seemed a particularly appropriate beneficiary. Eugene LaBrie suggested additional reasons as follows:

> Other likely factors are the primary nature of these industries, their chronically depressed state, their considerable importance both economically and numerically as voting taxpayers and the laudable independence and innate conservatism of taxpayers engaged in these industries–factors that sometimes prompt the statement that these forms of livelihood are not a source of income but a "way of life."[53]

Indeed, tax reform in 1972 and thereafter saw, despite the contrary recommendations of Carter,[54] a flurry of provisions enacted to deal with the "problem" of fluctuating income. These provisions, except for the block averaging legislation and the optional inventory adjustment, applied not only for farmers, but for other taxpayers. Since then, however, the trend has been to phase out most of the income-averaging provisions.[55] It should be mentioned that some of the other provisions referred to above, such as the farm rollover and the capital gains exemption, assist with the problem of fluctuating incomes in the special circumstances arising from sale or death.

One of the reasons that the income-averaging provisions have been gradually eliminated may be that, over time, the federal tax rates for individuals have tended to "flatten." Indeed, the 1987 reforms telescoped the ten federal rate brackets into three at 29 percent, 26 percent and 17 percent.[56] Today there are four federal rate brackets:

- 16 percent up to $31,677;
- 22 percent up to $63,354;
- 26 percent up to $103,000;
- 29 percent over $103,000.

The provinces generally set their own rates on income steps similar to the federal ones on a similar tax base. Alberta is an exception: it levies a 10 percent flat tax.[57] Québec has its own separate rate and tax base.[58] Corporate tax in Canada is generally a flat-rate tax. Canadian-controlled private corporations, however, face increased rates on active business income over CDN$200,000.[59]

The income-averaging provisions that are in place today – besides the optional inventory adjustment – include the rule regarding farm losses and the special regime developed to recognize agricultural income assistance programs. The farm loss rule permits both individual farmers and farm corporations to carry back business losses three years but forward ten years (an extra three years more than the seven years generally allowed.)[60]

The taxation of agricultural stabilization programs and compensation programs

For political and trade reasons, there is no comprehensive income stabilization program for aquaculture. In contrast, the federal government, in conjunction with the provinces, has provided and continues to provide income stabilization for primary commodity producers other than aquaculturalists. In December 2003, the Minister of Agriculture announced[61] that the required two-thirds of provinces representing 50 percent of Canada's agricultural production had agreed to implement the new Canadian Agricultural Income Stabilization Program (CAIS),[62] which will replace the Net Income Stabilization Account (NISA) program[63] and the Canadian Farm Income Program (CFIP).[64] The new program will not incorporate the tax preference, which is a characteristic of the NISA program. Under that program, two funds are established for producers. A producer contributes to fund number one and governments contribute to fund number two. In low-income years, the producer can withdraw monies from the funds. Amounts from fund number one are not included in the producer's income. However, interest on the funds, and "matching" amounts contributed to fund number two by governments are included in the producer's income but – and this is the tax preference – not until they are withdrawn.[65] Under the CAIS program, the producer is required to provide an amount on deposit. Interest is included in income when earned. In low-income years, producers can withdraw non-taxable amounts from their accounts. There are no government funds on deposit. Rather, taxable government assistance is paid out separately and directly to producers on the basis of a pre-established "insurance" formula related to the producer's amount on deposit.

Despite considerable efforts, the aquaculture industry has not been successful in obtaining income support(s) similar to other farmers from either or both levels of government. Indeed, they have not obtained the more modest goal of a compensation/income support program to moderate the impact of diseases in which major costly measures to producers, including stock destruction, are necessary to reduce pathogen levels and the risk of disease spread. At the present time, the industry is frustrated by the failure of the federal government to implement a tripartite initiative by itself, the Department of Fisheries and Oceans, and provincial governments concerning a national fish health program that would include some compensation arrangements.[66]

To date, compensation for industry disasters has been provided on an *ad hoc* basis. The most significant "bail-out" of the aquaculture industry occurred from 1996 to 2001 in New Brunswick during the first recorded outbreak of infectious salmon anemia (ISA) in Canada. The payout was under the auspices of the provincial Disaster Relief Fund.[67] Over CDN$14 million of taxable compensation was paid to fish farmers who had to dispose of their inventory.

Research and development costs

The income tax system in Canada permits taxpayers who engage in research and development to claim tax credits.[68] Although these are not directly related to income fluctuations, it can be expected that new enterprises in beginning low-income years might be likely to incur these types of expenses. Further, the "new" aquaculture sector characteristically has embraced innovation and new technology. However, in tandem with other claimants, fish farmers have found it difficult, if not impossible, to make use of the credits. The smaller operators, in particular, do not have the expertise, time or money to document the basis for a claim. This is a familiar complaint that the revenue authorities have attempted to address on numerous occasions. Nonetheless, difficulties remain, and are exacerbated by the frequent legislative changes in the area.

Property taxes

Property taxes are levied in all the provinces of Canada. Property taxes are calculated as a percentage of the assessed value of real property, although in some circumstances tax may also be imposed on business machinery and equipment. The taxes have a long history and were originally imposed to support "local" services, especially schools and the responsibilities normally assumed by municipal governments. In some jurisdictions, such as New Brunswick[69] and Prince Edward Island,[70] there is a local tax at a rate set by the particular municipality, in addition to a provincial tax. In Nova Scotia,[71] British Columbia,[72] and Newfoundland and Labrador,[73] the tax is levied and the rate designated by each municipality.

In most provinces, a distinction is made between residential and non-residential land with residential land, generally being taxed at a lower rate. Thus, in New Brunswick the provincial property tax rate for residential property is CDN$1.50/$100.00 of assessed value but 1.5 times that amount

– that is, CDN\$2.25/\$100.00 – for non-residential uses.[74] Distinctions are also made among other uses of real property, and in British Columbia, for example, there are nine classes of land with different rates.[75] Significantly, agricultural land is treated separately and preferentially.

The reason for the special position of agriculture under the various property tax systems rests on the characterization of property taxes as primarily, but not exclusively, benefits-based taxes. The argument is that if property tax rates are set at levels that capture as closely as possible the cost of services consumed, then, as between farm and residential property of equal value, the effective tax rate on the farm property should be lower. The farm covers (usually) more land, it is true, but it has fewer people and does not require the same level of services. Other reasons for agricultural preferences may include the force of past practice and the desire to slow the rate of conversion of agricultural land to urban uses. Tax concessions may also be extended to agricultural property as a form of economic assistance.

In the five provinces considered in this chapter, British Columbia explicitly provides for the tax treatment for property tax purposes of aquaculture on a comparable basis with agriculture. In that province, primary agricultural production for the purposes of the *Assessment Act* includes aquaculture.[76] In Nova Scotia, on the other hand, farm property is generally exempt from taxation under the *Assessment Act*[77] but aquaculture is taxed.[78] In New Brunswick, aquaculture operations do not qualify for tax deferral under the Farm Land Identification Program administered under the *Real Property Tax Act*.[79] So too in Newfoundland, the Real Property Tax Exemption Program for Agricultural Land[80] does not apply to aquaculture. Finally, Prince Edward Island's legislation makes no reference to aquaculture. "Farm property" is defined as cleared arable land.[81]

It is noteworthy that the types of concessions generally extended to agricultural concerns in rural areas have not been typically extended to other commercial activities in those areas, though resource-based operations may enjoy separate tax relief.

Finally, it should be mentioned that there are other property taxes that may affect aquaculture operations. For example, in British Columbia the province levies a school tax under the *School Act*[82] that is based on the nine categories of property authorized in the *Assessment Act*. The *School Act* exempts 50 percent of the assessed value of property assessed as farmland.[83] The school rate is 0.68 percent (mill rate of 6.8) for farm property, 0.41–0.45 percent for residential property and 0.99 percent for light industry and business in each municipality.[84]

Value added (GST) and provincial sales taxes

The federal government imposes a 7 percent multi-stage or value added sales tax (GST) on goods and services consumed in Canada.[85] Three provinces – New Brunswick, Nova Scotia and Newfoundland – also piggyback on the federal tax so that the 7 percent tax is increased to 15 percent in those juris-dictions.[86] Every seller along the chain of production and distribution of goods and services must collect the tax, but each taxpayer, except the final non-commercial consumer, is entitled to claim a credit (input tax credit) against taxes owing equal to the taxes already paid by them on their particular inputs.

The production from agriculture and aquaculture is, generally, effectively exempt from GST because basic groceries are zero-rated; that is, the rate of tax on the sale to the final consumer is zero percent.[87] Further, since GST is exigible only on goods or services consumed in Canada, produce destined for the export market is not subject to tax. Though produce is zero-rated, farmers may claim input tax credits for GST paid in carrying on their operations. They may, however, experience cash flow difficulties because they do not collect GST on sales to consumers, but are required to pay GST on their inputs. Many farm inputs are, however, zero-rated, and farmers may register to recoup, on a monthly basis, an amount equal to their input tax credits.

The policy rationale for the zero rate of tax on agricultural and aquaculture produce consumed in Canada arises because of consumer, not producer, concerns, and is based on the politically expedient view that some human essentials should not be taxed. The reason for the non-taxation of exports is to encourage them.

British Columbia[88] and Prince Edward Island[89] impose retail sales tax. Neither tax is applicable to food for human consumption.

License and leasehold fees

In marked contrast to the private property basis of agricultural operations, an important aspect of aquaculture operations in the Atlantic provinces and British Columbia is a direct reliance on the use of public (marine) resources for private purposes. The rent charged for the use of these public resources is generated in the form of leasehold fees for leases or occupation licenses of marine acreage. Leasehold fees and license fees (which are necessary whether or not there is a marine leasehold) are not, of course, taxes, but will, as already indicated, be briefly addressed in this chapter.

Each province administers its own system of licenses and leases except Prince Edward Island, which is administered by the federal government.[90] The fees charged in each of British Columbia and the Atlantic provinces vary but it is fair to say that, overall, the amounts exacted tend to be quite modest (see Table 7.1). The different regimes are outlined below. British Columbia has the highest fees. This figure can be put in some perspective by the observation that an "average" total area of a salmon aquaculture site is generally about 10 hectares. In contrast to British Columbia and New Brunswick, Nova Scotia, Prince Edward Island and Newfoundland charge only nominal fees.

Table 7.1 Provincial lease and license systems[92]

British Columbia

License		*Lease*
Finfish	Intensive – 7.5% of Zone Value ($500 min.)	Intensive – 8% of Zone Value ($500 min.)
	Extensive – 7.5% of half the Zone Value ($500 min.)	Extensive – 8% of half the Zone Value ($500 min.)
Shellfish	New tenues – 4% of $4940/ha ($600 min.)	New tenues – 5% of $4940/ha ($600 min.)
	Replacements – 4% of double the assessed land value on file with LWBC ($600 min.)	Replacements – 5% of double the assessed land value on file with LWBC ($600 min.)

New Brunswick

License		*Occupational permit*	
Commercial	$50		$100
Private	$10	*Lease*	
Institutional	$20	Marine site, commercial license	
		Finfish	$250/ha
		Mollusks	$20/ha ($100 min.)
		Crustaceans	$250/ha
		Inland site, Commercial license	
			$20/ha
		Lease fee to holder of private license	
			$100
		Lease fee to holder of institutional license	
			$100

Newfoundland

License	$100/site

Nova Scotia

License		*Lease*	$10/ha
U-Fish	$200		
All Others	$300		
Ten-year lease/license	$300		

Prince Edward Island

License	None	*Lease*	$10/acre or $4.05/ha

Sources: This table was compiled from the following sources:

British Columbia: Land and Water British Columbia Inc., *Land Use Programs*, Vol. 3:3, *Agricultural and Aquacultural Land Use* (2001) at Appendix 1. Online. Available http://lwbc.bc.ca/applying_for_land/aquaculture/aqua_policy.pdf (accessed 18 August 2003). In an appendix, the document provided a map displaying the various "fee" zones that fall within the issuing body's jurisdiction. The zones are priced as follows:

Zone A Value $7,031/ha
Zone B Value $6,375/ha
Zone C Value $5,156/ha
Zone D Value $4,875/ha
Zone E Value $4,325/ha (*Ibid*. at Appendix 4).

Additionally, the document defined the following terms:

- **"Intensive area"** – The area of Crown land used for aquaculture activities and related improvements directly associated with the production of finfish, shellfish or marine plants. The intensive area will include net-cages, netting, float camps, net storage, docks and mort sheds as well as a thirty-meter buffer around these structures. The thirty-meter buffer is mandatory and is intended to cover the area where anchor lines are likely to pose a restriction to navigation owing to the scope and angle of lines closest to the structures. Outside of the thirty-meter buffer, the lines are generally at a suitable depth to allow safe passage of a boat; however, any anchor lines beyond the thirty-meter buffer that restrict access or hamper navigation will also be included as part of the intensive area.
- **"Extensive area"** – The area of Crown land used for anchoring structures outside of intensive areas that do not impede navigation or access to lands beyond. (*Ibid*. at 4–5).

New Brunswick: *Aquaculture Act*, S.N.B. 1988, c. A-9.2; N.B. Reg. 91–158. The Regulation establishes and defines the two categories of aquaculture sites, and the three classes of licenses as follows:
- **"Inland aquaculture site"** is a class of aquaculture site that is situated in non-tidal waters or on land.
- **"Marine aquaculture site"** is a class of aquaculture site that is situated in tidal waters.
- **"Commercial aquaculture license"** is a class of aquaculture license that permits a licensee to conduct aquaculture for commercial gain.
- **"Institutional aquaculture license"** is a class of aquaculture license that permits a licensee to conduct aquaculture for the purposes of research outside a laboratory or an aquarium, or for use in public fishery enhancement activities, and not for the purposes of commercial gain.
- **"Private aquaculture license"** is a class of aquaculture license that authorizes a licensee to carry on aquaculture for private use and not for commercial gain (*ibid*. s. 2).

Newfoundland: Department of Fisheries and Aquaculture. Online. Available http://www.gov.nl.ca/fishaq/Aqua/Licencing.stm#Cost (accessed 26 March 2004).
Nova Scotia: Department of Agriculture and Fisheries. Online. Available http://www.gov.ns.ca/nsaf/aquaculture/application/aqua_fees.htm (accessed 26 March 2004).
Prince Edward Island: Interview with Dale Smith, Chief, Aquaculture Division, Department of Fisheries and Oceans (18 August 2003).

In a review of the lease/license fees extant in New Brunswick prepared for the Licensing and Inspection Branch of the Department of Fisheries and Aquaculture in New Brunswick,[91] the authors concluded that while the New Brunswick fees were higher than those of most of the other east coast provinces, they were lower than those in most of the eastern seaboard states. It should be mentioned, however, that many provinces and states periodically review their fee structures.

It is noteworthy that in other resource sectors, Canadian governments have sometimes insisted on sharing, or indeed appropriating, the Ricardian rent derived from the use of public resources for private purposes. Thus, governments earn substantial royalties from the exploitation of minerals and fossil fuels. On the other hand, they have, for whatever reason, sometimes elected to forgo any share in this surplus such as in the capture fisheries. This is a debate that so far has not been much pursued in considering these license and leasehold fees.

Conclusion

As I have demonstrated, the agriculture model of taxation in Canada has often been applied to the aquaculture sector. Fish farmers have approved of this development, not the least reason being that the model provides tax benefits. Further, the tax treatment of aquaculturalists as farmers supports other initiatives by the aquaculture industry to have government supports and incentives similar to those for agriculture extended to aquaculture. There seems little doubt that most aquaculturalists consider themselves as farmers, not fishers, and, indeed, different from other small and large rural businesses.

This chapter has compared the taxation of aquaculture with that of agriculture in the three main categories of taxation. It found that under the income tax system in Canada, fish farmers are generally taxed in the same way as other farmers. On the other hand, except in British Columbia, aquaculture operators are not treated the same as farmers for property tax purposes, although they may enjoy some tax concessions. Finally, the chapter considered the sales taxes regimes of the federal government and the provinces. Aquaculturalists are treated similarly to other farmers. It is difficult to derive any useful policy insights from comparisons in this last category.

Two questions arise when considering the policy implications of the application of the agricultural model to aquaculture. The first-order question is whether the special rules for agriculture are defensible when applied to traditional farmers. The second question, assuming the special rules can be justified, is whether, given the similarities and differences between the two sectors, the rules should be extended to fish farmers.

While the agricultural model was considered in relation to all the main categories of taxes, this chapter concentrated on the income tax system(s).

The income tax system draws more completely on the agricultural model in the taxation of aquaculture and ultimately seems to yield more insights. The special rules for agriculture, generally extended to aquaculture, were examined from four functional perspectives. They were rules designed to:

- enhance administration and compliance;
- assist in intergeneration succession;
- help farmers provide for their retirement; and
- alleviate the burden of fluctuating incomes.

It is fair to say that many of the special rules relating to agriculture were developed in an economic and social context that no longer exists. Thus, their continued application to both agriculture and aquaculture is, from a policy perspective, suspect. For example, the rules relating to the administration of the *Income Tax Act* (cash-basis accounting), which permitted unsophisticated and unschooled farmers (among others) to calculate and report income more easily, have morphed into a labyrinth of complex rules and elections. Their extension to the aquaculture industry, which has a substantial representation from a younger, better-educated generation (although this is not true across the board), and where a substantial component of the industry comprises large corporations, is not convincing.

Other special rules can be criticized, even though one accepts the stated or implicit policy reasons advanced for their existence. For example, the goal of the special rules relating to intergenerational succession is to preserve "local" and "family" farms. The special rules relating to intergeneration succession may not operate effectively, however. There is no guarantee, even after the tax-free transfer, that the next generation will stay in the community and continue to farm. They may simply take the (unwarranted) tax advantages and sell off farm and assets. Further, the rollover rules may actually create additional barriers to new entrants to the aquaculture industry. So too, the CDN\$500,000 capital gains exemption (which also generally applies to qualified shares of small business corporations that are not farm corporations) goes far beyond providing tax assistance comparable to that provided to the general population to save for retirement. It helps farmers who sell out to retain more of the proceeds, but there are no stipulations that the amounts must be deposited into a retirement fund. In addition, in order to limit the provision to "real" hands-on farmers, government efforts to target its benefits have generated a maze of overly complex rules that may not fulfill their intended purpose but certainly constitute a trap for the unwary and unadvised, or both.

Finally, there may be insufficient recognition of the circumstances under which the aquaculture industry operates. For example, most of the special rules in the Act that attempt to ameliorate the tax effects of fluctuating incomes have been eliminated over the past decade or more. Further, fish farmers generally do not qualify for income support programs extended to other farmers, including the Net Income Stabilization Account (NISA) program, which incorporates some tax preferences and the new Canadian Agricultural Income Stabilization (CAIS) program, which generally does not because of its different structure.

In the property tax area, the agricultural model is sometimes and sometimes not applied to aquaculture, depending on the jurisdiction and particular tax provision. While a benefits-based justification of property taxes (and historical precedent) may support reduced rates of tax on agriculture, it is not so compelling when applied to aquaculture. From a property tax perspective, aquaculture may more appropriately be compared with other rural industries, including industries in the resource sector.

Finally, it should be noted that fish farmers in the marine sector generally pay modest leasehold fees (which are not taxes) for their marine acreage. The system of leasehold interests is in significant contrast to the full ownership rights of most farmers. A list of leasehold fees for the five marine provinces has been included in this chapter.

Notes

1 *Income Tax Act*, R.S.C. 1985, c. 1 (5th Supp.), as amended [ITA]. This chapter purports to reflect the law as of September 2003. Note that, in its 2 May 2006 budget, the Canadian fedeal government proposed extending the intergenerational rollover and the CDA$500,000 capital gains deduction to fishers with certain kinds of qualified fishing property. Online. Avaliable http://www.fin.gc. ca/budget06/bp/bpa3ae.htm#fishers.

2 *Ibid.* The federal government taxes individuals and corporations. For 2003, the tax on individuals is 16 percent to CDN$31,677, 22 percent to CDN$63,354, 26 percent to CDN$103,000 and 29 percent thereafter. Corporate rates vary depending on the type of corporation and the type of income. The general corporate rate is 28 percent, but Canadian-controlled private corporations (CCPCs) are taxed at about 12 percent on certain amounts and types of income. Corporate dividends received by individuals are also subject to tax but a dividend tax credit is credited against tax payable to partially or, in the case of certain income of CCPCs, completely recognize the corporate tax already paid.

   Part 1.3 of the Act imposes a special tax on corporations on their taxable capital employed in Canada in excess of their capital deduction for the year. The tax operates as a minimum tax applicable to larger corporations. The Part 1.3 tax is gradually being phased out and will be fully eliminated in 2008. Until that time, the capital deduction will be increased from CDN$10 to CDN$50 million for taxation years ending after 2003 and the rate gradually reduced from 0.225 percent to 0.0625. The tax has been of considerable concern to aquaculture operations.

3 *Excise Tax Act*, R.S.C. 1985, c. E-15, Part IX.

4 *Income Tax Act*, R.S.B.C. 1996, c. 215; *New Brunswick Income Tax Act*, S.N.B. 2000, c. N-6.001; R.S.N.S. 1989, c. 217; *Income Tax Act*, R.S.P.E.I. 1988, c. I-1; *Income Tax Act, 2000*, S.N. 2000, c. I-1.1. Provincial tax rates vary among provinces. Both individuals and corporations are taxed. For 2003 in British Columbia, the tax on individuals is 6.05 percent up to CDN$31,124, 9.15 percent to CDN$62,249, 11.7 percent to CDN$71,470 and 14.7 percent over CDN$86,785. Corporate rates depend on the type of corporation and the type of income. The general corporate rate is 13.5 percent but Canadian-controlled

private corporations are taxed at 4.5 percent on certain amounts and types of income. Corporate dividends received by individuals are also subject to tax, but a dividend tax credit is credited against tax payable to partially or, in the case of certain income of CCPCs, completely recognize the corporate tax already paid. The tax base for provincial individual and corporate taxes generally approximates the federal base.

New Brunswick and Nova Scotia levy provincial corporate capital taxes. *New Brunswick Income Tax Act*, S.N.B. 2000, N-6.001,s. 62(1); *Income Tax Act*, R.S.N.S. 1989, c. 217, as amended by S.N.S. 1997, c. 3, s. 8. In New Brunswick, the annual rate is 0.3 percent with the first $5 million non-taxable. The annual rate in Nova Scotia is generally 0.25 percent.

5 *Social Service Tax Act*, R.S.B.C. 1996, c. 431; *Revenue Tax Act*, R.S.P.E.I. 1988, c. R-14.

6 *Excise Tax Act*, R.S.C. 1985, c. E-15, Part IX, Div. VIII; *Federal–Provincial Fiscal Arrangements Act*, R.S.C. 1985, c. F-8; *Harmonized Sales Tax Act*, S.N.B. 1997, c. H-1.01; *Sales Tax Act*, S.N.S. 1996, c. 31; *Tax Agreement Act*, S.N. 1996, c. T-0.01.

7 See, generally, Canada, Agriculture and Agri-Food Canada, *Agricultural Property Tax Concessions and Government Transfers to Agriculture* (Ottawa: Agriculture and Agri-Food Canada, 2000). Online. Available http://www.agr.gc.ca/spb/rad-dra/publications/agprop/agprop_e.php (accessed 26 March 2004) [*Agricultural Property Tax*]. See the "Property taxes" section (p. 257).

8 See the "License and leasehold fees" section (p. 259).

9 Statistics Canada, "Aquaculture Statistics," (17 October 2002) *The Daily*. Online. Available http://www.statcan.ca/Daily/English/021017/d021017b.htm (accessed 26 March 2004).

10 *Les Immeubles Dramis Inc.* v. *M. N. R.* [1981] C.TC. 2319; 81 D.T.C. 512 [*Les Immeubles* cited to C.T.C.]. See also C.C.R.A., Interpretation Bulletin, IT-322R, "*Income Tax Act* Farm Losses" (25 October 1978) at para. 8.

11 *Les Immeubles, supra* note 10 at 2574.

12 See, generally, Gary H. Munro and Kurt Oelschlagel, *Taxation of Farmers and Fishermen*, looseleaf (Toronto: Carswell, 2002).

13 Canada, Department of Finance, "Tax Expenditures and Evaluations 2002," at Table 1, "Personal income tax expenditures." Online. http://www.fin.gc.ca/toce/2002/taxexp02_e.html (accessed 26 March 2004).

14 Canada, *Report of the Royal Commission on Taxation*, Vol. 4 (Ottawa: Queen's Printer, 1966) at 440 (Chair: Kenneth Le Mesurier Carter) [Carter Report].

15 *Ibid.*

16 ITA, *supra* note 1, s. 28.

17 Carter Report, *supra* note 14 at 441. Carter would have permitted farmers who earned less than CDN$10,000 of gross revenue in the year to use the cash method of accounting.

18 *Ibid.*

19 ITA, *supra* note 1, para. 28(1)(b).

20 ITA, *supra* note 1, para. 28(1)(c).

21 New Brunswick, Agriculture, Fisheries and Aquaculture, *Bay of Fundy Marine Aquaculture Site Allocation* (Fredericton: Agriculture, Fisheries and Aquaculture, 2000).

22 ITA, *supra* note 1, subsecs. 70(9) (transfer of farm property on taxpayer's death to child), 70(9.2) (transfer of shares of family farm corporations and interests in family farm partnerships on taxpayer's death to child). Intergenerational transfers may be made during the parent's lifetime but income and capital gains attribution rules may apply in some circumstances. See ITA, *supra* note 1, subsecs. 73(3) and (4).

23 ITA, *supra* note 1, subsec. 70(10).

24 *Ibid.* subsec. 70(10).

25 *Ibid.*

26 Carter Report, *supra* note 14 at 28.

27 Canada, Department of Finance, *Proposals for Tax Reform* (Ottawa: Queen's Printer, 1969) at 36. It should be noted that the taxation of depreciable property was changed at tax reform so that tax-free transfers of that type of property between non-arm's-length individuals were eliminated unless the taxpayers could take advantage of a specific rollover provision such as the farm rollover.

28 ITA, *supra* note 1, subsec. 70(9.4) repealed by S.C. 1986, c. 6, s. 33(3).

29 See "Tax treatment of inshoremen unfair – CCPFH," (1 October 2003) *The Sou'Wester* at 1.

30 ITA, *supra* note 1, subsec. 70(9).

31 The parent could roll over the property to several children who on their subsequent divestment of their share of the property might, in some circumstances, each be able to claim the CDN$500,000 capital gains exemption. But see ITA, *supra* note 1, subsec. 69(11).

32 See Stanley S. Surrey, *Pathways to Tax Reform: The Concept of Tax Expenditures* (Cambridge, MA: Harvard University Press, 1973). Most tax theorists have embraced, to a greater or lesser extent, Surrey's description of tax expenditures in his seminal work. He describes deviations from a normative tax base as tax expenditures. In Surrey's account, most of the special farm provisions, including the intergenerational rollover and the CDN$500,000 capital gains exemption, qualify as tax expenditures.

33 There is a uniform comprehensive limit on tax-deductible contributions to tax-assisted retirement savings vehicles. The two principal vehicles are registered pensions plans, ITA, *supra* note 1, s. 147.1, and registered retirement savings plans (RRSPs), ITA, *supra* note 1, s. 146. Generally, the earnings of registered retirement savings vehicles are not subject to tax, but any withdrawals are taxed in the hands of the withdrawing taxpayer. For the 2005 taxation year, the limit for deductible contributions to an RRSP, assuming the taxpayer has not otherwise accumulated tax-sheltered savings benefits, is the lesser of 18 percent of earned income or $16,500. The taxpayer may carry unused contribution room forward indefinitely. The dollar limit will be increased to CDN$18,000 in 2006 and thereafter indexed by the growth in the average wage. See also ITA, *supra* note 1, s. 147.1 for the rules regarding registered pension plans.

34 Canada Customs and Revenue Agency, *Income Statistics 2001–1999 tax year: Final Basic Table* 9. Online. http://www.ccra-adrc.gc.ca/tax/individuals/stats/gb99/pst/final/pdf/table9-e.pdf (accessed 26 March 2004).

35 ITA, *supra* note 1, s. 110.6 added by 1986, c. 6, s. 58 generally applicable to 1985 *et seq.*

36 See, for example, the critique by William R. G. Lawlor, "Surplus Stripping and Other Planning Opportunities with the New $500,000 Capital Gains Exemption," *Proceedings of the Thirty-seventh Tax Conference*, 1985 Conference Report (Toronto: Canadian Tax Foundation, 1986) at 8:1–8:10.

37 The ability of an individual to claim the capital gains exemption is limited by their "annual gains limit," "cumulative gains limit" and "cumulative net investment loss." See ITA, *supra* note 1, subsec. 110.6(1). The annual gains limit and the cumulative gains limit, among other things, reduces a taxpayer's net capital gains, and therefore his or her capital gains exemption, by net capital losses claimed. The cumulative gains limit is the amount, if any, by which a taxpayer's investment expenses (e.g. interest) exceeds investment income.

38 ITA, *supra* note 1, subsec. 110.6(1) introduced in the 1985 Budget, *supra* note 35.

39 *Ibid.*
40 *Ibid.*
41 The capital gains inclusion rate was increased by the 1988 Budget to two-thirds in 1988 and 1989 and then to three-quarters in 1990. The 2000 Budget decreased the rate (with some transition provisions) to the one-half rate that presently applies.
42 ITA, *supra* note 1, subsec. 110.6(3) repealed by S.C. 1995, c. 3, s. 32(3) with some election provisions for 1994 and 1995.
43 ITA, *supra* note 1, subsec. 110.6(2.1) added by S.C. 1988, c. 55, s. 81(6).
44 ITA, *supra* note 1, subsec. 110.6(1).
45 ITA, *supra* note 1, subsec. 110.6 added by 1988, c. 55, subsec. 81(4), applicable with respect to dispositions of shares after 17 June 1987.
46 *Technical Committee on Business Taxation Report* (Ottawa: Department of Finance, 1998).
47 *Ibid.* at 7.18.
48 *Ibid.* at 7.18–7.19.
49 *Ibid.* at 1.5.
50 *Ibid.*
51 *Ibid.* at 7–18.
52 *Ibid.*
53 Eugene La Brie, *The Principles of Canadian Income Taxation* (Toronto: C.C.H. Canadian, 1965) at 388.
54 Carter Report, *supra* note 14.
55 The block averaging provision was the longest-lasting income-averaging provision dedicated solely to farmers and fisherman. It was introduced in 1944 and was finally eliminated in the tax reform of 1987, which was accompanied by the rate flattening already referred to. Former ITA, *supra* note 1, sec. 119. Block averaging was available to both individuals and corporations, and permitted a taxpayer to average income over a five-year period. It was by far the most generous income-averaging provision used by farmers and fisherman.

Other income-averaging provisions instituted and then eliminated after tax reform included the general averaging provisions, former ITA, s. 118, income-averaging annuity contracts, former ITA, sec. 61, and forward averaging, former ITA, s. 123. These provisions were applicable to all taxpayers, including farmers, but only if they were unincorporated. The general averaging provisions and the income-averaging annuity contracts (IAACs) were eliminated in the November 1981 budget. General averaging did not give most taxpayers significant benefits, was applicable only in the year income rose (not fell) and was usually automatically done by the tax collector. The IAAC provisions permitted the taxpayer to buy an annuity for an amount equal to certain kinds of income she or he received in the year. The annuity had to be paid commencing within ten months, and in the years it was paid, both capital and interest were taxed. The IAAC was of particular significance to farmers because it permitted the taxpayer to buy an annuity for cash basis inventory in a year the farming ceased.

For a brief period, from 1981 to 1987, the provisions abolished were replaced by the forward averaging refundable tax, former ITA, s. 110.4. As the name implies, income over a certain amount in the year is subject to tax, but the income may be reincluded in income in a future year at a lower tax rate. The difference between the tax paid and the tax payable in the future year on the income is refunded.
56 ITA, *supra* note 1, s. 117 as amended by S.C. 1988.
57 *Alberta Personal Income Tax Act*, R.S.A. 2000, C.A-30, s. 4.
58 *Taxation Act*, R.S.Q., c. I-3.

59 ITA, *supra* note 1, subsec. 125(1). The small business income threshold will be increased for the next three years by $25,000 a year to a maximum of $300,000 on 1 January 2006.

60 ITA, *supra* note 1, s. 111.

61 Agriculture and Agri-Food Canada, New Release, "Minister Vanclief Officially Launches New Farm Income Program" (11 December 2003).

62 Agriculture and Agri-Food Canada, *Canadian Agricultural Income Stabilization (CAIS) Program Handbook*. Online. http://www.agr.gc.ca/producerassistance2003/cais03hb.html (accessed 26 March 2004).

63 *Farm Income Protection Act*, S.C. 1991, c. 22.

64 CFIP ended in 2002.

65 ITA, *supra* note 1, ss. 12(10.2) and (10.3).

66 See *A Business Case in Support of a National Aquatic Animal Health Program*, March 2002. Available from New Brunswick Salmon Growers Association, St. Georges, New Brunswick.

67 *Emergency Measures Act*, S.N.B. 1978, c. E-7.1, s. 20.

68 ITA, *supra* note 1, s. 127(5).

69 *Real Property Tax Act*, R.S.N.B. 1973, c. R-2, s. 4, as amended by S.N.B. 1994, c. 93, s.7; *ibid.* s. 5(1)–(2), as amended by S.N.B. 1982, c. 56, ss. 4, 5.

70 *Real Property Tax Act*, R.S.P.E.I. 1988, c. R-5, ss. 4(1), 8.

71 *Municipal Government Act*, S.N.S. 1998, c. 18, s. 72(6).

72 *Local Government* Act, R.S.B.C. 1996, c. 323, s. 359(1). *Taxation (Rural Area) Act*, R.S.B.C. 1996, c. 448; but see *School Act*, R.S.B.C. 1996, c. 412, Div. 4.

73 *Municipalities Act, 1999*, S.N. 1999, c. M-24, s. 112. *St. John's Assessment Act*, R.S.N. 1990, c. S-1, ss. 3, 4.

74 *Real Property Tax Act*, *supra* note 69, s. 5(1).

75 *Assessment Act*, R.S.B.C. 1996, c. 20, s. 19; B.C. Reg. 205/84; B.C. Reg. 438/81.

76 R.S.B.C. 1996, c. 20; B.C. Reg. 411/95, Sch. A.

77 R.S.N.S. 1989, c. 23, s. 46. as amended by S.N.S. 1990, c. 19, s. 14; S.N.S. 1996, c. 5, s. 3; S.N.S. 1998, c. 13, s. 2; S.N.S. 1998, c. 18, s. 547.

78 *Assessment Act*, R.S.N.S. 1989, c. 23, s. 2(1)(s), as amended by *Municipal Law Amendment (2000) Act*, S.N.S. 2000, c. 9, s. 2.

79 *Farm Land Identification Regulation – Real Property Tax Act*, N.B. Reg. 84–75. *Agricultural Land Protection and Development Act*, S.N.B. 1996, c. A-5.11, s. 1.

80 *Municipalities Act*, S.N. 1999, c. M-24, s. 118(j). Government of Newfoundland and Labrador, *Real Property Tax Exemption Program for Agricultural Land*. Online. Available http://www.gov.nl.ca/agric/soils/sl170.htm (accessed 26 March 2004).

81 *Real Property Assessment Act*, R.S.P.E.I. 1988, c. R-4, s. 4(1).

82 *School Act*, R.S.B.C. 1996, c. 412, ss. 119–120.

83 *Ibid.* s. 130.

84 These rates vary among school districts and from year to year. See *Agricultural Property Tax*, *supra* note 7 at A-3.

85 *Supra* note 3.

86 *Supra* note 6.

87 *Excise Tax Act*, *supra* note 3, ss. 123(1), 165(3).

88 *Social Service Tax Act*, R.S.B.C. 1996, c. 431, s. 70.

89 *Revenue Tax Act*, R.S.P.E.I. 1988, c. R-14, s. 12(1).

90 *Fisheries Act*, R.S.B.C. 1996, c. 149, s. 13(5); *Aquaculture Act*, R.S.N.B. 1988, c. A-9.2, ss. 5(1), 25(1); *Fisheries and Coastal Resources Act*, S.N.S. 1996, c. 25, s. 64; *Aquaculture Act*, R.S.N.L. 1990, c. A-13, s. 4.

91 *A Review of Aquaculture Lease/License Fees in New Brunswick and Other Jurisdictions* (May 1999).

Part III

Aboriginal title and rights in aquaculture

8   The potential impact of aboriginal title on aquaculture policy

Diana Ginn

Introduction

This chapter discusses the potential impact of aboriginal property rights on the development of aquaculture policy by considering whether such rights could provide a basis for First Nation peoples to participate in aquaculture or to manage the participation of others in this industry. The purpose of the chapter is to describe the relevant law as it now stands, to identify issues that have not yet been decided and to consider how the courts might approach such issues in the future.

There are two categories of property-based[1] claims which First Nation peoples might consider using in relation to areas in which aquaculture is or could be carried out: claims based on aboriginal title and claims based on common law riparian rights. In an aboriginal title claim, a First Nation would argue that because of its historical use of an aquaculture area, it holds that area by way of aboriginal title. Given that the doctrine of aboriginal title has developed in relation to dry land, the first question to be addressed is whether the courts are likely to apply the doctrine to water areas. The first part of this chapter outlines the current law on aboriginal title; discusses the applicability of that law to rivers, lakes and marine coastal areas, reaching the tentative conclusion that the doctrine of aboriginal title could apply; and considers how recognition of aboriginal title in such areas might affect aquaculture policy. The second part considers whether property-based arguments might be made based on the common law concept of riparian rights to the land beneath rivers. At English common law,[2] the owner of land bounded by the non-tidal portion of a river or stream was presumed to own the waterbed to the center line of the river, while the Crown was presumed to own the land beneath the tidal portion of rivers. The latter part of this chapter outlines the issues that would have to be decided if a First Nation attempted to use the concept of riparian rights to claim a portion of a riverbed. There are so many unanswered questions in this area that it is difficult to predict how courts would respond to such a claim; however, aquaculture policy-makers should be aware of the issues and watch how the law develops in this area.

Aboriginal title

The doctrine of aboriginal title

Section 35(1) of the *Constitution Act 1982*[3] states, "Existing treaty and aboriginal rights are hereby recognized and affirmed." The source of treaty rights is self-explanatory: these are rights that have been recognized in a treaty (whether a historic agreement or a modern land claims agreement) between the Crown and a particular aboriginal nation or community. Aboriginal rights, however, do not have their source in any document or agreement; instead, these rights arise from the occupation of what is now Canada by aboriginal nations at the time of the British assertion of sovereignty. "Aboriginal rights" is thus an umbrella term that includes both activity rights (such as rights to hunt, fish or gather) and aboriginal title to land.[4]

The most thorough discussion by the Supreme Court of Canada on aboriginal title is the 1997 decision of *Delgamuukw* v. *British Columbia*.[5] Combining what is said in *Delgamuukw* with discussion in several other cases, we can say that aboriginal title, as currently conceptualized by Canadian courts, has dual sources: first, historic use and occupation of the land by First Nations,[6] and, second, the relationship between the common law and pre-existing aboriginal systems of law.[7] Aboriginal title is more than simply a right to carry out certain activities on the land: it is title to the land itself.[8] This form of title is *sui generis* (that is, unique),[9] communal and inherent.[10] Furthermore, aboriginal title is an exclusive form of title; that is, it carries with it the right to exclude others from using or occupying the area covered by aboriginal title.[11] Aboriginal title confers the right to use land for a variety of activities. Thus, a First Nation holding aboriginal title is not limited to traditional uses of the land. There is, however, one inherent limit: aboriginal title land cannot be used in ways that are "irreconcilable with the nature of the attachment to the land which forms the basis of the particular group's aboriginal title."[12] Aboriginal title exists in conjunction with underlying or radical Crown title[13] and can be alienated (transferred) only to the federal Crown.[14] With the protection afforded by s. 35(1) of the *Constitution Act*, existing aboriginal title can now be extinguished only with the consent of the First Nation involved, by way of a land claims agreement or other bilateral instrument.[15] Prior to 1982, federal jurisdiction over "Indians and lands reserved for the Indians"[16] was seen as permitting the federal government to extinguish aboriginal title unilaterally as well, so long as the intent to do so was "clear and plain."[17] Finally, because aboriginal title represents a relationship between common law principles and aboriginal systems of law, it "must be understood by reference to both common law and aboriginal perspective."[18]

Application of the doctrine of aboriginal title to water areas

If a First Nation makes a claim of aboriginal title to a terrestrial area, it is clear that the doctrine is applicable, and the question becomes one of evidence: what evidence is there of historic use of the area and what evidence is there of extinguishment? At present, however, no Canadian case law directly addresses the question of whether the doctrine of aboriginal title can be applied to water areas.[19] First Nations have made several aboriginal title claims to rivers and portions of the sea but the issue has not yet been decided by Canadian courts. The best-known such claim is probably the one filed by the Haida Nation in 2002[20] claiming aboriginal title to land, internal waters and a portion of the seabed and seas off the coast of British Columbia. However, the Haida case has not yet come to trial. In other cases where a decision has been rendered, it has been on other (usually preliminary or procedural) grounds.[21] Thus, the question of whether aboriginal title can exist in water areas has still not been resolved.

Courts in both Australia and New Zealand have heard aboriginal title claims to the seabed. In 2002, the High Court of New Zealand heard an appeal from a 1997 decision of the Maori Land Court, which recognized the possibility that the foreshore and seabed of the Marlborough Sounds could be held by way of customary Maori title.[22] The High Court reversed this finding.[23] With respect to the seabed,[24] the Court focused on the wording of s. 7 of the *Territorial Sea and Fishing Zone Act* (now the *Territorial Sea and Exclusive Economic Zone Act*, 1977), which states:

> Subject to the grant of any estate or interest therein (whether by or pursuant to the provisions of any enactment or otherwise, and whether made before or after the commencement of this Act), the seabed and subsoil of submarine areas bounded on the landward side by the low-water mark along the coast of New Zealand (including the coast of all islands) and on the seaward side by the outer limits of the territorial sea of New Zealand shall be deemed to be and always to have been vested in the Crown.[25]

According to the High Court of New Zealand, this title was vested in the Crown for the benefit of all subjects and could not be granted to anyone, including Maori, in fee simple, as that fee simple would conflict with the public right of navigation.[26]

In 2003, the Court of Appeal of New Zealand reversed the High Court,[27] and found that the Maori Land Court did have jurisdiction to determine whether portions of the foreshore or seabed were held by Maori customary title. The Court of Appeal cautioned that the outcome was not a finding of customary title,[28] only a finding that "the Maori Land Court can enter into the substantive inquiry."[29] The Court of Appeal rejected the idea that property law principles applicable to terrestrial land were inapplicable to the foreshore and seabed,[30] and the argument that legislation such as the *Territorial Sea and Exclusive Economic Zone Act* precluded a claim of customary title.[31]

In a 2001 decision, the High Court of Australia considered a native title claim to "the seas in the Croker Island region of the Northern Territory."[32] The case was heard in the first instance by Olney J of the Federal Court, who held that native title extended only to the low-water mark, although he recognized the claimants as having a right to fish, hunt and gather in the claimed area, to travel through it, and to "visit and protect places within the claimed area which are of cultural or spiritual importance [and] to safeguard the cultural and spiritual knowledge of the common law holders."[33] On appeal to the full court of the Federal Court, a majority of the court upheld Olney J's decision. Both the claimants and the defendants appealed to the High Court.

The High Court of Australia held that the idea of title – even radical title in the Crown – was not an appropriate concept for the seabed. The Court considered the wording of the *Sea and Submerged Lands Act, 1973*, s. 11 of which states:

> The sovereign rights of Australia as a coastal state in respect of the continental shelf of Australia, for the purpose of exploring it and exploiting its natural resources, are vested in and exercisable by the Crown in right of the Commonwealth.[34]

The court concluded that this "did not amount to an assertion of ownership of or radical title in respect of the seabed or superadjacent sea in that area, whether as a matter of international law or municipal law."[35] The court then held that "[a]s a matter of international law, the right of innocent passage is inconsistent with any international recognition of a right of ownership by the coastal state of territorial waters."[36] Further, the existence of title in the seabed was negated by "the recognition of public rights of navigation and fishing."[37] Thus, exclusive native title in the seabed could not be recognized because it would be inconsistent with the right of innocent passage and the rights of public navigation and fishing.[38] Olney J's order recognizing non-exclusive native interests was upheld.

Leaving aside issues of evidence, how is a Canadian court likely to respond to aboriginal title claims to water areas? Given that Canadian jurisprudence characterizes aboriginal title as a "burden" on the underlying radical title of the Crown, the first step in considering whether aboriginal title could exist in water areas is to ask whether the Crown holds title to those areas. The second step is to ask whether there are aspects of the doctrine of aboriginal title (particularly the concept of exclusivity) that would be more problematic in the context of water areas than on dry land. Before we address these issues, however, a word about terminology: should aboriginal title claims in aquatic areas be described as claims to the subaquatic land, or to the water and waterbed as a whole?

The common law position in Britain and Canada is that running or per-colating water cannot be owned, although the land beneath it can.[39] Thus, at common law, the concept of proprietary rights is applied to the underlying land, rather than the water itself. On the other hand, it is certainly possible that some First Nations may conceive of a river or marine area as a unified resource. Since aboriginal title has been described by the Supreme Court of Canada as based in both aboriginal legal systems and the common law,[40] aboriginal perspectives on use of and control over water areas would have to be taken into account and should be reflected in the development of the law of aboriginal title in aquatic areas.

In this chapter, however, I focus on title to the waterbed, for three reasons. First, it cannot be assumed that all First Nations would view water resources in exactly the same way, and therefore, in the context of a claim, the perspective of the particular First Nation making the claim would have to be explored, rather than trying to factor in some sort of generic "aborigi-nal perspective." Second, Lamer CJC's wording makes it clear that the common law will still have to be considered, although not privileged, so it is necessary to consider how an aboriginal title claim to submerged land would fit with Canadian law generally. Third, it may be that the distinction between title to the waterbed alone versus title to the bed and the water together would not actually have much impact on the consequences of a suc-cessful claim. If it were established that a First Nation held unextinguished aboriginal title to a riverbed or a portion of the seabed, use of that sub-merged land would seem inevitably to carry with it use of the water flowing over it – in other words, use of the water resource as a whole.

Crown title to submerged land

As has been noted already, aboriginal title – at least as currently conceptual-ized by Canadian courts – coexists with underlying fee simple in the Crown. Therefore, to consider whether aboriginal title might be found to exist in submerged lands, one must first inquire as to the existence of Crown title in those areas.

The English common law distinguished among land beneath non-tidal waters, land beneath tidal waters to the low-water mark, and land below the low-water mark. Land beneath non-tidal waters was presumed to lie with the owners of the adjacent river or stream bank, while land beneath the tidal portions of rivers lay, prima facie, with the Crown. According to the common law, the territory of the realm extended only to the low-water mark, so no one owned the seabed. The common law position was adopted, with some variations, in Canada, but, more to the point here, has been largely overridden by legislation. Most jurisdictions in Canada have passed legislation vesting the ownership of waterbeds within the province in the provincial Crown. In the case of the territories, title now lies with the federal Crown. Arguably, Canada also holds title to the land beneath

Canada's 12-mile territorial seas, by virtue of ss. 7 and 8 of the *Oceans Act*, which state:

> 7.   For greater certainty, the internal waters of Canada and the territorial sea of Canada form part of Canada.
>
> 8(1) For greater certainty, in any area of the sea not within a province, the seabed and subsoil below the internal waters of Canada and the territorial sea of Canada are vested in Her Majesty in right of Canada.[41]

Where the fee simple to subaquatic land in Canada lies with the Crown,[42] whether federal or provincial,[43] it seems possible that, as with dry land, this title could be subject to aboriginal title. In fact, one of the *sui generis* aspects of aboriginal title is that it exists as a burden or limitation on the underlying Crown title. Certainly, existence of Crown title is likely to be seen by courts as a precondition for any consideration of aboriginal title in submerged land. If, as has been suggested here, the Crown holds title to most waterbeds within Canada, as well as the bed of Canada's territorial seas, this precondition has been met. The next question to consider, then, is whether the various aspects of aboriginal title, as described by the courts, raise any greater or different conceptual problems in relation to submerged lands, as compared to terrestrial areas.

The nature of aboriginal title

Aboriginal title has been described as flowing from historic use and occupation and from the relationship between aboriginal systems of law and the common law. Nothing about this aspect of the doctrine seems inherently inconsistent with a First Nation being able to claim aboriginal title in a riverbed or a portion of the seabed.[44] Nor would Lamer CJC's comments in *Delgamuukw* concerning the purposes for which aboriginal title land can be used seem to cause any greater interpretational difficulties for subaquatic lands than for terrestrial lands.

It seems likely that the Supreme Court of Canada's characterization of aboriginal title as exclusive might well be the most problematic issue, at least with regard to those portions of rivers that are at common law subject to public rights of fishing and navigation, and those portions of the seabed, that are subject to the international right of innocent passage. As noted above, the High Court of Australia held in the Croker Island case that the recognition of any title – even Crown title – in the seabed is irreconcilable with the right of innocent passage, and that aboriginal title could not coexist with common law rights of public fishing and navigation. It is this author's position however, that these conclusions should not be seen as persuasive by Canadian courts, partly because of differences in the legal framework, but also because of flaws in the court's reasoning.

INNOCENT PASSAGE

Thinking first about the right at international law for ships of one nation to make innocent passage through the territorial waters of other nations, it is unclear why this would negate the possibility of title in the seabed. Certainly, the language of ss. 7 and 8 of the *Oceans Act*, cited on p. 276, would seem to indicate that Canada intended to acquire property rights over the bed of its territorial sea. The fact that international law places some limits on a title-holder's power to exclude others (in this case, ships of other nations) is not irreconcilable with the existence of title. As to the impact of the right of innocent passage on the possible existence of aboriginal title, the short answer would seem to be that if Crown title can coexist with such a right, so could aboriginal title, and it, like Crown title, would be subject to the international right. Where the exclusivity of the underlying Crown title is curtailed by international law, it seems logical that any aboriginal title which exists as a burden on that title would be similarly curtailed. While the right of innocent passage would limit the rights otherwise associated with aboriginal title, it should not, however, be seen as preventing courts from recognizing aboriginal title in the bed beneath Canada's territorial seas.

PUBLIC RIGHTS

Public rights of fishing and navigation in the tidal portions of rivers have been entrenched in English common law since the Magna Carta. This is based on an interpretation of s. 47 of Magna Carta, the modern translation of which states: "All forests that have been created in our reign shall at once be deforested. River-banks that have been enclosed in our reign shall be treated similarly."[45] In Canada, the Supreme Court of Canada has held that rivers "so far as the ebb and flow of the tide" are

> open to public use and enjoyment freely by the whole community, not only for the purposes of passage, but also for fishing, the Crown being restrained by Magna Charta from the exercise of the prerogative of granting a several fishery in that part of any river.[46]

This does not mean that public rights of fishing and navigation are sacrosanct; only that if the Crown wishes to curtail such rights, it must do so through legislation rather than through an exercise of the royal prerogative. Nor has the common law viewed the existence of such rights as irreconcilable with the concept of title. As noted earlier, the prima facie assumption is that the title to land beneath the tidal portions of rivers lies with the Crown. Thus, at common law, title exists to the riverbed but the otherwise exclusive character of that title is subject to Magna Carta rights. Where there is a conflict between the rights associated with title, and the public rights of fishing and navigation, the latter prevail.[47]

Extrapolating from this, it seems logical that aboriginal title could coexist with the common law public rights of navigation and fishing. In fact, there is even a possible argument that if a court recognized aboriginal title with relation to the bed of the tidal portion of a river, this would oust any rights based on Magna Carta.[48] The argument here would be that Magna Carta is irrelevant when the title being claimed arises not from a Crown grant but from use and occupation of the land before British sovereignty. Based on the Supreme Court of Canada's approach in *R. v. Gladstone*[49] to reconciling aboriginal fishing rights with Magna Carta rights, it seems doubtful, however, whether courts would currently be willing to accept such an argument. In *Gladstone*, Lamer CJC stated:

> [T]he aboriginal rights recognized and affirmed by s. 35(1) exist within a legal context in which, since the time of the Magna Carta, there has been a common law right to fish in tidal waters that can only be abrogated by the competent legislation. . . . While the elevation of common law rights to constitutional status obviously has an impact on the public common law rights to fish in tidal waters, it was surely not intended that, by the enactment of s. 35(1), those common law rights would be extinguished in cases where an aboriginal right to harvest fish commercially existed. . . . [I]t was not contemplated by *Sparrow* that the recognition and affirmation of aboriginal rights should result in the common law right of public access in the fishery ceasing to exist with respect to all those fisheries in respect of which exist an aboriginal right to sell fish commercially. As a common law, not constitutional, right, the right of public access to the fishery must clearly be second in priority to aboriginal rights; however, the recognition of aboriginal rights should not be interpreted as extinguishing the right of public access to the fishery.[50]

On the other hand, while courts may be unwilling to see aboriginal rights as extinguishing the common law rights of navigation and fishing, clearly these common law rights should not be seen as negating the possibility of aboriginal title. It is noteworthy that in the Marlborough Sounds case, the New Zealand Court of Appeal rejected the notion that "public interests" in navigation would "make private property interests somehow unthinkable."[51] Even if a court felt compelled to construct the rights associated with aboriginal title in such a way that the common law public rights were not completely ousted, *Gladstone* makes it clear that constitutionally protected rights would have some degree of priority over those grounded only in the common law – which is very different than saying that aboriginal title cannot exist in areas subject to public rights.[52]

Implications for aquaculture policy

A 1982 publication, *Aquaculture: The Legal Framework*, noted the relevance of aboriginal title to aquaculture in Nova Scotia:

> Aboriginal land claims are of particular interest to the aquaculturist as they claim a usufructory interest, i.e. a right to use the land and resources as they had historically. These uses and the area where they are carried on are pertinent to aquaculture, as they include taking shellfish and marine plants along the marine foreshore and salmon along the rivers of the province.[53]

Thus far, the potential significance of aboriginal title for aquaculture policy does not appear to have been explored in the academic literature. However, the potential interplay between aboriginal title and aquaculture has become even more significant than would have been the case in 1982. As noted above, in *Delgamuukw* the Supreme Court of Canada confirmed that aboriginal title is more than simply a right to use the land in traditional ways; it is title to the land itself and carries with it a right to use the land for a broad range of activities and the right to exclude others. Furthermore, there are now constitutional restraints on the government's ability to infringe aboriginal title, either directly through its own actions or by permitting others to engage in activities that would interfere with a First Nation's title to land.

If aboriginal title were recognized in a riverbed or a portion of the seabed, could the First Nation holding aboriginal title decide to use the area for aquaculture? If so, would the First Nation be required to obtain a license under, or otherwise adhere to, the aquaculture legislation in that jurisdiction? Could a First Nation prohibit others from carrying on aquaculture in the aboriginal title area, or decide to permit but regulate such activities? These questions raise two issues: first, absent any government regulation, what rights would flow from the recognition of aboriginal title? And second, to what extent could government restrict those rights?

Rights associated with aboriginal title

Would aboriginal title give a First Nation the right to engage in aquaculture or to control whether or how others could engage in aquaculture? As noted, the only limit that the Supreme Court of Canada has put on the use of aboriginal title land is that the uses must not be irreconcilable with the community's attachment with the land. There is already disagreement in the academic literature as to exactly how this limitation should be interpreted, but it seems unlikely that this would preclude all First Nations from using aboriginal title land for aquaculture. If such activity were challenged, a court would have to decide, on a case-by-case basis, whether the particular form of aquaculture being practiced was irreconcilable with the attachment of the particular First Nation to its land.

Assuming that aquaculture could take place on aboriginal title land, could the First Nation regulate how its own community members practiced aquaculture, prohibit outsiders from coming into the area for aquaculture purposes, or decide to allow outside involvement, but regulate it? Subject to what is said below regarding justified infringements of aboriginal rights, authority to do each of these things would seem to flow from the fact of holding title. With regard to regulating its own aquaculture activities, where land is held by a community (as is the case for aboriginal title land), presumably the government structure of that community has the authority to regulate members' use of the land.[54] With regard to outsiders engaging in aquaculture, the Supreme Court of Canada has made it clear that aboriginal title, like other title to land, includes a right to exclude others, so (again subject to what is said below) presumably it would be up to the First Nation to decide whether an individual or corporation outside the community should be allowed to engage in aquaculture in the aboriginal title area. The corollary of the power to exclude is the power to invite others in, and to regulate the conduct of those so invited. Therefore, absent valid legislation limiting the rights flowing from aboriginal title, it seems that First Nations holding land by way of aboriginal title might well, depending on the nature of their historic connection to the land, have the right to engage in aquaculture. They would also have the authority to prohibit or permit such activities by those outside the community and to regulate any aquaculture that was permitted.

Justified infringement of aboriginal rights

It is possible, however, that governments might be able to restrict the rights referred to above, whether by prohibiting the First Nation from carrying on aquaculture, by allowing it to do so but requiring adherence to a federal or provincial regulatory scheme, or by controlling any aquaculture activities by those outside the First Nation.

Although aboriginal rights, including aboriginal title, are recognized and affirmed by the Constitution, courts have held that these rights are not absolute. They can be infringed by both the federal and provincial governments,[55] provided that the infringement can be justified. The Supreme Court of Canada has set out a two-part analysis on the issue of justification.[56] First, an aboriginal nation wishing to challenge legislation as infringing its aboriginal rights bears the onus of proving prima facie infringement; not every application of legislation will be seen as an infringement of aboriginal rights. To determine whether there is a prima facie infringement, the courts must consider various factors, including whether the legislation limits the exercise of the right, whether the limitation is unreasonable, and whether it imposes undue hardship.

If the First Nation is able to show infringement using these tests, the second step of the analysis comes into play. It is up to the Crown to show that applying the legislation in this context could be justified. The Crown would have to prove that the legislation in question was "enacted according to a valid objective,"[57] and that the infringement of the aboriginal right can be justified in terms of the "honour of the Crown."[58] According to the Supreme Court of Canada in *Delgamuukw*, this might mean that the government would have to show that it accommodated the participation of Aboriginal peoples in resource development or that Aboriginal peoples had been involved in decision-making in respect of their lands.[59] *Delgamuukw* also stated that there would always be a duty to consult where the government is seeking to justify the infringement of aboriginal rights; however, the nature and extent of the consultation required could differ significantly from case to case.[60] The Supreme Court of Canada also held that since aboriginal title "has an inescapably economic aspect . . . fair compensation would ordinarily be required when aboriginal title is infringed."[61] Therefore, rights flowing from aboriginal title could be limited by aquaculture legislation if the Crown could show that the relevant provisions of the Act met the test for justification.

Common law riparian rights

Introduction

This section considers whether a First Nation might be able to claim proprietary rights in a riverbed by virtue of the fact that it holds aboriginal title to the adjacent land; that is, whether rights could be claimed through the application of common law principles regarding riparian rights[62] without the need to prove aboriginal title to the riverbed itself. As was noted in the previous section, the English common law drew a distinction between tidal and non-tidal waters in determining the ownership of the waterbed. Ownership to the beds of tidal waters lay prima facie with the Crown, and since the time of Magna Carta, there has been a public right to fish in such waters, which could only be restricted by an Act of Parliament. When non-tidal water runs in a definite stream or channel, there is at common law a presumption that the owner of land bounded by the river or stream owns the submerged land to the centre of the riverbed (*usque ad medium filum aqua*).[63] This presumption could be rebutted by a contrary expression in the grant or conveyance, or by evidence that the grantor of the abutting lands had not intended to convey the stream bed as well.[64]

At common law, fishing rights in non-tidal waters, unless at some time separated and conveyed as a *profit a prendre*, go with ownership of the riverbed.[65] Thus, there existed in England an exclusive common law right to fish in non-tidal waters, which right was an incident of ownership of the submerged land beneath. Finally, at English common law the soil beneath a lake or pool as well as the water[66] in it belonged to the person whose land surrounded it.[67] Where one's property abuts rather than surrounds a lake, the law was for a time less clear,[68] but according to Cheshire and Burns it is likely that the same rules would apply as to non-tidal waters.[69]

Application in Canada

In Canada, the *ad medium filum* presumption has been applied more narrowly in some jurisdictions than was the case in England, and most provinces and territories have passed legislation placing the ownership of watercourses and waterbeds in the Crown. While English common law distinguished between tidal and non-tidal waters, with the *ad medium filum* presumption applying to non-tidal waters, in the western provinces courts have drawn the distinction between navigable and non-navigable waters such that the presumption applies only to waters that are both non-tidal and non-navigable. Similarly, in Ontario the *Beds of Navigable Waters Act*[70] ousts the presumption with regard to navigable waters. In these provinces, then, riparian rights to the waterbed could at most apply to those parts of rivers or streams that are non-tidal and too small for navigation.

In Atlantic Canada, however, the English approach has been followed, so that the key issue is whether water is tidal, not whether it is in fact navigable.[71] The Canadian situation was summarized by the Supreme Court of Canada in a 1992 decision, *Friends of the Oldman River*:

> Except in the Atlantic provinces, where different considerations may well apply, in Canada the distinction between tidal and non-tidal waters was abandoned long ago. Instead the rule is that waters are navigable in fact whether or not the waters are tidal or non-tidal.[72]

Even more significant is that fact that most provinces have, by legislation, appropriated ownership of all river and stream beds to the provincial Crown.[73] Similar legislation has been passed by the federal government in relation to the Yukon, the Northwest Territories and Nunavut.[74] Thus, of the common law provinces and territories, common law riparian rights would seem only to exist in Ontario, New Brunswick, Prince Edward Island, and Newfoundland and Labrador, and even in several of these provinces the common law rights have been significantly limited.[75]

Potential application of common law riparian rights to aboriginal title lands

A consideration of whether the common law presumption that the owner of riparian lands also owns a portion of the riverbed could be applied to riparian lands held by way of aboriginal title is, at best, highly speculative, given the lack on jurisprudence on this issue. The Supreme Court of Canada has been willing to assume without deciding that the presumption might apply to reserve lands,[76] and a 1985 decision of the British Columbia Supreme Court, *Pasco* v. *C.N.R.*,[77] held that that there was "a serious question to be tried" with regard to an Indian band's claim of riparian rights attached to a reserve.[78] Given comments by the Supreme Court in both *Guerin*[79] and *Delgamuuk*[80] regarding the similarity of the interest held by aboriginal communities in reserve lands and aboriginal title lands, these cases at least leave open the possibility of arguing that the presumption is relevant. However, there is very little else to guide the discussion or to suggest how courts might respond to the series of questions that would have to answered if the riparian rights presumption were held to be applicable. All that this section attempts to do, therefore, is to outline the questions and subquestions that would arise, depending on how a court reasoned at each stage of the inquiry, and to sketch in some of the factors that might be taken into account.

In considering the question "do common law riparian rights attach to aboriginal title land?" a key issue may well be whether the application of the *ad medium filum* presumption is limited to the interpretation of a grant – in other words, whether the existence of title created by grant is a necessary precondition for the application of the presumption. From the perspective of aboriginal title, it matters greatly whether the rule is seen to mean "if you are granted or conveyed property fronting on a non-tidal portion of a river, it is assumed, absent evidence to the contrary, that you were also granted or conveyed half the riverbed" or whether it means "if you hold title to property fronting on a non-tidal portion of a river, it is assumed, absent evidence to the contrary, that you also hold title to the half the riverbed." Only the latter would permit the argument that riparian rights could attach to aboriginal title land, given that aboriginal title is founded on historic use, rather than on a Crown grant. *Halsbury's* simply states that "By a presumption of law, and in the absence of any evidence to the contrary, the ownership of a bed of a non-tidal river or stream belongs in equal halves to the owners of the riparian land."[81] Thus, the focus in *Halsbury's* seems to be on the fact of ownership rather than on the source of that ownership. However, *Anger and Honsberger Law of Real Property* refers specifically to a Crown grant: "[T]he rule respecting non-tidal waters applies only if there is a Crown grant extending to the centre of the water. If the bed is not included in the grant, the land extends only to the water's edge."[82]

As was noted above, the Supreme Court of Canada has left for another day the issue of how or whether the presumption might operate in the context of reserve lands. In both *R.* v. *Nikal*[83] and *R.* v. *Lewis*,[84] it was argued that a fishing by-law passed by the band applied as far as the midpoint of a river adjacent to the reserve, by virtue of the *ad medium filum* presumption. In *Nikal*, the court held that the river in question was navigable and therefore the presumption did not apply. In *Lewis*, the court decided against the claimants both because of the navigability of the river and because of historical evidence regarding the government's intention in creating the particular reserve. However, in *Nikal* the British Columbia Court of Appeal saw the absence of a grant or conveyance as fatal to a claim that riparian rights attached to reserve land:

> The creation of the Moricetown reserve did not involve a grant or conveyance of title or ownership to the Gitksan Wet'suwet'en people. It affirmed their right to use and occupation of the land. The English property rule in question applies to the interpretation of grants and conveyances of land. It has no application to circumstances which do not involve a grant but which recognize or affirm existing rights.[85]

In *Lewis*, despite the statement from the Supreme Court of Canada that for the purposes of the appeal it would assume without deciding that the presumption could apply to reserve lands, Iaccobucci J, speaking for the court, raised the question of whether title in reserve land would be seen as ownership for the purposes of the presumption: "At the outset, it should be noted that since the *ad medium filum aquae* presumption related to ownership of land, the question remains as to whether it applies to Indian reserves."[86] If a court held that riparian rights could attach to aboriginal title land, the court would then have to consider the nature of the estate acquired by way of the presumption: would the First Nation hold the riverbed by way of fee simple, as would be the case in a wholly common law context, or by way of aboriginal title? While it seems logical that the presumption, if applicable, would replicate the title to which it attached (meaning that the First Nation would hold the riverbed by way of aboriginal title), there is no case law on this point.

If riparian rights could attach to aboriginal title land and if the title in the riverbed so acquired were characterized as aboriginal title, rather than fee simple, the question then arises as to whether the title to the riverbed would have the same protection as other aboriginal title. On the one hand, it might be argued that since title to the riverbed is not based on historic occupation and use of the riverbed itself,[87] it is not deserving of the constitutional protection provided to aboriginal title in other contexts. The opposing argument would be that s. 35(1) of the *Constitution Act 1982* simply refers to aboriginal title and so there is no basis for creating different categories of aboriginal title, with different levels of protection. Again, there is no case law to guide this discussion. However, one author has suggested that "since riparian rights are not Aboriginal rights, they do not need to meet the requirements of infringement or extinguishment set out in the doctrine of Aboriginal rights."[88]

If this were the case, then aboriginal title in a riverbed acquired by way of the *ad medium filum* presumption could be restricted without governments having to meet the test for justification, and could be unilaterally extinguished by the federal government. The argument might be made that such a different species of aboriginal title, arising solely from the application of the common law and therefore having little inherently "aboriginal" about it, could even be extinguished by provincial governments. If the court adopted this approach, and if a claim arose in a province or territory where title to waterbeds is, by legislation, vested in the Crown, then the court would have to consider whether such legislation should be seen as expressing sufficiently clear and plain intent to extinguish the aboriginal title.

In some jurisdictions, this question might be answered by the wording of the legislation itself. If the Act was to the effect that after a certain date any grant or conveyance of riparian lands could not extend past the high-water mark, this would not seem to affect any aboriginal title based on the *ad medium filum* presumption, since aboriginal title does not depend on any grant or conveyance for its existence. Where legislation simply states that the title to the beds of all watercourses lies with the Crown, a court would have to decide whether this meant that the Crown intended to acquire the unencumbered fee simple and whether it had made that intention sufficiently plain to extinguish aboriginal title. An alternative argument might be that since aboriginal title coexists with Crown title, the fact that the riverbed is held by the Crown would be no impediment to a claim based on riparian rights.

If a court held both that the riparian presumption could attach to aboriginal title lands, creating aboriginal title in the riverbed, and that all aboriginal title, however acquired, has the same constitutional protection, then the title in the riverbed could only be infringed by legislation meeting the tests for justification. Furthermore, it could only be extinguished by the federal Crown, and could not be unilaterally extinguished after 1982. In this scenario, legislation placing the ownership of waterbeds in the provincial Crown would certainly not extinguish such rights, as provincial governments cannot extinguish aboriginal title.[89] With regard to the impact of legislation vesting waterbeds in the federal Crown, then the earlier discussion on the wording of such legislation and the possible coexistence of the Crown title with aboriginal rights would be relevant here as well. Interestingly, in *Lewis* and *Nikal* the Supreme Court of Canada rejected claims based on riparian ownership on the grounds that the river was navigable, and therefore the *ad medium filum* presumption would not apply. The court did not state that all such rights have been ousted in British Columbia by the *Water Act*.

Conclusion

The first section of this chapter considered whether the doctrine of aboriginal title, as developed thus far by Canadian courts, could apply to rivers, lakes and the seabed. While there is no Canadian jurisprudence directly on point, it is at least arguable that the doctrine of aboriginal title might apply in these areas. Aboriginal title carries with it the right to use the land for a variety of purposes, the right to exclude others and the right to regulate use of the land. In areas where aboriginal title claims might be made out, careful thought will therefore have to be given to accommodating aboriginal title rights within aquaculture policy, and considering the extent to which attempts to limit such rights would be upheld as justifiable.

The second part of this chapter asked whether the common law *ad medium filum* presumption could be used by an aboriginal community holding river-bank land by way of aboriginal title in order to claim title to the riverbed. Given the lack of relevant case law and the number of questions that would have to be worked through by any court tackling this question, it is difficult to draw even tentative conclusions on this question. However, the possibility of riparian rights claims cannot be rejected out of hand, and so any developments in this area of the law should be watched closely by those responsible for developing and implementing aquaculture policy.

Notes

I would like to thank the AquaNet Project for providing funding for research assistance. I would also like to thank Anne Tardif (LL.B., Dalhousie University, 2005) for her research assistance on this chapter.

1 While there are other categories of rights that might be relevant to the issue of First Nations and aquaculture – for instance, treaty rights or aboriginal fishing or gathering rights – these are not discussed here. My focus is solely on aboriginal or common law rights to the land itself.
2 Each of the provinces and territories of Canada (with the exception of Québec) incorporated English law as of a certain date (ranging from the mid-eighteenth to the early nineteenth century, depending on the jurisdiction) into its law. Thus, English common law is relevant to the issue of property rights in Canada. It must be remembered, however, that the common law can evolve over time. In some instances, Canadian courts have specifically altered the English common law to take account of different circumstances pertaining to Canada. Further, the common law can be changed or abrogated by statute.
3 *Constitution Act 1982*, s. 35(1), being Schedule B to the *Canada Act 1982* (U.K.), 1982, ch. 11.
4 The range of rights encompassed by "aboriginal rights" was discussed by the Supreme Court of Canada in *R. v. Adams* [1996] 3 S.C.R. 101 (an aboriginal fishing rights case). The Supreme Court of Canada stated that "claims to title to the land are simply one manifestation of a broader-based concept of rights" (para. 25), and that "fishing and other aboriginal rights can exist independently of a claim to aboriginal title" (para. 3). This was again discussed in *Delgamuukw* v. *British Columbia* [1997] 3 S.C.R. 1010 [*Delgamuukw*]:

> The picture which emerges from *Adams* is that the aboriginal rights which are recognized and affirmed by s. 35(1) fall along a spectrum with respect to their degree of connection with the land. At one end, there are those aboriginal rights which are practices, customs and traditions that are integral to the distinctive aboriginal community of the aboriginal culture of the group claiming the right. However, the "occupation and use of the land" where the activity is taking place is not "sufficient to support a claim of title to the land." ... Nevertheless, those activities receive constitutional protection. In

the middle, there are activities which, out of necessity, take place on land and indeed, might be intimately related to a particular piece of land. Although an aboriginal group may not be able to demonstrate title to the land, it may nevertheless have a site-specific right to engage in a particular activity.... At the other end of the spectrum, there is aboriginal title itself. As *Adams* makes clear, aboriginal title confers more than the right to engage in site-specific activities which are aspects of the practices, customs and traditions of distinctive aboriginal cultures.... What aboriginal title confers is the right to the land itself (para. 138).

5 *Delgamuukw, supra* note 4 at para. 111
6 The 1973 case of *Calder* v. *British Columbia (Attorney General)* [1973] S.C.R. 313 [*Calder*], which forms the foundation for the modern law on aboriginal title, recognized that such title is inherent. Judson J stated, "[w]hen the settlers came, the Indians were there, organized in societies and occupying the land as their forefathers had done for centuries" (328). In *Delgamuukw, supra* note 6, the Supreme Court of Canada described "the prior occupation of Canada by Aboriginal peoples" (para. 114) as one of the sources of aboriginal title. In *Delgamuukw*, the court also set out the following requirements for the establishment of aboriginal title:

 i   the land must have been occupied prior to sovereignty,
 ii if present occupation is relied on as proof of occupation pre-sovereignty, there must be a continuity between present and pre-sovereignty occupation, and
 iii at sovereignty that occupation must have been exclusive (para. 143).

7 *Delgamuukw, supra* note 4 at para. 114.
8 *Ibid.* at para. 111.
9 *Ibid.* at paras. 112–114, and *Canada* v. *Guerin* [1984] 2 S.C.R. 335 at 336 [*Guerin*]. See John Borrows, "Listening for a Change: The Courts and Oral Tradition," (2001) 39 *Osgoode Hall Law Journal* 1, for commentary on this.
10 *Delgamuukw, supra* note 4 at para. 115.
11 *Ibid.* at paras. 116, 117.
12 *Ibid.* at para. 111. For commentary on this restriction, see Nigel Bankes, "Delgamuukw, Division of Powers and Provincial Land and Resource Laws: Some Implications for Provincial Resource Use," (1998) 32 *University of British Columbia Law Review* 317 [Bankes, "*Delgamuukw*"]; Richard H. Bartlett, "The Content of Aboriginal Title and Equality before the Law," (1998) 61 *Saskatchewan Law Review* 377 [Bartlett, "Content"]; Brian Burke, "Left Out in the Cold: The Problem with Aboriginal Title under s. 35(1) for Historically Nomadic Aboriginal Peoples," (2000) 38 *Osgoode Hall Law Journal* 1; William Flanagan, "Piercing the Veil of Real Property Law: *Delgamuukw v. B.C.*," (1998) 24 *Queen's Law Journal* 279; and Kent McNeil, *Defining Aboriginal Title in the 90s: Has the Supreme Court Finally Got It Right?* (Toronto: Robarts Centre for Canadian Studies, York University, 1998) at 117–118 [McNeil, *Defining*].
13 *Calder, supra* note 6 at 353, and *Delgamuukw, supra* note 4 at para. 145.
14 *Delgamuukw, supra* note 4 at para. 113.
15 *R.* v. *Van der Peet* [1996] 2 S.C.R. 507 at para. 28, and *Delgamuukw, supra* note 4 at para. 35.
16 *Constitution Act 1867* (UK), 30 & 31 Victoria, c. 3, s. 91(24).
17 *Delgamuukw, supra* note 4 at para. 180.
18 *Ibid.* at para. 112.
19 There are, however, several authors who argue that the doctrine is applicable. Terence P. Douglas has suggested that, given the wording in *Calder, supra* note 6, "[a]lthough there is an absence of jurisprudence specifically relating to a

claim of Aboriginal title to specific bodies of water of water, the available case law permits the inference that such a claim is compatible within the context of Aboriginal title." See Terence P. Douglas, "Sources of Aboriginal Water Rights in Canada," at para. 17 [Douglas, "Sources"]. Online. Available http://www.firstpeoples.org/land_rights/canada/summary_of_land_rights/water_rughts.htm (accessed 14 April 2004). Even more emphatically, Bartlett, "Content," *supra* note 12, has stated, "A right to water is . . . an integral part of aboriginal title."

20 Action No. L020662, filed in the Supreme Court of British Columbia, 6 March 2002.
21 These cases include the following:

In *Kainaiwa/Blood Tribe* v. *Canada* [2001] F.C.J. No. 1502 (F.C.T.D.), aboriginal groups claimed use and ownership of rivers and riverbeds adjoining their reserve, by way of "existing aboriginal rights, treaty rights, or as riparian owner" (para. 6). However, the interim decision of the Federal Court focused solely on the jurisdiction of the court to hear a claim against the province of Alberta.

In 1999, the Association of Mi'kmaq Chiefs of Nova Scotia opposed the grant of a license allowing Maritime and Northeast Pipeline to build a natural gas pipeline over Crown lands on the grounds that "[t]he Mi'kmaq did not surrender Aboriginal title to the provincial Crown lands, including watercourse lands, along the pipeline corridor" (application reproduced in *Union of Nova Scotia Indians* v. *Nova Scotia (Attorney General)* [1999] N.S.J. No. 270 (N.S. S.C.) at para. 30). There has not been any decision on this issue.

In *Oregon Jack Creek Band* v. *CNR* [1989] B.C.J. No. 211, the Band claimed aboriginal title in a "river system." However, a 1989 decision of the British Columbia Court of Appeal dealt solely with the issue of whether the trial judge had been correct in refusing to allow the plaintiffs to amend their statement of claim so as to make a claim not only on behalf of three bands, but on behalf of the members of three First Nations.

22 *Re Marlborough Sounds*, unreported, Maori Land Court, 22A Nelson Minute Book I, 22 December 1997.
23 *Ngati Apa* v. *Attorney-General*, [2002] 2 N.Z.L.R. 661 [*Marlborough Sounds*]. The matter came before the High Court as stated case on questions of law.
24 *Ibid.* With regard to the foreshore, the High Court held, in keeping with the decision of *In Re Ninety-Mile Beach* [1963] N.Z.L.R. 461, that "the foreshore cannot be customary land unless the adjoining land is also customary land, as the rights to the foreshore go with the dry land" (para. 36).
25 *Territorial Sea and Exclusive Economic Zone Act 1977* (N.Z), 1977/28.
26 *Marlborough Sounds*, *supra* note 23 at para. 16. The focus of the Maori Land Court in New Zealand is to investigate claims to title, and, where such claims are substantiated, to convert the customary title to fee simple. This is very different from the situation in Canada, where the Supreme Court of Canada has made it clear that aboriginal title is not a form of fee simple. The High Court did state that its holding regarding title would "not preclude Maori from establishing customary rights over the foreshore, the seabed and the waters over them short of a right of exclusive possession" (para. 52).
27 *Ngati Apa* v. *Ki Te Tan Ihu Trust* [2003] N.Z.C.A. 117.
28 *Ibid.* at para. 8.
29 *Ibid.* at para. 12.
30 *Ibid.* at para. 51.
31 *Ibid.* at para. 63.
32 *The Commonwealth* v. *Yarmirr* [2001] H.C.A. 56 at para. 1 [*Yarmirr*].
33 *Ibid.* at para. 2.
34 *Seas and Submerged Lands Act 1973* (Cth.).

35 *Yarmirr, supra* note 32 at para. 54.

36 *Ibid.* at para. 57.

37 *Ibid.* at para. 61.

38 On appeal, the claimants had responded to the primary judge's refusal to recognize exclusive native title by "acknowledging the existence of the public rights to navigate and to fish and the right of innocent passage and contending that a determination of native title should be made subject to a qualification recognizing those rights" (*ibid.* at para. 94). However, the majority of the High Court responded to this by stating (*ibid.* at para. 98):

> [T]here is a fundamental inconsistency between the asserted native title rights and interests and the common law public rights of navigation and fishing, as well as the right of innocent passage. The two sets of rights cannot stand together and it is not sufficient to attempt to reconcile them by providing that the exercise of native title rights and interests is to be subject to the other public and international rights.

39 Lord Hailsham of St. Marylebone, ed., *Halsbury's Laws of England*, 4th ed., Vol. 49(2) (London: Butterworths, 1973) at para. 86 [*Halsbury's*].

40 *Delgamuukw, supra* note 4 at para. 112.

41 S.C. 1996, c. 31. The issue of Crown and aboriginal title in the seabed is discussed in greater detail in Diana Ginn, "Aboriginal Title and Oceans Policy in Canada," in D. Rothwell and D. VanderZwaag, eds., *Towards Principled Oceans Governance: Australian and Canadian Approaches and Challenges* (London: Routledge, 2006).

42 Where it is argued that submerged land belongs to a private, non-aboriginal owner rather than the Crown, presumably the first question would be whether any aboriginal title to the area had been lawfully extinguished, as this would seem to be a precondition to the Crown having the authority to grant the land to others. If aboriginal title did exist in the area and there had been no valid extinguishment, then it would seem the underlying radical title would still lie with the Crown and would be burdened by the aboriginal title. If aboriginal title existed, and was extinguished without consent, then compensation to the First Nation might well be in order, since Lamer CJC stated in *Delgamuukw, supra* note 4 at para. 168, that, given the inescapably economic aspect of land, loss of title would ordinarily require compensation.

43 Whether the underlying fee is federal or provincial should not affect the possibility of aboriginal title existing. Litigation might be necessary to determine which Crown holds the unencumbered fee simple once aboriginal title has been surrendered or extinguished, but that presumably is irrelevant to the question of whether aboriginal title exists.

44 It is worth noting that the language used in the Nunavut Land Claims Agreement (Ottawa: published under the joint authority of the Tungavik and the Honourable Tom Siddon, PC, MP, Minister of Indian Affairs and Northern Development, 1993) seems to recognize the possibility that the Inuit held aboriginal title to water areas:

> 2.7.1 In consideration of the rights and benefits provided to Inuit by the Agreement, Inuit hereby:
>
> (a) cede, release and surrender to Her Majesty The Queen in Right of Canada, all their aboriginal claims, rights, title and interests, if any, in and to lands *and waters anywhere within Canada and adjacent offshore areas within the sovereignty* or jurisdiction of Canada. [emphasis added]

45 *Alford* v. *Canada (Attorney General)* [1997] B.C.J. No. 251 (S.C.) at para. 17.

46 *Canada* v. *Robertson* (1882) 6 S.C.R. 52 at 88.
47 Gerard V. LaForest, *Water Resources Study of the Atlantic Provinces* (Ottawa: Department of Regional Economic Expansion, 1968) at 27–28 [LaForest].
48 See Bankes, *"Delgamuukw,"* *supra* note 12 at para. 2; Peggy Blair, "Solemn Promises and Solemn Rights: The Saugeen Ojibeway Fishing Grounds and *R.* v. *Jones and Nadjiwon,"* (1996–1997) 28 *Ottawa Law Review* 125; and Mark Walters, "Aboriginal Rights, Magna Carta and Exclusive Rights to Fisheries in the Waters of Upper Canada," (1998) 23 *Queen's Law Journal* 301.
49 *R.* v. *Gladstone* [1996] 2 S.C.R. 723.
50 *Ibid.* at para. 67.
51 *Supra* note 27 at para. 50.
52 This is discussed further in Ginn, *supra* note 41.
53 B. Wildsmith, *Aquaculture: The Legal Framework* (Toronto: Emond-Montgomery Ltd., 1982) at 165.
54 In *Delgamuukw, supra* note 4 at para. 115, Lamer CJC noted that:

> [a] further dimension of aboriginal title is the fact that it is held communally. Aboriginal title cannot be held by individual aboriginal persons; it is a collective right to land held by all members of an aboriginal nation. Decisions with respect to that land are also made by that community. This is another feature of aboriginal title which is *sui generis* and distinguishes it from normal property interests.

55 For commentary on provincial power to infringe, see J. Lambert, *"Van der Peet* and *Delgamuukw:* Ten Unresolved Issues," (1998) 32 *University of British Columbia Law Review* 249 at 226; and McNeil, *Defining, supra* note 12 at 25.
56 This test was first developed in *R.* v. *Sparrow* [1990] 1 S.C.R. 1075 [*Sparrow*], an aboriginal fishing rights case. In *Delgamuukw, supra* note 4, Lamer CJC applied the *Sparrow* test in the context of aboriginal title.
57 *Sparrow, supra* note 56 at para. 64.
58 *Ibid.*
59 *Delgamuukw, supra* note 4 at para. 168.
60 *Ibid.*
61 *Ibid.* at para. 169.
62 The term "riparian rights" is sometimes used to encompass both rights of water usage, which were not dependent on ownership of the underlying waterbed, and ownership of the riverbed. At common law, the owner of land abutting a river or stream or the sea has certain rights to use the water, including:

- the right of access to the water;
- the right of drainage;
- rights relating to the flow of water;
- rights relating to the quality of water (pollution);
- rights relating to the use of water;
- the right of accretion (LaForest, *supra* note 47 at 32–33).

The focus of this chapter, however, is rights of ownership; that is, title to the land acquired by way of riparian rights, rather than on rights of usage.
63 *Halsbury's, supra* note 39 at para. 101.
64 *Ibid.* at para. 106. La Forest, *supra* note 47 at 82, expresses the English rule as follows:

> That the owner of the land through which a non-tidal stream flows owns the bed of the stream unless it has been expressly or impliedly reserved, and if the stream forms the boundary of lands owned by different persons, each proprietor owns the bed of the river *ad medium filum aquae* – to the centre thread of the river.

65 Sir Robert E. Megarry and Sir William Wade, *The Law of Real Property*, 4th ed. (London: Stevens, 1975) at 30–56.

66 Standing water was the exception to the general common law rule that water could not be owned.

67 *Halsbury's, supra* note 39 at para. 107.

68 *Ibid.*

69 Cheshire and Burns, *Modern Law of Real Property*, 15th ed. (London: Butterworths, 1994) at 165, and LaForest, *supra* note 47 at 86.

70 *Beds of Navigable Waters Act*, R.S.O. 1990, c. B-4, s. 1:

> Where land that borders on a navigable body of water or a stream or on which the whole or part of a navigable body of water or stream is situate, or thru which a navigable body of water or stream flows has been or is granted by the Crown, it shall be deemed, in the absence of an express grant of it, that the bed of such body of water was not intended to pass and did not pass to the grantee.

71 Thus La Forest, *supra* note 47 at 94, states, "There is no instance in any of the Atlantic provinces where rivers have been considered navigable unless they were tidal and the courts have throughout acted on the basis that the English law prevails."

72 *Friends of the Oldman River* v. *Canada* (Minister of Transport) [1992] 1 S.C.R. 3 at 54.

73 For instance, in what are now the three prairie provinces, the common law presumption was initially ousted by federal legislation, the *Northwest Irrigation Act 1894*, 57 & 58 Vict., c. 30., which vested "property in and the right to use" of all watercourses in the Crown. In 1930, this was transferred to the provincial Crowns, as part of the transfer of ownership of natural resources.

74 The *Northwest Territories Waters Act*, S.C. 1992, c. 39, s. 4, the *Nunavut Waters and Nunavut Surface Rights Tribunal Act*, S.C. 2002, c. 10, s. 8(1) and the *Yukon Waters Act*, R.C. 1992, c. 40, s. 4, all state:

> Subject to any rights, powers or privileges granted pursuant to the *Dominion Water Power Act* or preserved under that Act, the property in and the right to the use and flow of all waters are vested in Her Majesty in right of Canada.

75 For instance, LaForest, *supra* note 47 at 122, states with regard to New Brunswick:

> [W]ith limited exceptions, private interests in land abutting on the above named rivers, deriving from Crown grants after 1884, and private interests in lands abutting on any river or lake deriving from a Crown grant after 1927, are subject to riparian ownership reserved to the Crown.

76 *R.* v. *Lewis* [1996] 1 S.C.R. 921 [*Lewis*]; *R.* v. *Nikal* [1996] 1 S.C.R. 1013 [*Nikal*].

77 *Pasco* v. *C.N.R.* [1985] B.C.J. No. 2818 (B.C. S.C.) [*Pasco*]. In this case, the Indian band had sought an interim injunction to prevent the Canadian National Railway from building tracks along a river that ran beside the reserve, construction of which would involve adding some fill to the river. The band argued that this would interfere with their fishery, and also claimed ownership in the riverbed, based on the *ad medium filum* presumption.

78 *Ibid.* at para. 29.

79 *Guerin, supra* note 9.

80 *Delgamuukw, supra* note 4.

81 *Halsbury's, supra* note 39 at para. 101.

82 A. H. Oosterhoff and W. B. Rayner, *Anger and Honsberger Law of Real Property*, 2nd ed., Vol. 2 (Aurora, ON: Canada Law Book, 1985) at 987.

83 *Nikal, supra* note 76.

84 *Ibid.*

85 *R. v. Nikal* [1993] B.C.J. No. 1399 at para. 58 (C.A.).

86 *Lewis, supra* note 76 at para. 57.

87 If the test for aboriginal title would be met, it would seem more straightforward to attempt to claim the submerged land directly by way of an aboriginal title claim, rather than through the riparian rights route.

88 Douglas, "Sources," *supra* note 19 at para. 7.

89 *Pasco, supra* note 77, raised the question of whether provincial legislation vesting the title to waterbeds in the provincial Crown is of any effect when the adjacent land is held as an Indian reserve. In *Pasco*, the Court posed the following question with regard to the British Columbia *Water Act*, R.S.B.C. 1986, c. 483: "Does the province have the legislative competence to deny riparian rights to the federal Crown in connection with an Indian reserve? . . . could such a provincial power infringe on federal rights in respect of Indians and fisheries?" (paras. 28–29).

9 Aquaculture law and policy in Canada and the duty to consult with Aboriginal peoples

Ronalda Murphy, Richard Devlin and Tamara Lorincz

Introduction

In November 2003, a Mi'kmaq elder from the community of Eskasoni launched a court action seeking to stop seismic testing in the waters off Cape Breton. He claimed that the government of Nova Scotia had failed to consult with his First Nation before issuing an approval to allow the testing by Corridor Resources, as part of its oil and gas exploration program.[1] Aboriginal communities throughout Canada assert they must be consulted before governments or corporations make decisions that could impair the constitutional rights of Aboriginal peoples. Invocation of the duty to consult as an independent source of legal entitlement is increasing. This chapter explains why these claims are being made and how they may operate in the specific context of aquaculture. Our objective is to illuminate the historical and political context of the duty to consult, and to canvass the responses of courts and governments to date. Unfortunately, the law is at a nascent stage of development, as the courts are struggling to give substance and structure to this novel doctrine.[2] Nonetheless, it is possible to identify basic themes and to outline the contours of the duty to consult as it presently is being defined by courts in response to urgent and complicated litigation.

Our conclusions can be briefly stated. Aboriginal communities may or may not have a right to develop, or refuse to develop, aquaculture projects. If they claim such a right, and the government has knowledge, real or constructive, of the potential existence of aboriginal right or title and the contemplated government conduct might adversely affect Aboriginal peoples, then the duty to consult is triggered. If the duty to consult exists, the consultation process may reveal a further duty to accommodate Aboriginal concerns by making changes to proposed actions. Because these duties exist and are triggered once a prima facie case has been made out, they afford aboriginal communities and governments the opportunity to draft plans for development that are responsive to the needs and concerns of Aboriginal peoples. Importantly, however, the duty to consult is distinct from the substantive right that triggers its application. We explain the significance of this distinction in the discussion of the case law. Our emphasis is on the holdings of the Supreme Court of Canada, though we make mention of a few significant decisions from lower courts.

The chapter is organized as follows. First, we sketch the overall legal context for aboriginal rights and the relationship between Aboriginal peoples and the governments of Canada. Then we explain the role of s. 35 of the *Constitution Act*, the provision that guarantees the rights of Aboriginal peoples, and the legal tests required for establishing a constitutional aboriginal right. With this in place, we introduce the legal regime for the duty to consult in the context of aquaculture and survey the responses of various governments to this issue. Finally, we outline the specific contours of the duty to consult as it has been articulated to date. The law can be quickly grasped by examining the following four questions: Who has the duty? What is the nature and intensity of the duty? Where does the duty arise? When does it get triggered? We conclude by addressing the question of why the duty exists, and we offer our tentative views regarding its potential impact on aquaculture development in Canada.

Historical legal context of the duty to consult

An understanding of the duty to consult requires an inquiry into the history and nature of aboriginal rights. The starting point is the colonization of what we now call North America and claims to sovereignty over Aboriginal peoples, in particular Britain's claim in Canada as explicitly stated in the Royal Proclamation of 1763.[3] In 1867, when four of the colonies agreed to the terms of the *British North America Act 1867* (BNA Act), they divided the legislative powers between themselves and the newly created federal government. The judiciary was, and remains, responsible for ensuring compliance with the division of powers in the Constitution. The federal government exercises legislative authority or jurisdiction over "Indians and Lands reserved for the Indians."[4] The policies and laws applied to "Indians" over the next hundred-plus years can only be described as vacillating from explicitly racist to effectively racist. Aboriginal peoples had very little political power, and nothing in the rule of law seemed capable of responding to the unjust treatment they received at the hands of the Canadian state.[5]

Indeed, with respect to law, under the principles of parliamentary (federal) or legislative (provincial) supremacy, any aboriginal rights recognized at common law could be abrogated, even extinguished. Validly enacted federal or provincial law could therefore eradicate any protection the common law afforded to Aboriginal peoples,[6] and, more shockingly, even extinguish treaty rights.[7] While the provincial government cannot legislate with respect to "Indians and the lands reserved to them," provincial laws can apply to Aboriginal peoples in one of two ways. First, the provincial law can have an incidental effect on Aboriginal peoples provided the law is otherwise validly enacted – in other words, it is a law that in pith and substance is one regarding a subject matter assigned to the provinces under s. 92 of the BNA

Act. Second, a provincial law of general application can apply by virtue of s. 88 of the *Indian Act*.[8]

The general principle of parliamentary or legislative supremacy operated until 1982, when the Canadian Constitution was amended to add the Canadian Charter of Rights and Freedoms ("the Charter"). Under the new legal regime, Canada became a constitutional democracy. Legislative enactments are now subject to the rights and freedoms enshrined in the Charter, or any rights established elsewhere under the *Constitution Act 1982*. The rights and freedoms protected in the Charter are subject to such reasonable limits as are prescribed by law and justified in a free and democratic society, which is set out in s. 1 of the Charter. The constitutional entrenchment of aboriginal rights is found in s. 35 of the *Constitution Act 1982*. It is not part of the Charter, and is not subject to s. 1 of the Charter. Section 35 provides that "the existing aboriginal and treaty rights of the Aboriginal peoples of Canada are hereby recognized and affirmed." While s. 35 is the "source" of current litigation and of central significance to Aboriginal peoples, a proper understanding of the law requires an examination of pre-s. 35 legal developments.

Recall that "Indians" and *lands* belonging to them are "assigned" to the federal level of government. In Canada, courts have identified key differences between title to aboriginal lands held by Aboriginal people, and other types of property rights with respect to land.[9] Aboriginal title exists in conjunction with radical or ultimate title flowing to the Crown, and aboriginal title can be sold only to the Crown. While aboriginal title can be put to a wide variety of uses, it cannot be put to a use that is inconsistent with the nature of the title being held by Aboriginal people as a communal interest. This is because of the origin of the title: it arises as a result of historical use and occupation of the land by Aboriginal people with a special relationship to their land. All of this makes the interest unique, or *sui generis*.

The Supreme Court of Canada ruled in the case of *Guerin* v. *The Queen*[10] that the federal government (in its capacity as the Crown) was required to deal with aboriginal land only in a manner that was for the benefit of that community. In reaching this conclusion, the Court interpreted various legislative clauses regarding the creation and sale of reserve lands, and the common law rules governing aboriginal title. Under those rules, aboriginal title lands cannot be sold directly to any third parties. They can be sold or "surrendered" only to the Crown, which in turn generates a "distinctive fiduciary obligation on the part of the Crown to deal with the land for the benefit of the surrendering Indians."[11] The rationale for this discretionary power in the Crown, as stated in the Royal Proclamation of 1763 and confirmed in the provisions of the *Indian Act*, was to "interpose the Crown between the Indians and the prospective purchasers or lessees of their land, so as to prevent the Indians from being exploited." The Court characterized this as a "historic responsibility" undertaken by the Crown, and a relationship that courts would supervise as a fiduciary obligation, albeit a *sui generis* one based on the "unique character both of the Indians' interest in land and of their historical relationship with the Crown."[12]

In this particular case, Crown agents negotiated the surrender of reserve lands so the lands could be leased to a golf club. However, they acted improperly at law by accepting contractual terms from the golf club that were worse than those agreed to by the aboriginal community. In engaging in that course of action, the federal agents were not acting in a manner consistent with the honor of the Crown. This honor is expressed as the Crown owing Aboriginal peoples a "fiduciary duty," or an equitable obligation to deal with the land for the benefit (not the disadvantage) of the aboriginal community involved. The Court explained that a breach of this duty occurred when the Crown agents failed to consult with the aboriginal community about the terms of the lease. It stated, "In obtaining, without consultation, a much less valuable lease than that promised, the Crown breached the fiduciary obligation it owed to the Band. It must make good the loss suffered in consequence."[13]

This case affirmed the existence of a fiduciary relationship and underlines the role that consultation plays within it. At this point, it is critical to note that the case was not a constitutional case, and there was no legislative action. Rather, there was a legislative framework in place (the *Indian Act*) that circumscribed what the Crown was entitled to do upon surrender of aboriginal land in light of the *historical relationship* between the Crown and Aboriginal peoples. Consultation was characterized as a feature of a court-imposed fiduciary relationship, and in this case proved to be a major obstacle to the Crown's attempts to deal with the land as it pleased. Nonetheless, it should not be forgotten that the court would not have refused to enforce a validly enacted law that altered or even eradicated aboriginal rights without consultation.

One example suffices to make this critical point. A series of numbered treaties were entered into between the federal government and a wide variety of aboriginal communities to facilitate settlement of non-Aboriginal people in the western region of Canada between 1871 and 1923. The treaties typically protected the right to commercial hunting by Aboriginal peoples. In the 1930s, however, the *Natural Resources Transfer Agreement* (NRTA) was entered into by each of the three western provinces of Alberta, Saskatchewan and Manitoba and the federal government. In paragraph 12 of those Agreements, the commercial right to hunt was extinguished and replaced with a right to hunt for food, on an extended land base. In a non-s. 35 case in 1990, *R. v. Horseman*, it was argued on behalf of the Aboriginal litigant that precedent affirming the loss of the treaty right to commercial hunting should not be followed. The claim was that paragraph 12 of the NRTA was passed without the consultation, much less the consent, of the affected communities. Further, counsel asserted that allowing the Crown to unilaterally change and derogate the treaty rights granted earlier "could only lead to the dishonor of the Crown."[14] The Supreme Court of Canada effortlessly rejected the argument and, in doing so, revealed the true limits of the fiduciary duty concept in the context of a legislative enactment:

> These contentions cannot be accepted.[15] ... In addition, although it might be politically and morally unacceptable in today's climate to take such a step as that set out in the 1930 Agreement without consultation and concurrence of the Native peoples affected, nonetheless the power of the Federal Government to unilaterally make such a modification is unquestioned and has not been challenged in this case.[16]

In sum, when it was not inconsistent with government intent as expressed in law, the Supreme Court of Canada imposed a powerful fiduciary duty on the Crown to act in the aboriginal community's best interest and in a manner that reflected the "honor" of the Crown.[17] Nonetheless, absent the status of constitutional rights, treaties did not serve to insulate Aboriginal peoples from government laws that unilaterally altered, or even extinguished, the rights of Aboriginal peoples even though they were protected in their treaties with the Crown. Any common law protection for Aboriginal or non-Aboriginal peoples is always vulnerable to legislative changes. What remains to be determined is how the recognition of aboriginal rights as "constitutionally protected" creates new vigor to the duty to consult, and in what manner. This leads us to s. 35.

Contemporary legal framework for the duty to consult

Canadian courts use a fairly straightforward – though not necessarily wise[18] – approach to defining s. 35 rights and the conditions under which they can be violated. Section 35 has repeatedly been held by the Supreme Court of Canada to exist for the purpose of recognizing the prior occupation of North America by Aboriginal peoples, and the reconciliation of prior occupation by Aboriginal peoples with the assertion of Crown sovereignty.[19]

Section 35 protects three classes of rights. First, and in many ways the strongest, are rights to aboriginal title. As we have noted, the nature of aboriginal title is quite unusual. It is a collective concept, attaching to the community for use of traditional lands.[20] Second, there are treaty rights. Treaties are not exactly like contracts, but they do create binding obligations and rights. Canadian courts now interpret treaty provisions generously, and in a manner that recognizes the unequal bargaining relationship that often characterized the negotiation process prior to treaty formation.[21] The treaties vary; some are compared with peace and friendship, some involved the "surrender" of land; and they are both quite old and very new.[22] Third, there are ancestral rights (often confusingly called aboriginal rights as well).[23] These protect traditional practices, traditions and customs that were, prior to contact with Europeans, integral to the distinctive culture of the specific Aboriginal group. There must also be continuity between those historic activities and the particular practice, tradition or custom relied upon by aboriginal communities today. These activities can be manifested in contemporary form – bows and arrows are not required to exercise an aboriginal right to hunt – and they may or may not exist alongside of an aboriginal title claim.[24] Finally, the right claimed must be infringed.[25] In addressing this requirement, the court considers whether the limitation is unreasonable, whether it imposes undue hardship or whether it denies rights holders their preferred means of exercising their rights.[26]

Interestingly, it will not suffice for the state to keep silent with respect to aboriginal rights holders and wait for claims of infringement to be made. In *R. v. Adams*, the court was confronted with a legislative scheme that allowed for sport and commercial fishing, but had no provisions for aboriginal food fishing except when a special permit was granted at the discretion of the relevant minister. It had been previously established that regulatory schemes that allow for ministerial discretion are not *per se* violative of the Charter; rather, the court will simply require the discretion to be exercised in a manner that accommodates the Charter.[27] Yet the court refused to follow earlier precedents and distinguished aboriginal rights as requiring more than this:

> I am of the view that the same approach should not be adopted in identifying infringements under s. 35(1) of the *Constitution Act, 1982*. In light of the Crown's unique fiduciary obligations towards aboriginal peoples, Parliament may not simply adopt an unstructured discretionary administrative regime which risks infringing aboriginal rights in a substantial number of applications in the absence of some explicit guidance. If a statute confers an administrative discretion which may carry significant consequences for the exercise of an aboriginal right, the statute or its delegate regulations must outline specific criteria for the granting or refusal of that discretion which seek to accommodate the existence of aboriginal rights. In the absence of such specific guidance, the statute will fail to provide representatives of the Crown with sufficient directives to fulfil their fiduciary duties, and the statute will be found to represent an infringement of aboriginal rights under the *Sparrow* test.[28]

Clearly, then, s. 35 is of great significance. However, it is important to note that although s. 35 is not subject to s. 1 of the Charter, in its first s. 35 decision, *R. v. Sparrow*, the Supreme Court of Canada held that no rights are absolute, and thus it was essential to subject these rights to limits that were justified.[29] Compelling and substantial government interests would be justifiable.[30] A valid objective is conservation and safety. This approach protects laws designed to preserve the existence of a natural resource (such as wildlife) or ensure safe practices in an activity that might otherwise expose individuals to danger (i.e. weapons restriction).[31]

In some cases, where the right in issue is one to sustenance, the state must accord that right a priority over other users of the resource who are not rights holders, such as sport and commercial fishers, subject to the overriding requirement of conservation of the resources itself.[32] When the court recognizes a "commercial" right to a certain resource, as in *R.* v. *Gladstone*, s. 35 does not operate to secure absolute or unlimited priority to the Aboriginal peoples; instead, if there is no internal limit to the right (i.e. limited to food), then

> the doctrine of priority requires that the government demonstrate that, in allocating the resource, it has taken account of the existence of aboriginal rights and allocated the resource in a manner respectful of the fact that those rights have priority over the exploitation of the fishery by other users.[33]

Finally, and most important for our purposes, the court held that the manner in which governments seek to achieve their "compelling and substantial" objectives must "uphold the honour of the Crown," in light of the Crown's fiduciary obligations to Aboriginal peoples. In a later case, the court explained: "The fiduciary duty, where it exists, is called into existence to facilitate supervision of the high degree of discretionary control gradually assumed by the Crown over the lives of Aboriginal peoples."[34] Section 35 is intended in part to limit the degree of that control. In *Sparrow*, the court thus expanded the fiduciary duty to extend to s. 35 rights. It stated:

> The *sui generis* nature of Indian title, and the historic powers and responsibility assumed by the Crown constituted the source of such a fiduciary obligation. In our opinion, *Guerin*, together with *R.* v. *Taylor and Williams*, ground a general guiding principle for s. 35(1). That is, the Government has the responsibility to act in a fiduciary capacity with respect to aboriginal peoples. The relationship between the Government and aboriginals is trust-like, rather than adversarial, and contemporary recognition and affirmation of aboriginal rights must be defined in light of this historic relationship.[35]

Of critical note, however, is that the duty to consult is not limited to s. 35 rights or to existing reserves, though it must still be tied to specific Indian interests rather than "a plenary Crown liability covering all aspects of the Crown–Indian relationship."[36]

Delgamuukw, the seminal case explaining the ubiquitous role that consultation plays in the Court's understanding of the nature of aboriginal/state relations, is one involving aboriginal title, where the court stated:

> There is always a duty of consultation. Whether the aboriginal group has been consulted is relevant to determining whether the infringement

of aboriginal title is justified, in the same way that the Crown's failure to consult an aboriginal group with respect to the terms by which reserve land is leased may breach its fiduciary duty at common law: *Guerin*. The nature and scope of the duty of consultation will vary with the circumstances. In occasional cases, when the breach is less serious or relatively minor, it will be no more than a duty to discuss important decisions that will be taken with respect to lands held pursuant to aboriginal title. Of course, even in these rare cases when the minimum acceptable standard is consultation, this consultation must be in good faith, and with the intention of substantially addressing the concerns of the aboriginal peoples whose lands are at issue. In most cases, it will be significantly deeper than mere consultation. Some cases may even require the full consent of an aboriginal nation, particularly when provinces enact hunting and fishing regulations in relation to aboriginal lands.[37]

It is inconsistent with the honor of the Crown to not consult with Aboriginal peoples when their interests are being affected in a manner that violates their rights. In the very first case, *Sparrow*, involving ancestral rights to fish for food, social and ceremonial purposes, the court explained:

Within the analysis of justification, there are further questions to be addressed, depending on the circumstances of the inquiry. These include the questions of whether there has been as little infringement as possible in order to effect the desired result; whether, in a situation of expropriation, fair compensation is available; and, *whether the aboriginal group in question has been consulted* with respect to the conservation measures being implemented. The aboriginal peoples, with their history of conservation-consciousness and interdependence with natural resources, would surely be expected, at the least, to be informed regarding the determination of an appropriate scheme for the regulation of the fisheries.[38]

Similarly, in *Gladstone*, a case that recognized commercial fishing rights as a traditional feature of the particular aboriginal community involved, the court required the state to engage in consultation as a component of the justification analysis: "questions relevant to the determination of whether the government has granted priority to aboriginal rights holders are those enumerated in *Sparrow* relating to consultation and compensation."[39] The court sent the case back to the trial court because of the Crown's failure to lead evidence as to justification in support of the infringements the court identified. Further, in *R. v. Nikal*, the court, again confronting a case where the Crown had not led any evidence to justify its infringement of aboriginal rights, commented on the essential role of consultation in the assessment of whether aboriginal rights are being legitimately considered by the state:

So long as the infringement was one which in the context of the circumstances presented could reasonably be considered to be as minimal as possible then it will meet the test. The mere fact that there could possibly be other solutions that might be considered to be a lesser infringement should not, in itself, be the basis for automatically finding that there cannot be a justification for the infringement. So too in the aspects of information and consultation the concept of reasonableness must come into play. For example, the need for the dissemination of information and *a request for consultations cannot simply be denied. So long as every reasonable effort is made to inform and to consult,* such efforts would suffice to meet the justification requirement. This is no more than recognizing that regulations pertaining to conservation may have to be enacted expeditiously if a crisis is to be avoided. On occasion, strict and expeditious conservation measures will have to be taken if potentially catastrophic situations are to be avoided. The nature of the situation will have to be taken into account in assessing the conservation measures taken. The greater the urgency and the graver the situation the more reasonable strict measures may appear.[40]

The jurisprudential history of the concept of consultation in aboriginal law was traced in recent decisions of the Supreme Court of Canada in *Haida*[41] and *Taku*.[42] These cases held the duty to consult and accommodate aboriginal interests flows from the legal requirement that the Crown act honorably in their relationship with Aboriginal peoples. The court distinguished "honor" from the "fiduciary duty" and explained the latter was limited to situations when a specific aboriginal interest has been established as requiring the Crown to act in the aboriginal group's best interest in exercising discretionary control over the subject matter of the defined or proven aboriginal right or title.[43] Honor, by contrast, is ever-present as an obligation. This means that the duty to consult can be triggered relatively simply by the assertion of aboriginal claims that meet a minimum threshold of legitimacy. The consultation must be meaningful and may generate a duty to accommodate by governments, though it does not necessarily mean that as a result of the process an agreement will or must be reached.[44] While there is no aboriginal veto over government decisions based on asserted rights, there may well be a veto with respect to established rights that are shown to be validated.[45]

The foregoing is the birth story of the constitutional duty to consult with Aboriginal peoples. It was analytically located within the justification stage of s. 35 cases involving proven violations, but its conceptual basis is the broader "honor of the Crown," and thus it is triggered by the assertion of claimed rights of which the government has knowledge, whether actual or constructive. Before turning to the topic of aquaculture, and our review of important operational aspects of the duty to consult cases, we offer a quick assessment of the outcomes in s. 35 cases.

The language of s. 35 requires that rights claimed under it be "existing." Any rights previously existing, but extinguished by state action prior to 1982, cannot be revived.[46] Of the various s. 35 claims, title cases are the strongest in law. But title is hard to prove,[47] and such cases more likely to succeed in the more recently settled parts of Canada than on the east coast.[48] The second set of cases involves treaties, and these have been quite successful as a source of rights protected under s. 35. Indeed, the most controversial s. 35 decision from the Supreme Court of Canada involved interpretation of a 1751 treaty as supporting a contemporary right to access and trade traditional resources in order to earn a moderate livelihood.[49] As for ancestral rights, which we will call aboriginal rights to maintain consistency with the jurisprudence, these are very difficult cases. It is hard to find evidence to prove a practice "at the time of contact," again more so in the parts of the country where there was early exposure to European settlement. However, when that evidence exists, it can support economically significant rights, including commercial rights to exploit resources.[50] The concept of priorities then applies to protect aboriginal access to rights (though if it is a commercial right it will not be exclusive access) after conservation goals are secured.

The key point is that before there is a duty to consult, there must be a rights claim, whether based on title, treaty or on aboriginal customs, practices and traditions. There is no "at large" duty to consult. More importantly, there is no free-standing right to engage in – or not engage in – aquaculture development. This cannot be addressed in the abstract. If there is an established aboriginal title claim, then aquaculture development will be protected, as will the right to refuse to allow it on aboriginal lands. If there are no aboriginal title claims, and the claim depends on a treaty provision, then that provision must support the interpretation that the treaty incorporates aquaculture development.[51] If the source of the right is based on ancestral practices, cultures and traditions, then there must be evidence to support the view that fish farming was an activity that was a defining feature of the aboriginal community in question at the time of contact with Europeans and that has continued.[52] Finally, levels of development have to be considered. The level of development of the resource completely turns on the evidence in the specific case. The evidence may or may not support production to the levels of large-scale commercial trade (e.g. *Gladstone*), but rather be limited to food or sustenance (e.g. *Sparrow*) or to a moderate livelihood (e.g. *Marshall No. 1*).

Having provided this historical and analytical overview of the larger framework within which the duty to consult needs to be understood, we are now in a position to delineate the specific dimensions of the duty to consult. However, before we do, it will be necessary to consider the degree to which contemporary government policies on aquaculture manifest an understanding of their duty to consult with Aboriginal peoples.

Federal and provincial policy on aquaculture and the duty to consult

Introduction

Aquaculture is a very new and evolving area of law and policy in Canada, and operates within a complex legal and regulatory system. At present, a patchwork of statutes, regulations and policies federally and provincially administer this expanding industry.[53] The basis of federal and provincial legislative jurisdiction over the subject is complicated, and beyond the scope of this chapter. At present, there are memoranda of understanding providing for shared jurisdiction between the federal government and several of the provinces.[54]

We have already noted that provincial laws cannot, in pith and substance, be directed to Aboriginal people, as the topic is reserved to the federal government. The legal context for aquaculture involves a variety of laws – fisheries, industry, environmental – as well as laws that relate specifically to Aboriginal peoples as such. It suffices to note that as far as Aboriginal people are concerned, both levels of government may be pursuing aquaculture projects that have the effect of violating aboriginal rights. The relationship between aquaculture law and the duty to consult is essentially undeveloped.[55] While some governments have recently drafted policy documents on consultation with Aboriginal peoples in the context of natural resource development (often on land), they have not specified aquaculture. In this section, we briefly review the available material on this topic. Because of a lack of information, we are unable to offer any assessment of the status of aquaculture in the Yukon, Northwest Territories and Nunavut.

Federal

The main federal[56] statutes that currently frame aquaculture are the *Fisheries Act*, the *Canadian Environmental Assessment Act*, the *Canadian Environmental Protection Act 1999*, the *Navigable Waters Protection Act* and the *Oceans Act*.[57] There are also a number of secondary federal statutes such as the *Canada Shipping Act*, the *Coastal Fisheries Protection Act*, the *Food and Drugs Act*, the *Pest Control Products Act* and the *Feeds Act*[58] that relate to specific aspects of aquaculture. However, there is no federal aquaculture act at the present. In 1995, Cabinet endorsed the Federal Aquaculture Development Strategy (FADS) and designated Fisheries and Oceans Canada (DFO) as the lead federal agency for aquaculture development. In 1997, the *Oceans Act* expanded the DFO's role in the development of ocean resources premised on three guiding principles: sustainable development, precaution and integrated management.[59] Sections 32 and 33 of the *Oceans Act* refer to cooperation or consultation between the minister and aboriginal communities. Section 33 states:

In exercising the powers and performing the duties and functions assigned to the Minister by this Act, the Minister

(a) shall cooperate with other ministers, boards and agencies of the Government of Canada, with provincial and territorial governments and with *affected aboriginal organizations*, coastal communities and other persons and bodies, including those bodies established under land claims agreements.[60]

In 1999, the Prime Minister appointed a Commissioner for Aquaculture Development to advise the Minister of Fisheries and Oceans.[61] The Office of the Commissioner for Aquaculture Development (OCAD) is mandated to determine the most appropriate and effective role of the federal government in the development of the industry.[62] In 2000, the DFO launched the Program for Sustainable Aquaculture and developed a six-point action plan to increase industry competitiveness and public confidence.[63]

The need for confidence became apparent when, later in 2000, the Auditor General of Canada found that DFO was not fully meeting its responsibilities to conserve and protect wild fish from the effects of salmon farms in British Columbia.[64] The 2000 DFO program focuses on the research and development related to biological science, the environment, human health, and improving the regulatory and management framework of aquaculture in Canada, and is backed by a CDN\$75 million investment by the federal government for five years.[65] The objective of the plan is to "develop a sound and integrated policy and regulatory environment [that] contributes to a stable and supportive business climate that will enable the industry to continue contributing to Canada's social and economic development."[66] In particular, the Ministers of Aquaculture and Fisheries agree to "resolve current legal and regulatory impediments and put in place a harmonized legal framework that enables development of the aquaculture sector in Canada."[67] To assist in attaining this goal, they also commit to "seek opportunities to enable Aboriginal and other stakeholders participation in aquaculture development."[68] The Office of the Commissioner for Aquaculture Development subsequently released the *Legislative and Regulatory Review of Aquaculture in Canada* in March 2001. In this document, thirty-six detailed measures were recommended as "urgently" needed to serve the development of aquaculture. However, we highlight that aboriginal interests, not to mention aboriginal rights, are barely mentioned in this otherwise comprehensive review of the existing and proposed legislative environment for aquaculture.[69]

In 2002, in response to FADS, the principles of the *Oceans Act* and the Aquaculture Program and Action Plan, DFO prepared an Aquaculture Policy Framework (APF). In it, DFO agrees to "clearly convey to other federal government departments, the provincial and territorial governments, the aquaculture industry, *Aboriginal groups*, and stakeholders the framework within which DFO is committed to taking action."[70] The framework also states, "DFO will respect constitutionally protected Aboriginal and treaty rights and will work with interested and affected Aboriginal communities to facilitate their participation in aquaculture development."[71]

Finally, in April 2003 the Standing Committee on Fisheries and Oceans, chaired by Member of Parliament Tom Wappel, submitted a report titled *The Federal Role in Aquaculture in Canada*. The report is a comprehensive review of the federal and provincial regulatory, political and legal developments in aquaculture. The report also includes a list of recommendations for the federal government. These include passage of a federal aquaculture Act and regulations that will "provide a clear set of standards for operators, other stakeholders and the public,"[72] and "promote communications between stakeholders, reduce and mitigate potential user conflicts, and enhance public awareness of the social and economic benefits of the industry."[73] There are, once again, no recommendations that refer specifically to Aboriginal peoples or their rights, much less consultation. The report does note the opposition to fish farming by some west coast First Nations communities,[74] and we identify a recent case on this point (*Heiltsuk Nation* v. *British Columbia*) in the last section of this chapter.

This review indicates that the federal government is aware of aboriginal interests but the attention is sporadic rather than systemic, as we believe to be required by the Constitution and in light of the "honorable" relationship between the federal Crown and Aboriginal peoples of Canada. There is no detail as to how DFO policies will work with respect to aboriginal and treaty rights. It is unclear whether DFO and the Commissioner for Aquaculture appreciate that the interests, indeed rights, of Aboriginal peoples are of a different order from those of commercial stakeholders who cannot claim any constitutional source for their desire to participate in resource development. It is also obvious that the federal initiative is strongly in favor of aquaculture development; not every aboriginal community shares that view. Owing to the novelty of this framework, it is too early to judge its effectiveness in engaging aboriginal communities on issues involving aquaculture but, at a minimum, current federal policy reflects an opening for the articulation of aboriginal interests and rights in this area of development and law.

British Columbia

British Columbia (BC)[75] is the largest producer of aquaculture products in Canada.[76] The two main subdivisions of BC aquaculture are the salmon farming and shellfish farming industries. In British Columbia, fish farms provide more than 3,500 direct and indirect jobs, mostly in coastal communities. According to the provincial Ministry of Agriculture, Food and Fisheries, the export value of farmed fish is approximately CDN$370 million, which accounts for 40 percent of the value of all BC seafood.[77] A license is required to operate a fish farm in the province.[78] Among First

Nations in British Columbia, support for aquaculture is divided. There are some First Nations, such as the Tlowitsis, Quatsino, Kwakiutl, Gwa'Sala-Nakwaxda'xw, Kitasoo and Qwe'Qwa'Sot'Enox, that want aquaculture in their coastal communities for economic development.[79] However, there is also vociferous opposition against aquaculture by First Nations, such as the Heiltsuk, Nuxalk, Namgis, Tsawataineuk, Kwicksutaineuk-ah-Kwah-ah-Mish, and the Gwawaenuk tribes, who fear that farmed salmon are causing fatalities among wild salmon from sea lice.[80] In the recent past, there have been many protests and acts of civil disobedience organized by the aboriginal groups opposing salmon aquaculture in the province.[81]

The British Columbia government has had numerous studies and policies on aquaculture over the past decade.[82] A controversial scientific study in 1997[83] led to a new policy in 1999 and the establishment of a committee to bring together stakeholders, including aboriginal groups. First Nations have been extensively involved in the governmental and public debates concerning the development of aquaculture in that province.[84] A moratorium was imposed in 1995.[85] No new farms sites were permitted until April 2002, when the moratorium was lifted.[86]

In October 2002, British Columbia released its Provincial Policy for Consultation with First Nations.[87] The policy applies to all provincial ministries, agencies and Crown corporations. This is a comprehensive effort by the British Columbia government to respond to the various court decisions discussed in the second part of this chapter. It requires government to engage in a "pre-consultation assessment" to determine whether a particular activity or decision requires consultation and lists a series of factors to aid in that inquiry while noting that it is rare that consultation would not be required on the assertion of an aboriginal claim.[88] Importantly, the document recognizes that consultation is always required whenever there is a "sound claim" of aboriginal interests, and does not only arise when those claims have been upheld in a court of law. This is consistent with the decisions from that jurisdiction's appeal court, which we will discuss in the next section. Finally, the document notes that the nature of consultation will vary depending on the strength or "soundness" of the claim being asserted.[89] We explain shortly why this is an appropriate understanding of the law on the duty to consult. Two final points are worthy of note. First, this is a policy document, not a legislative enactment. Second, there is no specific mention of aquaculture.

It is clear that the government of British Columbia acknowledges that Aboriginal peoples are readily implicated in the aquaculture industry and require engagement. In April 2003, the federal and British Columbia governments launched a "Salmon Aquaculture Forum" that is designed to bring all parties together to help "strengthen fisheries management" at the request of "governments, stakeholders, First Nations and concerned British Columbians."[90] The accompanying press release stated that the Forum "will provide an opportunity for interested parties to discuss current issues in aquaculture as they arise, including scientific and public policy questions."[91] Consultation with respect to aboriginal rights is now clearly stated as necessary in s.8.1.8 of the most recent Aqualculture Land Use Policy document, dated 5 October 2005.[92]

Alberta

In Alberta, aquaculture is regulated by the Alberta Agriculture, Food and Rural Development (AAFRD) department.[93] The main cultured freshwater species in the province are rainbow trout, tilapia, goldfish, koi and Arctic char.[94] The provincial *Fisheries Act* provides the licensing provisions for cultured fish. There is nothing in the *Fisheries Act* or the *Fisheries Regulations*[95] that requires public hearings or consultation with affected stakeholders, such as Aboriginal peoples in Alberta. In 2000, the provincial government released a report, entitled *Strengthening Relationships: The Government of Alberta's Aboriginal Policy Framework*, to address socio-economic disparities between aboriginal and non-aboriginal communities. In terms of consultation with aboriginal communities on resources development, the report states:

> The Government of Alberta encourages a "good neighbour" approach based on respect, open communication and co-operation. It expects those who propose natural resource developments to consult with and consider the views, values and experiences of communities and people that could be affected by their developments.[96]

This may require consultation with aboriginal communities on aquaculture projects, but that is not specified in the *Strengthening Relationships* report. Unlike British Columbia, it seems as if Alberta goes out of its way to minimize any special significance attaching to the presence and rights of Aboriginal peoples with respect to resource development.

Saskatchewan

In Saskatchewan, aquaculture is regulated by the Ministry of Agriculture, Food and Rural Revitalization.[97] There is nothing in the *Fisheries Act* or the *Fisheries Regulations* that requires public hearings or consultation with affected stakeholders, such as First Nations. In 2000, the Saskatchewan Environment and Resource Management (now Saskatchewan Environment) released its Aboriginal Consultation Guidelines.[98] This document provides a basic overview of the case law on the duty to consult, and in 2001 the same department issued the *Aboriginal Consultation Field Guide*.[99] This document details the consultative obligation of the government in a practical manner, covering a wide range of activities while not mentioning any specific interests of the Aboriginal peoples in the province or aquaculture.[100] As with

British Columbia, this initiative represents a good effort by the government to honor its obligations to Aboriginal peoples through a clear and public commitment to the duty to consult whenever aboriginal interests may be adversely affected. Again, however, these documents do not have the force of law and function rather as policy determinations.

Manitoba

The aquaculture industry in Manitoba is small and dominated by the production of rainbow trout.[101] The industry is governed by the Ministry of Conservation and is regulated by the *Fisheries Act* and the *Fisheries Regulations*.[102] A license is required to operate a fish farm in the province.[103] There is nothing in these pieces of legislation that requires public hearings or consultation with affected stakeholders such as First Nations. A recent report prepared for Manitoba's Aboriginal Justice Implementation Commission, entitled *Consultation with Aboriginal Peoples*, presented results from a survey conducted with Aboriginal peoples in that province:

> [A] series of questions was about whether respondents [aboriginals] had noticed any weaknesses in the province's capacity to communicate, administer programs and develop policies in relation to Aboriginal peoples. There was a resounding yes to all segments of this question from all respondents.[104]

Ontario

Ontario is the leading producer of rainbow trout in Canada.[105] In 2001, Ontario produced 4,100 tonnes of rainbow trout worth CDN$16,900,000. This represents approximately 63 percent of the entire amount of rainbow trout produced across Canada.[106] The Ministry of Agriculture and Food (OMAF) oversees aquaculture development in the province. The principal legislation is the *Fish and Wildlife Conservation Act*, and a license is required to culture fish in the province.[107] There is nothing in the Act that requires public hearings or consultation with affected stakeholders such as First Nations.

Québec

In Québec, the main species used for commercial aquaculture are also trout, with certain regions dominating the production.[108] Production in Québec was valued at CDN$5,299,000 million for 1,267 tonnes of aquaculture products for the year 2001.[109] The industry is regulated under the *Act Respecting Commercial Fisheries and Aquaculture* (Loi sur les pêcheries et l'aquaculture commerciales) and the *Commercial Aquaculture Regulation*.[110] There is nothing in the recently passed act or the regulations that requires public hearings or consultation with affected stakeholders, including Aboriginal peoples.

Newfoundland and Labrador

The Ministry of Fisheries and Aquaculture governs the highly developed aquaculture industry in Newfoundland and Labrador. The industry produced 4,263 tonnes of aquaculture products, valued at CDN$18.9 million in 2001.[111] Newfoundland and Labrador is one of the few provinces with a dedicated *Aquaculture Act* and *Aquaculture Regulations*.[112] However, there is no reference to Aboriginal peoples or First Nations in the legislative scheme, although the law contemplates regulations that prescribe procedures by which "neighbouring land owners, municipalities, other affected or interested persons and the general public may participate in helping the minister in his or her decision whether to grant an aquaculture license."[113] Current regulations are, however, silent on this point.

New Brunswick

In New Brunswick, the Department of Agriculture, Fisheries and Aquaculture administers the aquaculture industry. Currently, there are approximately fourteen hatcheries, ninety-seven marine sites and nine processing plants. The salmon aquaculture industry alone provides 3,005 direct and indirect jobs.[114] The industry is regulated by the *Aquaculture Act* and its accompanying *Aquaculture Regulations*.[115] Under the Act, public consultation is anticipated;[116] however, nothing in the Act or the regulations refers to Aboriginal peoples.

Prince Edward Island

Prince Edward Island is the leading producer of blue mussels in Canada and the famous Malpeque oysters.[117] The Department of Fisheries, Aquaculture and Environment oversees the aquaculture industry.[118] The *Fisheries Act* gives the Minister power to set regulations for the aquaculture industry.[119] There is nothing in the act that requires public hearings or consultation with Aboriginal peoples or other affected stakeholders.

Nova Scotia

The aquaculture industry in Nova Scotia is overseen by the Department of Agriculture and Fisheries. Nova Scotia has one of the most diverse industries in Canada. There are approximately 370 issued aquaculture sites in the province that culture various species.[120] The industry is regulated by the *Fisheries and Coastal Resources Act* and the *Aquaculture Licence and Lease Regulations*.[121] A license is required to operate a fish farm in the province. Under the act, the minister may call for public hearings.[122] There is a specific reference to mandatory consultations:

> 47 Before making a decision with respect to the application, the Minister
>   (a) shall *consult* with
>   i   the Department of Agriculture and Marketing, the Department of the Environment, the Department of Housing and Municipal Affairs and the Department of Natural Resources, and
>   ii any boards, agencies and commissions as may be prescribed; and
>   (b) may refer the application to a private sector, regional aquaculture development advisory committee for comment and recommendation.[123]

In addition, the statute states:

> 56 (1) The Minister may, with the approval of the Governor in Council, . . .
>   (2) Before recommending designation of an aquaculture development area or conditions or restrictions to be applicable thereto, the Minister shall
>   (a) *consult* with
>   i   the Department of Agriculture and Marketing, the Department of the Environment, the Department of Housing and Municipal Affairs and the Department of Natural Resources, and
>   ii any boards, agencies and commissions as may be prescribed.[124]

There is, however, nothing specifically related to aboriginal communities in the act or the regulations.

Conclusion

The federal, British Columbia and Saskatchewan governments recognize their legal obligation to consult with Aboriginal peoples. The frameworks in place appear to map onto prevailing law that we reviewed at the beginning of this chapter. Aquaculture in particular is a topic of concern at the federal level and, most acutely, in British Columbia, and it will be useful to watch developments in these jurisdictions closely. Our survey of law and policy in the remainder of the provincial governments reveals a stunning lack of attention to the issues of aboriginal consultation generally, or aquaculture and aboriginal rights specifically. This stance may be useful politically, but it is not sustainable in law. In fact, it is illegal. In the next section, we outline the emerging jurisprudence in various courts across Canada, which give more detail to the pronouncements on the duty to consult from the Supreme Court of Canada discussed in the second part of this chapter. This analysis will suggest that the challenges posed by the duty to consult are complex but also mandatory: governments must obey the constitutional requirement to consult.

Recent developments on the duty to consult: who, what, when, where and why?

To whom does the duty apply?

Once a case is litigated, and a court concludes that an aboriginal right is infringed, the onus of proof is on the Crown to demonstrate that the infringement is justified. It will fail to establish justification if the Crown did not "consult" with Aboriginal peoples.[125] The duty to consult applies to the Crown – that is, both the federal and provincial governments – and in particular it applies to civil servants, who, as agents of the Crown, are exercising governmental authority that infringes aboriginal rights.[126] In cases where there is more than one ministry or government department involved, the obligation will fall on the "responsible authority."[127] Courts are not subject to such a duty, nor are all creatures of the executive branch of government.[128] Finally, as noted earlier, the duty does not mean the government can only address aboriginal interests in making decisions;[129] other interests and rights, such as those held by non-Aboriginal Canadians, also merit consideration, but some form of priority must be accorded to aboriginal rights.[130] As we have already explained, the duty does not exist at large, but rather depends upon a triggering claim of a cognizable aboriginal interest.[131] In the context of aquaculture, such a claim to participate in the industry, or a claim to prevent aquaculture developments, must be based on a s. 35 protected title, treaty or aboriginal right.

Cases may involve developers, and it is important to identify the existing case law on the relationship between the Crown, aboriginal communities and developers. This is a hotly contested political, economic and legal concern, and current case law is in a state of fragility. We do know that the Crown cannot delegate, devolve or divest its duty to consult onto interested private parties.[132] Consulting by such third parties does not relieve the Crown of its duty under s. 35(1).[133] Governments can, however, require private developers whose projects might have an impact on aboriginal rights to have direct consultations with affected First Nations and, in assessing the extent of these consultations, courts can factor in the conduct of the private developers.[134] This is just an attribute of state power to condition its approval of any activity.[135] These lower court opinions on these points have now been held to be correct by the Supreme Court of Canada.[136]

In 2002, the British Columbia Court of Appeal radically expanded the potential scope of the duty to consult to private corporations but this was rejected by the Supreme Court of Canada on further appeal. In *Haida Nation No. 1*, the Haida Nation applied for a declaration that the Minister of Forests had breached his fiduciary duty when he renewed Tree Farm Licence 39 (T.F.L. 39) for Weyerhaeuser Company Ltd. on the Queen Charlotte Islands/Haida Gwaii pursuant to s. 29 (now s. 36) of the *Forest Act* without adequately consulting the Haida people, who claimed to hold aboriginal title and ancestral rights over this land. Lambert JA, speaking for a unanimous court, declared that both the provincial Crown and the logging company (as well as MacMillan Bloedel, Weyerhaeuser's predecessor) were subject to an enforceable legal and equitable duty to consult with and accommodate the Haida with regard to their economic and cultural claims.[137] He did not expressly address the reasons for this apparent expansion, except to note in passing that "Weyerhaeuser [were] aware of the Haida claims to aboriginal title and aboriginal rights ... through evidence supplied to them by the Haida people and through further evidence available to them on reasonable inquiry, an inquiry which they were obliged to make."[138] But Lambert JA never explicitly explained the source of such a corporate obligation. Thus, it was unsurprising that Weyerhaeuser, and several intervenors,[139] petitioned the Court to reconsider its holding on this point. To their undoubted dismay, in *Haida No. 2*, the British Columbia Court of Appeal, although this time only by a two to one majority (again authored by Lambert JA),[140] reaffirmed that third parties might owe a duty to consult in good faith and endeavor to seek workable accommodations, and that this is a duty separate from that owed by the Crown to Aboriginal people.[141]

The Court of Appeal justified its decision on three grounds: the existence of provisions within both the *Forest Act* and T.F.L. 39 that could be interpreted as requiring consultation with Aboriginal peoples; the equitable doctrines of constructive trust and knowing receipt; and a pragmatic argument that in the day-to-day management of the resource the real and effective capacity to consult and accommodate the concerns of Aboriginal peoples rested with Weyerhaeuser.[142]

Rather than following the pragmatic and grounded analyses of the British Columbia Court of Appeal in *Haida No. 2*, the Supreme Court of Canada retreated to a formalistic public/private dichotomy in reaching the conclusion that the duty to consult does not attach to third parties at all. This flows from the court's view that the Crown alone is responsible for third-party conduct that affects Aboriginal peoples, and the Crown's honor cannot be delegated (though informational aspects of the consultation itself may be delegated, and often are, to the industry developers).[143] The court was crystal clear on this point: "The remedy tail cannot wag the liability dog. We cannot sue a rich person, simply because the person has deep pockets or can provide a desired result."[144] However, the court also rejected the assumption that fed the holding on this point in the Court of Appeal below, namely that if Weyerhaeuser were not held to the duty there would be no remedy for the Haida Nation, pointing to recent legislation in British Columbia that claws back 20 percent of all forest licensees' harvesting rights partly to make land available to Aboriginal peoples. It is hard to predict the effect of this reasoning, but much seems to turn on the assumption that the state is willing to learn and meet aboriginal concerns, and not perpetuate the long-established patterns of the past. The court sounds optimistic. It is too early to tell whether that optimism is well placed. It requires a new relationship with the state, and a state willing to seriously devote itself to that task by reining in the legal power of corporations to unfairly undermine aboriginal claims to resource distribution by the state.

Just as important as the question of who has the duty to consult, is the question of "with whom must the Crown consult?" Aboriginal communities can be complex political units with difficult issues of representative legitimacy.[145] There can be personal and political differences within a band, between band members and non-band members, between bands, between First Nations, between local bands and regional organizations, etc. In any given litigation, only certain parties may be involved, but other aboriginal groups may have rights that would nonetheless be affected by decisions.[146]

To date, the courts have had little to say on this important issue, but what they have said reflects the broader proposition that the components of the duty to consult are highly fact dependent. In other words, the answer to the "with whom" question is "it depends." For example, in *R. v. Jack*, the Court of Appeal of British Columbia indicated that is not necessary that there be a vote by the band council, or that there be unanimous consent by a band to a particular governmental policy, but it also indicated that there may be situations where a vote will be essential.[147] In *R. v. Sampson*, the same court seemed to suggest that conversations with a band manager may be inadequate consultation. Lambert JA, in his dissenting opinion in *R. v. Gladstone*, indicated that consultations with the Native Indian Brotherhood rather than the Heiltsuk band itself were not sufficient.[148] In another case, *Alberta Wilderness Association v. Canada (Minister of Fisheries and Oceans)*, a regional umbrella organization representing various First Nations was held to be entitled to consultation.[149] It is a complex issue. Clearly, however, the prudent course of action is for the state to identify all aboriginal communities with a potential right, and to ascertain the legitimate authority within each community before engaging in consultation proceedings that seek to avoid infringements. The duty only falls on those infringing rights, but once it attaches, it requires good-faith participation by everyone involved, including aboriginal communities. In *Haida*, the Supreme Court of Canada stressed this point:

> At all stages, good faith on both sides is required. The common thread on the Crown's part must be "the intention of substantially addressing

[Aboriginal] concerns" as they are raised ... through a meaningful process of consultation. Sharp dealing is not permitted. However, there is no duty to agree; rather, the commitment is to a meaningful process of consultation. As for Aboriginal claimants, they must not frustrate the Crown's reasonable good faith attempts, nor should they take unreasonable positions to thwart government from making decisions or acting in cases where, despite meaningful consultation, agreement is not reached. Mere hard bargaining, however, will not offend an aboriginal people's right to be consulted.[150]

The nature of the duty to consult

In this section, we address the question of intensity: *what* must a government do to fulfill the standard of legally sufficient consultation? The short answer is, once again, "it depends." Conventionally, legal rights have been understood as either procedural or substantive, the right to a fair trial being an example of the former, freedom of expression being an example of the latter. In the context of aboriginal claims, the courts have, in our opinion, resisted this dichotomy and instead created a hybrid right, one that is more than procedural but, in most cases, less than a veto right.

The starting point for an inquiry into the nature and intensity of the duty to consult is the famous dictum in *Delgamuukw*, where Lamer CJ announced a context-sensitive proportionality test, which we reproduced in the second part of this chapter. The British Columbia Court of Appeal has developed the analysis further to suggest a twofold "adequate and meaningful" standard of consultation: (1) "a positive obligation to reasonably ensure that Aboriginal peoples are provided with all necessary information in a timely way so that they have an opportunity to express their interests and concerns;" and (2) "to ensure that their representations are seriously considered and, whenever possible, demonstrably integrated into the proposed plan of action."[151] More recently still, the same court has invoked a number of other descriptors that appear to intensify the obligations of governments even further: "ensur[ing] the substance of the concerns are addressed" and determining whether the "needs" and "concerns have been met or accommodated";[152] and seeking "accommodation" and "workable accommodation."[153] In *Haida Nation*, the Supreme Court of Canada has employed similar language: "accommodate," "stringent duties," "deep consultation," "finding a satisfactory interim solution," "reconciliation" and "balance and compromise."[154]

This is a relatively full-bodied conception of consultation. However, this raises a critical point, namely: what is the difference between a duty to consult and a substantive right to a particular outcome – in other words, a right to veto the state action in issue? As noted, several courts have held that the duty to consult does not give a veto right to aboriginal claimants,[155] and this is surely correct as otherwise there is little point in characterizing something as consultation, and little incentive for the state to engage in any process at all. Equally, however, a duty to consult cannot mean that no weight is given to the fact that the claims being made are not merely "interests," but actual constitutional "rights" that exist independently of any specific consultative process. Thus, the confusion is over whether the duty to consult is procedural only, or also substantive. If it is only procedural, it is not sensitive to the fact that there are rights being claimed. If it is fully substantive, and as such requires recognition of the rights claims themselves, then it is difficult to see what there is to "consult" about.

Haida clears up the law in this area. In a case where aboriginal rights are asserted rather than proven, the fiduciary duty is not yet triggered so as to require the Crown to act in the best interests of the aboriginal community in the Crown's exercise of discretionary control over the subject or the right claimed.[156] The notion of "honor" nonetheless dictates that even potential rights must be "determined, recognized and respected" through a process of consultation and, depending on the circumstances, reasonable accommodation.[157] The court stressed, "To unilaterally exploit a claimed resource during the process of proving and resolving the Aboriginal claim to the resource, may be to deprive Aboriginal claimants of some or all of the benefit of the resource. That is not honourable."[158] The duty is triggered as soon as the Crown knows, or ought to know, of the potential existence of an aboriginal claim and contemplates conduct that might adversely affect those rights or title claim.[159]

The court emphasizes, as we do, that the intensity of the obligations to consult and accommodate will depend upon the factual matrix of each case. As we have noted in this chapter, however, good faith requires engagement, not denial or deference, and this is exactly what the Supreme Court of Canada concluded. There is no veto, but in situations where there is a strong case that a significant right may be infringed, and the risk of non-compensable damage is high, "deep consultation, aimed at finding a satisfactory interim solution, may be required,"[160] and in the process the Crown is entitled to consider other societal interests.[161] The holding in *Haida* implicitly acknowledges the hybrid nature of the rights of consultation and accommodation, noting specifically that while the purpose of s. 35 rights is the reconciliation, "[r]econciliation is not a final legal remedy in the usual sense. Rather it is a process flowing from the rights guaranteed by s. 35(1),"[162] and thus the duty is intended to require parties to make a good-faith effort to understand and address each other's concerns.[163]

The idea that the duty to consult is a hybrid right that is more than procedural, but less than fully substantive, makes sense of the judicial dicta that impose reciprocal aboriginal obligations. Several courts have made it clear that the duty to consult is a two-way street.[164] While the primary obligation is on the Crown to consult, there is also a reciprocal obligation on Aboriginal peoples to participate fully and in good faith in the consultation process.[165] Courts have held, as we have noted, that the duty to consult cannot be used to give Aboriginal people a veto power; nor does it necessarily require the "agreement," "consensus" or "informed consent" of Aboriginal peoples.[166] Aboriginal peoples "cannot frustrate the consultation process by refusing to meet or participate, or by imposing unreasonable conditions"[167] or by making unreasonable demands for further information.[168] Nor can they, "in good faith, refuse to actively participate in the consultation process and then complain that [they have] not been consulted."[169] If they initially participate in a process, they cannot abandon it and then complain of a lack of consultation.[170] In *Haida*, the court explained, "As for Aboriginal claimants, they must not frustrate the Crown's reasonable good faith attempts, nor should they take unreasonable positions to thwart government from making decisions or acting in cases where, despite meaningful consultation, agreement is not reached."[171]

Courts are articulating consultation as a right and duty to authentically participate in solving the problem at issue. To the extent that this recognizes that the duty to consult is a "democratic right,"[172] this is clearly a welcome development, as long as courts are sensitive to the realities of that engagement. Often the cases involve disputes over scientific data or require very specialized expertise. This is obviously true with respect to aquaculture developments. Many aboriginal communities simply lack the resources to hire the experts necessary to meaningfully participate in the discussion over whether or not a proposed development has an adverse effect on aboriginal rights and interests. In *Kelly Lake Cree Nation* v. *British Columbia (Minister of Energy and Mines)*, the Ministry of Energy and Mines refused to fund a specialist to advise a First Nation on the consultation process, but instead offered to make available a member of the ministry to advise on the technical aspects of the project.[173] A First Nation may legitimately not have confidence that its interests will be adequately represented by a member of a bureaucracy whose very purpose is to promote development of resources – and this is very much the case with the Office of the Commissioner for Aquaculture. *R.* v. *Aleck* held that the bands would have to prove that funding for an independent analyst was "necessary," but this may impose a significant, perhaps insurmountable, burden on First Nations communities.[174] However, in *Taku River* the Supreme Court of Canada noted with approval the government's decision to retain an independent assessment by an expert approved of by the aboriginal community, and provide financial assistance to allow the community to participate meaningfully in the consultation process.[175]

The reciprocal nature of the duty to consult may cause concern for First Nations for other reasons as well. As Lilles CJ of the Territorial Court in *R.* v. *Joseph* has noted, "Native people are afraid that any information provided could be used against them in the future."[176] This may explain why, not infrequently, Aboriginal peoples state that their conversations and meetings are not "consultations."[177] Their point is that they do not want to legitimatize or be complicit in a "consultation" record-keeping process that is not about meeting aboriginal concerns as much as relevant actors (state and corporate) simply documenting compliance with a legally imposed duty to consult.

It is impossible to predict what view a court will take of a specific consultation process, though so far in the case law we can trace a fairly serious effort by courts to ensure that the consultation is seriously undertaken, with an openness to outcomes. Courts are displeased by aggressive aboriginal positioning on consultation, and remind communities that consultation is not a veto.[178] At the same time, courts are unwilling to tolerate the all too common Crown stance of a refusal to allow aboriginal claims to consultation to derail government approval of development projects. As one court stated recently, "On a legal basis, the shortness of time and economic interests are not sufficient to obviate the duty of consultation."[179]

Whether or not it makes sense to equate the position of the state and aboriginal communities by imposing a reciprocal requirement to participate, it is critical that the specificity of the relationship remain central to the elaboration of consultative duties. A clear and, we believe, appropriate recognition of this fact can be observed in the *Mikisew* case. There, the Crown argued that because it had provided opportunities for public consultations that are open to all stakeholders, the First Nations had a duty to participate in such fora and not frustrate the consultative process. In *Mikisew*, however, this argument was rejected and the judge went so far as to argue that "[a]t the very least, [the First Nation] is entitled to a distinct process if not a more extensive one."[180] This is important because it emphasizes the uniqueness of aboriginal rights in Canadian legal and political discourse, and their priority over generalized commercial development interests such as that of the emerging aquaculture industry. In *Taku River*, the Supreme Court of Canada explained that the key is that aboriginal concerns be taken seriously, and this must be done, whether in a specialized process or otherwise. The court explained:

> The Province was not required to develop special consultation measures to address TRTFN's [Taku River Tlingit First Nation's] concerns, outside of the process provided for by the Environmental Assessment Act, which specifically set out a scheme that required consultation with affected Aboriginal peoples.[181]

When is the duty triggered?

At what time does the duty to consult kick in: early, at the moment when Aboriginal peoples *assert* an aboriginal right, or much later, only after Aboriginal peoples have *proved* they have such a right? For twelve years (1990–2002), the dominant view was that the duty to consult arose quite late, only after the Aboriginal claimants had demonstrated an aboriginal right. A decision from the Ontario Court of Appeal illustrates this point. In

TransCanada Pipelines v. *Beardmore (Township)*,[182] plans were introduced to amalgamate several rural municipalities over the objection of two local First Nations bands who were claiming a violation of s. 35 rights to a particular territory. The Ontario Court of Appeal rejected the argument that the Ontario government owed a duty to consult on the basis of the *potential* claim by the First Nations to aboriginal rights. In reaching this conclusion, Borins JA drew on the logic of the three-point *Sparrow* test and held that the duty to consult came into being "only after" the First Nations had *established* (1) the requisite treaty right, and (2) infringement of such a right. The Ontario Court of Appeal characterized the s. 35 claim as "speculative" and therefore insufficient to trigger the duty to consult.[183] The court's concern was that a requirement of consultation prior to *proof* of an aboriginal right would impose too great a burden on government, and would encourage frivolous claims.

In *Haida*, the Supreme Court of Canada has rejected this analysis and authoritatively held that the duty exists early and is triggered relatively easily:

> Neither the authorities nor practical considerations support the view that a duty to consult and, if appropriate, accommodate arises only upon final determination of the scope and content of the right. . . . To limit reconciliation to the post-proof sphere risks treating reconciliation as a distant legalistic goal, devoid of the "meaningful content" mandated by the "solemn commitment" made by the Crown in recognizing and affirming Aboriginal rights and title. . . . It also risks unfortunate consequences. When the distant goal of proof is finally reached, the Aboriginal peoples may find their land and resources changed and denuded. This is not reconciliation. Nor is it honourable. . . . There is a distinction between knowledge sufficient to trigger a duty to consult and, if appropriate, accommodate, and the content or scope of the duty in a particular case. Knowledge of a credible but unproven claim suffices to trigger a duty to consult and accommodate. The content of the duty, however, varies with the circumstances, as discussed more fully below. A dubious or peripheral claim may attract a mere duty of notice, while a stronger claim may attract more stringent duties. The law is capable of differentiating between tenuous claims, claims possessing a strong prima facie case, and established claims. Parties can assess these matters, and if they cannot agree, tribunals and courts can assist. Difficulties associated with the absence of proof and definition of claims are addressed by assigning appropriate content to the duty, not by denying the existence of a duty.[184]

Where and why does the duty exist?

This chapter has explained that the Crown has an obligation to engage in *some* degree of consultation upon the assertion of a prima facie aboriginal right, but the greater the potential soundness of the right, the higher the standard of consultation. We also know that the duty originates with the honor of the Crown. The particular content of the duty and extent of any attendant obligations will vary with the nature and importance of the right claimed by Aboriginal people; it is not a general indemnity.[185] We can state, then, that the duty attaches wherever there are Aboriginal peoples who have rights that are being threatened. That means that any decision-maker involved in any area of development (including aquaculture) may find themself subject to the duty to consult and, thereafter, the duty to accommodate.

In our view, the purpose of the duty is to ensure that the parties to disputes involving aboriginal rights claims learn to take each other seriously and, in particular, to require the state to begin to treat Aboriginal peoples with respect and equality. Courts have been somewhat progressive in this area of law, at least in demanding engagement and meaningful communication. Some cases have gone further in holding the parties to a substantive standard of addressing the needs and concerns of legitimately advanced aboriginal claims to protection under s. 35. To the extent that those involved in the aquaculture industry ignore these judicial developments, they will likely find themselves legally vulnerable.

Conclusion

The duty to consult holds potential and perils for Aboriginal peoples in the context of aquaculture. On the one hand, it recognizes strong obligations that might be owed to Aboriginal peoples by governments, both federal and provincial. This review has indicated that to date many governments have not realized the significance of this legal duty. On the other hand, it is important for those concerned about the future of aboriginal communities not to buy into false optimism. The idea, and ideal, of a duty to consult is contingent upon a number of quite complex and technical legal relationships that require an ability to dovetail constitutional vision with quite context-specific facts. It may even have the possibility of intensifying the challenges faced by Aboriginal peoples. In short, the duty to consult with Aboriginal peoples in the context of aquaculture is still up for grabs. While recent decisions from the Supreme Court of Canada have clarified several aspects of the doctrine, we believe that its application will remain polyvalent: everything will depend upon the context.

Postscript

In March 2004, the Commissioner for Aquaculture Development issued a Report to the Minister of Fisheries and Oceans entitled *Recommendations for*

Change.[186] Filled with promises of jobs, money and stories of economic salvation, four of the fifty-seven pages are devoted to "Aboriginal Opportunities in Aquaculture." However, the Report misses a central argument of this chapter: that Aboriginal peoples in Canada have a potential constitutional right of consultation that is distinct from that owed to other Canadians. For example,

- The report calls for "enhanced participation in the aquaculture sector by Aboriginal peoples in a manner consistent with native culture, values and traditions"[187] but fails to elaborate upon what participation might mean, and it manifestly does not contemplate participation as objecting to aquaculture development.
- The Report acknowledges that "proto-aquacultural activities are believed to have been practiced by Aboriginal peoples,"[188] which may support claims for the constitutional obligation of substantive consultation. Such claims are ignored by the Report.
- In Part III, "Getting the Policy Right," the Report develops a "Vision for Sustainable Aquaculture Development in Canada" with scarce mention of the particular rights of Aboriginal peoples. Recommendations 7 and 8, dealing with "negotiat[ing] a New Aquaculture Framework Agreement" and "funding" and briefly propose to "[d]evelop a national strategy for Aboriginal aquacultural development with input from the provinces/territories"[189] with no reference to "input," much less "consultation" with the Aboriginal peoples themselves.

In 2005 in *Homalco Indian Band* v. *British Columbia (Ministry of Agriculture, Food and Fisheries)*[190] the British Columbia Supreme Court applied the principles outlined above. The British Columbia government had amended Marine Harvest Canada's existing aquaculture licence to allow them to raise Atlantic rather than Chinook salmon. The Homalco Indian Band claimed, *inter alia*, that they had not been consulted. Following the *Haida Nation* and *Taku River* decisions from the Supreme Court of Canada, the Court agreed. However, the Court refused the Homalco's application for an order that all of the Atlantic salmon at the site be removed until consultation and, if necessary, reasonable accommodation of their concerns had been achieved. This case confirms the application of the duty to consult and, if necessary, accommodate Aboriginal concerns in the context of aquaculture. However, it also raises further issues as to the most appropriate remedy for breach of the duty.

Notes

1 C. Macdonald, "Elder Loses Bid to Stop Seismic Testing off C.B.," *Chronicle Herald*, 12 December 2003, A3.
2 For example, as we will illustrate in our ensuing discussion, while there are significant points of consensus between appeal courts and the Supreme Court of Canada, there are also important differences. Compare, for example, *Haida Nation* v. *British Columbia (Minister of Forests)* (2002) 99 B.C.L.R. (3d) 209

(B.C.C.A.), [*Haida Nation No. 1*] with [2004] 3 S.C.R. 511; and *Taku River Tlingit First Nation* v. *Tulsequah Chief Mine Project* (2002) 211 D.L.R. (4th) 89 (B.C.C.A.) and [2004] 3 S.C.R. 550 [*Taku River*].

3 There are other relevant documents; this is just the most famous. The relevant passage is:

> [S]uch Parts of Our Dominions and Territories as, not having been ceded to or purchased by Us, are reserved to them [Aboriginal peoples] or any of them, as their Hunting Grounds. . . . We do, with the Advice of our Privy Council strictly enjoin and require, that no private Person do presume to make any purchase from the said Indians of any Lands reserved to the said Indians ... but that, if at any Time any of the Said Indians should be inclined to dispose of the said Lands, the same shall be Purchased only for Us, in our Name.

See discussion in *Delgamuukw* v. *British Columbia* [1997] 3 S.C.R. 1010 at para. 203.

4 *British North America Act 1867* (U.K.), 30 Vict., c. 3, s. 91(24).

5 Although, in 1973, the Supreme Court of Canada in *Calder* v. *Attorney General of British Columbia* [1973] S.C.R. 313 held that aboriginal interests in ancestral lands constituted a legal claim that predated and survived European settlement. See also *Mitchell* v. *MNR* [2001] 1 S.C.R. 911 at paras. 141–146.

6 Federal law on Indians is authorized by s. 91(24) and does not present a division of powers problem. Provincial law is vulnerable, however, and thus application is usually by virtue of s. 88 of the *Indian Act*, R.S.C. 1985, c. I-5. Note that treaties could also be extinguished: the *Natural Resources Transfer Agreement, infra*, is an example, as we will discuss in the second part of this chapter.

7 *R.* v. *Badger* [1996] 1 S.C.R. 771.

8 See *Kitkatla Band* v. *British Columbia* (2002) 210 D.L.R. (4th) 577 and, more recently, *Paul* v. *British Columbia* (Forest Appeals Commission) [2003] 2 S.C.R. 585.

9 See Diana Ginn, Chapter 8 this volume.

10 *Guerin* v. *The Queen* [1984] 2 S.C.R. 335.

11 *Ibid.* at 382.

12 *Ibid.* at 387.

13 *Ibid.* at 389.

14 *R.* v. *Horseman* [1990] 1 S.C.R. 901 at 932.

15 *Ibid.*

16 *Ibid.* at 934. See also *R.* v. *Badger* [1996] 1 S.C.R. 771 at 815: "The Federal government, as it was empowered to do, unilaterally enacted the *NRTA*. It is unlikely that it would proceed in the same manner today."

17 The power of the fiduciary duty concept can be readily appreciated in complicated case of *Blueberry River Indian Band* v. *Canada* [1995] 4 S.C.R. 344. The Supreme Court of Canada held that the Crown violated the fiduciary duty by selling potentially valuable mineral rights of a reserve land and not reacquiring the rights upon realizing a mistake had been made.

18 See, generally, P. Macklem, *Indigenous Difference and the Constitution of Canada* (Toronto: University of Toronto Press, 2001).

19 *R.* v. *Sparrow* [1990] 1 S.C.R. 1075 [*Sparrow*]; *R.* v. *Gladstone* [1996] 2 S.C.R. 723 at paras. 71, 72 [*Gladstone*]; *R.* v. *Van der Peet* [1996] 2 S.C.R. 507 at para. 39; *R.* v. *Adams* [1996] 3 S.C.R. 101 at para. 57 [*Adams*]; *Haida Nation No. 1, supra* note 2 at paras. 16, 17.

20 *Delgamuukw* v. *British Columbia* [1997] 3 S.C.R. 1010 [*Delgamuukw*]. See also Diana Ginn's chapter in this volume (Chapter 8).

21 An excellent summary of the principles of treaty interpretation can be found in the dissenting (not on this point) opinion of Chief Justice McLachlin in *R.* v. *Marshall No. 1* [1997] 3 S.C.R. 456.

22 See, for example, maritime versus the numbered treaties in western Canada, and new treaties in British Columbia, Québec and Labrador.

23 The leading case on ancestral rights remains *R.* v. *Van der Peet* [1996] 2 S.C.R. 507.

24 *Adams, supra* note 19 at 117: "[W]hile claims to aboriginal title fall within the conceptual framework of aboriginal rights, aboriginal rights do not exist solely where a claim to aboriginal title has been made out."

25 See *Sparrow, supra* note 19 at 1100; and *R.* v. *Nikal* [1996] 1 S.C.R. 1013 at 1057 [*Nikal*].

26 Sometimes this is a fairly easy step. For example, in *Sparrow* the government net length restriction for fishing was found to violate the ancestral right to fish. However, mere licensing may not constitute an infringement. See *Nikal, supra* note 25 at 1059, where it did not, but the court noted that the conclusion may turn on whether the license is one that could only be obtained at great difficulty or expense. If this is the situation, an infringement may be established.

27 As in *Slaight Communications Inc.* v. *Davidson* [1989] 1 S.C.R. 1038.

28 *Adams, supra* note 19 at para. 54.

29 For example, in *Nikal, supra* note 25 at para. 92, the court rejected the argument that no government regulation was legitimate once a right was established:

> It has frequently been said that rights do not exist in a vacuum, and that the rights of one individual or group are necessarily limited by the rights of another. The ability to exercise personal or group rights is necessarily limited by the rights of others. The government must ultimately be able to determine and direct the way in which these rights should interact. Absolute freedom in the exercise of even a Charter or constitutionally guaranteed aboriginal right has never been accepted, nor was it intended. Section 1 of the Canadian Charter of Rights and Freedoms is perhaps the prime example of this principle. Absolute freedom without any restriction necessarily infers [*sic*] a freedom to live without any laws. Such a concept is not acceptable in our society.

30 Initially the test seemed rigorous: *Sparrow, supra* note 19 at 1113, announced:

> If a *prima facie* interference is found, the analysis moves to the issue of justification. This is the test that addresses the question of what constitutes legitimate regulation of a constitutional aboriginal right. The justification analysis would proceed as follows. First, is there a valid legislative objective? Here the court would inquire into whether the objective of Parliament in authorizing the department to enact regulations regarding fisheries is valid. The objective of the department in setting out the particular regulations would also be scrutinized. An objective aimed at preserving s. 35(1) rights by conserving and managing a natural resource, for example, would be valid. Also valid would be objectives purporting to prevent the exercise of s. 35(1) rights that would cause harm to the general populace or to aboriginal peoples themselves, or other objectives found to be compelling and substantial.

But in several more recent cases, the Supreme Court of Canada has indicated an incredibly expansive list of potentially valid government objectives. See *Gladstone, supra* note 19 at para. 64, and *Marshall No. 2* [1999] 3 S.C.R. 532 at para. 41 and again in *Delgamuukw, supra* note 20. In the last-mentioned case, the court stated at para. 165 (citations omitted):

The general principles governing justification laid down in *Sparrow*, and embellished by *Gladstone*, operate with respect to infringements of aboriginal title. In the wake of *Gladstone*, the range of legislative objectives that can justify the infringement of aboriginal title is fairly broad. Most of these objectives can be traced to the reconciliation of the prior occupation of North America by aboriginal peoples with the assertion of Crown sovereignty, which entails the recognition that "distinctive aboriginal societies exist within, and are a part of, a broader social, political and economic community". In my opinion, the development of agriculture, forestry, mining, and hydroelectric power, the general economic development of the interior of British Columbia, protection of the environment or endangered species, the building of infrastructure and the settlement of foreign populations to support those aims, are the kinds of objectives that are consistent with this purpose and, in principle, can justify the infringement of aboriginal title. Whether a particular measure or government act can be explained by reference to one of those objectives, however, is ultimately a question of fact that will have to be examined on a case-by-case basis.

At least in one case, however, the court was clear that not everything will fall within the list. In *Adams, supra* note 19 at para. 58, the court declined the opportunity to find that sport fishing constituted a valid state objective:

> While sports fishing is an important economic activity in some parts of the country, in this instance, there is no evidence that the sports fishing that this scheme sought to promote had a meaningful economic dimension to it. On its own, without this sort of evidence, the enhancement of sports fishing accords with neither of the purposes underlying the protection of aboriginal rights, and cannot justify the infringement of those rights. It is not aimed at the recognition of distinct aboriginal cultures. Nor is it aimed at the reconciliation of aboriginal societies with the rest of Canadian society, since sports fishing, without evidence of a meaningful economic dimension, is not "of such overwhelming importance to Canadian society as a whole" (*Gladstone at* para. 74) to warrant the limitation of aboriginal rights.

31 *Sparrow, supra* note 19.
32 *Ibid.*
33 *Gladstone, supra* note 19 at para. 62.
34 *Wewaykum Indian Band* v. *Canada* (2002) 220 D.L.R. (4th) 1 at para. 79 [*Wewaykum*]. But note also at para. 96:

> When exercising ordinary government powers in matters involving disputes between Indians and non-Indians, the Crown was (and is) obliged to have regard to the interest of all affected parties, not just the Indian interest. The Crown can be no ordinary fiduciary; it wears many hats and represents many interests, some of which cannot help but be conflicting.

35 *Sparrow, supra* note 19 at 1108 (citations omitted).
36 See *Wewaykum, supra* note 34 at paras. 79–81, and *Haida Nation No. 1, supra* note 2 at para. 18.
37 *Delgamuukw, supra* note 20 at 1113.
38 *Sparrow, supra* note 19 at 1119 (emphasis added).
39 *Gladstone, supra* note 19 at 64.
40 *Nikal, supra* note 25 at 110 (emphasis added).
41 *Haida Nation No. 1, supra* note 2.
42 *Taku River, supra* note 2.
43 *Haida Nation No. 1, supra* note 2 at para. 18.

44 *Ibid.* at para. 48.
45 *Ibid.* at para. 48.
46 *Sparrow, supra* note 19 at 1091.
47 See, for example, *Delgamuukw, supra* note 20.
48 See *R.* v. *Bernard* (2003) 230 D.L.R. 4th 57 [*Bernard*]; *R.* v. *Marshall* (2003) 218 N.S.R. (2d) 78 [*Marshall*]; *R.* v. *Bernard* and *R.* v. *Marshall* [2005] 2 S.C.R. 200.
49 *Marshall No. 1* [1999] 3 S.C.R. 456. But note the retreat in *Marshall No. 2* [1999] 3 S.C.R. 532.
50 See, for example, *Gladstone, supra* note 19. Usually a non-commercial context is involved.
51 See also *Bernard* and *Marshall, supra* note 48.
52 That it is possible to find such evidence is indicated by the Report of the Standing Committee on Fisheries and Oceans, Chaired by Tom Wappel, M.P., in 2003. At 5, under the heading "Part 1 – Aquaculture in Canada," the report states, "There is anecdotal evidence that basic aquaculture was first practiced in Canada by aboriginal peoples who transferred fish between streams and rivers." *The Federal Role of Aquaculture in Canada*, Report of the Standing Committee on Fisheries and Oceans (Ottawa: House of Commons of Canada, April 2003) [Standing Committee Report 2003]. Online. Available http://www.parl.gc.ca/InfoComDoc/37/2/FOPO/Studies/Reports/foporp03-e.htm (accessed 24 March 2004).
53 "A New Way to Feed the World?" *The Economist*, 9 August 2003 at 9.
54 Standing Committee Report 2003, *supra* note 52 at 16.
55 Federal and provincial environmental assessment legislation for aquaculture projects may require public hearings with affected stakeholders, which may include First Nations, but this area is beyond the scope of our chapter. Nor do we discuss the myriad of administrative law issues that can arise in the context of project approvals for resource development. See discussion in David VanderZwaag *et al.*, Chapter 3, this volume.
56 The following websites provide information regarding federal policy on aquaculture and Aboriginal peoples: DFO Sustainable Aquaculture, online http://www.dpo.mpo.gc.ca/aquaculture/main_e.htm; Office of the Commissioner for Aquaculture Development (OCAD), online http://www.ocad-bcda.gc.ca; Canadian Aquaculture Industry Alliance (CAIA), online http://www.aquaculture.ca; and Aboriginal Canada Portal, online http://www.aboriginalcanada.gc.ca/abdt/interface/interface2.nsf/engdocBasic/11.html (accessed 22 March 2004).
57 *Fisheries Act*, R.S. 1985, c. F-14; *Canadian Environmental Assessment Act 1992*, c. 37; *Canadian Environmental Protection Act 1999*, 1999, c. 33; *Navigable Waters Protection Act*, R.S. 1985, c. N-22; and *Oceans Act*, 1996, c. 31.
58 *Canada Shipping Act*, R.S. 1985, c. S-9; *Coastal Fisheries Protection Act*, R.S. 1985, c. C-33; *Navigable Waters Protection Act*, R.S. 1985, c. N-22; *Food and Drugs Act*, R.S. 1985, c. F-27; *Pest Control Products Act*, R.S. 1985, c. P-9; and *Feeds Act*, R.S. 1985, c. F-9.
59 *Oceans Act, supra* note 57 at s. 30.
60 *Ibid.* s. 33. Online. Available http://laws.justice.gc.ca/en/O-2.4/88102.html (accessed 22 March 2004) (emphasis added). The DFO was contacted and asked if there was anything further explaining the scope and nature of engaging aboriginal groups on ocean resource development, such as aquaculture, but no reply to our inquiry was received.
61 Standing Committee Report, *supra* note 52 at 21.
62 In November 2001 both the mandate of the Commissioner and the current Commissioner were extended for a further two years, until 31 March 2004.

According to the Standing Committee Report, the extension was "intended to allow the Commissioner to develop a 10 year vision for aquaculture development in Canada," *supra* note 52 at 23.

63 DFO, *Program for Sustainable Aquaculture*. Online. Available http://www.pac. dfo-mpo.gc.ca/aquaculture/susaqua_e.htm (accessed 22 March 2004); DFO, *Aquaculture Action Plan*. Online. Available http://www.dfo-mpo.gc.ca/aquaculture/response_details_e.htm (accessed 22 March 2004).

64 Chapter 30 of the December 2000 Report to Parliament, *The Effects of Salmon Farming in British Columbia on the Management of Wild Salmon*, Auditor General of Canada, cited in Standing Committee Report, *supra* note 52 at 14.

65 News Release, "Dhaliwal announces $75 million for sustainable and environmentally sound aquaculture in Canada," 8 August 2000. Online. Available http://www.dfo-mpo.gc.ca/media/newsrel/2000/hq-ac71_e.htm (accessed 6 January 2005).

66 *Aquaculture Action Plan, supra* note 63 at 1.

67 *Program for Sustainable Aquaculture, supra* note 63 at 2.

68 *Ibid.* at 3.

69 The Commissioner noted that in the context of federal lease approvals for aquaculture, the federal government "must also assess the impact of proposed aquaculture sites with respect to … native rights and land claims." OCAD, *Legislative and Regulatory Review of Aquaculture in Canada* (Ottawa: Fisheries and Oceans Canada, 2001) at 16.

70 DFO, *Aquaculture Policy Framework* at 3. Online. Available http://www.dfo-mpo.gc.ca/aquaculture/policy/pg001_e.htm (accessed 24 March 2004) (emphasis added).

71 *Ibid.* at 4. DFO's view of the matter is further elaborated at 26: "Consistent with the general trend toward increased Aboriginal self-employment, Aboriginal communities are also playing an instrumental role in the development of Canada's aquaculture sector through the creation of their own community owned and operated aquaculture companies."

72 Standing Committee Report 2003, *supra* note 52, Recommendation 2 at 18.

73 *Ibid.*, Recommendation 10 at 24–25.

74 *Ibid.* at 4 and 25: "A number of major concerns emerged during the Committee's hearings regarding the siting of salmon farms: … on the West Coast, particular concerns about the infringement of Aboriginal title and rights through the placement of farms on 'Aboriginal' waters."

75 Useful information may be found at the following websites on aquaculture in British Columbia: British Columbia Aquaculture, Online: http://www.agf.gov.bc.ca/fisheries/index.htm; Coastal Alliance for Aquaculture Reform, Online: http://www.farmedanddangerous.org/about.htm; BC Salmon Farmers Association, Online: http://www.salmonfarmers.org/ (accessed 24 March 2004).

76 The primary species used in commercial production are Atlantic salmon, Chinook salmon and Pacific oysters. Secondary species include Coho salmon, trout, Manila clams, scallops, blue mussels, sea cucumbers, sea urchins, geoduck clams, abalone and marine plants. See Canadian Aquaculture Industry Association, "Industry Profile." Online. Available http://www.aquaculture.ca/English/IndustryProfile/ (accessed 22 July 2003) [CAIA Industry Profile].

77 "Aquaculture Forum to Link Interests, Seek Common Ground," News Release, 2003AGF0013-000413, 30 April 2003, Governments of Canada and British Columbia at 2. Online. Available http://www2.news.gov.bc.ca/nrm_news_releases/2003AGF0013-000413.htm (accessed 24 March 2004).

78 The major pieces of legislation informing aquaculture development are the

Fisheries Act, the *Environmental Assessment Act*, the *Aquaculture Regulation* and the *Aquaculture Waste Control Regulations*.

79 "B.C. Salmon Farmers Association Proposed Salmon Farming Relocation Sites Have Full First Nations support," News Release, 21 June 2000. Online. Available http://www.salmonfarmers.org/media/2000/6-21-00.html (accessed 6 January 2005).

80 Coastal Alliance for Aquaculture Reform (CAAR), "Salmon Farming and First Nations." Online. Available http://www.farmedanddangerous.org/farm_nations.htm (accessed 6 January 2005).

81 J. Stackhouse, "Trouble in Paradise," *Globe and Mail* (19 November 2001) at A8.

82 These are summarized in a document entitled "B.C. Salmon Aquaculture Policy Backgrounder" dated 31 January 2002. On file with the authors.

83 *Ibid.* at 5.

84 See, for example, the lengthy document entitled "Salmon Aquaculture Review: First Nation Perspectives," Vol. 2, August 1997. BC Environmental Assessment Office. On file with authors. Online. Available http://www.eao.gov.bc.ca/epic/output/html/deploy/epic_document_20_6046.html (accessed 24 March 2004).

85 Friends of Clayoquot Sound, "Moratorium in place since 1995," News Release, Online. Available http://www.docs.cc/newsreleases/020131.htm (accessed 24 March 2004).

86 "Expansion of industrial salmon farming will hurt BC. Liberals lift moratorium on salmon farm expansion," News Release, 31 January 2002. Online. Available http://www.focs.ca/lnewsreleases/020131.htm (accessed 6 January 2005).

87 *Provincial Policy for Consultation with First Nations* (British Columbia, October 2002). Online. Available http://www.gov.bc.ca/tno/down/consultation_policy_fn.pdf (accessed 24 March 2004).

88 *Ibid.* at 23–24.

89 *Ibid.* at 25: "The depth of consultation and the degree to which workable accommodations should be attempted will be proportional to the soundness of that interest."

90 *Aquaculture Policy Framework*, *supra* note 70. See also Pacific Fisheries Resource Conservation Council press release, John Frazer, MP, "Salmon Aquaculture Forum Consultations Begin," 12 May 2003. Online. Available http://www.fish.bc.ca/reports/pfrcc_salmon_aquaculture_forum_consultations_begin.pdf (accessed 24 March 2004).

91 *Ibid.*

92 Ministry of Agriculture and Lands: Crown Land Use Operational Policy: Aquaculture (Government of British Columbia: 2005) at 17. Online. Available http://www.lwbc.bc.ca/ollwbc/policies/policy/land/aquaculture.pdf.

93 Online. Available http://www.agric.gov.ab.ca/livestock/aquaculture/aqua_news/0211h.html (accessed 24 March 2004). See, generally, CAIA Industry Profile, *supra* note 76.

94 *Ibid.*

95 *Fisheries (Alberta) Act*, F-16 R.S.A. 2000; *General Fisheries (Alberta) Regulation*, 203/1997; *Fisheries (Ministerial) Regulation*, 220/1997.

96 *Strengthening Relationships: The Government of Alberta's Aboriginal Policy Framework* at 15. Online. Available http://www.aand.gov.ab.ca/PDFs/final_strengthrelations.pdf (accessed 24 March 2004).

97 Online. Available http://www.agr.gov.sk.ca/docs/statistics/livestock/aquaculture.asp (accessed 24 March 2004). The chief aquaculture species is trout. In 2001, 875 tonnes of farmed trout was produced, with an estimated value of CDN$4 million.

98 On file with the authors.
99 On file with the authors. See also "Saskatchewan Environment Consultation Requirements," undated. On file with authors.
100 See, generally, Manitoba Conservation, Online. Available http://www.gov. mb.ca/conservation (accessed 24 March 2004).
101 Approximately 16 tonnes, worth CDN$62,000, was produced in 2001, a significant increase from the 7 tonnes worth CDN$27,000 produced in 2000. See, generally, CAIA Industry Profile, *supra* note 76.
102 *Manitoba Fisheries Act*, R.S.M. 1987, F-90, and *Manitoba Fishing Licenses Regulations*, 124/97.
103 *Ibid.*, s. 14(3)(1).
104 "Final Report: Consultation with Aboriginal Peoples," 8 May 2001 by Therese Lajeunesse at 25.
105 See generally, Ontario Aquaculture Association, Online. Available http://www.aps.uoguelph.ca/~ontaqua/ (accessed 24 March 2004).
106 CAIA Industry Profile, *supra* note 76.
107 S.O. 1997 c. 41, s. 47(b). *Fish and Wildlife Conservation Act*, O. Reg. 664/98.
108 CAIA Industry Profile, *supra* note 76. Rainbow and brook trout mainly, but other species include brown trout, Arctic char, lake trout, perch, mussels and oysters. The regions of highest productivity are l'Estrie, l'Outaouais, Chaudière-Appalaches, la Mauricie and the Laurentides. In 1998, these regions were responsible for 72 percent of the entire production of fish in the province. See also Ministère de l'Agriculture, des Pêcheries et de l'Alimentation du Québec, online. Available http://www.agr.gouv.qc.ca/pac/ (accessed 24 March 2004).
109 CAIA Industry Profile, *supra* note 76.
110 *An Act Respecting Commercial Fisheries and Aquaculture*, L.R.Q. C.P-9.01, 1987, and *Commercial Aquaculture Regulations*, D. 1311-87, 1987 G.O. 2 5677.
111 CAIA Industry Profile, *supra* note 76. The most prominent species cultured are Atlantic salmon, steelhead trout and blue mussels. Other species cultured include brook trout, Arctic char, wolffish, halibut, yellowtail flounder, witch flounder, Atlantic cod, eels, giant scallops, sea urchins and soft shell clams.
112 *Aquaculture Act*, R.S.N.L. 1990 C. A-13 as amended, and *Aquaculture Regulations* (O.C. 96-939). See also Department of Fisheries and Aquaculture, online. Available http://www.gov.nf.ca/fishaq/ (accessed 24 March 2004).
113 *Aquaculture Act, ibid.* s. 11 (cc).
114 CAIA Industry Profile, *supra* note 76. Aquaculture is the most important agrifood product in the province. Atlantic salmon production represents 94 percent of all aquaculture production in New Brunswick, the majority of which is produced in the Bay of Fundy.
115 *Aquaculture Act, supra* note 112 Chapter A-9.2; *Regulations* are O.C. 91–806. See also Department of Agriculture, Fisheries and Aquaculture, online. Available http://www.gnb.ca/0027/index-e.asp (accessed 24 March 2004).
116 See *Aquaculture Act, supra* note 112 at s. 35: "The Minister shall undertake such public consultation in relation to aquaculture as the Minister considers appropriate or as is required by or in accordance with the regulations." Also, s. 42 authorizes regulations on this topic.
117 CAIA Industry Profile, *supra* note 76.
118 See, generally, Department of Fisheries, Aquaculture and Environment, online. Available http://www.gov.pe.ca/af/faa-info/index.php3 (accessed 24 March 2004).
119 Sections 4 and 7(b) *Fisheries Act* C. F-13.01.
120 CAIA Industry Profile, *supra* note 76. The major species cultured commercially are Atlantic salmon, steelhead trout, rainbow trout, blue mussels, American and European oysters, and sea scallops.

121 *Fisheries and Coastal Resources Act*, S.N.S. (1996) C-25, and *Aquaculture Licence and Lease Regulations Act*, N.S. Reg. 125/2000. See, generally, Department of Agriculture and Fisheries, online. Available http://www.gov.ns.ca/nsaf/aquaculture/ (accessed 24 March 2004).

122 *Ibid.* s. 48(c).

123 *Ibid.* s. 47 (emphasis added).

124 *Ibid.* s. 56 (emphasis added).

125 *Sparrow*, *supra* note 19. Aboriginal peoples do not have to prove that the government did not adequately consult them. *Mikisew Cree First Nation* v. *Canada (Minister of Canadian Heritage)* [2002] 1 C.N.L.R. 169 (F.C.) at para. 157 [*Mikisew*].

126 *Haida Nation No. 1*, *supra* note 2 at paras. 57–59. *Halfway River First Nation* v. *British Columbia (Ministry of Forests)* (1999) 64 B.C.L.R. (3d) 206 (B.C.C.A), paras. 55, 86 [*Halfway River*]. But also see the spirited objection by Southin JA who argues that this imposes too high a burden on government administrators who have no legal knowledge or training in *Halfway River First Nation*, paras. 227–231.

127 *Union of Nova Scotia Indians* v. *Canada (A.G.)* 1997 138 F.T.R. 103 at 328 (F.C.T.D.).

128 For example, quasi-judicial tribunals (such as the National Energy Board) are independent from the state and litigants, and thus the duty does not apply to them. In *Quebec (Attorney General)* v. *Canada (National Energy Board)* [1994] 1 S.C.R. 159, 184, it was stated that the National Energy Board's quasi-judicial nature "is inherently inconsistent with the imposition of a relationship of utmost good faith between the Board and a party appearing before it." See also *Treaty 8 Tribal Association* v. *Alliance Pipeline Ltd.* (1998), [1999] 4 C.N.L.R. 257 (N.E.B.) [*Treaty 8 Tribal Association*].

129 *Wewaykum*, *supra* note 34 at para. 96; *Haida Nation No. 1*, *supra* note 2 at paras. 45, 50; *Taku River*, *supra* note 2 at para. 2.

130 See *Sparrow*, *Adams* and *Gladstone*, *supra* note 19.

131 *Wewaykum*, *supra* note 34 at para. 85; *Tsartlip Indian Band* v. *Canada* (1999) 181 D.L.R. (4th) 730 (F.C.A.) at para. 35.

132 *Treaty 8 Tribal Association*, *supra* note 128 at paras. 9, 21.

133 *Mikisew*, *supra* note 125 at para. 156.

134 *Kelly Lake Cree Nation* v. *British Columbia (Ministry of Energy and Mines)* (1998) [1999] 3 C.N.L.R. 126 (B.C.S.C.) [*Kelly Lake*], paras. 154, 164.

135 Unless there is a right to engage in an act, the state can regulate it as it pleases, subject to the rule of law. There is no right to property or commercial development *per se* in the Canadian Constitution.

136 See our discussion of *Haida*, *infra*.

137 *Haida Nation No. 1*, *supra* note 2 at paras. 48, 52, 58, 60, 61, 62.

138 *Ibid.* at para. 49.

139 Of the five intervenors, four supported Weyerhaeuser: the Council of Forest Industries, the Business Council of British Columbia, the British Columbia Chamber of Commerce and the British Columbia Cattlemen's Association. Only one supported the Haida Nation, namely the Squamish Indian Band.

140 *Haida Nation* v. *British Columbia (Minister of Forests)* (2002), 5 B.C.L.R. 4th 33 [*Haida Nation No. 2*]. There was a dissent in *Haida Nation No. 2*, but it is tied to the procedural history of the case and does not address the substance of the majority's analysis.

141 *Ibid.* at para. 103.

142 *Ibid.* at paras. 48–92, 108. For a further analysis of these arguments, see R. Devlin and R. Murphy, "Recent Developments in the Duty to Consult," (2003) 14 *National Journal of Constitutional Law* 167.

143 *Haida Nation No. 1*, *supra* note 2 at para. 53.

144 *Ibid.* at para. 55.

145 *R.* v. *Ned* [1997] 3 C.N.L.R. 251 (B.C. Prov. Ct. Crim. Div.) [*Ned*] at 268. For illustrative examples of how differing aboriginal groups and interests can come into play, see *Perry* v. *Ontario* (1997) 33 O.R. (3d) 705 (Ont. C.A.); *Kelly Lake*, *supra* note 134; *Cheslatta Carrier Nation* v. *British Columbia (Environmental Assessment Act, Project Assessment Director)* [1998] 3 C.N.L.R. 1, 53 B.C.L.R. (3d) 1 (B.C.S.C.) [*Cheslatta*]; *Makivik Corp.* v. *Canada (Minister of Canadian Heritage)* (1998), [1999] 1 F.C. 38 (F.C.T.D.); and *R.* v. *Aleck* [2000] B.C.J. No. 2581 (B.C. Prov. Ct.) [*Aleck*].

146 This was a major concern of Justice Gerow in *Heiltsuk Nation* v. *British Columbia* [2003] B.C.S.C. 1422 [*Heiltsuk*]. See para. 60: "Although the petitioners argue that I should ignore the claims of the Nuxalk [who were not a party], I am of the view that making any findings regarding the Heiltsuk claim of rights and title which could potentially impact the overlapping claim of the Nuxalk in this proceeding is inappropriate."

147 *R.* v. *Jack* (1995) 16 B.C.L.R. (3d) 201 (C.A.) at para. 81 [*Jack*].

148 *Gladstone*, *supra* note 19 at para. 97; *Mikisew*, *supra* note 125 at para. 131.

149 (1998) 146 F.T.R. 257 (F.C.T.D.).

150 *Haida Nation No. 1*, *supra* note 2 at para. 42.

151 *Halfway River*, *supra* note 126 at paras. 160, 191.

152 *Taku River*, *supra* note 2 at paras. 193, 202.

153 *Haida Nation No. 1*, *supra* note 2 at paras. 51, 52, 60.

154 *Ibid.* at paras. 37, 44, 45.

155 *Ibid.* at para. 48; *Jack*, *supra* note 147 at 223; *Ned*, *supra* note 145 at 268; *Heiltsuk*, *supra* note 146 at para. 114: "No authority has been provided to me to support the proposition that the right to consultation carries with it the right to veto a use of the land."

156 *Haida Nation No. 1*, *supra* note 2 at para. 18.

157 *Ibid.* at para. 25.

158 *Ibid.* at para. 27.

159 *Ibid.* at para. 35.

160 *Ibid.* at para. 44.

161 *Ibid.* at para. 50.

162 *Ibid.* at para. 32.

163 *Ibid.* at para. 50.

164 *Cheslatta*, *supra* note 145.

165 *Haida Nation No. 1, supra* note 2 at paras. 42, 36; *Taku River*, *supra* note 2 at paras. 34, 36; *Ryan* v. *British Columbia (Ministry of Forests, District Manager)* [1994] B.C.J. No. 194 (C.A) at para. 6; *Halfway River*, *supra* note 126 at para. 161; *TransCanada Pipelines Ltd.* v. *Beardmore (Township)* (2000) 186 D.L.R. (4th) 403 (Ont. C.A.) at 455; *Heiltsuk*, *supra* note 146 at para. 118. In *Kitkatla*, the Court stated, "There is, however, no duty on First Nations to consult with the Crown. And there is no correlative right in the Crown to compel consultation." See *Kitkatla Band* v. *British Columbia (Minister of Small Business, Tourism and Culture)* (1998) 61 B.C.L.R. (3d) 71 (B.C.S.C.), para. 45.

166 *R.* v. *Sampson* (1995) 16 B.C.L.R. (3d) 226 (C.A.), 251 at 252; *Ned*, *supra* note 145 at 268; *Aleck*, *supra* note 145 at paras. 54–57. Note that where title is involved, consent will be required for surrender (see *Delgamuukw*, *supra* note 20 at 1113). Compare *Heiltsuk*, *supra* note 146 at para. 114, which suggests there is no veto over use of the land, which may be a misreading of *Delgamuukw*. *Haida Nation No. 1*, *supra* note 2 at paras. 48, 49.

167 *Halfway River*, *supra* note 126 at paras. 161, 182; *Aleck*, *supra* note 145 at para. 71.

168 *Cheslatta, supra* note 145 at para. 23.

169 *Vuntut Gwitchin First Nation* (1997) 138 F.T.R. 103 (F.C.T.D.), appeal dismissed [1988], F.C.J. No. 755 at para. 23; *Cheslatta, supra* note 145 at para. 23; *Kelly Lake, supra* note 134 at para. 159.

170 *Cheslatta, supra* note 145 at para. 73; *Kelly Lake, supra* note 134 at para. 164; *Aleck, supra* note 145 at para. 71. Indeed, there are dicta to indicate that Aboriginal peoples cannot even require that the consultations take place "on their own terms." *Cheslatta, supra* note 145 at para. 13.

171 *Haida Nation No. 1, supra* note 2 at para. 42.

172 *R. v. McIntyre* [1991] 1 W.W.R. 548 (Sask. Prov. Ct.) 569.

173 *Kelly Lake, supra* note 134 at para. 90.

174 *Aleck, supra* note 145 at para. 62.

175 *Taku River, supra* note 2 at para. 37.

176 *R. v. Joseph* [1991] N.W.T.R. 263 (Yukon Terr. Ct.) 272.

177 See, for example, *Heiltsuk, supra* note 146 at para. 108:

> I find on the evidence that prior to the petition the Heiltsuk have been unwilling to enter into consultation regarding any type of accommodation concerning the hatchery. This is apparent both from the position they have taken throughout the meetings where they have clearly indicated that they do not consider the meetings to be consultation and from correspondence between counsel in which the Heiltsuk have continued to express the view that no consultation has taken place.

178 *Haida Nation No. 1, supra* note 2 at para. 48; *Heiltsuk, supra* note 146 at 112:

> [T]he conduct of the Heiltsuk both in stating their position as one of zero tolerance to Atlantic salmon aquaculture and in attending meetings at which they stated they did not consider the meeting to be consultation indicates, in my view, an unwillingness to avail themselves of consultation.

> Compare with *Gitxsan First Nation* v. *British Columbia (Minister of Forests)* [2002] B.C.S.C. 1701, where Tysoe JA held that the First Nations acted reasonably in refusing to meet with the government until a response to their letter complaining of the state action had been received (at para. 89). The court agreed that the government's "consultation" was completely inadequate because it was not undertaken with a "genuine intention of substantially addressing the concerns of the petitioning first nations." (at para. 88).

179 *Gitxsan, ibid.*, para. 91, citing *R. v. Noel* [1995] 4 C.N.L.R. 78 (N.T.T.C.) at 95; *Mikisew, supra* note 125 at para. 132; and *Haida Nation No. 1, supra* note 2 at para. 55.

180 *Mikisew, supra* note 125 at para. 153.

181 *Taku River, supra* note 2 at para. 40.

182 (2000) 186 D.L.R. (4th) 403 (Ont. C.A.) leave to appeal to S.C.C. refused (2000) S.C.C.A. 264.

183 *Ibid.* at para. 121.

184 *Haida Nation No. 1, supra* note 2 at paras. 31, 33, 37.

185 *Ibid.*

186 *Recommendations for Change: Report of the Commissioner for Aquaculture Development to the Minister of Fisheries and Oceans Canada* (Ottawa: Office of the Commissioner for Aquaculture Development, 2004).

187 *Ibid.* at 4.

188 *Ibid.* at 6.

189 *Ibid.* at 44.

190 (2005) 39 B.C.L.R. (4th) 263 (S.C.).

10 Indigenous rights

Implications for aquaculture

Douglas Sanders

Introduction

Indigenous rights to resources in Australia, Canada, New Zealand and other jurisdictions have been major domestic issues over the past thirty years. The leading judicial decisions have dealt with land rights. Additionally, there are important decisions on hunting and fishing. Very recently, cases and settlements have dealt with rights to rivers, lakes, foreshore and offshore.[1] No significant decision on indigenous rights in the field of aquaculture has occurred, though there has been some activity on the issue in Canada and New Zealand. Any indigenous claims in relation to aquaculture would build on the judicial rulings on land, fish and water.

Land

In Canada, the changes in law and policy date from the 1973 *Calder* land rights decision,[2] which prompted the country to begin a land claims settlement process that continues more than thirty years later. The first comprehensive settlement of indigenous claims was the 1975 James Bay and Northern Quebec Agreement, recognizing, among other things, subsistence and commercial harvesting rights to fish and wildlife. The *Constitution Act 1982*[3] introduced s. 35, recognizing existing aboriginal and treaty rights. Substantive judicial recognition of rights began with the decisions in *Guerin* in 1984, *Sparrow* in 1990 and *Delgamuukw* in 1997.[4]

Australian developments date back to the 1976 *Aboriginal Land Rights (Northern Territory) Act*,[5] which began a land claims settlement process in the Northern Territory. Some extent of land rights was extended to the whole country by the 1992 *Mabo* decision.[6] A national system for determining surviving native title was established in the 1993 *Native Title Act*.[7]

Nordic developments began with the Alta dam fight in northern Norway in 1978. New Zealand judicial activism began with the *State Owned Enterprises* decision in 1987.[8] Malaysia applied Canadian and Australian decisions to Orang Asli traditional lands in the 1997 *Adong* decision.[9] The Inter-American Court of Human Rights upheld indigenous territorial rights on the Atlantic coast of Nicaragua in the *Awas Tingni* case in 2001.[10] Land rights were recognized in South Africa for the Khomani San (Bushmen) in an agreement in March 1999 prompted by the *Restitution of Land Rights Act 1994*. The Supreme Court of Appeal, on 24 March 2003, upheld a "customary law interest" in land based on a system of indigenous law for the Richtersveld Community, again giving force to the rights under the 1994 legislation, which was designed to redress the loss of land as a result of apartheid.[11] The Supreme Court of Appeal cited leading decisions from Australia and Canada, and the 1975 International Court of Justice decision in relation to Western Sahara.

Hunting and fishing

Rights to hunt and fish have featured in pioneering cases. In Canada, the lead fishing rights decisions are *Sparrow* in 1990, the commercial fishing rights trilogy in 1996[12] and the *Marshall* decision of 1999.[13] Also notable are the fishing rights provisions in the *James Bay Agreement* of 1975 and the *Nisga'a Treaty* of 2000.[14]

The first hunting case in Australia was the 1999 *Yanner* decision, which ended the idea that aboriginal rights there might be limited to land.[15] The New Zealand settlement of Maori fisheries claims was dramatic. It was upheld by the New Zealand courts[16] and by the United Nations Human Rights Committee in the *Mahuika* decision.[17] In the United States, the most important fishing rights litigation occurred in *United States* v. *Washington*,[18] upholding treaty-protected rights in Washington State both of a commercial and of a subsistence character (including fish stocks resulting from enhancement programs).

Water

Rights to inland, foreshore and offshore waters are much less clearly articulated. Fishing had been associated with ownership of waters in the early common law, but that link passed into history as governments regulated fishing by licensing systems. Provincial regulatory roles in fisheries in Canada, however, remain based on provincial claims to the ownership of the beds of lakes, rivers and coastal waters. Indian reserve boundaries may or may not include the property in lakes, rivers and foreshore. Such issues have not been treated in a consistent way in Canada over time, and very little litigation has occurred.

The pioneering claims to offshore waters in Canada were made by Inuit, who traditionally hunted and fished in coastal waters and on sea ice. Their claims have largely been settled in the Inuvialuit, Nunavut and Northern Quebec agreements. Current claims on the west coast include the long-standing Musqueam First Nation claim to the foreshore, the Nuu-Chah-Nulth claim to the offshore as far as the eye can see, and the Haida claim to the waters within Canada's territorial limit. There have been no judicial determinations of any of these claims.

In *Commonwealth* v. *Yarmirr*,[19] the Australian High Court upheld certain offshore rights, based on the 1993 *Native Title Act*. The later decision in *Risk* v. *Northern Territory*[20] ruled that such claims could not be handled under the *Aboriginal Land Rights (Northern Territory) Act* of 1976. Rights to waters, rivers and seas were upheld in *Lardil Peoples* v. *State of Queensland*.[21]

In New Zealand, the 2003 decision of the Court of Appeal in *Ngati Apa* v. *Attorney General* upheld the possibility of Maori ownership to foreshore and offshore areas in the Marlborough Sounds.[22] The litigation was triggered by a dispute over aquaculture. The matter was sent back to the Maori Land Court for a determination of rights under customary Maori law. Parliament, in a highly controversial move, declared any customary Maori rights to these areas to now be "public domain."

Aquaculture

We have only the beginnings of litigation on aboriginal rights in relation to aquaculture. One inconclusive judgment has been given in British Columbia, and important litigation in New Zealand on ocean rights, referred to above, was prompted by aquaculture initiatives. Litigation in the jurisdictions discussed in this chapter is recent and has tended to focus on "traditional" activities. It has not yet caught up with the rapid expansion of aquaculture.

Canadian developments

The Sparrow decision

The 1990 fishing rights decision of the Supreme Court of Canada in *Regina* v. *Sparrow*[23] interpreted the provision on "existing aboriginal and treaty rights" in s. 35 of the *Constitution Act 1982* for the first time. The joint judgment of Chief Justice Dickson and Mr. Justice LaForest began by referring to s. 35 as a "promise" to the Aboriginal peoples of Canada. Later in the judgment, s. 35 was described as "the culmination of a long and difficult struggle in both the political forum and the courts for the constitutional recognition of aboriginal rights." This language signaled that the court did not regard Indian rights cases as routine litigation, but as opening the possibility for redress for a regrettable history.

The key ruling held that the word "existing" in s. 35 refers to "unextinguished" rights, with a new rule that aboriginal and treaty rights could only be extinguished by state actions that were "clear and plain." A hundred years of virtually complete regulation of aboriginal fishing rights by federal legislation had not "extinguished" those rights. Now they could only be extinguished or limited by legislation that met certain standards. Clearly, there were now new ground rules for old rights. And the old rights were not "frozen," as of some particular date. They continued and evolved as aboriginal economies changed. This approach has obvious implications for rights in relation to aquaculture.

In the court cases, aboriginal plaintiffs first have to establish that they have an aboriginal right to fish, based on traditional usage, and that the right has been interfered with. Establishing the aboriginal fishing right was easy in the *Sparrow* case. An anthropologist had testified that for the Musqueam people, "the salmon fishery has always constituted an integral part of their distinctive culture."

If there is interference with an aboriginal right, then a three-part test determines whether the interference is justified.[24] There must be a valid legislative objective. In the context of fishing rights, conservation of the species is obviously a valid objective. Second, there must be a concern with the honor of the Crown and the special trust responsibility of the Crown. Finally, the infringement should be as limited as possible and should proceed only following consultation with the group in question. Any extinguishment of rights would require fair compensation. Previous decisions limiting aboriginal fishing rights because of *Fisheries Act* provisions were now "inapplicable" as a result of s. 35.

The 1996 commercial fishing trilogy

In three decisions in 1996, the Supreme Court of Canada ruled on an aboriginal right to sell fish in British Columbia.[25] There were strong reasons to assume that the Court would find in favor of upholding the aboriginal right. First, *Sparrow* had upheld aboriginal rights against a century of regulation of the food fishery, and all that was necessary now was to apply the same reasoning to the commercial fishery. Second, aboriginals in Washington State and Maori in New Zealand had gained commercial fishing rights through treaty rights. Third, Canada had already conceded that commercial fishing rights could come within land claims settlements, and had begun an "Aboriginal Fisheries Strategy" that authorized special commercial openings for Aboriginals (though there had been public protests over this development). Fourth, commercial hunting rights had been upheld by the Supreme Court of Canada as a treaty right in the *Horseman* decision,[26] reduced to subsistence rights only by express constitutional wording in the Natural Resources Transfer Agreement of 1930. Finally, commercial fishing rights would alleviate some of the poverty that still plagued aboriginal communities.

While victory seemed the most likely outcome, there were at least five factors that clouded the issue. First, the *Fisheries Act* had always recognized aboriginal food fisheries, but never expressly an aboriginal commercial fishery.[27] While there had been programs to promote the Indian presence in the commercial fishery, there had been nothing parallel to the Department of Fisheries policy to give priority to the Indian food fishery, an established policy that had been given legal force in *Sparrow*. Second, aboriginal rights argumentation naturally looks to rights established in the past. Chief Justice Lamer marshaled quotations from leading pro-aboriginal decisions and used those statements to restrict aboriginal rights to those enjoyed prior to contact with Europeans. It was the past that informed Lamer's analysis, not the forward-looking language that characterized the judgment in the *Sparrow* case. Third, management of the commercial fishery was probably much more complicated than management of the food fishery. Canada argued that there was no natural limit to an aboriginal right to a commercial fishery, but there obviously was to a food fishery. Fourth, the commercial fishery was experiencing difficult times, with reduced runs and management problems. Aboriginal commercial fishing rights would be at the expense of non-aboriginal commercial fishers. Fifth, Justices Dickson and Wilson, who wrote the key earlier decisions, were no longer on the court. Chief Justice Lamer, with a rather different approach in both substance and style, wrote the judgments.

Chief Justice Lamer limited aboriginal rights on what he held out to be a logical, objective basis. Aboriginal rights were limited to elements of pre-contact Indian societies. By Lamer CJ's rulings, aboriginal rights only include the commercial sale of fish if sale had been an integral part of the distinctive culture of the Aboriginal grouping at the point of contact with Europeans. While this may sound like an impossible standard to meet, in one of the cases, the *Gladstone* decision, it was held that the Indian tribe had the right to sell a large quantity of herring roe on kelp, a lucrative specialized product much sought after by the Japanese.

If the judgment in the *Sparrow* case had used expansive, celebratory language, the *Van der Peet*, *Gladstone* and *NTC Smokehouse* decisions were tightly structured and deliberate in their reasoning. In passages elaborating on the basic test, Lamer CJ ruled that it required more than demonstrating that a practice, tradition or custom was an aspect of, or took place in, the aboriginal society. The rationale behind the recognition of aboriginal rights was to reconcile aboriginal rights and Aboriginal peoples to the rights of the dominant society.

> Aboriginal rights have their basis in the prior occupation of Canada by distinctive Aboriginal societies. To recognize and affirm the prior occupation of Canada by distinctive Aboriginal societies it is *to what makes those societies distinctive* that the court must look in identifying Aboriginal rights. The court cannot look at those aspects of the Aboriginal society that are true of every human society (e.g. eating to survive), nor can it look at those aspects of the Aboriginal society that are only incidental or occasional to that society; the court must look instead to the defining and central attributes of the Aboriginal society in question. It is only by focusing on the aspects of the Aboriginal society that make that society distinctive that the definition of Aboriginal rights will accomplish the purpose underlying s. 35(1).[28]

The idea that fishing for salmon is peculiarly "Indian" seems ridiculous. Any people who live on the west coast of Canada, Aboriginal or non-Aboriginal, will fish for salmon. Not to fish would be peculiar.

Lamer CJ then went on to discuss the purpose of s. 35:

> [T]he Aboriginal rights recognized and affirmed by s.35(1) are best understood as, first, the means by which the Constitution recognizes the fact that prior to the arrival of Europeans in North America the land was already occupied by distinctive Aboriginal societies, and as, second, the means by which that prior occupation is reconciled with the assertion of Crown sovereignty over Canadian territory. The content of Aboriginal rights must be directed at fulfilling both of these purposes.[29]

This passage comes after eight pages of citations from judicial decisions in Canada, the United States and Australia, and references to the writings of several academics. All these references cite the prior existence of Aboriginal peoples as the basis for aboriginal rights. Those passages were written in support of aboriginal rights. Their description of the historical origins of the rights was not written to define limits, but that was how Lamer used the quotations.

The Marshall decision

In *Marshall*,[30] the Supreme Court upheld the commercial sale of eels as a treaty right. The decision was one of the most controversial decisions in the court's history. It triggered a native lobster fishery that pitted Aboriginals against non-Aboriginals. Bitter conflict and violence captured the national news for weeks.

In the 1760 treaty at issue in the Marshall case, the Mi'kmaq promised not to

> traffick, barter or Exchange any Commodities in any manner but such persons or the managers of such Truck houses as shall be appointed or Established by His Majesty's Governor at Lunenbourg or Elsewhere in Nova Scotia or Accadia.[31]

This recognized that the Mi'kmaq would sell to non-Aboriginals, but the particular arrangement, the use of "truck houses," was soon abandoned. Writing for the majority, Mr. Justice Binnie turned to the "underlying negotiations" and concluded that they supported a treaty right of the Mi'kmaq to a commercial trading right, sufficient to allow them to secure "necessaries," which would amount to gaining a "moderate livelihood." He observed that

> [t]he British certainly did not want the Mi'kmaq to become an unnecessary drain on the public purse of the colony of Nova Scotia or of the

Imperial purse in London, as the trial judge found. To avoid such a result, it became necessary to protect the traditional Mi'kmaq economy, including hunting, gathering and fishing. A comparable policy was pursued at a later date on the west coast where, as Dickson J. commented in *Jack* v. *The Queen*, [1980] 1 S.C.R. 294, at p. 311:

> What is plain from the pre-Confederation period is that the Indian fishermen were encouraged to engage in their occupation and to do so for both food and barter purposes.

The same strategy of economic aboriginal self-sufficiency was pursued across the prairies in terms of hunting: see *R.* v. *Horseman*, [1990] 1 S.C.R. 90.[32]

This approach gave legitimacy to upholding commercial rights. Mr. Justice Binnie was reversing the spirit and the reasoning of the 1996 commercial fishing decisions, but not, technically, the actual decisions.

Rights to waters

What of rights to the waters? Reserves in British Columbia do not include the foreshore or rights in relation to internal or adjacent waters. They do not include the beds of lakes, streams or rivers. But the reserves do not constitute a comprehensive settlement of land issues. Negotiations for treaties are ongoing in British Columbia.

Some dealings with coastal areas and coastal waters have occurred in connection with the James Bay and Northern Quebec Agreement and the Nunavut land claims settlement in the eastern Arctic. The *Nisga'a Treaty* in British Columbia did not depart from the tradition of excluding rights to rivers, streams and foreshore from recognized Indian lands. But it did define an important extent of fishing rights, both subsistence and commercial, for the Nisga'a people.[33]

The Nuu-Chah-Nulth people, on the west coast of Vancouver Island, assert claims to the territorial sea as far as can be seen from land. There has been no settlement of their claims, and litigation on the issue is not expected in the near future. The Haida have claimed Hecate Strait and the full territorial sea in the claim they have filed in British Columbia Supreme Court.[34] A trial on the issues in the case is probably a couple of years away.

Aquaculture

To date, the only Canadian judicial decision relating to aquaculture is that of Madam Justice Gerow of the British Columbia Supreme Court, a trial-level court, in *Heiltsuk Tribal Council* v. *British Columbia*.[35] The litigation challenged provincial licenses authorizing a hatchery designed to raise Atlantic salmon fry for use in aquaculture. Between 2001, when the

Heiltsuk Tribal Council first knew of the project, and March 2003, the company invested CDN$9.5 million in the hatchery site. Meetings had occurred between the company and the Heiltsuk. The Heiltsuk denied that those meetings constituted any "consultation." Madam Justice Gerow concluded that the Heiltsuk were opposed to Atlantic salmon aquaculture and were unwilling to consult or seek accommodation. She ruled that the hatchery would not interfere with Heiltsuk aboriginal rights, including their non-exclusive rights to use land and their rights to fish. The strident opposition of the Heiltsuk to the hatchery and aquaculture was seen as unreasonable. The decision upholding the licenses was not appealed.

Australian developments

The first significant Australian rights case was the 1970 decision in *Milirrpum* v. *Nabalco*,[36] challenging a bauxite mining permit on aboriginal reserve land in the Northern Territory. The judgment found a detailed system of land law on the part of Aboriginal people. But Justice Blackburn concluded that the doctrine of native title had never formed part of the law of Australia. He placed some reliance on Canadian judgments, in particular the decision of the British Columbia Court of Appeal in the *Calder* case (later decided on much different grounds by the Supreme Court of Canada).

Gough Whitlam and the Labor Party came to power in a national election in the latter part of 1972 promising a national land rights policy. Whitlam appointed Mr. Justice Woodward to conduct an inquiry on how to recognize native title in the Northern Territory, an area where the national government had full jurisdiction.[37] The *Aboriginal Land Rights (Northern Territory) Act* was enacted in 1976.[38] It gave Aboriginal people a veto over mining, though the veto could be overridden by the national government.[39] It also established an effective land claims process under the Aboriginal Land Commissioner, Mr. Justice Toohey (who later sat in the High Court).

The 1976 legislation established something of an Australian model on land rights, involving (1) the transfer of existing reserves to aboriginal ownership; (2) the creation of an aboriginal management body; (3) (sometimes) the establishment of a land claims process for unextinguished traditional land rights outside the reserves; and (4) (sometimes) the creation of a fund to acquire land for aboriginal communities. South Australia copied quite a bit of this model in the *Pitjantjatjara Land Rights Act* in 1981. The Pitjantjatjara scheme was challenged on the basis of racial discrimination, but upheld in 1985 by the Australian High Court in *Gerhardy* v. *Brown*.[40] New South Wales established a land fund and added a limited claims process. Also, Queensland introduced a statutory claims process.[41]

The 1992 High Court decision in *Mabo (No. 2)* v. *Queensland*[42] established some extent of land rights nationally. The *Native Title Act*[43] followed, designed to confirm non-aboriginal titles, and to provide an orderly process for determining where native title survived. As in Canada and New Zealand, a national land claims process had been established. The process has proven complex and slow, with the country being badly divided on a set of amendments to the *Native Title Act* in 1998 designed to mitigate somewhat the effects of native title rulings.

Rights to waters

In *Commonwealth* v. *Yarmirr*,[44] certain Aboriginal groups claimed native title to the seas and seabed around Croker Island, off the north coast of the Northern Territory. Croker Island and adjacent islands were recognized under the *Aboriginal Land Rights (Northern Territory) Act* of 1976, and were included in the Arnhem Land Aboriginal Land Trust for the benefit of the traditional owners. The claimants sought a determination for waters, reefs and lands that had not been granted under the 1976 legislation. Australia argued that no native title existed in relation to the sea and the seabed. Section 6 of the *Native Title Act* states that the legislation applies to a number of areas, including application "to the coastal sea of Australia and of each external Territory, and to any waters over which Australia asserts sovereign rights under the Seas and Submerged Lands Act 1973." The majority commented:

> The reference to "sea" in this definition, taken both with the other elements of the definition of "waters" and with the provision of s 6 of the Act, indicates clearly that the Act is drafted on the basis that native title rights and interests may extend to rights and interests in respect of the sea bed and subsoil beyond low-water mark and the waters above that sea bed.[45]

It was necessary for the High Court to determine what extent of sea rights fell within the traditional laws and customs of the claimants. A limited right to the sea and seabed was recognized. It was limited in two ways.

First, the court found that when sovereignty was asserted over the seas, some qualifications on any existing rights were imposed (para. 61):

> [T]he recognition of public rights of navigation and fishing and, perhaps the concession of an international rights of innocent passage. Those rights were necessarily inconsistent with the continued existence of any right under Aboriginal law or custom to preclude the exercise of those rights.

The various actions of Australia relating to the territorial sea, which led to these imposed rights held by others, were not, however, inconsistent with the common law of Australia recognizing native title rights in relation to the sea and seabed.

Second, the trial judge had ruled that under traditional law, the claimants had not enjoyed exclusive possession, occupation, use and enjoyment of the waters of the claimed area. Each clan allowed members of other claimant clans to use their areas, with permission, and this seemed to extend to other Aboriginal people from outside the claimed area. Evidence established that in the eighteenth and nineteenth centuries, large numbers of fishermen from Macassar, in what is now Indonesia, came for four to seven months each year to gather sea cucumber in the claimed area. The trial judge held that under aboriginal law and custom there was no right to exclude such people. Permission was required for other Aboriginal groups, but not for such outsiders.

In *Risk* v. *Northern Territory*,[46] claims had been filed for the seabed of bays and gulfs within the limits of the Northern Territory. The area claimed was large and included the port area of the city of Darwin, the largest settlement and capital of the Northern Territory. The legislation invoked was the *Aboriginal Land Rights (Northern Territory) Act* of 1976. Under it, extensive claims to lands outside reserves had been heard, and native title rights recognized for large areas. The Aboriginal Land Commissioner, established under the legislation to hear claims, had ruled that sea claims fell outside the Act and could not be considered. The majority of the Full Court of the Federal Court of Australia, with one dissent, affirmed that decision. That was affirmed by the High Court, which added that there might be a claim under the *Native Title Act*.

The reasoning was textual. The legislation dealt with claims to "land in the Northern Territory." An examination of the wording of the legislation led the various levels to conclude that it did not envisage claims to the seabed:

> First, there are strong textual indications in the *Land Rights Act* that "land in the Northern Territory" does not include the seabed. Secondly, the nature of the interest which is granted to a Land Trust suggests that the seabed is not "land in the Northern Territory". Thirdly, any remaining doubt about the matter is put to rest when regard is had to relevant extrinsic material and the legislative history which lies behind the *Land Rights Act*.[47]

The Northern Territory was authorized to create a 2-kilometer "buffer zone" of sea adjoining aboriginal land. This very provision, the High Court said, confirmed the distinction between land and sea in the legislation.

The ruling was not based on aboriginal law, which does not distinguish between land and waters in understanding traditional "country." There can be "sea country"; that is, coastal maritime areas that are part of native title according to aboriginal law.

In March 2004, the Full Federal Court decided the case of *Lardil Peoples* v. *State of Queensland*.[48] At issue were native title rights to rivers and seas in the Wellesley Island region. In a detailed judgment, the following rights were upheld:

1 The right to access the land and waters seaward of the high-water line in accordance with and for the purposes allowed by and under their traditional laws and customs.

2 The right to fish, hunt and gather living and plant resources, including the right to hunt and take turtle and dugong, in the intertidal zone and the waters above and adjacent thereto for personal, domestic or non-commercial communal consumption in accordance with and for the purposes allowed by and under their traditional laws and customs.

3 The right to take and consume fresh drinking water from freshwater springs in the intertidal zone in accordance with, and for the purposes allowed by and under, their traditional laws and customs.

4 The right to access the land and water seaward of the high water line in accordance with and for the purposes allowed under their traditional laws and customs for religious or spiritual purposes and to access sites of spiritual or religious significance in the land and waters within their respective traditional territory for the purposes of ritual or ceremony.

The rights upheld are circumscribed. They do not allow the exclusion of all others and do not extend to mineral or petroleum rights. They are subject to common law rights of fishing and navigation, and to the right of innocent passage.

Aquaculture

In November 2003, the Aboriginal native title holders in the area around Croker Island, whose rights had been the subject of the Yarmirr litigation, signed an agreement authorizing the establishment of three aquaculture pearl farms. The operators of the farms also held Crown leases for aquaculture under the *Northern Territory Fisheries Act*.[49]

New Zealand developments

The 1840 Treaty of Waitangi guaranteed the Maori the continuation of rights to their lands and fisheries. A special tribunal or court was established to handle the transfer of Maori lands to settlers. The direct enforceability of the treaty in domestic law was denied.

In response to the large 1975 land rights march to the capital, Wellington, the government established the Waitangi Tribunal as an advisory body to hear Maori claims. Gradually, with the support of certain judicial decisions, a land claims process came into being. The first comprehensive settlement was with the Tainui tribe in the Waikato area, south of Auckland. By June 2003, settlements had occurred with fourteen tribes (*iwi*), with three additional settlements awaiting legislative implementation.[50] Also, there has been a comprehensive settlement of commercial fisheries claims.[51]

A minority of Maori opposed the comprehensive settlement of commercial fisheries claims. A challenge to the settlement in the New Zealand courts failed.[52] The matter was taken to the United Nations Human Rights Committee in the *Mahuika* case.[53] The settlement was again upheld. By the settlement, some 23 percent of commercial fisheries quota in New Zealand is held by a Maori fisheries commission.

The Ngati Apa decision

The most significant recent litigation in New Zealand on Maori rights dealt with the waters of the Marlborough Sounds.

> In 1997 some iwi [sub-tribes] from the top of the South Island were concerned about the way marine farming was developing in the Marlborough Sounds and its impact on their customary fishing rights and what they considered their more general customary interests.
>
> They took a test case to the Maori Land Court asking it to determine that areas of the foreshore and seabed in the Sounds were Maori customary land. This case challenged the assumptions that there was little if any scope for customary rights that amounted to a full title to remain in the foreshore, and that there was no ability for a customary claim to the seabed.[54]

The case went to the Court of Appeal in *Ngati Apa* v. *Attorney General*, decided in June 2003, on preliminary questions of law. No determination has yet taken place on traditional Maori rights to the areas in question.

The Crown argued that the 1963 decision *In Re Ninety-Mile Beach* meant that Maori interests in the foreshore had been extinguished when customary rights ended for the adjacent onshore land (which would probably defeat all Maori claims in the Marlborough Sounds).[55] Second, the Crown argued that various statutes had extinguished any Maori customary property rights in the area by vesting rights to the foreshore and the seabed in the Crown. A judge in the High Court ruled that lands below the low-water mark were beneficially owned by the Crown at common law, blocking any Maori claims.[56] All five judges in the Court of Appeal rejected these arguments and ordered the Maori Land Court to proceed to hear the claims.

Chief Justice Elias of the Court of Appeal stressed the preliminary character of the court's ruling. While she upheld the Maori's legal arguments, she emphasized that the Court of Appeal could not determine whether any Maori customary land rights existed in the foreshore and seabed. There was no factual record in the case on which to make such a determination. Customary rights vary from case to case, depending on the customs and usages of the specific communities involved. Madam Justice Elias noted that such rights could vary from usufructory rights, on the one hand, to exclusive ownership, as recognized by the Supreme Court of Canada in *Delgamuukw* v. *British Columbia*.[57]

There are clear suggestions in the judgments of the Court of Appeal that any Maori rights to foreshore and seabed would probably be of a limited character, and would be subject to certain statutes. Only Mr. Justice Gault described the claims in what we may assume was the full strength of the plaintiff's assertions: "The claim is essentially that the whole of the foreshore and seabed of the Marlborough Sounds, extending to the limits of New Zealand's territorial sea, is Maori customary land as defined in the Act."[58] He was also the only judge to refer to the Marlborough Sounds Resource Management Plan of 1995, which perhaps was the government move that prompted the Maori litigation.[59]

The Treaty of Waitangi of 1840, unlike the North American Indian treaties, did not extinguish native title. Instead, Maori ownership was confirmed. The history of land dealings in New Zealand is the history of a specialized court with the statutory authority to convert Maori customary land rights into fee simple titles, bringing the land, parcel by parcel, onto the open market.[60] Non-indigenous settlers took over the land, as they did in North America, but through a different formal process.

Because the statutory scheme dominated, the common law recognition of customary land rights was, in Chief Justice Elias's words, "little developed."[61] Some fundamentals were clear, she ruled. For example, it was clear that the Crown's acquisition of sovereignty over New Zealand did not, in itself, affect customary property rights.[62] Those interests were preserved by the common law until extinguished in accordance with law, as held in rulings from New Zealand, the United States and Canada.

Over the years, Elias commented, there have been some good decisions and some bad decisions. The 1877 decision in *Wi Parata* v. *Bishop of Wellington*, which denied Maori customary land rights, was wrong, she declared.[63] Its reasoning had been rejected by the Judicial Committee of the Privy Council.[64] Yet it had influenced New Zealand thinking for most of a century. The 1963 decision in *Ninety-Mile Beach* was also wrong.[65] The idea that there were no enforceable Maori customary land rights had "proved hardy."[66] Elias CJ even found a culprit for the heresy: Sir John Salmon, who drafted the *Native Lands Act* of 1909.[67]

The modern shift in judicial and political attitudes meant that there was now a clear contradiction between the recognition of customary rights to land and the *de facto* lack of recognition of customary rights to the foreshore and subsurface. This contradiction gave rise to the Marlborough Sounds litigation.[68]

The Court of Appeal ruled that if Maori custom and usage recognized rights to the foreshore and seabed, that would prevail over the English common law rulings, which rejected any such private ownership in the United Kingdom. The case had to go back to the Maori Land Court for a determination on Maori customary rights, based on evidence of Maori customs and usage. Only after that determination was made could the legal question be addressed as to whether legislation had extinguished some or all of the rights involved.

A set of general national statutes was cited, dealing with harbors and territorial seas. While many provisions seemed to assume there were no Maori rights to foreshore or seabed in existence, the statutes failed to demonstrate any intention to extinguish pre-existing Maori customary rights. Madam Justice Elias examined the various statutes and concluded that no "confiscatory effect," "expropriatory purpose" or act of "appropriation" could be discerned in the legislation.[69] Mr. Justice Keith specifically quoted the rule of extinguishment now adopted in Canada and Australia in the leading cases of *Sparrow* and *Mabo*. The onus of proving extinguishment lies on the Crown, and the intent to extinguish must be "clear and plain."[70] Mr. Justice Tipping required extinguishment by express words or necessary implication. "Parliament would need to make its intention crystal clear."[71]

The Court of Appeal ruling, while preliminary, proved highly controversial. The government, after a year of public disputes, enacted legislation declaring any possible Maori interests in Marlborough Sounds to be "public domain."

International law developments

International law developments on indigenous rights include: (1) the Draft Declaration on the Rights of Indigenous Peoples (currently being considered by the United Nations Commission on Human Rights);[72] (2) the study on indigenous peoples and their relationship to land by Special Rapporteur Erica Irene Daes;[73] (3) the decision of the Inter-American Court of Human Rights in 2001 in *Awas Tingni* v. *Nicaragua*;[74] (4) the decision of the United Nations Human Rights Committee in *Mahuika* v. *New Zealand* in 2000;[75] and (5) the reports of the Special Rapporteur on Indigenous Issues, the first issued in 2002.[76]

These texts see indigenous rights not so much in specific historical-legal terms (the framework for domestic litigation) as in terms of a general history of the loss of economic base by various indigenous communities. They are concerned with the survival of indigenous collectivities. As part of that minority rights concern, they address the issues of the resources that Indigenous peoples need to gain or regain in order to survive and develop. The texts do not address aquaculture specifically, but would see it as a significant potential economic resource for Indigenous peoples in various parts of the world.

The decision of the United Nations Human Rights Committee in Mahuika *v.* New Zealand

In *Mahuika* v. *New Zealand*, the United Nations Human Rights Committee considered a settlement of Maori claims to fishing rights, under the provisions of the International Covenant on Civil and Political Rights.[77] The Treaty of Waitangi of 1840 guaranteed to the Maori the "full, exclusive and undisturbed possession of their lands, forests, fisheries and other properties."[78] But the treaty could not be directly enforced in New Zealand law, and most commercial fishing passed into non-Maori hands.

A settlement of Maori fishing rights claims occurred, placing around 23 percent of total commercial fisheries quota in the hands of a Maori fisheries commission. While this was a clear compromise of the provisions of the treaty, it was upheld by the New Zealand courts, having been put into place by effective settlement legislation.[79] The committee did not re-examine the settlement in terms of the Treaty of Waitangi promise, interpreting instead Article 27 of the Covenant on Civil and Political Rights:

> In those States in which ethnic, religious or linguistic minorities exist, persons belonging to such minorities shall not be denied the right, in community with the other members of their group, to enjoy their own culture, to profess and practise their own religion, or to use their own language.

A large minority of Maori, claiming to represent seven (out of eighty-one) *iwi* (tribal groups) challenged the settlement. If the *iwi* were considered as separate claimants, then an analysis would have to be undertaken of the impact of the settlement on each iwi. Instead, the committee proceeded to consider the Maori as a single minority group. The committee found that the Maori had been consulted and had agreed to the settlement. That consent prevailed over the interests of "individuals who claim to be adversely affected." The dissenting *iwi* were groups of individuals, not separate rights-holding collectivities. The committee did not discuss whether the proper minority entities were the eighty-one *iwi* or the Maori as a pan-*iwi*/pan-tribal entity.

The committee ruled, as it had before, that economic rights could come within Article 27. The exercise of fishing rights, the Committee said, was "a significant part of Maori culture."[80] It also confirmed its established position, especially in the case of Indigenous peoples, that the enjoyment of the right to one's own culture may require positive legal measures of protection by a state, and measures to ensure the effective participation of members of minority communities in decisions that affect them.

The committee noted that the settlement gave Maori "access to a great percentage of quota, and thus effective possession of fisheries was returned to them." The committee understood that the New Zealand government saw the settlement in developmental terms, not simply in terms of Treaty of Waitangi claims. In other words, the settlement was a good thing, which was to the economic benefit of Maori as whole, and should not be derailed by minority dissidents.

The committee ruled that the settlement did limit the rights of the Maori to enjoy their own culture, but held that the limitation was justified in the particular case, without discussing the factors that justified the limitation. The history of the non-recognition of Maori fishing rights had led to a situation where non-Maori were well established in the industry. It seems that the committee was not going to order the expropriation of all or part of those rights, but neither was it willing to discuss the problem of the non-indigenous rights, which now meant that only a compromise or settlement of Maori claims was possible, not a full vindication of treaty rights.

While the outcome supported indigenous economic rights, the reasoning is weak and, in parts, non-transparent. But given the limited background of the committee members on indigenous resource claims, the decision was commendable.

The Awas Tingni case in the Inter-American Court of Human Rights

In 1995, a petition was submitted to the Inter-American Commission on Human Rights by the Indian Law Resource Center, based in the United States, on behalf of the Awas Tingni Indian community on the Atlantic coast of Nicaragua.[81]

Timber harvesting rights had been granted to a Korean-owned company over traditional lands of the Awas Tingni without community consultation, consent or compensation. A challenge to the timber concession in the Nicaraguan courts was unsuccessful. The commission attempted friendly settlement procedures, but without success.[82] It issued a report on the case, ruling that Nicaragua was in breach of the American Convention on Human Rights in failing to demarcate the traditional lands of the Awas Tingni community and the other indigenous communities. In 1998, the commission submitted the matter to the Inter-American Court of Human Rights.[83] In August 2001, the court ruled in favor of the Awas Tingni community, ordering Nicaragua to begin the process of demarcating the Indian lands and issuing titles in accordance with local indigenous customary law.

The commission had gradually built up jurisprudence around the recognition of Indigenous peoples and their territorial rights. The decision of the Inter-American Court of Human Rights in the *Awas Tingni* case confirms this jurisprudence, and stands as the most important decision of an international court or tribunal on the issue of indigenous rights:

> 149. Given the characteristics of the instant case, it is necessary to understand the concept of property in indigenous communities. Among indigenous communities, there is a communal tradition as demonstrated by their communal form of collective ownership of their lands, in the sense that ownership is not centered in the individual but rather in the group, and in the community. By virtue of the fact of their very existence, indigenous communities have the right to live freely on their own territories; the close relationship that the communities have with the land must be recognized and understood as a foundation for their cul-

tures, spiritual life, cultural integrity and economic survival. For indigenous communities, the relationship with the land is not merely one of possession and production, but also a material and spiritual element that they should fully enjoy, as well as a means through which to preserve their cultural heritage and pass it on to future generations.

. . .

151. The customary law of indigenous peoples should especially be taken into account because of the effects that flow from it. As product of custom, possession of land should suffice to entitle indigenous communities without title to their land to obtain official recognition and registration of their rights of ownership.

The Court said that guaranteeing respect for territorial rights required the issuance and registration of formal titles to the land.

Conclusion

This chapter has surveyed developments on indigenous rights to lands, fisheries, foreshore and offshore to provide a context for analyzing indigenous claims to aquaculture. The analysis demonstrates the recent character of the present framework of indigenous rights. While national land claims systems exist in Australia, Canada and New Zealand, in each case they are no more than thirty years old and are far from completing their work. The process of settling claims and working out a new system of indigenous rights has proven more complex and much more time-consuming than observers expected. This is a work in progress. The same is true in relation to indigenous fishing rights, now being resolved in Canada largely in the context of land claims settlements. It is striking how recent the cases dealing with waters, foreshore and offshore are. The most important judicial decisions are in 2001 and 2003.

At the time of writing, we have only one trial level decision in Canada relating to aquaculture. The New Zealand litigation in *Ngati Apa* v. *Attorney General* was prompted by a dispute about aquaculture, though aquaculture is not directly addressed in the decision of the Court of Appeal. Further litigation is certain on these issues.

Postscript

After I had completed this chapter, three items came to my attention: (1) The Maori Commercial Aquaculture Claims Settlement Act of 2004 guaranteed 20 percent of commercial aquaculture space in New Zealand to Maori, in settlement of aquaculture claims. (2) The Aboriginal and Torres Strait Islanders Commission published *Water and Fishing: Aboriginal Rights in Australia and Canada* (2004), which covers a number of matters discussed in this chapter, but has no section titles specifically referring to aquaculture.

(3) The British Columbia Supreme Court has issued two decisions relating to indigenous objections to an aquaculture farm in Bute Inlet that planned to introduce Atlantic salmon.

Notes

The research assistance of Warren Pearson, past Research Assistant with Dalhousie's Marine and Environmental Law Institute, is gratefully acknowledged.

1 This chapter does not deal with claims to self-government, tribal sovereignty, self-determination or autonomy.
2 *Calder* v. *Attorney-General of British Columbia* [1973] S.C.R. 313.
3 *Constitution Act*, 1982, s. 35, being Schedule B to the *Canada Act 1982* (U.K.), 1982, ch. 11.
4 *Guerin* v. *The Queen* [1984] 2 S.C.R. 335; *Regina* v. *Sparrow* [1990] 1 S.C.R. 1075 [*Sparrow*]; *Delgamuukw* v. *British Columbia* [1997] 3 S.C.R. 1010 [*Delgamuukw*].
5 Commonwealth, 1976.
6 *Mabo (No. 2)* v. *State of Queensland* (1992) 66 A.L.J.R. 408 (High Court of Australia) [*Mabo No. 2*].
7 Commonwealth, 1993.
8 *New Zealand Maori Council* v. *Attorney General (State Owned Enterprises)* [1987] 1 N.Z.L.R. 641 (Court of Appeal).
9 *Adong bin Kuwau* v. *Johor* [1997] 1 *Malayan Law Journal* 418–436. The reasons for judgment of Gopal Sri Ram JCA for the three-person panel in the Court of Appeal were issued on 24 February 1998. The appeal to the Federal Court was dismissed without reasons.
10 Series C No. 79, I/A Court H.R., The Mayagna (Sumo) Awas Tingni Community Case, judgment of 31 August 2001. Online. Available http://www.corteidh.or.cr/seriecing/serie_c_79_ing.doc (accessed 22 March 2004).
11 *Richtersveld Community* v. *Alexkor Limited & the Government of the Republic of South Africa*, 2003 S. Afr. S.C. 48/2001.
12 *R.* v. *Van der Peet* [1996] 2 S.C.R. 507 [*Van der Peet*]; *R.* v. *NTC Smokehouse* [1996] 2 S.C.R. 672; *R.* v. *Gladstone*, [1996] 2 S.C.R. 723.
13 *R.* v. *Marshall* [1999] 3 S.C.R. 456 [*Marshall*].
14 *Nisga's Final Agreement Act*, S.C. 2000, c. C-7. See also Douglas Sanders, "'We Intend to Live Here Forever': A Primer on the Nisga'a Treaty," (1999) 33 *University of British Columbia Law Review* 103.
15 *Yanner* v. *Eaton* (1999) 201 CLR 351. This viewpoint had been around, in spite of language in s. 223 (2) of the *Native Title Act 1993* that made references to "hunting gathering, or fishing rights and interests," language cited in the *Yarmirr* decision, para. 8.
16 *Te Runanga o Wharekauri Rekohu* v. *Attorney-General* [1993] 2 N.S.L.R. 301 [*Te Runanga*].
17 CCPR/C/70/D/547/1993 (27 October 2000) [*Mahuika*].
18 *United States* v. *Washington*, 384 F. Supp. 312 (Western District Washington, 1974), aff'd 520 F. 2nd 676 (9th Cir. 1975), cert. denied, 423 U.S. 1086 (1976) (for subsequent opinion on many of the issues see *Washington* v. *Fishing Vessel Association*, 443 U.S. 658 (1979)).
19 (2001) 75 A.L.J.R. 1582.
20 (2002) 76 A.L.J.R. 845.
21 [2004] F.C.A. 298 (Federal Court of Appeal).
22 *Ngati Api & Ors* v. *Attorney-General & Ors*, 2003 NZCA 117 [*Ngati Api*].
23 *Sparrow, supra* note 4.

24 Section 35 is outside the Canadian Charter of Rights and Freedoms, Part I of the *Constitution Act*, 1982, *supra* note 3, being Schedule B to the *Canada Act 1982* (U.K.), 1982, c. 11. Therefore s. 1, which allows for limitations on rights, does not apply. The court, without reference to s. 1, created an equivalent justification test. *Sparrow*, *supra* note 4.

25 *Supra* note 12.

26 *R.* v. *Horseman* [1990] 1 S.C.R. 901 [*Horseman*].

27 *Fisheries Act*, R.S., c. F-14, s. 1.

28 *Van der Peet*, *supra* note 12 at para. 56.

29 *Ibid.* at para. 43.

30 *Marshall*, *supra* note 13.

31 *Ibid.* at para. 5.

32 *Horseman*, *supra* note 26 at para. 48.

33 See Sanders, *supra* note 14.

34 Council of the Haida Nation, Guujaaw, suing on his own behalf, and on the behalf of all members of the Haida Nation v. Her Majesty the Queen in Right of the Province of British Columbia and the Attorney-General of Canada. File no. L020662, Supreme Court of British Columbia, 14 November 2002.

35 *Heiltsuk Nation* v. *British Columbia* (2003) 19 B.C.L.R. (4th) 107.

36 (1971) 17 F.L.R. 141.

37 Austl., Commonwealth, Aboriginal Land Rights Commission, Parl. Paper No. 69 (1974).

38 Commonwealth, 1976.

39 *Aboriginal Land Rights (Northern Territory) Act 1976* (Cth.), s. 48 C (1)(b).

40 (1985) 159 C.L.R. 70.

41 For details of these processes, see Heather McRae, Garth Nettheim, Laura Beacroft and Luke McNamara, *Indigenous Legal Issues: Commentary and Issues* (Sydney: Lawbook Co., 2003).

42 *Mabo No. 2*, *supra* note 6.

43 Commonwealth, 1973.

44 *Commonwealth* v. *Yarmirr* (2001) 75 A.L.J.R. 1582 (H.C.A.).

45 *Ibid.* at para. 8.

46 *Risk* v. *Northern Territory* (2002) 76 A.L.J.R. 845 (H.C.A.).

47 *Ibid.* at para. 25.

48 [2004] F.C.A. 298.

49 See John Hughes, "Croker Seas Pearl Farming Agreement," (December 2003–January 2004) 5(29) *Indigenous Law Bulletin* 21.

50 Wayne Thompson, "Hapu Yearns for Crown Partnership," *New Zealand Herald*, 17 June 2003, A7. The article discusses claims in the Auckland area that are under negotiation.

51 Bill No. 90.1, *Maori Fisheries Bill 2003*, 1st Sess., 47th Parl., 2003. This bill has been introduced but has not come into effect. The bill is before the select committee.

52 *Te Runanga*, *supra* note 16.

53 *Mahuika*, *supra* note 17.

54 "Foreshore Report Guarantees Public Access" (an edited version of the Government's report on Maori claims to the seabed and foreshore released on 18 August 2003), *New Zealand Herald*, 19 August 2003, A14.

55 [1963] NZLR 461.

56 *Re: Ninety-Mile Beach* [1963] N.Z.L.R. 461.

57 *Delgamuukw*, *supra* note 4 at paras. 110–119. *Amodu Tijani* v. *Secretary, Southern Nigeria*, was cited for the same proposition.

58 *Ngati Api*, *supra* note 22 at para. 95. The act referred to is the statute establishing the Maori Land Court.

59   *Ibid.* at para. 94.

60   Additionally, there were confiscations after the land wars of the 1860s. The modern statutory scheme is the *Te Whenua Maori Act* of 1993.

61   *Ngati Api, supra* note 22 at para. 46. Equally, it was little developed in Australia and Canada until recent decades. While major US cases occurred in the first decades of the nineteenth century, their authority or applicability was subsequently undermined, then restored.

62   In para. 15, *supra* note 22, the classic statement on this point from *Amodu Tijani* v. *Secretary, Southern Nigeria* [1921] 2 A.C. 399 at 407, was quoted: "A mere change in sovereignty is not to be presumed as meant to disturb rights of private owners."

63   (1877) 3 N.Z. Jur. (NS) SC 72.

64   Elias notes that the *Wi Parata* decision was rejected by the Judicial Committee of the Privy Council in *Tamaki* v. *Baker* [1901] A.C. 561. The pioneering modern Canadian case, *Calder* v. *British Columbia, supra* note 2, modeled its pleadings on those in *Tamaki* v. *Baker*.

65   Mr. Justice Gault felt that the decision was fact dependent, rather than rejecting it categorically: *supra* note 22 at para. 123. Mr. Justice Keith, writing for himself and Anderson, finds it wrong on the rule for extinguishment: *supra* note 22 at para. 154.

66   *Ibid.* at para. 26.

67   *Ibid.* at para. 26 and 27.

68   *Ngati Api, supra* note 22.

69   For examples, see *ibid.* at paras. 60, 63 and 70. Of course, legislators do not bother to "extinguish" rights that they do not recognize as existing. The harshest periods of indigenous land loss are usually accompanied by a denial that there are any indigenous rights in play. Because the New Zealand statutes did not address the issue of indigenous rights, they were vulnerable to some reassessment after Maori rights regained legal recognition.

70   *Ibid.* at para. 148; *Sparrow, supra* note 4; *Mabo* v. *Queensland (No. 2)* (1992) 175 C.L.R. 1, 64, 111, 195–196.

71   *Ngati Api, supra* note 22 at para. 185.

72   United Nations Committee on Economic, Social and Cultural Rights (ESCOR), Commission on Human Rights, 11th Sess., Annex 1, U.N. Doc. E/CN.4/Sub.2 (1993).

73   Office of the United Nations High Commissioner for Human Rights, *Indigenous Peoples and Their Relationship to Land*, Final Working Paper, E/CN.4/Sub.2/ 2001/21 UN ESC (2001).

74   *Awas Tingni* v. *Nicaraguan* (2001) Inter-Am. Ct.H.R. (Series C) No. 79, *Annual Report of the Inter-American Commission of Human Rights*, 1999 OEA/Ser L/V/III.47/doc. 6 (2000).

75   *Apirana Mahuika et al.* v. *New Zealand* (2000), HRC Communication No. 547/1993, U.N. Doc. CCPR/C/70/D/547/1993.

76   Office of the United Nations High Commissioner for Human Rights, *Report of the Special Rapporteur on the Situation of Human Rights and Fundamental Freedoms of Indigenous People*, E/CN.4/2002/97 UN ESC (2002).

77   *Mahuika* v. *New Zealand*, Communication 547/1993, Views of 27 October 2000.

78   *The Treaty of Waitangi*, opened for signature 6 February 1840, Art. 1 (entered into force 6 February 1840).

79   *Te Runganga, supra* note 16.

80   *Ibid.* at para. 9.3.

81   See *supra* note 10 for the text of the court's decision. A considerable literature has developed about the case. See Patrick Macklem, "Indigenous Rights and

Multinational Corporations at International Law," (2001) 24 *Hastings International and Comparative Law Review* 475 at 478; Patrick Macklem and Ed Morgan, "Indigenous Rights in the Inter-American System: The Amicus Brief of the Assembly of First Nations in *Awas Tingni* v. *Republic of Nicaragua*," (2000) *Human Rights Quarterly* 569; Jennifer A. Amiott, "Environment, Equality and Indigenous Peoples' Land Rights in the Inter-American Human Rights System: Mayagna (Sumo) Indigenous Community of *Awas Tingni* v. *Nicaragua*," (2002) 32 *Environmental Law, Lewis and Clark Law School* 873; S. James Anaya and Todd Crider, "Indigenous Peoples, the Environment, and Commercial Forestry in Developing Countries: The Case of *Awas Tingni* v. *Nicaragua*," (1996) 18 *Human Rights Quarterly* 345; and S. James Anaya, "The Awas Tingni Petition to the Inter-American Commission on Human Rights: Indigenous Lands, Loggers and Government Neglect in Nicaragua," (1996) 9 *St. Thomas Law Review* 157.

82 *Awas Tingni* v. *Nicaragua* (2001) Inter-Am. Ct.H.R. (Series C) No. 79 at para. 13 [*Awas Tingni*]; *Annual Report of the Inter-American Commission of Human Rights*: 1999 OEA/Ser L/V/III.47/doc. 6 (2000).

83 *Awas Tingni, supra* note 82.

Part IV

International trade dimensions in aquaculture

11 Aquaculture and the multilateral trade regime

Issues of seafood safety, labeling and the environment

Ted L. McDorman and Torsten Ström

Introduction

Global aquaculture production in 2000 was estimated at 45.7 million metric tons (mt) and valued at US$56.5 thousand million.[1] Over 91 percent of the aquaculture production by weight and 82 percent by value originated from Asia. The top nine aquaculture-producing states are Asian: China, India, Japan, the Philippines, Indonesia, Thailand, South Korea, Bangladesh and Vietnam. Seven of these producing states are developing countries. The most spectacular growth in aquaculture production from 1995 to 2000 has been in Africa, where production has gone from 100,000 mt to nearly 400,000 mt with a value just shy of US$1 billion. Overall, since 1991 there has been a tripling of global aquaculture production.[2]

Numbers are lacking on the quantity and value of aquaculture products traded internationally, since farmed fisheries products are not segregated from non-farmed seafood, but the assumption is that at present the amount is small and concentrated on two products: shrimp and salmon.[3] Further, it is assumed that as aquaculture production grows and the harvest from wild fisheries declines, international trade in aquaculture products will expand.

Shrimp is the most traded seafood product internationally, with about 26 percent (1.1 million mt) coming from aquaculture.[4] The major markets are Japan and the United States, and other developed states. The major exporters of farmed shrimp are developing states such as Thailand, Ecuador, Indonesia, India, Mexico, Bangladesh and Vietnam.[5] The second most traded aquaculture product is salmon. The 2001 figures put the quantity of internationally traded farmed salmon at 1 million mt,[6] with most of that being Atlantic salmon and a small amount of Coho salmon. Norway is the principal exporter of Atlantic salmon, with the European Union the main market.[7] Chile is another major exporter, with Japan and the United States the principal markets. Other internationally traded aquaculture products of interest include crab, tilapia, mollusks, sea bass and sea bream. The major exporters of these farmed species tend to be developing states and the major importers tend to be developed states.[8]

Canada is not without an interest in aquaculture matters. Aquaculture production in Canada in 2002 was estimated at CDN$640 million.[9] This figure has been growing quickly and in 2000 constituted 22.8 percent of the total value harvested from living aquatic resources.[10] In 2002, Canada exported aquaculture product worth approximately CDN$474 million, of which nearly 90 percent was farmed salmon destined primarily for the US market.[11]

This snapshot of global production and trade volumes and values reveals that developing states, rather than developed states, have the most at stake economically, environmentally and socially (in terms of food supplies) regarding aquaculture matters. Approximately 91 percent of aquaculture production and 84 percent of value originate from low-income food deficit countries (LIFDCs).[12]

> Of particular significance is the fact that the growth of aquaculture pro-
> duction within developing countries and LIFDCs has been steadily
> increasing. In the last decade, the aquaculture sector within LIFDCs has
> been growing over seven times faster (over the period 1970 to 2000)
> than the aquaculture sector within developed countries.[13]

The purpose of this modest contribution is to look at one small part of the global aquaculture picture: the manner in which the rules and disciplines of the World Trade Organization (WTO)[14] apply to the international trade in aquaculture products. While the international trade in aquaculture products raises a host of issues, the principal focus here will be the application of the WTO trade regime to government laws and regulations affecting trade in aquaculture products. The key areas for examination are the general trade rules set out in the General Agreement on Tariffs and Trade, 1994,[15] which apply to environmental measures that affect international trade in aquaculture products; the Agreement on the Application of Sanitary and Phytosanitary Measures (the SPS Agreement),[16] which is the principal WTO agreement that deals with the health and safety issues that arise regarding internationally traded food/aquaculture products, including labeling directly related to food safety; and the Agreement on Technical Barriers to Trade (the TBT Agreement),[17] which deals with other forms of labeling. It is first useful, however, to provide some background regarding the WTO trade regime.

The World Trade Organization: institutional issues[18]

Structure and membership

The World Trade Organization is the international institution administering the rules and principles that constitute the multilateral trade law system. It was created at the conclusion of the Uruguay Round of multilateral negotiations and, on 1 January 1995, replaced the General Agreement on Tariffs and Trade (GATT), which had been the framework for inter-national trade law among its adherents since its inception in 1947.[19] There is continuity between GATT 1947 and the WTO as the principal set of rules to be applied by the WTO are still those originally set out in the 1947 GATT Agreement, although they have been recreated as part of the WTO Agreement and renamed GATT 1994.[20] Over thirty agreements and under-standings relating to trade in goods, services, trade-related intellectual prop-erty and dispute settlement that emerged as part of the Uruguay Round of Multilateral Trade Negotiations must be added to GATT 1994.[21] Together, GATT 1994 and the Uruguay Round agreements and understandings administered by the WTO constitute the multilateral trade law regime.

As with any international treaty, the rules and principles of the WTO are only applicable between those countries that are contracting parties to the WTO Agreement. Unlike most other international treaties, however, the WTO does not provide for automatic "membership" through direct state ratification. The WTO, like GATT 1947 before it, is an organization where the existing contracting parties determine "membership." Those countries that were contracting parties to GATT 1947 at the time of entry into force of the WTO in 1995 joined the WTO by designating their acceptance.[22] Other countries may accede to the WTO "on terms to be agreed between it and the WTO," with the accession agreement requiring a two-thirds major-ity vote of the members of the WTO.[23] As of April 2003, there were 146 members of the WTO.[24] The most recent high-profile member is the People's Republic of China (PRC), which became a WTO member in 2001. At present, there are twenty-five states in the process of negotiating their entry into the WTO.[25]

A key institutional component of GATT 1947 was the willingness of states to engage in intensive periods of negotiations in order to promote and enhance trade liberalization. These formal negotiating sessions, which can last years, are referred to as rounds. Including the initial negotiation of GATT 1947, there have been eight negotiating rounds. The most recent was the Uruguay Round, which commenced in 1986 and was completed in 1994. In 2001, the WTO launched a new round of comprehensive trade negotiations referred to as the Doha Round.[26]

Dispute settlement

The WTO has a well-defined dispute settlement process to assist countries to resolve differences arising from the interpretation or application of GATT 1994 and the other Uruguay Round Agreements.[27] A state that feels an obligation owed it under GATT 1994 or that the other Uruguay Round Agreements is being breached by another WTO member may, after a period of consultations, request a panel of experts to decide on the consistency of the alleged offending measure with the WTO rules. The request for a panel goes to the WTO Dispute Settlement Body (DSB), which is composed of the WTO membership. The DSB is to establish a panel of experts to examine a complaint unless "the DSB decides by consensus not to establish a panel."[28] Where a panel is established, and it issues its findings and recommendations, it is possible for a party to the dispute to appeal those findings and recommendations to the WTO Appellate Body.[29] The Appellate Body, an innovation introduced as part of the Uruguay Round, is a seven-person permanent court-like structure with jurisdiction to review issues of law and legal interpretations arising within panel decisions.[30] It has become the practice of states to refer almost all or parts of panel reports to the Appellate Body. Unappealed findings and recommendations of panels and results from the Appellate Body only become binding on the disputing parties when they are adopted by the DSB. Both panel reports and Appellate Body reports must be adopted unless "the DSB decides by consensus not to adopt the report."[31]

The WTO dispute settlement process recognizes that it is not always feasible for countries to implement immediately the results of adopted panel or Appellate Body reports. States are to be given a "reasonable period of time" to comply with adopted decisions.[32] Where it is not possible for the disputing states to reach agreement on what constitutes a "reasonable period of time," binding arbitration is used to make the temporal determination.[33]

A failure by a WTO member to implement the findings and recommendations of an adopted panel or Appellate Body report can lead to the adversely affected state seeking compensation from the recalcitrant state.[34] The aggrieved state may also seek the approval of the DSB to impose proportionate retaliatory trade measures against a state that does not implement the results of an adopted panel or Appellate Body report.[35]

Two additional comments are useful regarding the WTO dispute settlement process. First, unless disputing parties agree to the contrary, the jurisdiction of the panels and the Appellate Body is to interpret and apply the WTO-administered treaties and not to adjudicate other questions of international law or consider other factors such as environmental politics.[36] The WTO dispute settlement process is not a general court of international law. However, the WTO Appellate Body in the *Turtle/Shrimp Report*[37] made extensive reference to both the relevant regional treaty on turtle protection and the work undertaken as part of the United Nations Conference on Environment and Development (UNCED) in contextualizing the trade dispute. Second, the WTO dispute settlement process is not required to listen to or provide access to non-state actors.[38] Nevertheless, the WTO has opened a dialogue with non-governmental organizations, particularly those with an environmental focus, in order to meet mutual concerns.[39] Moreover, in the *Turtle/Shrimp Report* the Appellate Body decided that briefs originating from non-governmental organizations could be appended to the written submissions of the principals in an appeal.[40] Finally, in the *Asbestos* appeal, the Appellate Body interpreted its own rules of procedure to enable it, as a matter of discretion, to accept and consider unsolicited briefs submitted by non-governmental organizations and others.[41] Accordingly, it established an *ad hoc* procedure whereby prospective "friends of the court" could apply for leave to submit such briefs. Although the Appellate Body ultimately declined to grant leave to any of the seventeen applications received, its initiative touched off a firestorm of protest among some WTO members, mainly from developing countries.[42]

International trade law and the environment: issues for aquaculture

Main issues

The environmental concerns raised by aquaculture production and the international trade in farmed product are of either a local or a regional nature (degradation of land, water, vegetation, fish diseases, genetics) or transnational in nature (alien species, fish diseases, genetics). Without unduly simplifying the issues, it can be said that the environment/trade law issues raised by internationally traded aquaculture products are encapsulated by three concerns:

- first, the trade law consistency of measures taken by an importing state to protect *its* environment from possible adverse effects from imported aquaculture products such as the introduction of alien species, fish diseases or unwanted genetically modified marine species;
- second, the trade law consistency of measures taken by an importing state against imported aquaculture products because the production of the aquaculture products is having an adverse environmental effect *within* the producing state such as land, water and vegetation degradation and undesirable fish diseases or genetically modified marine species; and
- third, the trade law consistency of measures taken by an importing state against imported aquaculture products because the production of the aquaculture products is having an adverse environmental effect on the global commons beyond the national jurisdiction of both the producer and the importer, such as the introduction of undesirable fish diseases or genetically modified marine species.

It is the second and third concerns that attract the most attention. One spin on these concerns is that trade measures (market access) can and should be used to promote sustainable aquaculture development. A second, more common spin is that a producer in one state has to meet more stringent environmental concerns than the producer in another state and that the unlevel playing field requires or justifies trade measures against the foreign product in order to address concerns about "unfair" competition. Another, more conspiratorial spin is that importing states use such measures to interfere in sovereign decisions by exporting states as regards their domestic environmental policies. In less polite terms, this is sometimes referred to as eco-imperialism or eco-colonialism.[43] The "Plan of Action" adopted at the United Nations World Summit on Sustainable Development held in South Africa in 2002 indicates the concern about such unilateral trade measures in paragraph 101:

> States should cooperate to promote a supportive and open international economic system that would lead to economic growth and sustainable development in all countries to better address the problems of environmental degradation. Trade policy measures for environmental purposes should not constitute a means of arbitrary or unjustifiable discrimination or a disguised restriction on trade. Unilateral actions to deal with environmental challenges outside the jurisdiction of the importing country should be avoided. Environmental measures addressing transboundary or global environmental problems should, as far as possible, be based on an international consensus.[44]

For these and other environmental matters that arise from internationally traded aquaculture products, it is necessary to turn to the obligations within GATT 1994.

Primary obligations of GATT 1994

Within GATT 1994, which deals with trade in goods, the principal obligations that place constraints on a state party's actions as regards both the import and the export of goods are found in Articles I, III and XI. The main exceptions to these obligations are set out in Article XX.

Article I contains the obligation on state parties to respect the principle of most favored nation (MFN). MFN means that an importing state is not to discriminate by the use of tariffs or other measures between like products coming from different WTO members.[45] The essence of this is that an exporter's product is guaranteed the same treatment as another exporter's goods in access to a market of a WTO member, provided the goods are like products, as regards tariffs and similar measures. More simply put, an importer cannot discriminate in its use of tariffs or market access on the basis of the country of origin of a product.

Article XI contains the obligation on state parties not to use quotas or embargoes (quantitative measures) on either the import or the export of products. As regards imports, while a state may impose tariffs, provided non-discrimination between countries of origin exists under Article I, a state may not utilize import embargoes as a means to provide protection to domestic producers.

Article III contains the complex obligation on state parties of national treatment that means that a state is not to utilize internal measures (for example, a tax) which result in discrimination against foreign-produced products to the benefit of locally produced like products. Together, Articles III and XI are designed to prevent a state from implementing measures that impede the import of foreign goods in order to benefit domestically produced like goods.

While there a number of thresholds as regards the application of Articles I and III, one of the most critical is that of "like products." Under Article I, an importing state is not to discriminate in tariff application as between "like products" from differing countries. Under Article III, once a foreign product is "over the border," the importing state must not treat imports less favorably than it treats domestically produced products. While there is no definition of "like product" within GATT 1994, dispute settlement decisions generally have confined "like product" analysis to the physical properties of the goods and the substitutability of the products, and declined to be seduced by arguments regarding attributes not directly related to the goods in question.[46]

While Articles I, III and XI favour non-discrimination and unrestricted market movement of goods, Article XX establishes a series of exceptions that provide state parties with the justification to take measures inconsistent with their GATT 1994 obligations. The two most important exceptions are Article XX(b), respecting health and safety, and Article XX(g), respecting exhaustible natural resources:

> Subject to the requirement that such measures are not applied in a manner which would constitute a means of arbitrary or unjustifiable discrimination between countries where the same conditions prevail, or a disguised restriction on international trade, nothing in this Agreement shall be construed to prevent the adoption or enforcement by any contracting party of measures:
>
> (b) necessary to protect human, animal or plant life or health;
>
> . . .
>
> (g) relating to the conservation of exhaustible natural resources if such measures are made effective in conjunction with restrictions on domestic production or consumption.

Basics of the trade law and environment relationship[47]

Using trade measures to protect one's own environment

It has long been understood, and the GATT–WTO trade dispute settlement panels have endorsed the notion, that the Article XX exceptions allow a state to utilize trade measures, otherwise inconsistent with Articles I, III or XI of GATT 1994, that are bona fide for the purposes of the protection of the health, safety and protection of its citizens and environment.

In direct terms, Article XX assures that a state can prevent or condition the imports of goods where there is a bona fide risk that the goods will be or are harmful to the environment or to human, animal or plant life or health.

Using trade measures to further sustainable development, to ensure a level playing field or as eco-imperialism

The controversial issues of environment/trade law arise when a state utilizes trade measures against a product from another state where the justification is that the product is being produced in an environmentally unfriendly manner or that the production methods have environmentally unfriendly consequences either within the producing state or for the global commons.

As a generality, the application of multilateral trade law is to the products in the flow of international trade. It is understood that differing producers utilize differing methods for the production of goods. Economists often characterize this as an element of the concept of comparative advantage, one of the central theoretical principles that underlie international trade. As noted above, one of the thresholds in Articles I and III is the treatment of "like products" and that importing states are not to discriminate as between countries of origin where the products are like, and, once having entered the flow of domestic trade, like products are to be treated alike regardless of their domestic or foreign origin. The like product analysis generally has been about the good itself and not engaging the product process that created the good, except to the extent that the production process affects the good itself. Thus far, the argument that a measure which discriminates between a good produced in a more environmentally friendly manner than another is consistent with Article I or III has not been successful, primarily because looking behind the good would require an inquiry into the issues of national policy and thus be an interference in national sovereignty.

Article XI does not rely on like products but is categorical in that quantitative controls (quotas, embargoes) are prohibited. The high-profile environment/trade cases before the GATT and WTO dispute settlement panels have involved US embargoes against tuna and shrimp because of concerns over dolphin and turtles.[48] In these cases, attention was placed on the Article XX environmental exception.

Summarizing the "state of the law" of Article XX, subparagraphs (b) and (g), is a challenging undertaking.[49] Clearly, the trade measure in question must "fit" the wording of subparagraphs (b) and (g), and then the chapeau wording must be applied. The key words in the "fit" examination of (b) is that the measure in question must be "necessary" to protect human, animal or plant life or health and in the case of (g) that the measure must be "relating to" conservation of exhaustible natural resources. The WTO Appellate Body has looked at the qualifiers "necessary" and "relating to" on several

occasions[50] and tended not to give either term a restrictive or narrow interpretation. For the purposes of this contribution, the key bedevilling issue is
whether (b) and (g) are restricted only to measures that apply to protect
one's own environment or whether (b) and (g) can be applied so as to protect
the health of citizens and environment beyond the borders of the enacting
state. Subparagraphs (b) and (g)

> had been traditionally applied to rule out environmental laws that pro-
> tected the environment outside the enacting country's borders, though
> the 1998 WTO Appellate Body ruling in the shrimp–turtle case may
> have changed this by requiring merely a "sufficient nexus" between the
> law and the environment of the enacting state. Although the ruling did
> not fully explore what constituted a sufficient nexus, it appears that
> transboundary impacts on air and water, or impacts on endangered and
> migratory species, for example, might provide such a nexus.[51]

The second step of Article XX application involves the chapeau. The
three tests in the chapeau to be met are whether, in its application, the
measure is arbitrarily discriminatory, is unjustifiably discriminatory or constitutes a disguised restriction on trade. The clearest statement to date on
these tests in an environmental context comes from the 1998 shrimp–turtle
case. Although the Appellate Body did not try to define these terms, it
arguably defined a number of criteria for *not* meeting the tests, including, for
example, the following:

- A state cannot require another state to adopt specific environmental
  technologies or measures; different technologies or measures that have
  the same final effect should be allowed.
- When applying a measure to other countries, regulating countries must
  take into account differences in the conditions prevailing in those other
  countries.
- Before enacting trade measures [with an extraterritorial reach], countries
  should attempt to enter into negotiations with the exporting state(s).
- Foreign countries affected by trade measures should be allowed time to
  make adjustments.
- Due process, transparency, appropriate appeals procedures and other
  procedural safeguards must be available to foreign states or producers to
  review the application of the measure.[52]

The Appellate Body in the 1998 *Turtle/Shrimp Report* concluded that the US
measure in question did not meet the wording of the chapeau and thus was
not covered by the Article XX wording.[53] Without altering the US legislation, the United States undertook actions to implement the law in a different manner, which, according to the Appellate Body in 2001, met the
criteria of the chapeau.[54] A reasonable conclusion is that states seeking to use

trade measures directed at environmental concerns beyond its borders may have scope for action but the hurdles for such action are significant and include attempting to reach international accords and maintaining a non-discriminatory approach. In sum, although the final outcome of the dispute demonstrated that it is possible for states to impose trade-related measures to protect exhaustible natural resources beyond their territorial limit, the turtle–shrimp results cannot be read as providing *carte blanche* to states to utilize trade measures against states that have differing environmental standards from themselves.

Subsidies[55]

The only direct mention of fisheries in the 2001 Doha Ministerial Declaration launching the new round of multilateral trade negotiations is respecting fisheries subsidies, an issue that has been seen as an important component of the overcapacity of the world's fishing fleets and the consequent decline in marine catches.[56] Subsidy questions have frequently been raised as regards aquaculture products with a pre-WTO case involving a Norwegian challenge of US countervailing duties applied against subsidized farmed salmon.[57] In addition to the Norwegian situation, Chilean farmed salmon was targeted unsuccessfully, in the late 1990s by US countervailing duty legislation[58] and there are current rumors of a case regarding farmed shrimp being on the horizon.[59] The European Union has also used countervailing duties (see below) against Norwegian farmed salmon.[60]

All states provide subsidies in some manner. It is recognized that in most cases, subsidization causes trade distortions for the competitors of the products subsidized. Balancing these two realities has always been a major challenge for the international trade regime. It is important to recognize that part of this balance is that a subsidy for the purposes of international trade law is not to be confused with what is referred to in common parlance as a subsidy, and that only certain categories of subsidies are subject to scrutiny under international trade law.

The key WTO document on subsidies is the Uruguay Round Agreement on Subsidies and Countervailing Measures.[61] The Uruguay Round Subsidies Agreement and its GATT precursors recognize that in certain situations there is a domestic remedy available to states faced with the importation of subsidized products. The domestic remedy is countervailing duties, which, in certain circumstances, can be applied against products that have benefitted from subsidization where the imported subsidized goods cause material economic injury to the producers of like domestic goods. The Uruguay Round Subsidies Agreement also creates an international remedy attainable through the WTO dispute settlement process where states utilize certain categories of subsidies.

The Uruguay Round Subsidies Agreement has created different categories of subsidies. First is the category of prohibited subsidies. In simple terms, a prohibited subsidy is a financial benefit[62] available contingent on the product in question being exported.[63] Second is the category of actionable subsidies. An actionable subsidy is any governmental financial benefit available to a *specific* company or industry[64] that results in injury to the domestic industry of another country or that creates "serious prejudice" to the interests of another member of the WTO.[65] A state can only utilize a countervailing duty against an imported product that has received either a prohibited or an actionable subsidy. If the government financial benefit does not fit either of these two categories, then no countervail can be levied. Where the alleged subsidized product is not being imported – another condition for the application of a countervailing duty – but it is a product that has had the benefit of a prohibited subsidy or an actionable subsidy, then the complaining state must make use of the WTO dispute settlement process in order to have the subsidy removed.

While it is usually taken as axiomatic that reduction and removal of subsidies in the marine fisheries sector would enhance both the maximum economic return for harvested fish and resource conservation, it has been argued that for certain small developing states, fisheries subsidization is critical in the development of a viable economy and necessary to ensure that a coastal state reaps the benefits of the living resources within its 200-nautical mile economic zone.[66] While the argument would be slightly different for aquaculture development, the so-called infant industry justification – subsidizing until the industry develops a certain strength – has both an economic and a political resonance.

Another issue regarding subsidies is whether a state that has lower environmental standards than another state can be said to be providing a subsidy to the produced products. It is easy to argue that the existence of lower environmental standards in one country amounts to a subsidy on products entering a state with higher environmental standards, and therefore a countervailing duty may be appropriate against the imported product.[67] However, the Uruguay Round Subsidies Agreement definition of a subsidy requires that a government take positive action to provide a financial benefit or to forgo the collection of revenue.[68] Differing environmental standards would not easily meet the requirement of the giving of a financial benefit or the forgoing of government revenue.[69] Ultimately, the principal reason differential environmental standards should not be considered as a subsidy is that states have the right to establish their own domestic environmental measures and standards, and must have this autonomy without fear of trade action by another state.

The agreement on sanitary and phytosanitary measures, food safety and aquaculture products[70]

The general picture

The trade issues regarding food safety standards are easily understood. Countries, consumers and producers rely upon the free flow of food and food products across national borders, a flow that benefits both cost-efficient producers and cost-conscious consumers. In theory, if not always in practice, the free flow of food and food products makes more food available to more people at lower prices. Countries and consumers desire that food be safe to consume, and therefore countries must be able to impose food safety requirements to protect their consumers. Such food safety standards, while clearly legitimate, can interfere with the free flow of food and food products. There is also the concern that national food safety standards, in some situations, are or can be adopted to protect not domestic consumers from health concerns, but domestic producers from unwanted foreign competition. Aquaculture products raise many of the same consumer safety issues as wild fish, plus some different ones because of transmittal of unwanted substances through feed and the use of veterinary drugs.[71]

It is the Sanitary and Phytosanitary Measures Agreement that is most directly relevant to internationally traded aquaculture products because it applies to government measures (including import requirements) whose purpose is to protect human and animal health from food-borne risks.[72] For example, the SPS Agreement applies to national regulations that address microbiological contamination of food, pesticide levels, permitted food additives, Hazard Analysis and Critical Control Point (HACCP) requirements, and packaging and labeling requirements directly related to the safety of the food.

The goal of the SPS Agreement is to impose certain disciplines on the development and implementation of food safety-related requirements "so as to ensure that these measures can continue to be properly employed to protect human, animal or plant life and health but cannot be employed as unjustifiable non-tariff barriers, protecting domestic production from import competition."[73] Simply put, in the food context the SPS Agreement is about trying to ensure consumer protection while preventing unwarranted market protection.

While the SPS Agreement is related to obligations in GATT 1994, the WTO US/Canada/EU beef hormone dispute determined that the SPS Agreement stands alone from the obligations in GATT 1994.[74] Thus, an import requirement that falls within the scope of the SPS Agreement must be consistent with the obligations set out in the SPS Agreement. The measure must also be consistent with GATT 1994, although a measure that meets the requirements of the SPS Agreement is also very likely to meet the applicable disciplines in GATT 1994.

Application of the SPS Agreement

Article 2.1 of the SPS Agreement provides that states are free to impose SPS requirements for the protection of human, animal and plant life and health subject to the measures not being inconsistent with the SPS Agreement. This raises two critical questions: (1) when are measures consistent with the SPS Agreement; and (2) who has the burden of showing the consistency or inconsistency of a measure?

Consistency with the SPS Agreement: Codex international standards

Article 3.2 indicates that an import requirement covered by the SPS Agreement "which conforms to" international standards, guidelines or recommendations is deemed to be consistent with the SPS Agreement and thus permissible. Respecting food safety, it is the work of the Codex Alimentarius Commission (hereinafter Codex)[75] that is to be taken as the international standards, guidelines and recommendations set out in Article 3.2.[76] As one author has summarized, "The objectives of the Codex programme are to protect the health of consumers, to ensure fair practices in the food trade and to promote the coordination of all food standards work undertaken by national governments."[77]

In this regard, Codex has had some remarkable successes and it is the principal international organization mandated to develop common food requirements. However, Codex is both a voluntary organization and one that seeks voluntary adoption of its guidelines, codes and standards (voluntary harmonization). Moreover, the Codex "common requirements" may be seen as being set at the lowest common denominator (the minimum standard of seafood safety) and may be aspirational rather than technical, the result being that states have some degree of flexibility in applying the Codex "common requirements." Thus, state conformity with the internationally agreed requirements is far from uniform. Also, there have been concerns expressed that developing countries have been unable to participate and have their views expressed and considered at Codex because of a lack of funds, the multiplicity of committees and lack of technical capacity.[78] Nevertheless, the SPS Agreement gives prominence to Codex by taking the work of Codex to be the internationally agreed-upon requirements respecting food. This was clear in the WTO US/Canada/EU beef hormone dispute, where both the dispute settlement panel and the Appellate Body accepted that it was the Codex hormone standards which were the internationally agreed-upon requirements.[79]

It should be noted that there is *no obligation* on states to have import requirements that conform to Codex. States are not obligated to have food safety requirements at all. The SPS Agreement does not elevate the Codex voluntary guidelines and recommendations to ones with obligatory force and effect.[80] At best, Article 3.1 of the SPS Agreement directs that states should seek to harmonize their import requirements on the basis of the work of Codex, which is not the same as conformity with Codex.[81] Nevertheless, the work of Codex is a "benchmark" for the application of the SPS Agreement.

Consistency beyond or in the absence of international standards

The SPS Agreement specifically acknowledges that a state can impose food requirements that exceed or result in a higher level of human, animal or plant protection than that afforded by internationally agreed standards. Moreover, states can impose food requirements where no internationally agreed standards exist. For import requirements in these situations to be in compliance with the SPS Agreement, the import requirement must be based on "risk assessment."[82] The "risk assessment" is to take into account scientific evidence and demonstrate the "potential for adverse effects on human or animal health" of the risk.[83] While the importing state has the opportunity to determine what is its "appropriate" level of protection from the risk of the hazard and the appropriate measure to accomplish that level of protection (risk management), the state is also to take into account the objective of minimizing trade effects.[84]

Codex has achieved a level of agreement of the formalized definitions of risk assessment and risk management within the context of food safety risk analysis. The larger challenge is to develop principles and guidelines for risk analysis (risk assessment, risk management and risk communication) such that the food safety risk analysis done is based on good science and achieves transparency and consistency.[85] A common understanding of the manner in which risk assessment is to be carried out and the range of risk management choices associated with the risk would assist in addressing the key SPS Agreement issue of whether a food import requirement was "scientifically justified" or a disguised trade barrier.

Overall, one is left with the situation under the SPS Agreement that import requirements related to food must have a relationship with or be based upon a "real risk" of adverse health effects as demonstrated by scientific evidence. As one author summarized, "As long as there is a scientific justification for a particular ... [import requirement], a member is free to choose its own level of protection after determining that the health or safety risk is genuine."[86] The Appellate Body in the Japan/US fruit dispute stated:

> [T]he obligation ... that an SPS measure not be maintained without scientific evidence requires that there be a rational or objective relationship between the SPS measure and the scientific evidence. Whether there is a rational relationship between an SPS measure and the scientific evidence is to be determined on a case-by-case basis.[87]

Burden of showing that an import measure is consistent/inconsistent

The manner in which the SPS Agreement is worded *appears* to indicate that it is the importing state that has the burden of showing that an import requirement is consistent with the SPS Agreement where the import requirement is not based upon an internationally agreed standard. The dispute settlement panel in the WTO US/Canada/EU beef hormone dispute accepted that this was correct. However, the Appellate Body indicated that the panel was incorrect on this point and that both the complaining state and the importing state had evidentiary obligations. The burden is first on the complainant to prove a prima facie case that the import requirement is *not* consistent with the SPS Agreement. If this can be shown, then the burden shifts to the importing state to disprove the complainant's case and show that the import requirement is consistent with the SPS Agreement.[88] From this, one can infer that import requirements covered by the SPS Agreement are presumed valid unless it can be shown to the contrary. This burden is consistent with the burden that applies generally to other WTO Agreements such as GATT 1994.

In the seafood trade context, the burden issue is supportive of import requirements (developed countries) at the expense of exporters (developing countries) and puts an additional impediment on the use of the SPS Agreement as a means of questioning possible trade distorting measures. From a consumer protection perspective, it allows an importing state more latitude in imposing seafood safety requirements.

Minimal scientific information (precautionary approach)

Respecting food, it is not an unreasonable argument that a particular import requirement is important since the state wants to be cautious in protecting consumer health. The idea of a precautionary approach to justify import restrictions has a certain appeal. The European Union characterized its concern respecting hormones in beef as involving reliance on precaution,[89] as did the Japanese respecting quarantine and test measures for imported fruit and the codling moth. However, the Appellate Body in the WTO US/Canada/EC beef hormone dispute did not accept the precaution argument of the European Union and, as yet, has not had to comment on a more robust articulation of precaution as a justification for stopping the import of certain food products.[90] The Appellate Body noted that there was a relationship between the precautionary principle and the SPS Agreement, and that it was both Article 3.3 and Article 5.7 of the SPS Agreement where the precautionary principle "finds reflection" in the SPS Agreement.[91]

WTO jurisprudence on the SPS Agreement

Thus far, there have been four dispute settlement cases involving the SPS Agreement:[92] the US/Canada/EU beef hormone dispute,[93] the Japan/US fruit dispute,[94] the Canada/Australia salmon dispute[95] and the Japan/US apples dispute.[96] In all four cases, the import requirements under review were found to be inconsistent with the SPS Agreement. Only the US/Canada/EU beef hormone dispute was directly concerned with human health issues; the other three disputes were concerned with transported diseases or pests that might affect local plants or animals.

In the US/Canada/EU beef hormone dispute, there were Codex standards for the hormones in question. However, the EU measures sought a higher level of protection than that afforded by the Codex codes. One reviewer summarized the final result as follows:

> The EC measures on hormone-treated beef failed to satisfy the SPS Agreement on two counts: 1) all available scientific evidence, as well as experts consulted by the panel, stated that the hormones in question are safe when used in accordance with good practice; and 2) the EC failed to conduct a risk assessment that satisfied the provisions of the SPS Agreement.[97]

On the last point, it appears from the record that no risk assessment based on scientific information existed, or, at least, information arising from such an assessment was not presented to the WTO dispute settlement bodies. As a consequence, the dispute settlement bodies had no recourse but to conclude that the import requirement was "not based on a risk assessment" as required by Articles 5.1 and 5.2 (and Article 2.2) of the SPS Agreement.

The requirement that a proper risk assessment be carried out was the central question in the Appellate Body decision in the Canada/Australia Salmon Dispute. The import regulation concerned salmon from Canada, and the risk at issue was the risk of diseases for native Australian salmon. As concluded by one writer,

> Because the 1996 Final Report [risk assessment report done by the Australian government] did not contain an evaluation of the likelihood of entry, establishment, or spread of the diseases of concern nor an evaluation of the likelihood of entry, establishment, or spread of the diseases according to the SPS measures which might be applied, the 1996 Final Report could not qualify as a risk assessment.[98]

Finally, in the Japan/US fruit dispute, one of the issues was that there did not appear to have been a risk assessment undertaken as required under Article 5.1 of the SPS Agreement.[99] In the Japan/US apples dispute, the panel identified a number of inadequacies in the risk assessment methodology used by Japan, and concluded that the pest risk analysis did not meet the requirements of Article 5.1 taking into account the definition of risk assessment found in Annex A of the SPS Agreement.[100]

The four Appellate Body cases demonstrate the legal requirement to support import regulations, which are not based on international standards and where the health and safety risk is not obvious, with risk assessment studies. It can be surmised in all four cases that the inability to produce adequate risk assessment studies by three states or groups of states (the European Union, Australia and Japan) was because there existed no scientific basis for the import requirements. The situation could be different for a developing state that may not have the resources for conducting its own of risk assessment and would, therefore, have to rely on risk assessment done by other countries or through international organizations. However, it appears from both the SPS Agreement and the four decided cases that the threshold for whether a risk exists is relatively low.

The technical barriers to trade agreement and eco-labeling[101]

The general picture

Correlated with food safety matters are issues concerning product standards, specifications, description and labeling. As regards the international trade in aquaculture products, the major issues in this area arise with respect to product labeling, particularly whether a fishery product is farmed or wild, organically produced, grown or harvested in an environmentally sound manner, or contains genetically modified organisms.[102]

The Technical Barriers to Trade Agreement is an important companion to the SPS Agreement regarding food-related issues and is, as regards labeling for non-food safety reasons such as environmental matters and other issues, the principal WTO agreement.

The TBT Agreement is based upon its predecessor agreement, the GATT Standards Code, which was created as part of the Tokyo Round and came into being in 1980.[103] The development of the Standards Code by the GATT Contracting Parties was principally a response to two developments. First, government regulation during the 1960s was increasingly focusing on product standards and specifications, in part to protect consumer safety and to guard against deceptive and fraudulent practices. Products were also becoming more complicated, and government-mandated standards were part of an effort to ensure uniformity and to preserve the integrity of the marketplace. Second, as tariffs declined as a result of successive GATT negotiating rounds, governments began to look for other means to control imports. Using product standards as a non-tariff barrier to trade became an effective substitute for the formerly high tariff walls.

Scope of the agreement

As has been mentioned, the TBT Agreement establishes disciplines that, among other things, govern the use of labeling measures other than those directly related to food safety. These would include both mandatory and voluntary eco-labeling schemes, although, depending on whether a labeling program is mandatory or voluntary, different rules will apply.

Article 1.3 specifies that the TBT Agreement covers all products, including industrial and agricultural products, but it does not apply to government procurement or to SPS measures.[104] It can, however, apply to products that are also subject to SPS measures, as the reference to agricultural products makes clear.[105] In any event, it is clear that the TBT Agreement extends to aquaculture products.

The TBT Agreement divides measures into "technical regulations" and "standards." These are defined, respectively, as

> Document which lays down product characteristics or their related processes and production methods ... with which compliance is *mandatory*. It may also include or deal exclusively with ... labeling requirements as they apply to a product, process or production method.[106]

> Document approved by a recognized body, that provides, for common and repeated use, rules, guidelines or characteristics for products or related processes and production methods, with which compliance is *not mandatory*. It may also include or deal exclusively with ... labeling requirements as they apply to a product, process or production method.[107]

In other words, measures – such as regulations – that fall within the definition of either a "technical regulation" or a "standard" are subject to the requirements of the TBT Agreement[108] and must be consistent with those requirements.

There are a couple of distinctions to note in the two definitions. First, technical regulations are measures that are mandatory, while standards are measures that are voluntary. "Mandatory" means that the requirements of the measure must be met in order for the product to be allowed market access. "Voluntary," on the other hand, means that the measure restricts the manner in which a product is marketed but does not preclude market access entirely. For example, a voluntary labeling measure might establish criteria that must be met in order for the product to carry a particular label, but the label itself is not a prerequisite for market access. In contrast, a mandatory labeling requirement would require the product to carry a particular label in order to have market access.[109]

Technical regulations are subject to Articles 2 and 3, while standards are subject to Article 4 and the Code of Good Practice set out in Annex 3 to the TBT Agreement. This distinction will be examined in greater detail later in the chapter.

The second distinction to note is the reference in the first sentence of the definition of a technical regulation to "their related processes and production methods," and the corresponding reference in the second sentence to "a product, process or production method." This may be an important difference in terminology, with respect not only to determining whether a particular measure meets the definition of a "technical regulation" or a "standard," but also to its potential implications in the context of the determination of "like products" in Article 2.1 of the TBT Agreement, particularly in relation to labeling programs. This issue will also be explored further in what follows.

Technical regulations and eco-labeling

Article 2 contains the main provisions that govern technical regulations. These include non-discrimination, avoidance of unnecessary obstacles to trade, the use of international standards, notification and transparency, and mutual recognition or equivalence.

Non-discrimination

Article 2.1, like Articles I and III of GATT 1994, requires technical regulations to accord MFN status and national treatment to imported products.[110] This means that such products must be treated no less favorably than "like products" of national origin, and "like products originating in any other country."

As with GATT 1994, a key issue is determining whether the imported and domestic products are "like."[111] The jurisprudence indicates that this is to be done on a case-by-case basis. It is conceivable that, in addition to a number of other considerations, whether two products are considered "like" will be influenced by the nature of the measure being challenged. For example, it may be that two products that are considered like – and therefore subject to the obligation in Article 2.1 – for the purposes of a technical specification relating to the products' performance characteristics may not be considered like for the purposes of a labeling regime. Each case must be considered on its own merits, taking into account all relevant factors.

Unnecessary barriers to trade

Article 2.2 imposes a requirement on all WTO members that their technical regulations not create "unnecessary obstacles to trade."[112] This requirement raises several considerations. First, it means that the purpose underlying the measure must be a "legitimate objective." Article 2.1 includes a non-exhaustive list of legitimate objectives, including the protection of human health or safety, animal or plant life or health, and protection of the environment. Second, it means that the measure chosen must be the "least trade-restrictive"; that is, it must obstruct trade as little as possible while also achieving the objective. Third, in preparing, adopting and applying the measure, the risks created by the non-fulfillment of the objective must be taken into account.[113]

Although no jurisprudence exists to confirm it, a reasonable interpretation of Article 2.2 suggests that, in contrast to the SPS Agreement, the TBT Agreement does not contain a legal obligation either to conduct a risk assessment or to base a measure on such an assessment. At the same time, the wording of Article 2.1, including a reference to the assessment of risks, suggests that a risk assessment may be a practical necessity in many cases – particularly for measures involving the protection of human health or the environment – in order to defend such measures effectively if a challenge were to arise.

International standards

Like the SPS Agreement, the TBT Agreement favors the use of international standards wherever possible and appropriate. To this end, Article 2.4 requires WTO members to use such standards as a basis for their technical regulations "except when such standards . . . would be an ineffective or inappropriate means for the fulfilment of the legitimate objectives pursued." This bias is also reflected in Articles 2.5 and 2.6. Article 2.5 states that a technical regulation that "is in accordance with relevant international standards" is presumed not to create an unnecessary obstacle to trade. Article 2.6 strongly encourages, without requiring, WTO members to take an active role in international standardizing bodies, particularly in relation to international standards for products where they have adopted, or expect to adopt, technical regulations.

Standards and eco-labeling

Disciplines on the use of standards are set out in Article 4 and the Code of Good Practice found in Annex 3 of the TBT Agreement. However, the obligation placed on WTO members regarding use of standards is slightly different as compared to technical regulations.

Article 4.1 requires WTO members to "ensure that their central government standardizing bodies accept and comply with the Code of Good Practice," but in the case of local and non-governmental standardizing bodies, members are only obliged to "take such reasonable measures as may be available to them" to ensure such acceptance and compliance. WTO members must also refrain from taking any measures that would require or encourage standardizing bodies to act in a manner inconsistent with the Code. In other words, standardizing bodies and agencies that the central government controls directly must comply, and the central government is obligated to ensure that they do so. However, with respect to all other standardizing bodies, the central government likely cannot be held responsible for their actions, although it must exert whatever authority might be available to it to ensure compliance.

The Code of Good Practice sets out several requirements that more or less parallel those found in Article 2, including non-discrimination,[114] avoidance of unnecessary obstacles to trade,[115] and usage of international standards.[116]

WTO jurisprudence on the TBT Agreement

There have been few WTO panel or Appellate Body reports that have dealt with the TBT Agreement in detail. In the EC asbestos dispute,[117] the Appellate Body examined the definition of a "technical regulation" in order to determine whether the measure in question in that dispute constituted a technical regulation. Although the Appellate Body found that the measure was a technical regulation, it declined to examine whether the measure met the requirements of Article 2 of the TBT Agreement on the grounds that it had an inadequate factual foundation upon which to base its analysis.

In the EC sardines dispute,[118] the measure in question was clearly a technical regulation in the form of a requirement that a particular trade description – the term "sardines" – be used only on the packaging of a particular species of fish.[119] However, the findings of both the panel and the Appellate Body focused almost exclusively on the use of international standards, in particular whether an international standard existed, and whether the EC measure met the requirements of Article 2.4 of the TBT Agreement.[120]

The outlook

As has been noted, the TBT Agreement does not apply to measures covered by the SPS Agreement. This raises the question of what types of measures relevant to aquaculture are likely to be considered TBT measures; that is, "technical regulations" or "standards." The simple answer to this question is that it will depend on the measure. Measures that prescribe product characteristics or process and production methods that affect product characteristics, or measures that prescribe labeling requirements,[121] are likely to be captured as technical regulations.

A central issue that has arisen in the context of the "like product" test is whether it is legitimate for regulators to distinguish between products – that is, treat them as being different rather than "like" – on the basis of how they are produced or manufactured even where the production or manufacturing process does not affect the characteristics of the products themselves (called non-product-related processing and production methods, or nprPPMs).[122] An example might be shrimp caught at sea versus farmed shrimp. Assuming that there are no differences with respect to taste, texture or the relative health risks arising from their consumption, is it nevertheless legitimate for the regulator to differentiate between these shrimp from a regulatory standpoint, and, if so, in what respects? The answer to these questions will likely depend on the type of measure that is involved, and the basis for the differentiation.

Another question has arisen, this time in the context of Article 2.2, concerning labeling requirements that are driven by the desire of the public for certain information about the product, in particular information relating to the manner in which the product was produced or manufactured. Is it legitimate for a government regulator to require a producer or importer to include information regarding the environmental impact caused by the harvesting, manufacture or production of the product?

Because the jurisprudence dealing with the TBT Agreement is sparse and fairly narrow in its consideration of TBT-related issues, there is currently little guidance on how these rules are likely to be interpreted in practice. In particular, the precise relationship between nprPPM-based measures and the "like product" test – be it for labeling or other types of regulations – has yet to be defined with any degree of precision. Likewise, the ambit of the term "legitimate objective," and the likelihood that it can encompass the idea of "consumer information," and, if so, under what conditions, is currently unclear and therefore the subject of much debate.

Conclusion

The international flow of trade in aquaculture products, which is increasing rapidly, provides important export markets and economic development possibilities for developing states. Thus, it should always be borne in mind that the application of the WTO international trade regime to aquaculture products is of utmost concern to developing states.

The purpose of this chapter has been to outline the general contours of the existing multilateral trade regime that falls under the administrative authority of the WTO. For international trade in aquaculture and the issues of seafood safety, labeling and the environment, the principal agreements, as discussed above, are GATT 1994, the SPS Agreement and the TBT Agreement. While the rules contained in these agreements are complex, the general thrust of them is not: discrimination for the purposes of market protection is not tolerated, and a state cannot dress up a protectionist measure as environmental or related to food safety as a means of evading the trade rules. Bona fide concerns of an importing state about potential food safety or domestic environmental degradation are generally respected by the trade rules. Unsubstantiated fear or unicultural ethical concerns about imported aquaculture products are not.

Notes

Ted L. McDorman, Professor, Faculty of Law, University of Victoria, Victoria, British Columbia, co-wrote this contribution during the period when he was Academic-in-Residence, Bureau of Legal Affairs, Department of Foreign Affairs and International Trade, Ottawa. The views expressed are those of the author.

Torsten Ström, Counsel, Trade Law Bureau, Department of Foreign Affairs and International Trade/Department of Justice, Government of Canada. The views expressed are the author's personal views and should not necessarily be ascribed to the Government of Canada.

1 The production and value data in this paragraph are drawn from FAO, *Yearbook of Fishery Statistics: Aquaculture Production*, Vol. 90/2 (2000), (Rome: FAO, 2002), Tables A-1, A-2, A-3 and A-4, and include production of aquatic plants. The data are summarized in Albert J. Tacon, "Aquaculture Production Trend Analysis" in FAO, *Review of the State of World Aquaculture* (Rome: FAO Fisheries Circular No. 886, Rev. 2, 2003) at 2–29. See also FAO, *The State of the World Fisheries and Aquaculture 2002* (Rome: FAO, 2003) at 26–28. Online. Available http://www.fao.org (accessed 25 February 2004).

2 Regarding projected aquaculture growth, see generally Sena S. De Silva, "A Global Perspective of Aquaculture in the New Millennium," in R. P. Subasinghe, P. B. Bueno, M. J. Phillips and C. Hough, eds., *Aquaculture in the Third Millennium* (Technical Proceedings of the Conference on Aquaculture in the Third Millennium) (Bangkok: Network of Aquaculture Centres in Asia-Pacific, 2001) at 431–459. *The State of the World Fisheries and Aquaculture 2002, supra* note 1 at 113, states:, "Aquaculture production is expected to double to 1.5 million tonnes in three decades."

3 Helga Josupeit, Audun Lem and Hector Lupin, "Aquaculture Products: Quality, Safety, Marketing and Trade," in Subasinghe *et al.*, *supra* note 2 at 252; and *The State of the World Fisheries and Aquaculture 2002, supra* note 1 at 36–38.

4 *The State of the World Fisheries and Aquaculture 2002, supra* note 1 at 36; and Josupeit *et al.*, *supra* note 3 at 252.

5 *Ibid.*

6 *The State of the World Fisheries and Aquaculture 2002, supra* note 1 at 36.

7 *Ibid.* See also Josupeit *et al.*, *supra* note 3 at 253.

8 *Ibid.* See also Josupeit, *supra* note 3 at 252–254.

9 See Canada, Office of the Commissioner of Aquaculture Development, which includes statistical information. Online. Available http://www.ocad-bcda.gc.ca (accessed 25 February 2004). Regarding production, see online, http://www.dfo-mpo.gc.ca/communic/statistics/aqua/aqua02_e.htm (accessed 25 February 2004); and *Northern Aquaculture Statistics 2001: The Year in Review*, prepared by PricewaterhouseCoopers, available on the Office of the Commissioner of Aquaculture website.

10 Canada, Department of Fisheries and Oceans, *DFO's Aquaculture Policy and Framework* (Ottawa: DFO, 2002) at 12. It is noted that the Canadian aquaculture sector employs in excess of 14,000 people and generates approximately CDN$1 billion annually in direct and indirect economic activity.

11 See Statistics Canada, Aquaculture Statistics 2002. Online. Available http://www.statcan.ca/english/freepub/2-22-XIE/free.htm (accessed 25 February 2004).

12 Tacon, *supra* note 1 at 14. See also at 27–29 data regarding the contribution of aquaculture to the global food supply.

13 *Ibid.*

14 Agreement Establishing the World Trade Organization, in GATT Secretariat, *The Results of the Uruguay Round of Multilateral Trade Negotiations: The Legal Texts* (Geneva: GATT, 1994) at 14 and reprinted in (1994) 33 *International Legal Materials* 1144. See generally the WTO website. Online. Available http://www.wto.org (accessed 25 February 2004).

15 See text accompanying *infra* note 20.

16 In *The Results of the Uruguay Round, supra* note 14 at 69 and on the WTO website. Online. Available http://www.wto.org/english/docs_e/legal_e/15-sps.doc (accessed 25 February 2004).

17 In *The Results of the Uruguay Round, supra* note 14 at 138 and on the WTO website. Online. Available http://www.wto.org/english/docs_e/legal_e/17-tbt.doc (accessed 25 February 2004). For an excellent article that analyzes numerous aspects of the GATT 1994, SPS Agreement and TBT Agreement and their relationships, see Gabrielle Marceau and Joel P. Trachtman, "The Technical Barriers to Trade Agreement, the Sanitary and Phytosanitary Measures Agreement, and the General Agreement on Tariffs and Trade: A Map of the World Trade Organization Law of Domestic Regulation of Goods," (2002) 36 *Journal of World Trade* 811–881.

18 This section is drawn, with modification, from T. L. McDorman, "Fisheries Conservation and Management and International Trade Law," in Ellen Hey, ed., *Developments in International Fisheries Law* (The Hague: Kluwer Law International, 1999) at 503–506.

19 The 1947 General Agreement on Tariffs and Trade, 55 *U.N.T.S.* 194.

20 Article II(4) of the WTO Agreement, *supra* note 14. Regarding the transition from the 1947 GATT to the WTO, see Edmond McGovern, *International Trade Regulation* (Exeter: Globefield Press, 1998 (loose-leaf)), Chapter 1.

21 See *The Results of the Uruguay Round, supra* note 14. Many, but not all, of the Uruguay Round Agreements are reprinted in (1994) 33 *International Legal Materials* 1125, and all are available on the WTO website, *supra* note 14. Article II of the WTO Agreement deals with the relationship between the WTO and the Uruguay Round Agreements, *supra* note 14. See also McGovern, *supra* note 20 at section 1.121.

22 Articles XI and XIV of the WTO Agreement, *supra* note 14.

23 Article XII of the WTO Agreement, *supra* note 14. See generally, McGovern, *supra* note 20 at section 1.312.

24 See the WTO website, *supra* note 14.

25 See the WTO website, *supra* note 14.

26 The Doha Ministerial Declaration was adopted on 14 November 2001 and is available on the WTO website, *supra* note 14, and (2002) 41 *International Legal Materials* 746.

27 Understanding on Rules and Procedures Governing the Settlement of Disputes ["Dispute Settlement Understanding"], in *The Results of the Uruguay Round, supra* note 14 at 404; (1994) 33 *International Legal Materials* 1226 and on the WTO website, *supra* note 14.

28 Article 6(1) of the Dispute Settlement Understanding, *ibid.*

29 *Ibid.* at Article 17(4).

30 *Ibid.* at Article 17(6).

31 *Ibid.* at Articles 16(4) and 17(14).

32 *Ibid.* at Article 21(3).

33 *Ibid.* at Article 21(3)(c).

34 *Ibid.* at Article 22(1) and (2).

35 *Ibid.* at Article 22(1) and (2).

36 The issue is one of jurisdiction of a WTO panel. Article 7(1) of the Dispute Settlement Understanding, *ibid.*, has created a standard jurisdiction clause that confines a panel to dealing with issues arising from treaties and agreements within the WTO framework. See generally McGovern, *supra* note 20 at sections 2.2321 and 2.2323.

37 *United States – Import Prohibition of Certain Shrimp and Shrimp Products*, Report of the Appellate Body, WTO Doc. WT/DS58/AB/R, 12 October 1998 [Turtle/Shrimp Report] and reprinted in (1999) 38 *International Legal Materials* 121. All the Panel and Appellate Body reports referred to in this contribution are available on the WTO website, *supra* note 14.

38 The state-centric nature of the WTO dispute settlement procedures has been

criticized by environmental non-governmental organizations. See, for example, Daniel C. Esty, *Greening the GATT: Trade, Environment, and the Future* (Washington, DC: Institute of International Economics, 1994) at 210–215.

39 See the WTO website, *supra* note 14.

40 See Turtle/Shrimp Report, *supra* note 37 at paras. 11–13.

41 See *European Community – Measures Affecting Asbestos and Products Containing Asbestos*, Report of the Appellate Body, WTO Doc. WT/DS 135/AB/R, 12 March 2001, reprinted in (2001) 40 *International Legal Materials* 1193 [EC–Asbestos].

42 See, generally, Geert A. Zonnekeyn, "The Appellate Body's Communication on *Amicus Curiae* Briefs in the *Asbestos* Case: An Echternach Procession," (2001) 35 *Journal of World Trade* 553–563, and Ernesto Hernandez-Lopez, "Recent Trends and Perspectives for Non-state Actor Participation in World Trade Organization Disputes," (2001) 35 *Journal of World Trade* 469 at 485–495.

43 See Frank Biermann, "The Rising Tide of Green Unilateralism in World Trade Law: Options for Reconciling the Emerging North–South Conflict," (2001) 35 *Journal of World Trade* 421 at 422:

> In recent discussions on establishing environmental standards within WTO law, developing countries often feel threatened by what they perceive as an emerging environmental unilateralism and even "eco-imperialism" of industrialised countries.

44 The Plan of Action adopted at the World Summit on Sustainable Development held in 2002 in Johannesburg, *Report of the World Summit on Sustainable Development* (UN Doc. A/Conf.199/20, New York, 2002) at para. 101. Online. Available http://www.johannesburgsummit.org/html/documents/summit_docs/131302_wssd_report_reissued.pdf (accessed 25 February 2004).

45 In fact, the obligation extends to treatment accorded to non-WTO members. In other words, WTO members are entitled to the same (or better) treatment as compared to the most favored treatment provided to any other country, whether a WTO member or not.

46 Four criteria have been identified for establishing "likeness": the physical properties of characteristics of the goods, end uses of the goods, tariff classification, and consumer tastes and preferences. See, generally, Won-Mog Choi, *"Like Products" in International Trade Law: Towards a Consistent GATT/WTO Jurisprudence* (Oxford: Oxford University Press, 2003); and Donald H. Regan, "Regulatory Purpose and 'Like Products' in Article III:4 of the GATT (With Additional Remarks on Article III:2)," (2002) 36 *Journal of World Trade* 443–478. The most recent WTO decision to deal extensively with like-product analysis is EC – Asbestos, *supra* note 41.

47 For a recent review of these issues and the manner in which they are being discussed at the WTO, see: Sabrina Shaw and Risa Schwartz, "Trade and Environment in the WTO: State of Play," (2002) 36 *Journal of World Trade* 129–154.

48 The US turtle/shrimp matter within the WTO dispute settlement process has a lengthy history. *United States – Import Prohibition of Certain Shrimp and Shrimp Products*, Report of the Panel, WTO Doc. WT/DS58/R, 15 May 1998, (1998) 37 *International Legal Materials* 834; Appellate Body Report, Turtle/Shrimp Report, *supra* note 37; *United States – Import Prohibition of Certain Shrimp and Shrimp Products, Recourse to Article 21.5 by Malaysia, Report of the Panel*, WTO Doc. WT/DS58/RW, 15 June 2001 [Turtle/Shrimp, Malaysian Review]; and *United States – Import Prohibition of Certain Shrimp and Shrimp Products, Recourse to Article 21.5 by Malaysia*, Report of the Appellate Body, WTO Doc. WT/DS58/AB/RW, 22 October 2001, (2002) 41 *International Legal Materials* 149.

49 For a detailed analysis of Article XX, see McGovern, *supra* note 20 at sections 13.111, 13.112, 13.113, 13.131–132 and 13.154.

50 See regarding Article XX(b) and "necessary": *EC – Asbestos*, *supra* note 41 and *Korea – Measures Affecting Imports of Fresh, Chilled and Frozen Beef*, Report of the Appellate Body, WTO Doc. WT/DS161/AB/R and WT/DS169/AB/R, 11 December 2000. As regards Article XX (g) and "relating to," see Turtle/Shrimp Report, *supra* note 37, and *United States – Standards for Reformulated and Conventional Gasoline*, Report of the Appellate Body, WTO Doc. WT/DS2/AB/R, 29 April 1996.

51 International Institute for Sustainable Development (IISD) and the United Nations Environment Programme (UNEP), *Environment and Trade: A Handbook* (Winnipeg: IISD, 2000), and a version is updated online at http://www.iisd.org/trade/handbook/ (accessed 25 February 2004). The cited material is from http://www.iisd.org/trade/handbook/3_4_1.htm (accessed 3 September 2003). The requirement of "sufficient nexus" between the law and the environment of the enacting state was applied by the Appellate Body in the US Turtle/Shrimp Report, supra note 37, only to subparagraph (g) leaving the question open as regards subparagraph (b).

52 *Ibid.*

53 Turtle/Shrimp Report, *supra* note 37.

54 Turtle/Shrimp, Malaysian Review, *supra* note 48.

55 This subsection is drawn, with modification, from McDorman, *supra* note 18 at 510–512.

56 The Doha Ministerial Declaration, *supra* note 26 at para. 28. There is a growing literature on fisheries subsidies. A good overview is provided by William E. Schrank, "Subsidies for Fisheries: A Review of Concepts," in FAO, *Expert Consultation on Economic Incentives and Responsible Fisheries* (Rome: FAO Fisheries Report No. 638, Supplement, 2001) at 11–39.

57 *United States – Imposition of Countervailing Duties on Imports of Fresh and Chilled Atlantic Salmon from Norway*, Report of the Panel adopted by the Committee on Subsidies and Countervailing Measures on 28 April 1994, GATT Doc. SCM/153. Note: Frank Asche, "Trade Disputes and Productivity Gains: The Curse of Farmed Salmon Production?" (1997) 12 *Marine Resource Economics* 67–73.

58 The US International Trade Administration determined that Chilean farmed salmon was not being subsidized. See ITA, Final Negative Countervailing Duty Determination: Fresh Atlantic Salmon from Chile, 63 *Federal Register* 31437–31447. An anti-dumping duty was imposed in 1998, although was revoked in 2003. See: ITA, Fresh Atlantic Salmon from Chile: Preliminary Results of Antidumping Duty Changed Circumstances Review and Notice of Intent to Revoke Order, 68 *Federal Register* 39058.

59 See D. Michael Kaye and Phyllis L. Derrick, "Legal Remedies for U.S. Shrimp Producers to Restrict Imports," (October 2002) 5(5) *Global Aquaculture Advocate* 12–14.

60 Countervailing duties were imposed in 1997 but avoided through undertakings given by Norwegian producers. The duties remain in place and the undertakings subject to regular adjustment. See Council Regulation (EC) No. 321/2003, 18 February 2003, *Official Journal of the European Union*, L. 47/3, 21 February 2003.

61 In *The Results of the Uruguay Round*, *supra* note 14 at 264, and on the WTO website, *supra* note 14.

62 See Article 1.1 of the Agreement on Subsidies and Countervailing Measures [Subsidies Agreement], Annex 1, *ibid.*

63 Article 3.1 of the Subsidies Agreement, *ibid.* A prohibited subsidy can also arise where the subsidy is contingent upon the use of domestic over imported goods in the production process.

64 Specificity is defined in Article 2 of the Subsidies Agreement, *ibid.*

65 Articles 5 and 6 of the Subsidies Agreement, *ibid.*

66 See Roman Grynberg and Martin Tsamenyi, "Fisheries Subsidies, the WTO and the Pacific Island Tuna Fisheries," (1998) 32(6) *Journal of World Trade* 128 and 133–134.

67 See, for example, Thomas K. Plofchan, "Recognizing and Countervailing Environmental Subsidies," (1992) 26 *International Lawyer* 763–780.

68 Article 1.1 of the Subsidies Agreement, *supra* note 62.

69 Twisting the wording to cover differential standards has been described as "tortured." Esty, *supra* note 38 at 164.

70 This section is drawn, with modification, from T. L. McDorman, "The International Context of Harmonizing Fishery Product Standards: With Special Reference to ASEAN," an unpublished report done in 2000 for the FAO.

71 For a recent overview of these concerns, see FAO Committee on Fisheries, Sub-Committee on Aquaculture, 2nd Session, 7–11 August 2003, "Strategies to Improve Safety and Quality of Aquaculture Products" (COFI: AQ/II/2003/6) at paras. 10–32. Online. Available http://www.fao.org/fi/meetings/cofi/cofi_aq/2003/default.asp (accessed 25 February 2004).

72 Annex A, definition 1 to the SPS Agreement, *supra* note 16.

73 Jeffrey S. Thomas and Michael A. Meyer, *The New Rules of Global Trade* (Toronto: Carswell Publishing, 1997) at 86.

74 *European Communities – Measures Concerning Meat and Meat Products (Hormones)*, Report of the Panel, WTO Doc. WT/DS26/R/USA, 18 August 1997 and *European Communities – Measures Covering Meat and Meat Products (Hormones)*, Report of the Appellate Body, WTO Doc. WT/DS26/AB/R and WT/DS48/AB/R, 16 January 1998.

75 The Codex Alimentarius Commission was established in 1962 to implement the Joint FAO/WHO Food Standards Programme. The main work of Codex is to agree upon and compile internationally accepted voluntary food standards. See the Codex website. Online. Available http://www.codexalimentarius.net (accessed 25 February 2004).

76 Annex A, definition 3 to the SPS Agreement, *supra* note 16.

77 M. Kenny, "International Food Trade: Food Quality and Safety Considerations," (1998) 21 *Food, Nutrition and Agriculture* 1 at 2.

78 See H. E. Prasidh, "Food Trade and Implementation of the SPS and TBT Agreements: Challenges for Developing Countries in Meeting the Obligations of the SPS and TBT Agreements and the Codex Alimentarius," paper prepared for the Conference on International Food Trade beyond 2000: Science-Based Decisions, Harmonization, Equivalence and Mutual Recognition, Melbourne, October 1999 at para. 32, Online. Available http://www.fao.org/docrep/Meeting/X2666E.htm (accessed 25 February 2004).

79 *EC/US Beef Hormone Report*, Panel Report and *EC/US Beef Hormone Report*, Appellate Body, *supra* note 74.

80 EC/US Beef Hormone Report, Appellate Body, *supra* note 74 at para. 165.

81 *Ibid.* at paras. 160–168.

82 A reading together of Article 3.3 and 5.1 of the SPS Agreement, see EC/US Beef Hormone Report, Appellate Body, *supra* note 74 at para. 177.

83 Annex A, definition 4 to the SPS Agreement, *supra* note 16. In 1995, a report of joint FAO/WHO expert consultation, *Application of Risk Analysis to Food Standards Issues*, defined risk as "a function of the probability of an adverse effect and the magnitude of that effect, consequential to a hazard(s) in food." Cited in Jørgen Schlundt, "Principles of Food Safety Risk Management," (1999) 10 *Food Control* 299 at 299.

84 See Articles 3.3 and 5.1 to 5.6 of the SPS Agreement, *supra* note 16, and see

also EC/US Beef Hormone Report, Appellate Body, *supra* note 74 at paras. 173–177.

85 See Steve Hathaway, "Management of Food Safety in International Trade," (1999) 10 *Food Control* 247 at 248.

86 Kevin C. Kennedy, "Resolving International Sanitary and Phytosanitary Disputes in the WTO: Lessons and Future Directions," (2000) 55 *Food and Drug Law Journal* 81 at 86.

87 *Japan – Measures Affecting Agricultural Products*, Report of the Appellate Body, WTO doc. WT/DS76/AB/R, 22 February 1998 [Japan – Agricultural Products] at para. 84.

88 See EC/US Beef Hormone Report, Appellate Body, *supra* note 74 at paras. 97–109.

89 The European Union argued that risk assessment required in the SPS Agreement had to be interpreted to take into account the precautionary principle as a norm of customary international law. See EC/US Beef Hormone Report, Appellate Body, *supra* note 74.

90 It appears that the Appellate Body left open the possibility that in the future, if the precautionary principle emerges as a principle of public international law, the relationship between precaution and the SPS Agreement may be different. See EC/US Beef Hormone Decision, Appellate Body, *supra* note 74 at paras. 123, 124.

91 EC/US Beef Hormone Decision, Appellate Body, *supra* note 74 at para. 124. Article 5.7 of the SPS Agreement provides that where scientific information is insufficient or uncertain, a state can utilize a temporary import requirement.

92 For a good overview, see Kennedy, *supra* note 86 at 91–99.

93 EC/US Beef Hormone Decision, Appellate Body, *supra* note 74.

94 Japan – Agricultural Products, *supra* note 87.

95 *Australia – Measures Affecting Importation of Salmon*, Report of the Appellate Body, WTO Doc. WT/DS18, AB/R, 20 October 1998.

96 *Japan – Measures Affecting the Importation of Apples*, Report of the Panel, WTO Doc. WT/DS245/R, 15 July 2003 [Japan – US Apples], as upheld by the Appellate Body in WTO Doc. WT/DS45/AB/R, 26 November 2003.

97 Kennedy, *supra* note 86 at 96.

98 *Ibid.* at 97.

99 Japan – Agricultural Products, *supra* note 87.

100 Japan – US Apples, *supra* note 96.

101 There is an extensive literature on eco-labeling generally and eco-labeling in the fisheries context. An interesting recent contribution to this area is Cathy R. Wessells, Kevern Cochrane, Carolyn Deere, Paul Wallis and Rolf Willmann, *Product Certification and Ecolabeling for Fisheries Sustainability* (Rome: FAO Fisheries Technical Paper, No. 422, 2001). See also Rex J. Zedalis, "Labeling of Genetically Modified Foods: The Limits of GATT Rules," (2001) 35 *Journal of World Trade* 301–347.

102 A brief review of these issues is undertaken in FAO Committee on Fisheries, Sub-Committee on Aquaculture, 1st Session, 18–22 April 2002, "Aquaculture Development and Management: Status, Issues and Prospects," (COFI:AQ/1/2002/2 at paras. 42–47. Online. Available http://www.fao.org/docrep/meeting/004/y3277e.htm (accessed 25 February 2004).

103 The formal title of the 1980 agreement, like that of the TBT Agreement, was *Agreement on Technical Barriers to Trade*. See *GATT: Basic Instruments and Selected Documents* (26th Supp.) at 8. The GATT Standards Code was a plurilateral agreement. This means that it applied only to those GATT contracting parties that signed it. In contrast, the TBT Agreement applies to all WTO members.

104 Articles 1.4 and 1.5 of the TBT Agreement, *supra* note 17.

105 For example, a food product might be subject to an SPS requirement dealing with food safety and a technical regulation governing the labeling of nutritional qualities that are not directly related to food safety.

106 Annex 1, item 1 of the TBT Agreement, *supra* note 17.

107 Annex 1, item 2 of the TBT Agreement, *ibid.* note 17.

108 The TBT Agreement also governs the preparation, adoption and application of conformity assessment procedures. These are processes – often of a technical nature – to determine whether a particular product meets the specifications or qualities attributed to it by the manufacturer, producer or processor. Such specifications, if required by the importing state, may constitute technical regulations or standards in themselves. This chapter does not address the TBT disciplines on conformity assessment procedures.

109 Of course, this also means that the product must meet the specifications dictated by the label content.

110 Article 2.1 of the TBT Agreement, *supra* note 17, reads as follows:

> Members shall ensure that in respect of technical regulations, products imported from the territory of any Member shall be accorded treatment no less favourable than that accorded to like products of national origin and to like products originating in any other country.

111 See discussion regarding "like goods" in text accompanying *supra* note 46.

112 Article 2.2 of the TBT Agreement, *supra* note 17, reads as follows:

> Members shall ensure that technical regulations are not prepared, adopted or applied with a view to or with the effect of creating unnecessary obstacles to international trade. For this purpose, technical regulations shall not be more trade-restrictive than necessary to fulfil a legitimate objective, taking into account the risks non-fulfilment would create. Such legitimate objectives are, *inter alia*: national security requirements; the prevention of deceptive practices; protection of human health or safety, animal or plant life or health, or the environment. In assessing such risks, relevant elements of consideration are, *inter alia*: available scientific and technical information, related processing technology or intended end-uses of products.

113 Article 2.2 of the TBT Agreement, *ibid.*, does not make it clear who is to take these risks into account, but it is reasonable to surmise that the directive is aimed at the government that is adopting the measure.

114 Code of Good Practice, paragraph D, Annex III, TBT Agreement, *supra* note 17.

115 Code of Good Practice, paragraph E, Annex III, *ibid.*

116 Code of Good Practice, paragraphs F and G, Annex III, *ibid.*

117 *EC – Asbestos, supra* note 41.

118 *European Community – Trade Description of Sardines*, Report of the Appellate Body, WTO Doc. WT/DS231/AB/R, 26 September 2002 [European Community – Sardines].

119 The central issue in the dispute, which pitted Peru against the European Community, was that EC law only allowed a particular species of fish – *Sardina pilchardus Walbaum* – to be marketed in the Community under the trade name "sardine." Peru, which exported tinned fish of the species *Sardinops sagax* as sardines, claimed that the Codex Alimentarius Commission had adopted an international standard in 1978 regarding which fish species could be marketed as sardines, and that the standard allowed for *Sardinops sagax* to be billed as a sardine. In other words, according to Peru, the EC law was inconsistent with the international standard and, therefore, a violation of Article 2.4.

120 There were a number of subordinate or preliminary arguments put forward by the disputing parties, but the crux of the matter, and the issue that both the panel and the Appellate Body focused on, was whether Article 2.4 had been violated. Both the panel and the Appellate Body expressly declined to examine arguments about possible violation of Articles 2.1 and 2.2 once a violation of Article 2.4 had been established, citing the principle of "judicial economy" as their rationale. See European Community – Sardines, *supra* note 118 at para. 313.

121 Unless they are directly related to food safety. See Annex A, SPS Agreement, *supra* note 16, which specifies that SPS measures include all "packaging and labeling requirements directly related to food safety."

122 This is a key issue, since if the products are not "like," no issue of discrimination arises, and the measure would not amount to a violation of the national treatment/MFN obligation in Article 2.1 of the TBT Agreement, *supra* note 17.

12 Food safety and farmed salmon

Some implications of the European Union's food policy for coastal communities

John Phyne, Richard Apostle and Gestur Hovgaard

Introduction

"Mad cow disease," "foot and mouth disease," dioxin scares and genetically modified organisms are but a sampling of the food concerns that have emerged over the past decade. Ulrich Beck argues that we have entered a period where science and the industrial products of science are impacting upon human social organization in unforeseen ways. Because of this, risk science and risk management have become intrinsic to the regulation of social life.[1]

The European Union (EU) is at the forefront of human institutions engaged in risk science and risk management. The EU has come a long way from a food policy based upon security to one that is increasingly based upon safety. In the process, the consumer has replaced the producer at the center of food policy agendas. The Common Agricultural Policy (CAP), a policy instrument that still is contentious, has been superseded in ways unimagined within the EU itself. Beginning in the early 1990s, the EU began reducing funding to the CAP in favor of structural measures aimed at diversifying rural Europe. A policy canopy that embraced measures as wide-ranging as eco-tourism, organic agriculture and aquaculture emerged as "long-term" alternatives to the subsidies that generated the notorious "mountains" of butter and beef.

However, the consequences of agricultural overproduction were visited upon European consumers in several ways in the 1990s. "Mad cow disease" was an early concern, and this problem emerged periodically for the balance of the decade. There were also concerns over dioxin in food products and "foot and mouth" disease; the latter devastated British agriculture and contributed to Continental demands for restrictions on British food exports. By this time, an integrated food policy was emerging to deal with the assessment and management of food risks. The EU also moved to protect consumers from the "risks" of food imports from non-EU nations.[2] In the process, the EU shifted its priorities from import duties used to protect "producers" to food safety laws to protect "consumers." Nevertheless, the former remain an important dimension of EU policy.

The central objective of this chapter is to trace the evolution of EU food policy and show its potential consequences for farmed salmon exports from coastal communities. To that end, we will focus upon coastal communities in Norway and the Faroes as case studies. Both nations stand outside the EU, but are significant suppliers of fisheries products to EU consumers.[3] Scottish and Irish salmon farming concerns do not produce enough to satisfy EU demand and, as a result, significant imports come from Norway and the Faroes. However, since the early 1990s, Scottish and Irish producers have lobbied the EU in Brussels for duties against farmed salmon imports from Norway. The Irish view the Norwegians as "price-makers" who are in a position to flood the EU with cheap imports and thereby undermine Irish and Scottish production.[4] While duties and minimum import prices on farmed salmon imports were central in the 1990s, the food scares mentioned above also entered the picture. By the turn of the century, these scares affected farmed salmon imports into the EU, in addition to other food commodities. For our purposes, concerns over dioxin and genetically modified organisms (GMOs) have begun to impact heavily upon farmed salmon imports from outside the EU.

In this chapter, we discuss the impact of dioxin and GMO regulations upon feed producers and fish farming operations in Norwegian and Faroese coastal communities. We also examine the impact of the Faroes' recent food and feed agreements with the EU. These regulatory concerns, in addition to pre-existing Hazard Analysis at Critical Control Points (HACCP) and Codex Alimentarius Commission regulations,[5] and the current pattern of economic concentration and globalization in the salmon farming industry, potentially have negative consequences for small-scale producers and marginal salmon farming regions.

This chapter is divided into three sections. First, we briefly review the EU's transition to a food safety regulator. This not only follows in wake of the food scares over "mad cow" disease, "foot and mouth" disease and dioxins in food products, but also dovetails with recent global trends in the role of food safety laws as substitutes for import duties in the trade in food products.[6] Second, we assess the impact of EU food safety laws for farmed salmon imports from Norway and the Faroes. We argue that international, national and firm-level food regulations structure the export of farmed salmon to the EU. The impact of regulations on animal feed – a critical input for farmed salmon – is emphasized. The final section of the chapter explores the ways in which food safety regulations reinforce "buyer-driven" chains in the global food industry. We argue that by acting as "lead drivers" in the setting of food safety standards, international, national and firm-level players can contribute to social stratification at the points of production and processing. Nevertheless, we point to the role of state regulations and community embeddedness in fostering the ability of coastal communities to survive the current intersection of food safety trends and economies of scale in the salmon farming industry. These are countervailing forces to the full impact of standardized globalization. We conclude with some reflections on the consequences of the EU's food safety policies for Canadian coastal communities.

From security to safety: evolution of the EU's food policy

The EU has become synonymous with the CAP. This policy is part of a package of measures used to redistribute wealth from richer to poorer regions within the EU. Although the CAP's expenditures became controversial in the 1980s (after the 1973 and 1985 rounds of expansion), the initial impetus of EU agricultural policy was to provide food security in post-World War II Europe.[7]

With the entry of Ireland (1973) and Spain and Portugal (1985) into the EU, the importance of agricultural funding increased. The CAP includes a Guarantee Section used to provide price supports for a number of agricultural commodities and a Guidance Section to improve agricultural productivity. Despite the decline in overall agricultural funding, by 1998 the CAP still constituted over 54 percent of the entire EU budget. Nevertheless, in an effort to reduce overproduction, the funding within the CAP has shifted from price supports to income assistance. In 1992, market supports and export assistance constituted 82 percent of CAP funding; by 1998 this figure was reduced to 29 percent.[8] The stereotype of the CAP supporting only small and inefficient agricultural units is belied by the fact that large-scale farmers and processors have achieved their economies of scale with the assistance of EU price supports.[9]

By the early 1990s, the EU had moved to diversify the structural measures that included the CAP. The objective is to deal with both the resistance for continuing agricultural supports that generated overproduction and the need for the diversification of rural areas. Together with reforms to the Common Fisheries Policy (CFP), rural areas witnessed a diversification of regional policy in the form of assistance to rural manufacturing, tourism and environmental initiatives. These measures, which fell under the European Regional Development Fund (ERDF), were also associated with training initiatives under the European Social Fund (ESF).[10]

By the 1990s, the industrialization of agriculture was bearing witness to environmental and food safety problems. Prominent among the latter were the scares over BSE (bovine spongiform encephalopathy), foot and mouth disease, and dioxins. The BSE or "mad cow disease" scare was particularly notable, as it affected British agricultural exports to other EU nations. Concerns emerged over the contamination of animal feed. The BSE crisis was related to the mixing of animal waste in feed provided for animals for human consumption.[11] In addition to this, a dioxin scare emerged in Belgium in the late 1990s. Dioxins are present in industrial products and waste and accumulate in the food chain. By the late 1990s, the EU was concerned about dioxins in animal feed.[12]

During the 1990s, the EU shifted its food policy from "security" to "safety" in order to deal with consumer fears and the retailer concerns over food safety. A key measure was the need for traceability of food from the farm to the table. This required checks at each of the points in which contamination could occur. Banks and Marsden argue that with the deregulation of the British dairy industry in the 1990s, a re-regulation occurred under the guise of market power. Processors developed closer linkages with producers in order to provide high-quality and traceable dairy products for British retail giants. The latter replaced home delivery as the main outlet for dairy goods.[13] The EU was also concerned with food imports from outside the common market. Thus, in the 1980s and early 1990s, controversies emerged with the United States over beef hormones and bovine somatotropin (BST). The latter is a naturally occurring substance used to increase milk production.[14]

In the wake of an early BSE crisis, the European Commission (EC) formed the Office of Veterinary and Phytosanitary Inspection and Control (OVPIC) in 1991 under the auspices of the Directorate General – Agriculture. Thirty inspectors were hired to conduct on-site inspections pertaining to food safety. However, OVPIC had to compete with other EC agencies for funding. As the EC was in the process of transforming OVPIC into an independent agency with its own source of funding, controversy erupted in 1996 over the British government's position that it could not rule out a connection between BSE and Creutzfeldt-Jakob disease in humans. The ensuing crisis over "mad cow disease" involved accusations by the British beef industry that the EU was attempting to ban British beef from EU markets.[15]

OVPIC remained inside the Directorate-General – Agriculture until the food scares of the late 1990s. Despite concerns in the 1990s over the growing power of the EC bureaucracy in Brussels, member states and the European Parliament did not oppose the continued maintenance of the OVPIC (renamed the Food Veterinary Office [FVO]) under the authority of the EC, especially in the aftermath of more food scares in the late 1990s. The FVO maintains a risk management function and operates to enforce enhanced food safety laws. In 1997, the number of inspectors in the FVO increased to 200 and the agency was transferred to the Directorate General – Health and Consumer Protection (DG – SANCO).[16] This removed the enforcement of food standards from the EC Directorate-General – Agriculture and separated the production and regulatory components of EU food law.

At this time, the food scares resulted in the development of a *White Paper on Food Safety* that proposed the establishment of a European Food Safety Authority (EFSA).[17] This agency was launched in 2002 as a risk assessment and communication agency under the control of the European Parliament. The *White Paper on Food Safety* states that the EC, "in establishing agreements with third countries that recognise the equivalency of food safety controls under the WTO/SPS agreement, calls on the FVO for an evaluation of the health situation in the third countries concerned."[18] This entailed the continuance of border inspections for non-EU states exporting goods to the EU. The *White Paper*'s concern with food safety is reiterated in the legislation that established the EFSA.[19] What is central is the need to preserve safety measures in feed for animals used for human consumption and to trace the nature of feed back to its point of origin. The end result is a traceability system that covers items ultimately bound for human consumption. The feed manufacturing sector would be subject to the same checks for dioxin as the human food sector. The EFSA would be responsible for further testing for BSE.[20] In addition, feed for animal consumption and food for human consumption must not exceed permissible levels of GMOs. The labeling and authorization of GMOs, a highly contestable issue for major GMO producers such as the United States, Canada and Argentina, became intrinsic to EU food policy.[21]

In September 2003, the EU passed legislation making the labeling of genetically modified food for human consumption and feed for animal consumption mandatory.[22] The only exception is for food "containing materials which consists of or is produced from GMOs in a proportion no higher than 0.9 percent of the food ingredients considered individually or food consisting of a single ingredient, provided that this presence is adventitious or technically unavoidable."[23] A similar exception exists for feed.[24] However, in cases where the food or feed exceeds that 0.9 percent threshold, mandatory labeling is necessary, and the legislation requires the use of clear labeling.[25]

While no explicit mention is given to genetically modified fish or fish feed, Regulation (EC) No. 1829/2003 is significant in that it effectively acts to "identify" any transgenic fish imports into the EU, and any fish feed that contains genetically modified products such as soybeans. Both are significant for our purposes because of the current move by feed companies to decrease their dependence on pelagic stocks by using more vegetable-based materials in their feed. In light of this situation, one of the world's largest fish feed companies announced that it will not use genetically based materials in its feed.[26] This is clearly an attempt to prevent the occurrence of significant traces of GM products in farmed salmon imports into the EU.

With EFSA authority for risk assessment science and communication and FVO authority for risk management, the EU has further extended its regulatory scope. Concerns over BSE, dioxins and GMOs have resulted in a regulatory regime that aims to provide a full traceability system for all food imports. Although the implications for salmon farming are not fully clear, we are in a position to provide a tentative evaluation. Norway and the Faroes stand outside the EU, which is the biggest farmed salmon export market for each nation. Salmon farming is a good test of the degree to which EU regulations can impact upon third countries.

Vogel argues that trade liberalization is associated with more, and not less, food safety and environmental regulations. Declining tariff levels are being replaced with increasing levels of product harmonization.[27] A key empirical question, for our purposes, is the degree to which the "retreat of the state" in the era of globalization is occurring in the context of an international salmon aquaculture industry and supranational food regulations. As we shall see, food regulations in the context of declining export prices and economic consolidation in the international salmon farming industry pose challenges for the salmon farming industries of Norway and the Faroes. Nevertheless, we show that Norway and the Faroes are not totally subject to international regulatory and market forces, and have some scope in structuring the continued participation of salmon farming companies in export markets.

Food safety and farmed salmon exports to the European Union

Norway

> Unlike Danish farmers, who enthusiastically advocated EEC membership, the rural populations of Norway saw EEC membership as a colossal threat to their survival and as a direct contradiction to the powerful localism and regionalism espoused on the periphery.
>
> (Esping-Anderson on the reaction of rural Norwegians to the 1973 referendum on EEC membership[28])

> Norway is not part of the EU, but it is part of the EU.
> (Observation by interviewees and contacts in Norway)

Despite Norway's rejection of EEC (1973) and EU (1994) membership, the EU remains its largest trading partner. In 2000, nearly 77 percent of Norway's exports went to the EU and nearly 62 percent of its imports were from the EU. Over 37 percent of Norway's GDP is based upon its exports to the EU.[29]

While oil and gas constitute the main engine of Norway's economy, fishery products are still intrinsic to the well-being of the country. During the 1990s, Norway and Chile emerged as the two largest producers of farmed Atlantic salmon. By 2000, 883,558 tonnes of farmed Atlantic salmon was produced on a global basis. Norway produced 49.4 percent of this total. Chile (18.9 percent), Great Britain (14.6 percent), Canada (7.7 percent) and the Faroes (nearly 3.2 percent) were the next four largest producers.[30] The EU stands as Norway's largest seafood export market. Seven out of the top ten export markets for Norway's farmed salmon in 2001 were EU countries. In 2002, Norway sent to the EU nearly 50 percent of its total production of 2,109,323 tonnes of seafood exports (farmed and harvested).[31]

Furthermore, in contrast to Chile, which moved quickly to establish economies of scale and vertical integration in salmon farming,[32] Norway had a late move towards vertical integration in salmon farming. This dates back to a concerted effort by the state to decentralize production in the 1970s and 1980s. Firms were restricted to one license and one site. While this allowed production to increase throughout the country, it hindered the establishment of large firms. In the early 1990s, overproduction and the cold-storage freezing strategy of the Fish Sales Organization (FSO) contributed to bankruptcies and the restructuring of the industry. The government changed the license rules enabling producers to have more than one license and one site.[33] By 2001, six companies had over 30 percent of all salmon farm licenses.[34]

During the past decade, food safety laws and various import restrictions have been used to structure Norwegian seafood exports to the EU. Farmed salmon production from Norway has acted as a lightning rod in trade disputes with the EU. In the mid-1990s, the small Irish salmon farming industry was concerned over the "dumping" of Norwegian salmon in the European salmon market.[35]

In 1992, the European Free Trade Association (EFTA) (which includes Norway as a member) signed a free trade agreement with the EU. This reinforced the abolition of income supports to the fishery that was agreed to within EFTA in 1989.[36] In the 1990s, a minimum import price (MIP) was imposed on Norwegian farmed salmon. This price, which is negotiated in salmon agreements between Norway and the EU, establishes a "market" price that has to be maintained in order for Norwegian imports to gain access to EU markets. Irish and Scottish salmon producers demanded this so that the Norwegians cannot "dump" low-priced salmon on EU markets. Nevertheless, the Norwegian Federation of Fish and Aquaculture Industries (NHK) feels that the current MIP is inflated and it would like it abolished. Reports in 2002 indicated that the EU would replace the MIP with a 14 percent duty.[37] One official with NHK noted in an interview that "the biggest problem is that we are outside the big trade blocs." This has hindered access to markets.

Import duties in the 1980s and 1990s may have also impacted upon vertical integration. Nearly 70 percent of farmed salmon is exported fresh.[38] For Norway's EU exports, a typical pattern is for value added processing (such as smoking) to occur in Denmark. From Denmark, Norwegian salmon travels to the large EU markets. Prior to being shipped to the EU, most Norwegian salmon is handled by an export agency that acts on behalf of a given number of producers. Sea Star International and Lerøy are two of the bigger Norwegian export agencies; the former markets directly for one of the largest producers in southern Norway. Some producers bypass Danish processors and deal directly with supermarket chains such as Carrefour in France and Mercadona in Spain.

Although the abolition of tariff barriers and the ongoing negotiations over the MIP concerned Norwegian salmon farmers during the 1990s, quality controls also became crucial to food products produced for consumption in the EU. The FVO inspects member states and third-party countries such as Norway. Although the 2001 FVO report shows meat and dairy product inspections for Norway,[39] this does not mean that fish products are free of regulatory concerns. EU policies on dioxins and GMOs impact upon the entire food chain. Maximum dioxin levels in animal flesh sold to EU countries came into effect in July 2002, and DG – SANCO wants to reduce dioxin levels by 25 percent.[40]

Since pelagic stocks are used in fish pellets for farmed salmon, feed producers have to take this into account. Although dioxins are diminishing in the food chain, it is possible for feed to be produced with dioxin levels above acceptable EU levels. Moreover, dioxins accumulate in fats, and, as a result, can concentrate in the oil of pelagic fish. A fish veterinarian with Norway's Directorate of Fisheries noted that fish from the Baltic Sea have high levels of dioxin, and the challenge for feed companies is in the reduction of such levels.[41] Marine oil can be purified in order to lower dioxin levels. In addition, marine oil can be partially substituted with vegetable oil in order to lower dioxin levels. Nevertheless, he noted that the dioxin levels in effluent fell by 70 percent from 1985 to 1995.

The concern over dioxin levels makes the current forward integration strategy of some large feed companies a risky proposition. The existence of high dioxin levels in feed has potential implications for one feed company's recent global acquisitions of grow-out sites in Norway.[42] Moreover, an official noted that the company must buy feed from other companies because it does not have enough feed to supply all of its grow-out sites.[43] One implication that we can draw from this is that if there is a convergence of diminishing stocks for fish feed and high dioxin levels in the remaining feed, the global strategies of this feed giant may be undermined. Grow-out sites may have to increasingly turn to vegetable-based feed supplies in order to maintain operations. And, with this, firms will have to ensure that vegetable oils in fish feed do not exceed GMO levels accepted by the EU.[44] An official with another large feed manufacturer noted that there may be a horizontal integration of terrestrial and sea-based feed companies in the future due to the need for fish farming companies to decrease their reliance on pelagic stocks.[45]

While the reliance on vegetable oils will not result in a "vegetarian salmon," a new soya–pelagic synthesis is being produced as a feed supply. The EU requires the harmonization of standards, coupled with the labeling of GMOs in feed and final food products.[46] The ultimate objective is to have significantly reduced levels of GMOs in the food supply.[47] Shortly after the EU announced its proposed food policy for GMOs, one of the world's largest feed companies stated that it would not use GMOs in its feed. An official from Norway pointed to the role of the major Japanese and European markets in structuring his firm's decision:

> Japan wants the salmon naturally fed. Japan and Europe have questions about GMOs. . . . So, the question is if the oil is coming from GMO

plants. [This is] difficult to measure. The oil profile is mirrored in the salmon. Protein doesn't change. Oil you do change. If we use vegetable protein in the feed, it will not be shown in the salmon, but the oil profile does [show in the salmon]. [In] the US and Canada, there are no questions at all. Vegetable oil is used in feed for salmon production. In Chile – in terms of producers – this question is not an issue at all. But, there the feed companies follow the rules and expectations of the super-markets in Europe. They want to keep the possibility to sell in Europe. In Europe they don't want any GMO or land animal products in the feed. In Chile and Peru the price level for fish oil is relatively cheap. Vegetable oil is expensive. In the future there will be a shift to veget-able oil in Europe.[48]

In addition to regulating the feed supply beyond its borders, the EU also impacts upon the harvesting and processing of farmed salmon. This is through measures that predate the FVO and EFTA, namely HACCP and the SPS Agreement of WTO. Nevertheless, these polices are endorsed by the EC and are taken into account by the Directorate of Fisheries – the regulatory body for the Norwegian fishing and fish farming industries.

While it is not surprising that recent EU legislation on food safety emphasizes the issue of feed,[49] HACCP regulations impact heavily upon the use of fish feed, as well as upon the harvesting and processing of salmon. In Norway, HACCP and other quality regulations are based upon the incorpo-ration of EU regulations into Norwegian law.[50] The "critical control points" in the HACCP are originally based upon US regulations, and these deal with various "points" from the "fertilized egg to the table" that can impact negatively upon human health concerns.[51] To that end, fish processing estab-lishments use a check guide to inspect for hazards at each of the "critical control points."[52] A fish veterinarian with the Directorate of Fisheries in Bergen noted the importance of abiding by the HACCP system, which is based upon the Codex Alimentarius rules adopted by the WTO.[53] Most per-tinent here is the SPS Agreement, which aims for harmonized global stand-ards in the assessment of food safety for items subject to international trade.[54] The fish veterinarian added that Norway's HACCP regulations have "more descriptive and stronger requirements than EU regulations."[55] While we have no direct evidence to confirm this statement, this regulatory pro-cedure is extremely plausible. If Norway does have more stringent HACCP requirements than the EU, this may give food inspectors leverage so that "permissible deviance" within Norwegian law will enable producers to adhere to EU rules.

Another issue that impacts upon salmon farming interests is the role played by retailing giants in the EU. In the western Norwegian community of Austevoll, two supermarket giants were noted as being major players.[56] Carrefour of France is one of the largest supermarket chains in the world. It has a contract with an Austevoll salmon farming company that is one of the largest fish farming companies in Norway. An official with this firm noted that in addition to EU concerns, the company had to deal with an eighty-five-page manual from Carrefour.[57] He stressed that this company has an inspector who visits their processing plant on a regular basis. This inspector is not a seafood specialist.[58] This particular individual inspects a wide variety of food products for Carrefour. An official with the salmon farming company indicated that it was "easier dealing with the Japanese" because, although the Japanese were quality conscious, they sent over people who specialized in fish products.[59]

Two Austevoll firms sell to Mercadona, one of Spain's largest retailing companies. One of these firms sells seventy tonnes of fresh salmon a week. The salmon is sent to Valencia and redistributed to the 1,200 stores in the Mercadona chain. The manager of this firm noted that "two to three times a year we have visits from this chain – delegations that come to talk about quality and preferences for salmon."[60] He added that although the EU market was a stable market, it will not be one where increasing prices can be expected. He noted that

> [T]he main challenge for the industry in the future will be to produce fish more cheaply and integrate towards the market ... to increase the margin it is important to produce cheap. It is more likely that we will be able to produce two kroner cheaper than selling the fish two kroner more expensively.[61]

The implication of this is that labor and input costs will have to be secured at a cheaper rate. We will return to this issue later.

In addition, some of our respondents noted that in the latter half of 2002, the strong Norwegian krone (NOK) hindered export growth in EU markets for many Norwegian salmon farming companies. Thus, the need to meet quality controls was converging with pricing and currency trends that cut down on export volume. A fish-sourcing manager for Carrefour had no sympathy for the plight of salmon farming companies facing low profit margins:

> [A]ccusations of retail giants squeezing margins doesn't really apply to *Carrefour*. I don't have much sympathy for salmon farmers who are finding their profits low at the moment. High prices a couple of years ago almost put a great deal of French smokers out of business and no one was complaining about profits then.[62]

In summary, the Norwegian salmon farming industry faces challenges on a number of fronts. In addition to the MIP, food safety regulations governing feed, harvesting and processing have acted to govern the nature of farmed salmon production. The power of retail giants in the European food chain also reinforces quality demands. These firms also influence the ex-farm gate price for salmon. Next, we consider the impact of EU food safety regulations upon farmed salmon exports from the Faroes.

The Faroes

> EU membership or an extended fisheries zone under Faroese jurisdiction were mutually exclusive options. The strong internal demand for Faroese self-determination in issues concerning access to the fisheries settled the issue.[63]

When Denmark joined the EU in 1973, the Faroes elected to remain outside of the EU. Like their Norwegian counterparts, the Faroese feared the entry of EU vessels into their waters.[64] Despite being a small nation in the North Atlantic, the Faroes share common ground with Norway on the EU. Like many Norwegians, Faroese residents oppose entry into the EU, but at the same time, these residents live in the midst of fishing and aquaculture industries that are heavily dependent on EU markets. Fisheries and aquaculture constitute 90 percent of the value of the nation's exports.[65] The Faroes are the fifth largest global producer of farmed Atlantic salmon.[66] Most of this production is exported to the EU.[67]

The early period of aquaculture development on the Faroes was witness to government attempts to use this new industry as a development strategy for the country as a whole. This resulted in an early spatial dispersion of the industry. Nevertheless, disease problems due to the overcrowded site locations in areas with poor water exchange, as well as bankruptcies, ushered in a period of social change in the industry. In 1990, the government allowed mergers, and the industry reduced from sixty-five grow-out sites in 1990 to thirty locations by 1999. At the same time, production increased from over 12,000 tonnes (round weight) to over 36,000 tonnes.[68]

The EU is the largest market for Faroese exports and, as is the case for Norway, much of the product enters the EU through Danish ports. While Faroese firms deal with Danish processors, a large Faroese government-owned marketing firm facilitates the entry of Faroese fish products into the EU. It provides a multiplicity of product lines and outlets for groundfish, shellfish and farmed salmon. The firm has linkages with processing plants, trawlers and small vessel owners in the Faroes.[69] It also has marketing outlets in the northern Danish port of Hirtshals, Boulogne-sur-Mer in France and Preston in the United Kingdom, and marketing arrangement with Rainbow Seafoods in Gloucester, Massachusetts.[70]

As is the case for Norway, the Faroes have to meet the food safety requirements of the EU. This applies to issues such as feed, the import and export of livestock and the slaughtering of diseased salmon at sea. In the case of feed, we noted above that feed must meet the EU's requirements on dioxin and GMO levels in order to gain entry to EU markets. In the case of the Faroes, a large local firm accounts for 36 percent of the feed, with 46 percent accounted for by Ewos (the world's second largest feed producer) and 18 percent by Skretting (a Nutreco concern, Nutreco being the world's largest feed producer). The Faroese feed company sources 80 percent of its feed requirements from local catches. In contrast, the Ewos plant brings feed from Norway.[71] In fact, approximately 50 percent of the feed used on the Faroes comes from Norway, Denmark and Iceland.[72] This raises the issue of the extent to which feed from non-Faroese sources (especially from the Baltic Sea near southern Norway and Denmark) has appropriate levels of dioxin for EU purposes. While there is no evidence to date of Faroese salmon being banned from EU markets because of high dioxin levels, the issue of dioxin levels in farmed fish may intensify as the EU collects more data and establishes new permissible levels in 2006.[73]

In February 2001, the Faroes signed a Veterinary and Hygiene Agreement with the EU.[74] This facilitated the importation of EU livestock into the Faroes, as well as the exportation of livestock from the Faroes to the EU.[75] However, in this case, "livestock" means fish, because the Faroes does not have a substantial agricultural industry. For the Faroes, this means that the country can no longer prevent the importation of EU-approved ova and smolts for grow-out sites. Nevertheless, (as of 2003) there has been "no smolt importation because local production suffices to meet Faroese demand."[76] The Faroes are concerned about importing diseases. Despite this, the Faroes may have to accept ova and smolts from the EU if increased production is going to occur. In addition, the Veterinary and Hygiene Agreement formalizes the border inspection process and thus enables the Faroes to meet EU standards when it comes to exports to that market.[77]

Beginning in 1996, the Faroes maintained a compensation fund that enabled producers to defray the costs of slaughtering diseased animals at sea. The salmon farming industry and the government each contributed 50 percent of the costs of the fund.[78] This has to be discontinued for farmed salmon if the Faroes want to maintain access to EU markets. It will be replaced by a privately funded insurance program.[79]

Keeping in line with the EU's food policy, Faroese producers must maintain checks throughout the entire production cycle for farmed salmon. In the event of diseases that require the slaughtering of salmon at sea, salmon farmers will have to draw from their stock insurance fund.[80] This is another example of EU legislation that requires the elimination of subsidies on the part of nations that export food to the EU.

As is the case for Norwegian salmon farmers, Faroese producers not only are witness to EU regulations, but also are facing national legislation that must abide by EU requirements in order to maintain access to EU markets. Moreover, they also depend upon European processors and retail giants in order to gain access to EU markets. Only 25 percent of Faroese farmed salmon are value added prior to export.[81] An official with a large Faroese salmon farming company has an arrangement with a Danish smokehouse. The smokehouse sells its product to an Italian supermarket chain that distributes smoked salmon to Italian and French restaurants. In commenting on food safety requirements and its relation to the Faroese salmon farming company, an official with the Danish smokehouse noted:

The big challenge for a company like [ours] is to manage to follow up on public demands on food safety and other regulations. This is the main reason that many of the small ones [smokeries] are forced to go down; they don't have the resources to follow these demands. Plus the fact [is] that customers and deliverers are growing bigger. . . . [W]e have integrated with [the Faroese firm] because of: (1) traceability, because there is easy access to their provider; and (2) to buy salmon for the prices of the day is difficult, and makes partners like [the Faroese firm] a good investment.[82]

The Danish smokehouse official added that while there is "free trade" between the Faroese salmon farming company and his firm, the EU's dumping investigation against the Faroes may change this. He was referring to the recent EU decision to investigate Norway, Chile and the Faroes for dumping farmed salmon on EU markets below the MIP.[83]

An official with the Faroese salmon farming firm (discussed above) noted the importance of the Danish smokehouse in order for his firm to have access to the EU market.[84] But he added that dependence on this market is lessening, owing to increasing levels of local consumption. He stated, "[N]ormally around 30 percent is sold domestically, but it is a more safe business to sell on the Faroes. You get better prices here compared to export, but only small changes on the market may change this."[85] This company currently sells more than 50 percent of its output on the Faroes. The company official noted this to be unusual.[86]

The Faroese salmon farming company in question also has no HACCP system in place. Every worker on each of the firm's grow-out sites is required to work according to the company's fish quality handbook.[87] The company official indicated that customers (such as the Danish smokehouse) know where the fish is coming from and do not ask for food quality inspections.[88] He added that, despite the EU food policy, he has not seen anyone at "any of his sites doing any kind of inspection."[89]

Nevertheless, EU inspections of Danish seafood processing plants would most likely involve farmed salmon imports from the Faroes. These comments by the Faroese salmon farming company official may point to some limitations in the reach of EU food policy, but an official with a seafood marketing company stated that their new processing plant in the town of Fuglafjørður is the first in the Faroes with HACCP approval.[90] He also noted that the firm is asked about dioxins and GMOs in salmon and trout, but such quotations are more likely to come from the Japanese.[91] The important point is that the entry of HACCP through this seafood marketing company may entail a greater push by the EU to use this as a standard required by other Faroese producers. Yet this remains to be seen.

In comparison to Norway, the absence of a greater EU presence in the Faroes may be due to the much smaller production that the Faroes places on EU markets. These comparative differences mean that we may need to question the degree to which the EU's supranational "state" is contributing to the erosion of national sovereignty as trade liberalization increases.[92]

Next we examine the contribution of the EU's food safety regulations to "buyer-driven" global food chains. While these pose structural constraints that act to confine producers, we provide evidence that shows how a combination of national regulations and resilient communities in Norway and the Faroes continues to provide possibilities as salmon farmers seek to maintain access to EU markets.

Food safety, global food chains and coastal communities

Food safety and global food chains

The food safety requirements in national, EU and WTO rules are not "neutral" criteria. While these requirements establish a "level playing field" in international trade, food safety legislation stands testimony to an axiom in the sociology of law that legal equality can have unequal consequences. Henson and Loader argue that SPS requirements impose harsh burdens on developing countries. They lack the resources necessary to meet food safety laws, and face the possibility of having their exports rejected in the markets of developed countries. Developing countries single out the EU as a particularly difficult market when it comes to food safety requirements.[93] To that end, the EU's new food safety legislation is taking into account the need to assist developing countries when it comes to fully incorporating food safety laws into their national legislation.[94] However, this does not entail that all of the players in industrialized nations necessarily have equal access when it comes to meeting food requirements. To that end, some firms and peripheral regions within industrialized states may need "upgrading assistance" in order to secure better access to markets such as the EU.[95]

The need to upgrade is not exclusively determined by national and international food standards. Such standards work in conjunction with other trends. First, as many analysts have noted, power has shifted to actors in the food chain further removed from direct production. This includes processors, but especially retailers. Global food chains are now "buyer driven." This means that pricing and food quality standards are determined by retail giants that transfer their demands downstream to processors and producers.[96] Retail firms and international standards such as HACCP, SPS and the EU's food policies act as "lead drivers" in the governance of food chains. These set standards that must be met by other actors in the chain. Such standards also include sanctions. National producer associations often act as "chain upgraders" by providing industry members with technical support in order to meet international requirements.[97]

The concentration in power upstream in global food chains often influences economies of scale among downstream producers and processors. This is especially true for the Chilean salmon farming industry, which operates in a neoliberal environment.[98] First, in the midst of low retail prices in the mid-1990s, the world's two largest feed companies moved forward in the farmed salmon commodity chain and acquired grow-out sites. One feed company absorbed firms in Norway, Scotland, Ireland, Canada and Chile. Another acquired one of Chile's largest salmon farming companies. This was matched by greater concentration among the owners of grow-out sites. For example, Fjord Seafoods acquired two large vertically integrated salmon farming companies in Chile.[99] It also made investments throughout Norway and made linkages with Pieters, a large Belgian food retailer. Pan Fish, a company formed in the aftermath of the collapse of the Fish Farmers' Sales Organisation (FOS) in 1992, also made huge investments in Norway and in other salmon farming jurisdictions.[100] Pan Fish invested in pelagic vessels in Norway and the Faroes.[101] Our Norwegian interviewees noted that Pan Fish also invested in a "failed" feed processing facility. The obvious objective in Pan Fish's strategies was to secure control of a salmon farming chain that connected pelagic vessels, fishmeal companies, feed plants and salmon farming firms in order to compete in global markets.

It appears that this combination of international and national food standards and growing concentration in "buyer-driven" food chains places both small-scale firms and more peripheral communities in a precarious position. Yet a combination of state supports and "embedded communities" act as countervailing trends to these globalizing forces.

Norway

In Norway, the state relaxed ownership requirements in the wake of the 1992 collapse of the FOS, but still maintained rules that discouraged excessive concentration. The acquisition strategies of Pan Fish and Fjord Seafoods are limited by rules that cap the concentration of licenses at both the national and county level. In an August 2002 interview, an aquaculture official with the Directorate of Fisheries noted:

> The maximum number of concessions for salmon and trout is 15 percent of the [national total]. There are actually three levels. If you want to have more than 10 percent of the maximum total you need a special application. It will be very difficult to get this through. But if that is OK, you need a new application. The government has to approve if you pass 15 percent. But you cannot have more than 50 percent in a county.[102]

As a result, salmon farming companies and concessions are dispersed throughout Norway. In 2000, 194 firms held rights to 854 grow-out sites in Norway.[103] Although fifteen firms controlled nearly 40 percent of all concessions, there were 154 firms with five or fewer concessions.[104] In addition, the less populous northern counties of Troms and Finnmark held nearly 17 percent ($n = 144$) of all grow-out sites.[105] According to two officials with the NHK, the dispersion of licenses to the northern counties is a deliberate strategy on the part of the Norwegian state.[106] Moreover, the official with the Directorate of Fisheries noted that in the northern counties there is more room, which facilitates the separation of sites by a distance of 3–5 kilometers.[107] In the south, the minimum distance is one kilometer, and this is often difficult to enforce, owing to the concentration of sites. For example, populous Hordaland County has 139 sites – almost the same number as the thinly populated northern counties.[108]

The rules on ownership and minimum site distance between salmon farms act as forces that can minimize the impact of globalizing food safety trends and economic concentration in the food industry. First, the already economically troubled fish farming giants have to limit their investments in each of the Norwegian counties. This means that they cannot completely undermine the efforts of small-scale producers. Second, the minimum distance requirements between salmon farms entail that farms further north have a better chance to avoid the diseases that come with overcrowding. Thus, the disadvantages of distance potentially can be overcome by having fewer disease problems and better fish quality. And given the lower prices of salmon in international markets, firms that concentrate on production may be better situated than those that pursue an integrated strategy.

But challenges do exist for communities in the northern counties of Troms and Finnmark. In an August 2002 interview, an official with the Directorate of Fisheries noted that these northern counties are where the bulk of forty new licenses would be allocated in the fall of 2002.[109] These cost nearly NOK5 million. The cost is NOK4 million in Troms and Finnmark counties.[110] An *IntraFish* reporter noted that ten licenses allocated to Finnmark county were "redeemed in the first round, and companies had to decline in the counties of Troms, Rogaland and Møre og Romsdal. Exorbitant license duties and high risks were given for the refusals." Officials with the Norwegian salmon farming industry did not approve of this latest round in license allocations. They were especially concerned with the expiry of the EU–Norway Salmon Agreement and the implications that a new round of licenses would have for the MIP in a new agreement.[111]

While consistent with Norway's decentralized aquaculture policy, these new investments may create more anxiety in EU markets over the "dumping" of salmon. The north suffered greatly during the overproduction crisis of the early 1990s, and many small-scale firms collapsed. Some attribute this to overproduction and the cold storage policy of FOS.[112]

In short, EU food safety laws in the context of the current crisis in the global salmon farming industry will make it difficult for new salmon farming companies to emerge in peripheral Norwegian communities. A better strategy may be enhancing the prospects of existing firms. Moreover, the "shakedown" in large companies that is on the horizon may create opportunities for more solvent smaller-scale players. But in the context of the regional divisions between north and south in Norway, the allocation of existing licenses is politically an expeditious strategy.[113]

In addition to state policies, embedded communities can act to buffer the consequences of globalization. Despite the pressures of economies of scale in Austevoll, Norway, most of the firms there are embedded in the local community. While these firms feel the effects of the "cost-price" and regulatory squeeze, the history of entrepreneurship in this community is a pivotal dimension that cannot be ignored. A municipal official, echoing Weber, speaks of the role of "Pietism" and learning about the fishing industry at a young age as being critical to Austevoll's well-being.[114]

In addition, there is a network of aquaculture, fishing, fish processing and fish supply firms. Aquaculture, in conjunction with pelagic fisheries, gives Austevoll a diversity of fish products that are marketed to places as wide-ranging as Spain, Poland and Russia. Further, two of the larger aquaculture firms have investments throughout Hordaland county.[115] These are coupled with investments abroad. Business operators espouse competition, but note that "informal ties" are important in that they assist other businesses during times of need. This pertains to issues such as the "exchange of equipment" necessary for day-to-day operations in the event that a competitor has equipment in need of repair. Another Austevoll salmon farmer noted that salmon farming companies also have a collective tradition of working together in negotiating a deal for fish feed in order to ensure that all firms received the best possible price. Yet he indicated that the impact of economic concentration by outside forces has resulted in changes to the community whereby the two largest local producers are no longer interested in cooperating with the smaller salmon farms in securing a common agreement on feed purchases.[116]

Thus, while EU and corporate food safety laws present challenges in the short term, the local business community may be embedded enough to survive in the long term. The question is the degree to which food safety laws in conjunction with the international restructuring of the salmon farming industry may undermine the balance between "competition" and "intracommunity" cooperation in Austevoll.[117]

Austevoll depends upon imported labor in processing plants. In Austevoll, young Swedish (largely female) workers come to work in the processing of pelagics. These processing plants operate on a seasonal basis. An official with a large salmon farming company stated,

> [I]t is pretty easy to find workers for the processing plant. It is more popular than for instance Gilde [the meat company], because you get the opportunity to work a lot of hours in the fishing industry. The payment per hour is about NOK100 [CDN$20], but there are a lot of extras like overtime.

Austevoll residents are more likely to work in farmed salmon processing plants that operate on a yearly basis.[118] In 2001, there were 2,254 individuals employed (ages 16–74) in Austevoll, and nearly 16 percent ($n = 358$) were from outside the community. Over one-third of all the latter ($n = 124$) came from another country.[119]

If the current trend of low market prices continues, and if there are additional costs associated with meeting food safety requirements, we may witness changes in the Austevoll labor market. As noted earlier, an owner of an Austevoll salmon farm indicated that it is easier to save on production costs than it is to increase market prices. Two of the main production costs in salmon farming are feed and labor. Given that the feed in Austevoll is purchased from one of the three large fish feed companies, this means that production savings are equivalent to reducing labor costs. If this is the case, the labor-saving costs will come from increasing "efficiencies" either on grow-out sites or in plants that process farmed salmon. One result could be the gradual substitution of higher-priced local labor with lower-paid immigrant labor. Thus, immigrant labor may increase in year-round salmon processing lines, as is the current case for seasonal pelagic processing lines. Another result could be the substitution of labor by increased automation in fish processing plants. If either scenario occurs (or both), this will change the nature of the labor market. While these are possibilities, the critical point is that food safety regulations combined with low market prices can have localized impacts on labor markets. Furthermore, there is consolidation in the Austevoll salmon farming industry, and with this may come the need for reduced labor costs. Firms may be able to adjust to these realities of globalization, but the adjustment may come at the expense of the life chances of local workers.

The Faroes

EU food safety laws and economic concentration in the international salmon farming industry arguably pose greater challenges for the Faroes. Faroese producers are more distant from EU markets. However, as we have argued above, EU regulations have not fully impacted upon Faroese production. In addition, the Faroese version of social democracy and community embeddedness underscores the resilience of this nation in the face of globalizing forces.

Like Norway, the Faroes imposes limitations on salmon farming concessions. Ownership rules restrict voting and "decisive influence" "in all grow-out sites to 25 percent of the grow-out licenses, as well as 25 percent of the total shares and voting rights in all grow-out licenses in the Faroes."[120] In addition, no outside firm is allowed more than one-third of the voting rights and shares in a Faroese fish farm.[121]

The end result is that while the number of firms declined after the economic crisis of the early 1990s, there are both local and external limitations on the concentration of capital. Since 1990, the restructured aquaculture industry has been characterized by a diverse spatial and industrial structure. This ranges from firms that are spatially concentrated and vertically integrated to ones that are spatially dispersed and horizontally integrated.[122] Thus, while mergers occurred after 1990, vertical integration became only one of a number of strategies. Nevertheless, one of Austevoll's largest firms and Pan Fish have investments in Faroese pelagic vessels.[123] These investments may be within the letter of Faroese law; however, in the long term the question is the leverage that a substantial concentration of minority ownership may have upon the Faroes.[124]

In the Faroes, aquaculture is an example of "embedded entrepreneurship," in that investments made are "nested" in the communities in question. The industry easily adapts to the traditional lifestyles of Faroese coastal communities. The vertical integration in the Faroese fishing and aquaculture industries reached limits in the early 1990s.[125] The result was the threat of widespread closures of fishing plants, in addition to rising unemployment and out-migration (especially to Denmark). However, a combination of institutional and community investments in the 1990s turned the Faroese economy around. In place of "Faroese Fordism," a more diversified industrial and service sector was created. Vertical integration was replaced by more flexible contractual relationships between harvesting, processing and marketing firms. Firms such as Faroe Seafoods link harvesters, fish farms and processors to EU markets.[126] Intracommunity ties, and networks with "outsiders," improved the life chances for residents on the islands of Sandoy and Vágoy.[127] Hovgaard argues that a networking economy is slowly emerging on the Faroes. This builds on a tradition "of community entrepreneurship and locally derived supporting structures . . . supplemented and strengthened with direct and multifaceted global linkages."[128] This networking is occurring in several Faroese coastal communities.[129]

What are the implications of this for meeting the challenges of food safety laws and economic concentration? Faroese law has acted to maintain local ownership over grow-out sites. Moreover, this control, combined with interfirm networking and linkages to global markets, shows that vertical integration is not a necessary condition for interfacing with the global economy. A major challenge may come with the need to accept the importation of ova and smolts from the EU. This will occur if Faroese farmed salmon production is to increase. The importation of ova and smolts from the EU raises the safety threshold for all producers. It is here that the Faroese government may have to intervene with resources to ensure that imported ova and smolts pose minimal risks for producers.

Faroese ingenuity weathered the economic storm of the early 1990s and there is no reason why it cannot address further challenges. One of these challenges is the need to mitigate disease problems. To date, Faroese producers have wisely monopolized the better grow-out sites, and the riskier sites (those with poorer water exchange and disease problems) have investments by foreign firms.[130] Nevertheless, all Faroese producers (local and with foreign investments) are facing the consequences of an outbreak of infectious salmon anemia (ISA). Officials with two local firms view this as one of their biggest challenges.[131] An official with a large salmon farming company noted that while his firm can withstand the current low price for farmed salmon (around CDN$5 in 2002), ISA is a serious problem that will negatively impact upon all salmon farming businesses in the Faroes.[132] An official with another firm indicated that "[t]he quality of Faroese salmon has decreased since the problem with the ISA started. The producers are forced to slaughter at certain times, which makes the quality various even within the same site."[133]

After our field research (2002 and 2003), the ISA crisis intensified and spread throughout the Faroese archipelago. This disease, which is not peculiar to the salmon farming industry, was unknown on the Faroes until 2001.[134] At the time of the outbreak, the Faroes was producing 30,000 tonnes; this rose from 56,000 to 58,000 tonnes in 2003. In 2005, the production is estimated to be 10,000 tonnes. One of the remaining success stories is the locally owned feed company Havsbrún. A Norwegian interest is currently (January 2005) attempting to acquire majority ownership of this company.[135]

Føroya Banki, the largest shareholder in Faroese aquaculture, expects salmon exports to further decline because of the continuing ISA crisis and low export prices. The bank anticipates that exports of salmon will be zero tonnes in 2006; a few thousand tonnes of trout are expected to be exported.[136] On the latter note, one of the Faroes' largest aquaculture companies (Vestsalmon) exports 2,000–3,000 tonnes of trout to Japan. A director with that company expects the salmon aquaculture industry to rebound once diseases have run their course. He noted that "[t]he Faroe Islands has a marine environment that is ideal for aquaculture, and the balance between supply and demand will improve in the future."[137] Despite the pessimism of Føroya Bank; and the optimism of Vestsalmon, in order for the Faroese salmon aquaculture industry to rebound, it needs to use a vaccine in order to minimize the impact of further outbreaks of ISA. The EC gave clearance to use that vaccine in August 2004.[138]

Finally, as is the case for Norway, the Faroes experienced low unemployment levels in the early part 2000s. At the time, the country depended on imported labor for some of its salmon processing establishments. An interesting irony here is that the southern island of Suðuroy, which experienced high unemployment and rivalry between its two largest towns in the early to mid-1990s,[139] imported female Thai workers to process farmed salmon. At the time, these women were employed elsewhere in the Faroes in the salt fish industry. Some of these women married locals and integrated into the day-to-day life on Suðuroy. However, with the rise in unemployment levels in the wake of the ISA crisis and other problems in the seafood sector, the "guest workers" have become part of the "political debate" on where the Faroes is go next. The fish factory that employed these women has closed. As a result, the long-term prospect for these immigrant women remains to be seen.[140] The key is how these immigrants will be treated in the midst of economic decline. The future of Faroese social democracy may be partially judged on how such workers are treated, as the nation's farmed salmon is processed for global markets.

Conclusion

We have argued that the EU's food safety laws have implications for Norway and the Faroes. Exports to EU markets are increasingly judged by food safety criteria that impose traceability throughout the food chain. As a result, the EU's international standards are incorporated into national legislation. On the surface, this seems to fit with arguments that growing trade liberalization is being matched by a greater integration of environmental and food safety standards. The end result is a further contribution to globalization.[141] This is matched by greater power in food chains by feed giants, grow-out giants and international food retailers.

Globalization is not simply an all-encompassing force that diminishes the power of nation-states and buttresses the power of supranational institutions. Nation-states still have considerable leverage in structuring both the shape of their civil societies and the ways in which actors participate in the global economy.[142] Norwegian and Faroese versions of social democracy instituted policies that have limited the powers of large-scale interests in the salmon farming industry. To be sure, there have been economic crises (especially in the early 1990s) that resulted in the decline in small firms and greater economies of scale, but a great degree of decentralization (geographic and economic) is part of the aquaculture industry in both nations. In fact, the recent problems of Pan Fish and Fjord Seafood bear testimony to the limits of economies of scale.

Firms that avoid the forward integration complexes of feed giants are in a better position to weather economic downturns. Economies of scale, coupled with international regulatory pressures and low prices on export markets, are a recipe for disaster. Firms with loose connections to feed companies and exporters avoid being caught in a food chain that leaves them with little leverage. A perpetual lesson in seafood industries is that monopolies based upon single product lines have a precarious existence.[143]

In addition, a closer examination of coastal communities shows that community entrepreneurship and interfirm networking gives communities the ability to maintain linkages with export markets even in the face of international food standards and economic concentration. This is true for both Norway and the Faroes. In fact, a more immediate challenge for the latter is the need to revitalize aquaculture in the aftermath of the current ISA outbreak. Food safety laws only matter if farmed salmon can survive the grow-out period.

The EU's food safety laws will present a constant challenge for Norway and the Faroes. The Norwegian fish farming industry favors EU membership, if only to avoid the constant negotiations of the MIP and border inspections to guarantee food safety. Furthermore, while the enlargement of the EU from fifteen to twenty-five members may temporarily stretch EU food inspection resources, the expanded EU includes nations with coastal resources (such as Estonia and Poland) that may want to challenge the seafood exports of Norway and the Faroes.

The precautionary principle at the heart of EU food policy has social costs that need to be attended to by Faroese and Norwegian policy-makers. If existing and new firms need to upgrade to HACCP and other food safety requirements, institutional assistance may be required. In the current context of low market prices, the absence of such assistance may mean that upgrading costs will be downloaded on the labor forces of coastal communities. If this becomes the case, the forces of globalization will undermine the redistributive benefits of social democracy. Nevertheless, Norwegian and Faroese policy-makers have faced similar challenges in the past, and one should not underestimate their ability to provide localized benefits in the midst of standardized globalization.

Prior to concluding this chapter, it is pertinent to ask about the possible consequences of the EU's food policy for coastal communities in other nation-states. The coastal communities of Canada serve as a good example. While the current mandatory labeling policy for GM foods is of more concern to Canadian agricultural producers, given Canada's position as one of the world's largest producers of GM foods,[144] the EU's GMO policies have consequences for Canadian seafood producers too. They may reinforce Canadian dependence on the US market.

Although Norwegian and Faroese farmed salmon producers have a significant presence in EU markets, Canadian production is overwhelmingly exported to the United States. In 2002, 70 percent (85,400 tonnes) of Canada's farmed salmon production (Atlantic and Pacific) was in British Columbia, and 95 percent of this was exported to the United States.[145] The remainder of Canada's production of farmed salmon (all Atlantic) is largely in New Brunswick (with smaller quantities in Nova Scotia, and Newfoundland and Labrador).[146] These provinces also send significant shipments to the United States. In addition, access to the US market is being structured by adherence to the *Anti-Terrorism Act*. This bio-security provision is of concern to Atlantic Canadian seafood exporters, given that the original wording included a twenty-four-hour notification period for seafood exports to the United States. This had negative implications for the just-in-time delivery process that is crucial for seafood exporters. This was reduced to two hours on 12 December 2003.[147]

Given this dependence on the US market, one would think that farmed salmon producers, especially those in eastern Canada, would have a large interest in securing access to EU markets. However, while food safety standards such as HACCP are emphasized in recent aquaculture policy proposals,[148] this may not be enough to secure entry into EU markets. Chile has signed a free trade deal with the EU, and farmed salmon from that country has already (albeit in small quantities) secured a foothold in EU markets.[149] Perhaps only increased expansion of farmed salmon consumption in the EU may provide an avenue for farmed salmon exports from Canada. Furthermore, with the possibility of transgenic fish being produced in commercial quantities in Canada in the next decade, such production will most likely be centered on the American market. This is the case even if fish farmers voluntarily adopt labeling.[150] While eastern Canada is best positioned geographically to enter EU markets, its colder waters make it a likely candidate for transgenic fish that can survive at subzero temperatures. If there is a rapid shift to such fish in order to enhance survival capabilities and thereby reduce losses from "flash freezes," this will reinforce a continuing dependence on the US market.[151]

In summary, while the Norwegian and Faroese states have the necessary leverage to continue exporting to EU markets, even in the midst of more stringent EU food policies, the same cannot be said for Canada. The future direction of farmed salmon production Canada will most likely reinforce that country's dependence on the US market. This is at a time when major farmed salmon producers such as Norway and Chile are actively diversifying their export markets.[152]

Notes

The authors acknowledge the research funding provided by AquaNet, a Network of Centres of Excellence program based at Memorial University of Newfoundland. This chapter is part of a larger research project entitled *The Institutional and Social Structure of Aquaculture: A Comparative Analysis of Norway, Chile, the Faroes and Japan* (Richard Apostle, Principal Investigator, Gene Barrett, Co-Principal Investigator and John Phyne, Co-Principal Investigator). Several individuals assisted us with the Norwegian and Faroese phases of our research: For Norway, thanks are extended to Moira McConnell, Faculty of Law, Dalhousie University, for facilitating contacts with Roger Bennett and Knut Lindkvist of the Department of Geography, University of Bergen, and Stig-Erik Jakobsen of the Foundation for Research in Economics and Business Administration. We also thank Bernt Aarset of the Foundation for Research in Economics and Business Administration. Atle Roalsvik of the Department of Geography, University of Bergen, assisted in our meeting Gard Hansen (now at the Department of Interdisciplinary Studies of Culture, Norwegian University of Science and Technology). Mr. Hansen was indispensable to our field research in Austevoll in August 2002 and March 2003. Ólavur Waag Høgnesen of the Center for Local and Regional Development in Klaksvík, the Faroes, also assisted with the August research in Austevoll. We also thank Sunniva Seterås of the Directorate of Fisheries in Bergen, who facilitated our March and August 2002 interviews with personnel at the Directorate of Fisheries. For the Faroes, we thank the staff at the Center for Local and Regional Development. Thanks are extended to all of those who gave their time to be interviewed for this project. In order to protect the confidentiality of our informants, especially those in the Faroes, which is a small society, we will not use the names of companies in the reporting of our interview data. As is the case for most social science research, policy developments often supersede published information. The data in this chapter reflect the state of affairs as of January 2005. Finally, we extend our appreciation to the editors, David VanderZwaag and

Gloria Chao, and to Susan Rolston for their assistance in bringing this chapter to publication.

1 Ulrich Beck, *Risk Society: Towards a New Modernity* (London: Sage, 1992).
2 Commission of the European Communities, *White Paper on Food Safety* (Brussels: Commission of the European Communities, 2000).
3 Although formally part of Denmark, the Faroes are autonomous in most areas, with the exception of foreign policy and judicial affairs. Denmark's entry into the EU in 1973 was not followed by the Faroes. The Faroes remain outside of the EU; yet, as we shall see, the Faroes and Norway depend upon Danish firms in order to get access to EU markets.
4 John Phyne, *Disputed Waters: Rural Social Change and Conflicts Associated with the Irish Salmon Farming Industry* (Aldershot, UK: Ashgate, 1999) at 51–53. This view is still held. For a detailed overview of EU salmon farmers' arguments concerning Norwegian dumping of farmed salmon on EU markets, see Commission of the European Communities, *Proposal for a Council Regulation Terminating the Anti-dumping and Anti-subsidy Proceedings concerning Imports of Farmed Atlantic Salmon Originating in Norway and the Anti-dumping Proceedings concerning Imports of Farmed Atlantic Salmon Originating in Chile and the Faeroe Islands (presented by the Commission)* (Brussels, 30 April 2003). Later in this chapter, we will discuss some of the views of this report with specific reference to Norway and the Faroes.
5 HACCP is one of the food safety laws endorsed by the Codex Alimentarius Commission – a body established in 1961 by the Food and Agriculture Organization (FAO). In 1963, the FAO and World Health Organization (WHO) adopted the statutes of the Codex Alimentarius Commission (hereafter referred to as Codex). Codex has as its objective the establishment of food safety laws while minimizing the impact of such laws on international trade. However, Codex's rulings are not binding on nation-states. Nevertheless, Codex's scientific findings are used in the resolution of trade disputes. In 1995, with the establishment of the World Trade Organization (WTO), the Sanitary and Phytosanitary (SPS) Agreement became the basis for balancing food safety with trade liberalization. It made reference to Codex as a basis for its decisions. For more on the history of Codex and its role in food safety and trade, see FAO and WHO, *Understanding the Codex Alimentarius.* Online. Available http://www.fao.org/docrep/w9114e/W9114e00.htm (accessed 1 March 2004). For the SPS Agreement (which includes its endorsement of the Codex Commission), see WTO, "Agreement on the Application of Sanitary and Phytosanitary Measures." Online. Available http/www.wto.org/english/docs_e/legal_e/15-sps.pdf (accessed 1 March 2004). For a discussion on the relevance of Codex and the SPS Agreement for international trade in aquaculture products, see Ted McDorman and Torsten Ström, Chapter 11, this volume. Douglas Moodie, in Chapter 13, this volume, also refers to the impact of Codex and the SPS Agreement on trade in aquaculture products, but with specific reference to the intersection between the Canadian and global regulatory venues. Prior to the WTO, food safety and trade issues were dealt with by Article XX of GATT. Moodie notes that while GATT 1994 (which includes Article XX) was incorporated into the WTO in 1995, in the long term the SPS Agreement will become more important for food safety and trade issues. This is also the implication of McDorman and Ström's analysis. Later in this chapter, we will briefly deal with the implications of the SPS Agreement in our consideration of the relevance of EU food safety laws for farmed salmon imports.
6 For a discussion of this issue, see David Vogel, *Trading Up: Consumer and*

Environmental Regulation in a Global Economy (Cambridge, MA: Harvard University Press, 1995) at 150–195.

7 See European Commission, *Ireland: A Region of the European Union* (Brussels and Luxembourg: CECA – CEE – CEEA, 1994).

8 European Commission, *Unity, Solidarity, Diversity for Europe, Its People and Its Territory: Second Report on Economic and Social Cohesion*, Vol. 1: January 2001 (Luxembourg: Office for Official Publications of the European Communities, 2001) at 81. Although it lies outside the scope of our analysis, it should be noted that with EU enlargement in 2004 the CAP will be as important to East European nations as it was to Ireland when that country joined the EU in 1973.

9 Hilary Tovey, "'Of Cabbages and Kings': Restructuring in the Irish Food Industry," (1991) 22(4) *The Economic and Social Review* 333.

10 For details on the provisions of the ERDF and ESF, and the structural funds in general, see "Council Regulation (EC) No. 1260/1999 of 21 June 1999 laying down general provisions on the Structural Funds," in European Commission, *Structural Actions 2000–2006: Commentary and Regulations* (Luxembourg: Office for Official Publications of the European Communities, 2000) at 33.

11 Nigel Evans, "The Impact of BSE in Cattle on High Nature Value Conservation Sites in England," in Hugh Millward, Kenneth Beesley, Brian Ilbery and Lisa Harrington, eds., *Agricultural and Environmental Sustainability in the New Countryside* (Truro: Rural Research Centre, Nova Scotia Agricultural College, 2000) at 92. In the spring of 2003, a single case of BSE in Canada resulted in negative implications for beef exports to the United States and Japan.

12 Commission of the European Communities, *supra* note 2 at 23–24. In a scientific report on dioxins in the food chain, the EC noted that

> [m]eat, eggs, milk, farmed fish and other food products may be contaminated above background by dioxins from feeding stuffs. Such contaminants may be due to a high level of local environmental contamination, for example from a local incinerator, such as the Belgian episode, where animal feeding stuffs were contaminated, or to a high content of dioxins in fish meal and fish oil. Wild fish from certain polluted areas may be highly contaminated.

See Health and Consumer Protection Directorate-General, Scientific Committee on Food, European Commission, *Opinion of the SCF on the Risk Assessment of Dioxins and Dioxin-Like PCBs in Food* (adopted on 22 November 2000). Online. Available http://europa.eu.int/comm/food/fs/sc/scf/out78_en.pdf (accessed 1 March 2004) at 40. The report provides information on dioxin levels in various food sources. Wild fish and freshwater-farmed fish are noted as having dioxin levels above those for other food sources (at 15–16). Later in this chapter, we will discuss EU legislation pertaining to maximum dioxin levels permitted for food and feed traded within the EU and imported into the EU.

13 Jo Banks and Terry Marsden, "Reregulating the UK Dairy Industry: The Changing Nature of Competitive Space," (2001) 37 (3) *Sociologia Ruralis* 383.

14 Vogel, *supra* note 6 at 152–174. McDorman and Ström, in Chapter 11 of this volume, discuss the implications of the beef hormone dispute as an example of how the EU was viewed as violating the SPS Agreement.

15 R. Daniel Kelemen, "The Politics of 'Eurocratic Structure' and the New European Agencies," (2002) 25(4) *West European Politics* at 105.

16 *Ibid.* at 107.

17 Commission of the European Communities, *supra* note 2 at 14. Although the BSE issue deals with terrestrial as opposed to aquatic animals, it arguably informed the EU debate over food in general. For our purposes, the issue of

dioxins in food and feed is more pertinent. The relevant legislation is *Commission Regulation (EC) No. 466/2001 of 8 March 2001 Setting the Maximum Levels for Certain Contaminants in Foodstuffs (Text with EEA Relevance). Official Journal of the European Communities* L 077, 16/03/2001. Article 1(1) at L77/1 stipulates, "The foodstuffs indicated in Annex 1 must not, when placed on the market, contain higher contaminant levels than those specified in the Annex." Annex 1: Section 3.1.4 at L 77/4 states that the muscle meat of fish must have a maximum level (mg/kg wet weight) of 0.2 for heavy metals. These fish include live fish, fresh, chilled and frozen fish, fish fillets, dried and salted fish, crustaceans and mollusks (as defined by *Council Regulation (EC) No. 104/2000 of 17 December 1999 on the Common Organisation of the Markets in Fishery and Aquaculture Products* at Article 1 (a), (b) and (c) at L 17/26). These categories most likely cover wild and farmed fish (such as salmon). *Regulation (EC) No. 466/2001* further stipulates that these categories of fish are also evaluated for maximum levels of cadmium (0.1 mg/kg wet weight) (Article 3.2.5 at L 77/10) and mercury (0.5 mg/kg wet weight) (Article 3.3.1 at L 77/11). The regulation also covers other fish and foodstuffs. This regulation came into force on 5 April 2002.

18 Commission of European Communities, *supra* note 2 at 15.

19 *Regulation (EC) No. 178/2002 of the European Parliament and the Council of 28 January 2002. Official Journal of the European Communities* L-31/1-24.

20 Commission of European Communities, *supra* note 2 at 22–29.

21 *Ibid.* at 22–29 and 32. Moodie, in Chapter 13 of this volume, notes that the United States, Canada and Argentina produce 96 percent of the world's GM food and are opposed to the mandatory labeling stance of the EU. For Moodie, the position of these three countries is that "in the absence of actual health and environmental risks . . . there is no compelling reason to mark GM foods, and several good reasons (all commercial) not to."

22 *Regulation (EC) No. 1829/2003 of the European Parliament and of the Council of 22 September 2003 on Genetically Modified Food and Feed (Text with EEA Relevance). Official Journal of the European Communities* L 268/1. In this text, the EU notes that the protection of human and animal health is at the core of its concerns over GMOs. It calls for a safety assessment for GMOs entering the EU. In the preamble to the main legislation dealing with labeling, it is noted (at L 268/1) that

> [i]n order to protect human and animal health, food and feed consisting of, containing or produced from genetically modified organisms (hereinafter called genetically modified food and feed) should undergo a safety assessment through a Community procedure before being placed on the market within the Community.

The legislation goes on to add (at L 268/1) that the procedure of substantial equivalence is not to be confused with food safety because

> [w]hilst substantial equivalence is a key step in the procedure for assessment of the safety of genetically modified foods, it is not a safety assessment in itself. In order to ensure clarity, transparency and a harmonized framework for the authorization of genetically modified food, this notification procedure should be abandoned in respect of genetically modified foods.

Moodie, in Chapter 13, notes that FAO and GM-food-exporting states such as the United States endorse "substantial equivalence" when it comes to comparing GM and non-GM foods. "Substantial equivalence essentially stands for the proposition that if a GM food product looks, feels, tastes and acts the same as its non-GM counterpart, no more need be done." Moodie adds that the problem of scientific standards is raised in the debate over "substantial equivalence."

23 *Ibid.* at Article 12(2) at L 268/11.

24 *Ibid.* at Article 24(2) at L 268/16.

25 *Ibid.* at Article 13 at L 268/11–12 outlines the labeling requirements for food. Article 25 at L 268/17 pertains to the labeling of feed. In the fall of 2004, the EFSA published a guidance document for applications for GMOs entering the EU. This document covered items under *Regulation (EC) No. 1829/2003* and *Directive 2001/18/EC* (deliberate release of GMOs). See European Food Safety Authority, "Guidance Document of the Scientific Panel on Genetically Modified Organisms for the Risk Assessment of Genetically Modified Plants and Derived Food and Feed," (2004) 99 *The EFSA Journal* 1.

26 In the next section, we will devote more attention to this company's position on fish feed that contains vegetable-based materials.

27 Vogel, *supra* note 6.

28 Gøsta Esping-Anderson, *Politics against Markets: The Social Democratic Road to Power* (Princeton, NJ: Princeton University Press, 1985) at 223. The Norwegian electorate rejected EEC membership in 1973 and EU membership in 1994. Over 52 percent of the electorate rejected EU membership in 1994; this was at a time when both Finland (57.0 percent) and Sweden (52.2 percent) voted to join the EU. Shaffer notes that in a 1993 national survey on EU membership, no region in Norway favored this option. The highest percentage in favor was in the Oslo Fjord region (43.8 percent in favor). Disapproval increased among the electorate as one moved further west and north in Norway. Despite overall electoral approval, and subsequent entry of their nations into the EU, the northern populations of Finland and Sweden rejected EU membership. See William Shaffer, *Politics, Parties and Parliaments: Political Change in Norway* (Columbus: Ohio State University Press, 1998) at 98–104. A September 2004 poll by the Sentio-Nortstat polling institute showed over 49.1 percent of Norwegians to be in favor of EU membership, 37.7 percent opposed to membership and 13.2 percent undecided. These results matched those of a poll administered in the summer of 2003. The chair of a pro-EU Norwegian group stated that despite the promising results, it would be better to wait another two years before a third referendum on EU membership was held. See "Norwegian Majority for EU Membership: Poll," (29 September 2004) *EU Business*. Online. Available http://www.eubusiness.com/afp/041029125640. dluopzfp (accessed 27 January 2005).

29 The Economist, *Pocket World in Figures* (London: Profile Books, 2003). Calculated from data at 178–179.

30 These data are based upon figures contained in "Table 1.5: Volume, Value and Average Price of World Aquaculture Production of Atlantic Salmon, 1995–2000" of Éric Gilbert, *The International Context for Aquaculture Development: Growth in Production and Demand, Case Studies and Long-Term Outlook* (Ottawa: Office of the Commissioner for Aquaculture Development, 2002) at 10.

31 The salmon export figures are found at Norwegian Seafood Export Council, "The Norwegian Seafood Export Statistics" (2003). The total seafood export figures are found at Norwegian Seafood Export Council. Online. Available http://www.seafood.no/facts/export (accessed 1 March 2004).

32 John Phyne and Jorge Mansilla, "Forging Linkages in the Commodity Chain: The Case of the Chilean Salmon Farming Industry, 1987–2001," (2003) 43(2) *Sociologia Ruralis* 108.

33 Stig-Erik Jakobsen, "Development of Local Capitalism in Norwegian Fish Farming," (1999) 23(2) *Marine Policy* 117.

34 Directorate of Fisheries, *Key Figures from Norwegian Aquaculture Industry* (Bergen: Directorate of Fisheries, 2001) at 4. The data are based upon statistics in the table entitled "Licenses on 31.12.00." There are still license limitations in place. These are discussed later in this chapter.

35 Phyne, *supra* note 4 at 51–53. In a May 1995 interview, an official with the Irish Salmon Growers Association noted that the Norwegian salmon feed and farmed salmon production records presented to a recent meeting of the EC on import duties did not match. He argued that there should have been greater farmed salmon production, and that the "real figures" were not being revealed. In reference to the mismatch between "feed" and "farmed salmon production," he noted sarcastically that the Norwegian salmon farmers must "be eating the feed themselves." In an August 2002 interview, an official with the Directorate of Fisheries noted that the current feed regulations in the Norway were informed by the EU in 1996. The official noted that "the EU imposed a penalty against Norway – dumping. But before this dumping case came up, Norway tried to convince the EU that Norway pushed on this feed quota, we developed it ourselves." In the fall of 2003, Norway moved to phase out the feed quota regime. See Knut Eirik Olsen and Aslak Berge, "Norwegian Government to Do Away with Feed Quota System," (13 August 2003) *IntraFish.* Online. Available http://www.intrafish. com/article.php?articleID=37240.

36 Richard Apostle, Gene Barrett, Petter Holm, Svein Jentoft, Leigh Mazany, Bonnie McCay and Knut Mikalsen, *Community, State and Market on the North Atlantic Rim: Challenges to Modernity in the Fisheries* (Toronto: University of Toronto Press, 1998) at 102. EFTA now includes Norway, Iceland, Switzerland and Liechtenstein. Switzerland is not part of the EFTA–EU agreement. In 1994, Austria, Finland and Sweden left EFTA to join the EU.

37 K. E. Olsen, "The Strategy That Could Have Cost Norway dearly," (26 April 2002) *IntraFish* at 6. Online. Available http://www.intrafish.com/article. php?articleID=22687. There are already duties on Norwegian farmed salmon exports of 2 percent for fresh and frozen exports and 13 percent for smoked salmon. Commission of the European Communities, *supra* note 4 at 49. The push for an increase in duties was part of the anti-dumping and anti-subsidy investigation conducted in 2002 and 2003. The European Commission concluded in April 2003 that proceedings would be terminated regarding the investigation of anti-dumping and anti-subsidy measures on the part of Norwegian producers, and anti-dumping measures on the part of Faroese and Chilean producers. See Commission of the European Communities, *supra* note 4 at Articles (1) and (2) at 54.

38 Norwegian Seafood Export Council, *supra* note 31.

39 European Commission, *Food and Veterinary Office: FVO: Annual Report* (European Commission, Health and Consumer Directorate – General: Food and Veterinary Office, 2001) at 12 and 17.

40 Fiona Cameron, "Strong EC Move to Cut Exposure to Dioxins," (3 February 2003) *IntraFish* at 24. Online. Available http://www.intrafish.com/article.php? articleID=31484. Cameron is referring to *Council Regulation (EC) No. 2375/2001 of 29 November 2001 Amending Commission Regulation (EC) No. 466/2001 Setting Maximum Levels for Certain Contaminants in Foodstuffs (Text with EEA relevance). Official Journal of the European Communities* L 321/1. That regulation states at L 321/2 (14) that

all operators in the food and feed chain continue to make all possible efforts and do all that is necessary to limit the presence of dioxins in feed and food, the maximum levels applicable should be reviewed within a defined period of time with the objective to set lower maximum levels. An overall reduction of at least 25 percent of the human exposure to dioxins should be achieved by the year 2006.

However, the regulation is cautious on the issue of dioxin levels in farmed fish. The regulation states at L 321/2 (17):

> [M]aximum levels in feeding stuffs for fish means that, farmed fish have significantly lower dioxin levels. Once more data is available, it may in the future be appropriate to lay down different levels of fish and fishery products or exempt categories of fish, insofar they are of limited significance from an intake point of view.

While the regulation notes that dioxin levels are high in the Baltic Sea, it provides for derogation for Sweden and Finland because fish such as Baltic herring and salmon may not comply with maximum levels; if such fish are excluded, this "may have a negative health impact on Sweden and Finland" [L 321/3 (18)]. Because of this, Article 1(1) at L 321/3 amends *Regulation (EC) No. 466/2001* by permitting Sweden and Finland to place on the market fish

> from the Baltic region, which is intended for consumption in their territory with dioxin levels higher than those set in 5.2 of section 5 of Annex 1, provided that a system is in place to ensure that consumers are fully informed of the dietary recommendations with regard to the restrictions on consumption of fish from the Baltic region by identified vulnerable groups of the population in order to avoid potential health risks.

The fish products in question are the same as those identified in *Council Regulation (EC) No. 104/2000, supra* note 17. The interesting point here is that domestic supplies of herring and salmon may have dioxin levels that exceed maximum levels, but be allowed for food consumption in Sweden and Finland, but farmed salmon from Norway and the Faroes partially fed with herring from the Baltic Sea could possibly be banned from EU markets if the salmon has dioxin levels in excess of those permitted in the regulations.

The debate over dioxins in farmed salmon intensified in 2004 in the wake of a publication that argued that the levels of PCBs in farmed salmon from Europe (including Norway and the Faroes) exceeded levels in that from British Columbia and Chile. The authors of the study provided consumption advice for farmed salmon (meals per month), with the lowest consumption advisories for farmed salmon from Scotland, the Faroes and Norway. See Ronald A. Hites, Jeffrey A. Foran, David O. Carpenter, M. Coreen Hamilton, Barbara A. Knuth and Steven J. Schwager, "Global Assessment of Organic Contaminants in Farmed Salmon," (2004) 303, 9 January *Science* 226. This study, especially its recommendations for restricted consumption levels, outraged the global salmon farming community. The newly struck Norwegian Food Safety Authority asked the National Institute of Nutrition and Seafood Research to look at the study. On the basis of a larger Norwegian database than was present in the Hites *et al.* (2004) study, the Norwegian Food Safety Authority argued that the study in *Science* had some weaknesses and that it was not necessary to restrict consumption of farmed salmon. See "The Norwegian Food Safety Authority De-dramatizes Findings concerning Farmed Salmon" (9 January 2004). Online. Available http:www.mattilsyne.no/portal/page?_pageid=34,33401&_dad=portal92&_schema=PO... (accessed 26 January 2005). In June 2004, EFSA held a scientific colloquium on the methodologies concerning the impact of PCBs in food designated for human consumption. The Hites *et al.* (2004) study informed some of the discussions. In fact, David Carpenter (one of the co-authors of the Hites *et al.* (2004) study) gave a presentation at the colloquium. The colloquium concluded that while there was agreement on the science of dioxin toxicology, there was uncertainty in data interpretation that needed to be clarified in future research in order to estimate, among other things, "tolerable intake guidelines." See Scientific Colloquium Series of the European Food Safety Authority, (2004) June No. 1. *Summary Report EFSA Scientific Colloquium: Methodologies and Principles for Setting Tolerable Intake Levels for Dioxins, Furans and Dioxin-Like PCBs,* 28–29 June 2004, Brussels. This more recent debate over dioxins will most likely inform future EC regulations on acceptable dioxin levels for food such as farmed salmon.

41 Interview, 9 August 2002.

42 Interview, 1 August 2002. This company has acquired the operations of several large salmon farming companies in Norway, Scotland, Ireland and Chile.

43 *Ibid.*

44 Another issue is that pelagic fishing companies in Norway are not totally dependent on feed companies for a market outlet. The pelagic fleet also has access to firms that process fishmeal production for pet food. Pelagic vessels also have the option of selling better-quality pelagic stocks (e.g. herring) for human consumption. One pelagic company (four trawlers) in Austevoll sells two-thirds of its catch for human consumption – largely to Poland. The other one-third (mostly blue whiting) is sold for fishmeal. Interview, 5 August 2002.

45 Interview, 8 August 2002. This official noted that his firm was the last to use vegetable protein and oil in fish feed.

46 *Regulation (EC) No. 1829/2003, supra* notes 23 to 26.

47 See Commission of the European Communities, *Proposal for a Regulation of the European Parliament and of the Council on Genetically Modified Food and Feed* (Brussels: Commission of the European Communities, 2001) for an extended discussion on the need to regulate GMOs. The use of risk assessment science is noted as having a critical role in the detection of GMOs. Also see *Regulation (EC) No. 1829/2003, supra* notes 22–25.

48 *Supra* note 42.

49 *Regulation (EC) No. 1829/2003, supra* note 22.

50 Directorate of Fisheries, *Own Check Guide: For Use in Establishments Producing Fish and Fishery Products* (Bergen: Directorate of Fisheries, Department of Aquaculture, 1999), and Directorate of Fisheries, *Norwegian Quality Regulations Relating to Fish and Fishery Products* (Bergen: Directorate of Fisheries: Department of Aquaculture, 1999). A fish veterinarian with the Directorate of Fisheries indicated that Norwegian food safety governance is in the process of being centralized into a new regulatory body. This body emerged in 2004 as the Norwegian Food Safety Authority. This body (like the Directorate of Fisheries) is under the jurisdiction of the Ministry of Fisheries and Coastal Affairs. The Norwegian Food Safety Authority is charged with the responsibility to "ensure that Norwegian seafood is produced in accordance with Norwegian obligations under international agreements" (at 8). See Ministry of Fisheries and Coastal Affairs, *The Riches of the Sea: Norway's Future* (Oslo: Norwegian Ministry of Fisheries and Coastal Affairs, 2004). A basic outline of the role of the Norwegian Food Safety Authority can be found on the organization's website. See "This Is the Norwegian Food Safety Authority." Online. Available http://www.mattilsynet.no/portal/page?_pageid=34,33401&_dad=portal92&_schema=PO. . . . (accessed 26 January 2005). Any future development in Norway's food safety policies for farmed salmon will most likely be informed by this authority.

51 *Ibid.* The HACCP requirements discussed in these documents are emerging as required international standards in the salmon farming industry in particular and the food industry in general. Intesal, the research arm of SalmonChile (formerly known as the Chilean Salmon and Trout Producers Association), is providing HACCP training to Chilean salmon farming companies so their employees can "upgrade" their skills in processing farmed salmon for the American market. See Phyne and Mansilla, *supra* note 32 at 115–116.

52 Directorate of Fisheries, *Own Check Guide, supra* note 50.

53 *Supra* note 41.

54 Vogel, *supra* note 6 at 184–195. Vogel argues that the SPS Agreement is contentious because many nations use food standards not for safety purposes, but in order to protect domestic markets for local producers. The SPS legislation is at the heart of some recent cases before the WTO. See McDorman and Ström, Chapter 11, this volume, note 6. Of particular interest here was the dispute between Canada and Australia over the fish health effects of live salmon exports from the former to the latter. We will discuss this case later in the chapter because of its possible pertinence to a recent agreement between the EU and the Faroes over the exportation of ova from the former to the latter.

55 *Supra* note 41.

56 Austevoll was the site of our field research in Norway in August 2002 and March 2003. Austevoll is a coastal community of 4,400 situated one hour by ferry south of Bergen. It is a series of island communities shaped like a semicircle. The islands are connected by fishing and aquaculture. There is also a developed service sector with links to both of these industries. Some Austevoll firms also have connections with the Norwegian oil and gas industry. For a fuller discussion of our research methodology and findings pertaining to Austevoll, see John Phyne, Gestur Hovgaard and Gard Hansen, "Norwegian Salmon Goes to Market: The Case of the Austevoll Seafood Cluster" (2006) 22(2) *Journal of Rural Studies* 190.

57 Interview, 5 August 2002.

58 *Ibid.*

59 *Ibid.*

60 Interview, 8 August 2002.

61 *Ibid.*

62 C. Richardson, "Mighty Buying Power" (March 2002), *IntraFish* at 30. Of course, given the power of Carrefour in global food chains, one can reasonably assume that "French smokers" were "squeezed" by a combination of high ex-farm gate prices for salmon and low prices from supermarket chains such as Carrefour; given their subordinate role in global food chains, "smokers" most likely rely upon lower ex-farm gate prices for salmon in order to increase their profit margins. In 2002, most Norwegian salmon farming companies lost money before taxation. Firms with fifteen or more licenses (6.7 percent), five to nine licenses (13.8 percent), two to four licenses (7.8 percent) and one license (10.4 percent) had significant losses. The smallest losses were for firms that had between ten and fourteen licenses (0.1 percent). The latter were also the only firms that had a positive return (2.6 percent) on total assets. See Directorate of Fisheries, "Preliminary Results from the Profitability Survey on Norwegian Fish Farms in 2002: Salmon and Trout" (2003). Online. Available http://www.fiskeridir.no/english/pages/statistics/index.html (accessed 1 March 2004).

63 Jøgvan Mørkøre, "Faroese Fishing Industry at the Crossroads: Staying outside the EU," in Lawrence Felt and Ragnar Arnason, eds., *The North Atlantic Fisheries: Successes, Failures and Challenges* (Charlottetown, PEI: The Institute of Island Studies, 1995) at 137. Mørkøre shows that between 1984 and 1992,

opposition to EU membership declined in the Faroes (at 141). Nevertheless, at the time of writing, there was no concerted movement in the Faroes in favor of imminent EU membership.

64 *Ibid.* at 138.

65 Gestur Hovgaard, *Coping Strategies and Regional Policies: Social Capital in the Nordic Peripheries – Country Report – Faroe Islands* (Stockholm: Nordregio, 2002) at 8.

66 Gilbert, *supra* note 30 at 10.

67 For an analysis of the origins, growth and restructuring of the Faroese salmon farming industry, see Richard Apostle, Dennis Holm, Gestur Hovgaard, Ólavur Waag Høgnesen and Bjarni Mortensen, *The Restructuration of the Faroese Economy: The Significance of the Inner Periphery* (Frederiksberg, Denmark: Samsfundslitteratur Press, 2002) at 125–153.

68 *Ibid.* at 133.

69 Richard Apostle and Gestur Hovgaard (2003), "Global Commodity Chains and Faroese Aquaculture: Economic Integration and Institutional Buffers," unpublished draft manuscript at 11–14. Available from the lead author at the Department of Sociology and Social Anthropology, Dalhousie University, Nova Scotia, Canada.

70 Faroe Seafood, *Faroe Seafood* (Tórshavn, Faroe Islands, n.d.).

71 Apostle and Hovgaard, *supra* note 69 at 4–7.

72 Ari Johanneson, *Fish Farming in the Faroe Islands: Measures Taken to Minimise the Impacts of Salmon Aquaculture on Wild Stocks* (Tórshavn: Ministry of Trade and Industry of the Faroe Islands, n.d.) at slide 17 of Power Point presentation.

73 *Council Regulation (EC) No. 2375/2001, supra* note 40 at L 321/2 (14).

74 Johanneson, *supra* note 72 at slide 27.

75 *Ibid.*

76 Apostle and Hovgaard, *supra* note 69 at 10. Later, the outbreak of infectious salmon anemia that entered the Faroes in 2001 spread, to negatively impact the entire salmon farming industry. We will briefly discuss this later in the chapter.

77 *Ibid.* Since this agreement deals with trade issues, the SPS Agreement within the WTO is relevant. Any attempts by the Faroes to restrict ova and smolts can run the risk of running foul of the SPS Agreement. McDorman and Ström, in Chapter 11 of this volume, refer to a dispute between Canada and Australia over the latter's refusal to accept live salmon imports from Canada on the grounds that diseases could be spread from Canadian salmon to Australian salmon. The 1996 risk assessment report completed by the Australian government was viewed to be contrary to SPS measures because it contained no scientific evaluation of the danger of entry, establishment or spread of diseases.

78 Johanneson, *supra* note 72 at slide 35.

79 *Ibid.*

80 *Ibid.* at slide 38.

81 *Ibid.* at slide 15.

82 Interview, 19 October 2002.

83 *Ibid.* On the dumping issue, see V. Solsletten, "Mighty Buying Power," (November 2002) *IntraFish* at 22.

84 Interview, 13 May 2003.

85 *Ibid.*

86 *Ibid.*

87 *Ibid.*

88 *Ibid.*

89 *Ibid.*

90 Interview, 13 May 2003.

91 *Ibid.*
92 Vogel, *supra* note 6. One of the implications of Vogel's analysis is that the internationalization of food safety standards will be reflected in national regulations.
93 S. Henson and R. Loader, "Barriers to Agricultural Exports from Developing Countries: The Role of Sanitary and Phytosanitary Requirements," (2001) 29(1) *World Development* 85.
94 *Regulation (EC) No. 178/2002, supra* note 19 at L 31/10, Article 13. In addition, the SPS Agreement contains measures to account for the need for developing countries to have a time frame within which to upgrade their food safety standards (Article 10: Special and Differential Treatment is the relevant section). See WTO, *supra* note 5 at 74.
95 See Raphael Kaplinsky and Jeff Readman, *Integrating SMEs in Global Value Chains: Towards a Partnership for Development* (Vienna: United Nations Industrial Development Organization, 2001). While the authors' arguments for "upgrading" assistance are applied to small and medium-sized enterprises (SMEs) in the "developing world," firms in peripheral regions in the "developed world" may also need "upgrading" assistance, especially since standards are increasingly globalized. When failures to "upgrade" at the points of production and processing are combined with transportation costs, these can converge to further marginalize SMEs in peripheral regions of even the richest "developed" countries.
96 There is a huge literature in this area. Some notable examples include Banks and Marsden, *supra* note 13; Peter Gibbon, "Agro-commodity Chains: An Introduction," (2001) 32(3) *IDS Bulletin* 60; John Humphrey and Hubert Schmitz, "Governance in Global Value Chains," (2001) 32(3) *IDS Bulletin* 19; Raphael Kaplinsky and Michael Morris, *A Handbook for Value Chain Research* (Brighton: Institute for Development Studies, 2002); and Kaplinsky and Readman, *supra* note 95. In addition, in recent years, *Rural Sociology*, *Sociologia Ruralis* and *World Development* have published articles dealing with the shifting nature of power in the food chain.
97 In the case of the salmon farming industry, industrial upgrading is occurring in Chile; see Phyne and Mansilla, *supra* note 32 at 115–116.
98 *Ibid.*
99 *Ibid.* at 117–118.
100 In the fall of 2002, Pan Fish was declared insolvent and taken over by a large Norwegian bank. Pan Fish was viewed as overextending itself in the late 1990s, especially in terms of its pelagic fishing investments, notably the consolidation of one of Norway's largest purse seine companies under Pan Pelagic. Berge notes that "[w]ith less than 50 percent of the shares in a long line of purse seiners the aquaculture company was caught without being able to harvest any profit from the fishing boats accounts at the same time as considerable capital was tied up." This was coupled with the fall in market prices for farmed salmon in 2001 and 2002. See Aslak Berge, "Bank's Financing Proposal Accepted without a Hitch," (January 2003) 2(1) *IntraFish* at 23. Since this time, Pan Fish has restructured to become a smaller firm with less global reach. The failure of Pan Fish and the difficulties experienced by Fjord Seafoods suggest that firms that have smaller production schedules, less geographical scope as well as access to export markets may be better positioned in the foreseeable future than large-scale players that simultaneously attempt a greater global reach and corporate consolidation of most aspects of the food chain.
101 Aslak Berge, *The Rise and Fall of the Pan Fish Empire: Industry Report* (Bodø, Norway: IntraFish Media AS, 2002) at 10. Pan Fish, in conjunction with an

Austevoll firm, is still trying to develop this feed processing facility. For more details, see Phyne *et al.*, *supra* note 56.

102   Interview, 9 August 2002.

103   Directorate of Fisheries, *supra* note 34 at 4. The data are based upon the statistics in the table entitled: "Licenses on 31.12.00."

104   *Ibid.*

105   *Ibid.*

106   Interview, 27 February 2002.

107   *Supra* note 102.

108   Directorate of Fisheries, *supra* note 103.

109   *Supra* note 102.

110   *Ibid.*

111   "New Salmon Licenses in Norway: Not the Best of Times," (31 January 2003) *IntraFish*. Despite the concerns at this time, the EC did not consider the production from these licenses to be of concern until 2004. Moreover, the EC indicated that additional production would be absorbed by emerging markets such as Russia and China. See Commission of the European Communities, *supra* note 4 at 22. Inevitably, given the size of the Norwegian industry, any additional licenses will still be held in suspicion by producers in Ireland and Scotland. As a result, one should not rule out further action by these nations against Norway. In March 2004, the Norwegian government stated that no new production licenses were to be allocated for the balance of 2004 or for 2005. See Bent Are-Jensen, "Norway's Bid to Soothe EU: No New Licence Allocations," (15 March 2004) *IntraFish*. Online. Available http://www.intrafish.com/print.php?articleID=42807 (accessed 15 March 2004).

112   For an analysis, see Petter Holm and Svein Jentoft, "The Sky Is the Limit?: The Rise and Fall of Norwegian Salmon Aquaculture," in Conner Bailey, Svein Jentoft and Peter Sinclair, eds., *Aquacultural Development: Social Dimensions of an Emerging Industry* (Boulder, CO: Westview Press, 1996) at 23.

113   Moreover, despite the decline of the Labor Party's hegemonic position in Norwegian politics, an ongoing lesson in Norway is that political parties that attempt to implement strictly center-right policies while in power may not govern very long. This was the fate of the Conservative Party in the 1980s. Furthermore, since the 1990s, the Christian Democrats (formerly Christian People's Party) and the Center Party have maneuvered between the traditional left- and right-wing power blocs in Norwegian politics. The end result is that the Norwegian welfare state and its redistributive policies command wide support from most parties along the political spectrum. See Shaffer, *supra* note 28.

114   See Max Weber, *The Protestant Ethic and the Spirit of Capitalism*, 3rd ed. (Los Angeles: Roxbury Publishing Company, 2002) at 80–89. Weber argues that among the Protestant sects, German Pietism was more developed than Lutheranism in providing a methodical orientation necessary for the "this-worldly" asceticism crucial for the development of capitalism. Yet he argues that, in contrast to Pietism, "the virtues cultivated by Calvinism appear to stand in a relationship of greater elective affinity to the restrained, strict, and active posture of capitalist employers of the middle class" (at 89). In addition to the municipal official's comments, similar observations were made by the owner of a pelagic fishing fleet on the importance of religious orientation to day-to-day business. Of course, more detailed research would be needed to establish the importance of religious orientation to entrepreneurship in Austevoll.

115   Interview, 10 March 2003 and *supra* note 57.

116   Interview, 2 August 2002. Jakobsen, *supra* note 33 at 125, also notes that

Austevoll firms collectively purchase inputs. Nevertheless, he argues that these firms pursue largely individualistic strategies.

117 This issue is further explored in Phyne *et al.*, *supra* note 56.

118 Interview, 6 August 2002.

119 Statistics Norway, "Employed 16–74 Years, in Commuting and Time, 1244 Austevoll," Statistics Norway. Online. Available http://www.ssb.no/English (accessed 15 March 2004). The manager of a fish processing plant noted that there are 100–110 individuals employed in the plant. Approximately 50 percent of these are full-time. Most of the full-time employees are "locals," but there are 8–10 from countries such as England, France, the Philippines, Thailand, Somalia and Ghana. Over 90 percent of the seasonal workers ($n = 50$) are from Sweden. Some plants in Austevoll process only pelagics, whereas other plants process farmed salmon and pelagic products.

The use of migratory labor is not uncommon in the Norwegian fish processing industry. In the northern Norwegian community of Botsfjord, Tamil workers are employed in the fish processing plants. See Gestur Hovgaard, *Globalisation, Embeddedness and Local Coping Strategies: A Comparative and Qualitative Local Study of Local Dynamics in Contemporary Social Change* (Roskilde, Denmark: Department of Social Science, University of Roskilde, 2001). In contrast to the fish processing industry, workers employed in the fish harvesting industry are almost exclusively from Nordic backgrounds. An owner of a pelagic fishing fleet noted that changes must be made to the Nordic labor agreement that gives priority to residents of Nordic countries in the area of fish harvesting work. He stated that this contributes to high labor costs. This individual would like to employ East Europeans. *Supra* note 44.

120 Apostle and Hovgaard, *supra* note 69 at 11.

121 *Ibid.* at 10.

122 Apostle *et al.*, *supra* note 67 at 140–151.

123 Austevoll Havfiske ASA, *Årsrapport 2001* (Storebø: 2002) at 2; Pan Fish ASA, *Annual Report 2000* (Ålesund, Norway, 2001) at 11 to 12. Pan Fish's pelagic investments are based upon a 19.9 percent holding in Austevoll Havfiske ASA. However, as noted earlier, the extent to which a restructured Pan Fish remains involved in such investments remains to be seen.

124 Apostle and Hovgaard, *supra* note 69 at 17.

125 *Ibid.* and Apostle *et al.*, *supra* note 67.

126 Apostle and Hovgaard, *supra* note 69 at 11–14.

127 Hovgaard, *supra* note 65 at 25–28.

128 *Ibid.* at 7.

129 Apostle *et al.*, *supra* note 67. This argument is a theme that runs through all of the case studies.

130 *Ibid.* at 170.

131 *Supra* notes 84 and 90.

132 *Supra* note 84.

133 *Supra* note 90.

134 Jógvan H. Gardar, "ISA: A Late Arrival to the Faroes That's Settled Well," (9 August 2004) *IntraFish*. Online. Available http://www.intrafish.com/print. php?articleID=46384 (accessed 19 January 2005).

135 This information was provided by the Center for Local and Regional Development, which has offices in Klaksvík on the island of Borðoy and in Vágur on the island of Suðuroy.

136 Jógvan H. Gardar, "Faroe Islands Salmon Adventure over by 2006," (9 August 2004) *IntraFish*. Online. Available http://www.intrafish.com/print.php?articleID=46384 (accessed 19 January 2005).

137 Jógvan H. Gardar, "Vestsalmon CEO: 'Faroe aquaculture recovering'," (4

January 2005) *IntraFish*. Online. Available http://www.intrafish.com/print. php?articleID=50349 (accessed 19 January 2005).

138 "Faroe Islands Vaccinating All Spring Smolt in 2005 against ISA," (9 November 2005) *IntraFish*. Online. Available http://www.intrafish.com/print. php?articleID=48890 (accessed 19 January 2005).

139 Apostle *et al.*, *supra* note 67 at 27–74.

140 Unfortunately, we do not have quantitative data to further verify this statement. This observation comes from knowledge gleaned by researchers at the Center for Local and Regional Development. Some of the unemployed workers have returned to Thailand.

141 Vogel, *supra* note 6.

142 This is also the argument of Adair Turner in his *Just Capital: The Liberal Economy* (London: Pan Books, 2002).

143 Richard Apostle, Gene Barrett *et al.*, *Emptying Their Nets: Small Capital and Rural Industrialization in the Nova Scotia Fishing Industry* (Toronto: University of Toronto Press, 1992).

144 Moodie, Chapter 13 of this volume.

145 Government of British Columbia, Ministry of Agriculture, Forestry and Fisheries, "Fisheries Statistics: Salmon Aquaculture in British Columbia." Online. Available http://www.agf.gov.bc.ca/fish_stats/aqua-salmon.htm (accessed 24 March 2004).

146 These are Canada's Department of Fisheries and Oceans data for 2002. In that year, British Columbia led the country, with 90,000 tonnes of salmonid production (Atlantic and Pacific), New Brunswick had 39,450 tonnes (over 98 percent Atlantic), Nova Scotia had 2,385 tonnes (80 percent Atlantic) and Newfoundland 2,870 tonnes (almost 56 percent Steelhead). Fisheries and Oceans Canada, "2002 Canadian Aquaculture Production Statistics (tonnes)." Online. Available http://www.dfo-mpo.gc.ca/communic/statistics/aqua/aqua02 _e.htm.

147 Alain Meuse, "Anti-Terrorism Bill: Amendments a Boon to Seafood Industry," (1 November 2003) 36(2) *The Sou'Wester* at 3.

148 Report of the Commissioner for Aquaculture Development, *Achieving the Vision* (Ottawa: Office of the Commissioner for Aquaculture Development, 2003) at 39.

149 Commission of the European Communities, *supra* note 4 at 49.

150 Moodie, Chapter 13, this volume.

151 In addition to transgenic fish, any major shift to the use of genetically modified vegetable oils for use in fish feed will act as a barrier to gaining access to EU markets. Transgenic fish partially raised on GMOs in feed may have good survival rates but limited markets.

152 As the Chileans are entering Europe, the Norwegians are diversifying their still small Asian export market by moving into China. For more on Norway's export strategies in China, see Gard Hopsdal Hansen, "Fortune Fish and Boomerang Internationalization: Norwegian Activity and Local Response in China." Unpublished draft manuscript, available from the author at the Department of Interdisciplinary Studies of Culture, Norwegian University of Science and Technology, Trondheim, Norway.

13 Transgenic fish

Some Canadian regulatory issues

Douglas Moodie

Introduction

We live in a world inundated with transformative technologies. One of those is the genetic modification of living organisms for food purposes. Huge increases in the acreage of genetically modified (GM) crops have occurred since the mid-1990s.[1] More than 175 million hectares worldwide have now had GM crops grown on them at one time or another.[2] This "first generation" of GM food crops has consisted of plants with few and relatively straightforward trait changes. But this is likely to change soon.[3] It has been estimated that "the global market for biotechnology applications will reach [CDN]$50 billion annually by 2005 ... and the strongest growth is projected for the agri-food sector."[4]

The research and production emphasis in the international sphere, as in the domestic one, has so far been heavily geared to GM plants and the human food and livestock feed derived from those plants. Work involving GM animals lags far behind that involving plants.[5] Technical issues, including the greater complexities of animal biological systems, are in part responsible for this. The development of transgenic fish[6] represents something of an exception to this situation. Research on transgenic fish commenced several decades ago and "has occurred at a very rapid pace."[7] By 2000, as many as thirty-five different species of fish had been the subject of genetic manipulation.[8] The first transgenic fish research in Canada was focused on an antifreeze protein gene to permit the rearing of Atlantic salmon in the icy waters off Canada's east coast.[9] The majority of Canadian research, however, has been on increasing growth rates, with almost all the effort directed at Atlantic and Pacific salmon.[10] This research has produced a genetically altered Atlantic salmon that has a growth rate reputed to be as much as 600 percent greater than its non-GM counterpart.[11] Technological breakthroughs hold out the promise of revolutionizing the commercial production of salmon, and perhaps numerous other species. So far, though, transgenic fish remain at the pre-commercial stage.[12]

GM foods have fueled significant discourse, in Canada and abroad, in recent years. Two major contemporary Canadian studies of biotech food regulation have resulted. First, in August 2002 the Canadian Biotechnology Advisory Committee (CBAC)[13] issued a report on the regulation of GM foods in Canada. The *CBAC Report* followed almost two years of research, stakeholder and public consultations, and public comments on specific draft recommendations presented in an interim report published in August of 2001. Second, the Royal Society of Canada (Royal Society)[14] in early 2000 assembled an Expert Panel (Royal Society Expert Panel) to provide advice on certain food biotechnology[15] regulatory issues. The Royal Society Expert Panel issued its report in January 2001. The federal government published an action plan[16] in response to the *Royal Society Report*, as well as six (to January 2005) progress reports[17] on that action plan.

The *CBAC* and *Royal Society Reports*, together with the action plan and progress reports, represent intensive Canadian government-sponsored initiatives to study the issues swirling around GM foods.[18] The heightened attention paid to GM foods in Canada in the past three years or so has been prompted by several factors. First, as already mentioned, recent leaps in biotechnology have the potential to make available a much greater array of GM foods, with more complex characteristics, than the world has previously known.[19] Second, there are growing economic[20] and social[21] forces urging the broad acceptance and usage of GM foods and food products. Third, there is a fast-evolving awareness among the consuming public of the presence of GM foods and food products in their diets, of the potential for significant increases in that presence, and of the health, environmental and social/ethical issues connected with GM foods and food products.[22]

The purpose of this chapter is to examine aspects of the Canadian regulation, present and prospective, of transgenic fish. The emphasis of the examination is on three interconnected regulatory areas: food safety, environmental protection and consumer choice.[23] Transgenic fish regulation in Canada represents a discrete and valuable area of study for several reasons. The first is that research organizations with strong Canadian ties have been at the forefront of global efforts to develop the commercial potential of transgenic fish.[24] The second is that Canada has an already significant aquaculture industry that is poised, some predict, to grow exponentially in future decades.[25] Third, the emergence of an explicit, broad discourse regarding GM foods creates a backdrop that inevitably will influence, one way or another, any integration of transgenic fish into Canada's traditional aquaculture industry. This combination of factors suggests that the Canadian experience, and the decisions ultimately made by Canada *vis-à-vis* commercial production of transgenic fish, will be both intrinsically interesting and of instructional value to other nations traveling similar paths.

The first part of this chapter considers the international background and the laws, policies and mechanisms developed and continuing to be developed by the global community to address food safety, environmental protection and consumer choice issues relating to GM foods (and, particularly, transgenic fish). The second part of the chapter looks at the domestic situation. What regulatory structures and procedures are currently in place? How have international initiatives and directives been incorporated, or hold promise to be incorporated, in the Canadian context? To what extent has the recent governmental and extra-governmental attention paid to GM food issues so far shaped Canada's regulatory structures and procedures? How might one expect those structures and procedures to be affected, both by the formal government-sponsored discourse and by broader societal forces and opinions? How might the answers to the foregoing "GM food" questions vary, if at all, for transgenic fish?

The chapter ultimately attempts some conclusions on the sort of regulatory regime commercial transgenic fish producers might expect to face in Canada in three to five years, and beyond. It also offers some thoughts on the critical factors that may make the difference between a future multi-billion-dollar industry and one that never gets off the proverbial drawing board.

International laws, policies and mechanisms

Overview

The international context of GM food regulation is a complex one, involving various organizations each with its own unique mandate and agenda. The *CBAC Report* identifies several international bodies that are currently involved in "the coordination and regulation of biotechnology products."[26] These include the Food and Agriculture Organization (FAO), the World Health Organization (WHO), the Codex Alimentarius Commission,[27] the World Trade Organization (WTO), the Organisation for Economic Co-operation and Development (OECD) and the Secretariat of the Convention on Biological Diversity.[28] These bodies "cover a spectrum of functions from ... set[ting] science-based standards to ... broader objectives such as food security, trade facilitation, environmental protection, and other social and political goals."[29] It is the stated view of the CBAC that Canada demonstrates active participation in all these international bodies, and "at times lead[s] the efforts to develop international consensus on matters of science, governance and/or policy."[30] Over the past three or four years, this plethora of international organizations, along with several expert panels and commissions in various states, has produced numerous reports on regulatory issues related to GM foods.[31] Of the several international organizations involved with GM food regulation, three seem to be particularly conspicuous. They are the FAO, WHO and WTO.

Food safety

The existing international regimes and procedures for monitoring and regulating food safety are extensive and well established, though they do not much differentiate between GM and non-GM foods. It can be anticipated, therefore, that transgenic fish will be treated, at the international level, in a manner similar to that applied to any novel food product. International food safety regulation seems to have two facets: one is centered on the development and dissemination of food standards and guidelines; the other has a strong trade orientation. The precise interaction of these two facets is not an easy thing to discern.

The Codex Alimentarius Commission (Codex) is at the heart of the food standards/guidelines facet of the international regulatory effort. Codex has been at its task for more than four decades. Its secretariat is provided by the FAO and WHO. Codex is an intergovernmental body with 165 member countries.[32] Its primary functions are stated to be "protect[ing] the health of consumers, ensur[ing] fair practices in food trade and promot[ing] the coordination of food standards."[33] The explicit "food trade" function indicates that the Codex Alimentarius Commission's mandate may blur into those of the WTO and its related bodies. It is not abundantly clear what Codex does, independent of the WTO regime, to ensure "fair trade practices in the food trade." It may be that this trade aspect of Codex's mandate is a holdover from a period before the WTO took primary jurisdictional control over trade-related matters in the international forum. On the other hand, with the involvement of Codex in the establishment of global food labeling rules,[34] there exists an inextricable connection with trade issues. Whether or not a product is labeled, and what the content of that labeling is, have a direct bearing on the international trade in that product.

The FAO and WHO have furnished to the international community "expert scientific advice on the food safety aspects of foods derived from biotechnology since 1991."[35] The FAO and WHO, through Codex in 1999 established the *ad hoc* Intergovernmental Task Force on Foods Derived from Biotechnology (the CAC Task Force). The stated objective of the CAC Task Force is to consider the health and nutritional implications of biotech foods and, in particular, to

> develop standards, guidelines or recommendations, as appropriate, for foods derived from biotechnology or traits introduced into foods by biotechnology, on the basis of scientific evidence, risk analysis and having regard, where appropriate, to other legitimate factors relevant to the health of consumers and the promotion of fair trade practices.[36]

Again one sees the "fair trade practices" angle occupying a central place. The CAC Task Force submitted reports to Codex in 2001, 2002 and 2003.[37]

The joint FAO/WHO work on the human health evaluation of GM foods uses for its scientific foundation a series of expert consultations on the safety and nutritional aspects of such foods. These expert consultations are stated to be "completely independent from the inter-government negotiation process, and treat the subject from a pure scientific perspective."[38] They aim to furnish scientific advice, to review existing GM food assessment strategies and to offer recommendations on further research and evaluation needs.[39]

The structure of Codex's approach to the assessment of biotech foods starts with science-orientated output from various expert consultative bodies. That output typically moves to the CAC Task Force, which uses it to help mold standards, guidelines and recommendations for ultimate consideration by the Commission as a whole. The overall process seems comprehensive, if somewhat lumbering. And it is directed at constantly moving targets, as different foods with different characteristics produced through different means stream onto the stage. Ultimately, Codex does no more than gather and disseminate information about the "potential benefits and possible risks associated with the application of modern technologies to increase plant and animal productivity and production."[40] It is left to the governments of individual states to create and implement their own policies regarding these technologies. They can wholeheartedly embrace the Codex's standards, guidelines and recommendations, disregard them completely or use them with whatever modifications they consider appropriate.

Notwithstanding the non-binding status of Codex's efforts, they are significant for a couple of reasons. First, the scientific information generated has a great deal of value for developing states that do not have the financial and/or technical means to do their own science. Even for states that do, the scientific information emanating from Codex can serve as the basis for policy decisions or, at the least, for further independent research efforts. Second, the standards and guidelines set by Codex have persuasive international force because of the multilateral process involved and the level of expertise drawn upon. States that set up standards that substantially exceed those recommended by Codex do so at the risk of incurring domestic and/or international criticism for "overdoing it." Conversely, states that have standards markedly below those suggested by Codex greatly increase the chances of domestic complaints and/or of actions abroad to block entry of their "substandard" products. To the extent, therefore, that the Codex food safety assessment process is viewed as scientifically legitimate and independent of distorting factors (such as political ones) – and that, by and large, seems to be the case at the present time – what it has to say about biotech foods will carry serious weight.

Environmental protection

From an environmental perspective, the difficulty with transgenic fish is that no one yet has much idea what impact they might have on natural ecosystems. Regulatory initiatives are therefore to a large extent "grasping at shadows." Some environmentalists suggest scenarios of catastrophic proportions. Greenpeace Canada, presently the most active Canadian environmental non-governmental organization (ENGO) on the topic of GM food, considers the genetic alteration of fish to be "high risk technology with potentially disastrous consequences if the GE fish escape into the environment."[41] Citing past experience with invasive species and specific research findings on transgenic fish, Greenpeace Canada has suggested that "even a single fertile GE fish could be sufficient to destroy a local population under certain circumstances."[42] The Royal Society Expert Panel, not so dramatically disposed as Greenpeace, nevertheless underscored the serious dearth of information available. It concluded that "[b]ased on the limited research that has been published to date ... there is little, if any, empirical basis upon which one can reliably predict the outcome of interactions between wild and GM fish."[43] The complicating factor with this issue is the difficulty involved in achieving a satisfactory level of assurance, given that the only real way of doing so is to engage in comprehensive field trials – inherently risky business.

The research to date on the overall effects of transgenesis of fish shows "deleterious consequences to fish morphology, respiratory capacity, and locomotion associated with the introduction of growth hormone (GH) gene constructs in some transgenic variants of salmonids, notably Pacific and Atlantic salmon."[44] The concern here is that transgenic fish that cannot swim, forage or breathe properly not only may have difficulty thriving in a fish farm setting, but, if released into the environment, intentionally or not, could have detrimental effects on native populations through the exchange of genetic material ("introgression"). Escapes from aquatic fish farms are common,[45] and so any rearing of transgenic fish in sea-based open net-cages can be anticipated to result in environmental incursions of significant numbers of those fish. In addition to the concerns surrounding introgression, escaped transgenic fish might also have more direct and immediate ecosystem impacts through "hyper-predation" (eating everything in sight) and migration to novel habitats.

The international concern with protecting the environment from adverse effects of GM fish has both general and specific aspects. On the general side, since at least 1992 the concept of "precaution" has guided the international approach to environmental stewardship. What is now commonly referred to as the "precautionary principle" or "precautionary approach" was enshrined as, and seemingly popularized by, Principle 15 of the Rio Declaration on Environment and Development.[46] While the Rio Declaration was not the first time the idea of a principle of precaution found its way into an international agreement,[47] it certainly seems to have been energized by Rio. The broad acceptance of the Rio Declaration among states and the steady percolation of its concepts and principles downward into the strata of citizenry of those states lends support to the suggestion that the Rio Declaration represented a kind of international "codification" of the precautionary principle. In the decade since Rio, reference to the precautionary principle/approach in the international environmental arena has become increasingly common. Academics have noted the precautionary principle's incorporation "in virtually every recently adopted treaty and policy document related to the protection and preservation of the environment."[48] Even some members of the international judiciary have taken "precaution" into their lexicon.[49]

Notwithstanding the broad international acceptance of the precautionary principle, there remains uncertainty as to precisely what it embodies. It is said to represent a "better safe than sorry" attitude and approach. It has generated voluminous literature, with legal scholars toing and froing on its precise meaning and utility. There is no consensus; indeed, the views of academia on the precautionary principle seem to span the spectrum. Some commentators have categorized it as a tool of great import in ongoing efforts to protect the environment; others have dismissed it as being too loose and ambiguous to be of any real use.[50]

Specific rules and standards aimed at protecting the environment from possible negative ramifications of transgenic fish have emerged and/or evolved in the international forum. Bodies charged with creating and promulgating these rules and standards have adopted "precaution" as their mantra. The North Atlantic Salmon Conservation Organization (NASCO),[51] the International Council for the Exploration of the Seas (ICES),[52] and the FAO have all become directly involved in the business of safeguarding the seas from unwanted infiltration by GM fish.

In 1997, the Council of NASCO, its governing body, agreed to certain guidelines relating to the development of transgenic salmon.[53] The NASCO Guidelines obligate convention parties to, among other things, "take all possible actions to ensure that the use of transgenic salmon, in any part of the NASCO Convention Area, is confined to secure, self-contained, land-based facilities" and to "advise the NASCO Council of any proposal to permit the rearing of transgenic salmonids and provide details of the proposed method of containment and other measures to safeguard the wild stocks."[54]

In September 1995, ICES enacted its Code of Practice on the Introductions and Transfers of Marine Organisms,[55] which requires that "wherever feasible, initial releases of GMOs be reproductively sterile in order to minimize impacts on the genetic structure of natural populations."[56] The *Royal Society Report* noted that

> the Working Group of ICES that deals specifically with transgenic organisms issued, in 1997, a qualifying recommendation that "[u]ntil there is a technique to produce 100 percent sterilization effectiveness, GMO[s] should not be held in or [in] connection with open water systems."[57]

The ICES Code also imposes a requirement to notify member countries prior to the intentional environmental release of a genetically modified organism (GMO), encourages establishment of strong domestic measures to protect ecosystems from harmful effects of releases of GMOs and urges members to undertake research to evaluate the effects on the environment of released GMOs.

The Code of Conduct for Responsible Fisheries[58] (FAO Code) was adopted on 31 October 1995 by the Conference of the United Nations Food and Agriculture Organization. A segment of the FAO Code relates specifically to aquaculture development. Article 9.3.1 speaks directly to the issue of protection of the environment from the potentially adverse effects of genetically altered organisms.[59]

The NASCO Guidelines, ICES Code and FAO Code constitute non-binding general obligations on the Canadian government to defend ecosystems and biodiversity from any possible negative implications of transgenic fish. They do not have "the force of law" behind them in the sense they would if they were embodied in a convention. That said, there exists not insignificant moral compunction on Canada to adhere to these rules and guidelines in shaping its domestic approach to regulating transgenic fish.

The Biodiversity Convention[60] and the Cartagena Protocol made pursuant to the convention,[61] unlike the NASCO Guidelines, ICES Code and FAO Code, upon ratification by a state do constitute legally binding obligations on that state.[62] The primary purposes of the convention, as set out in Article 1, are "the conservation of biological diversity, the sustainable use of its components and the fair and equitable sharing of the benefits arising out of the utilization of genetic resources." The convention, in Article 8, imposes several general obligations that impact on possible future releases of transgenic fish into the environment. Paragraph 8(d) commits convention parties to "[p]romote the protection of ecosystems, natural habitats and the maintenance of viable populations of species in natural surroundings," and paragraph 8(h) has parties agreeing to "[p]revent the introduction of, control or eradicate those alien species which threaten ecosystems, habitats or species." It is paragraph 8(g), however, that is a "direct hit" in terms of transgenic fish and the environment. It reads, in part:

> Each Contracting Party shall, as far as possible and as appropriate . . . [e]stablish or maintain means to regulate, manage or control the risks associated with the use and release of living modified organisms resulting from biotechnology which are likely to have adverse environmental impacts that could affect the conservation and sustainable use of biological diversity, taking into account the risks to human health.

This provision, directly on topic as it may be, is nevertheless so vague as to be capable of multiple interpretations and a wide range of implementatory measures.

The Biodiversity Convention, being (and intended to be) a framework agreement, is noteworthy, at least as far as the present discussion is concerned, not so much for itself as for the biosafety protocol developed under it. The objective of the Cartagena Protocol, stated in Article 1, is to try to insulate the environment from genetically engineered organisms that might cause harm to it.[63] To what extent does the protocol affect transgenic fish, particularly their development and production within Canada? The impact is not, at least not directly, that significant. In the view of this author, the protocol is something of a "red herring" when it comes to transgenic fish. First, the protocol is a trade-based agreement, aimed at controlling the "transboundary movement ... of all living modified organisms that may have adverse effects on ... biological diversity."[64] The preamble of the protocol, at paragraph 9, specifically acknowledges that "trade and environment agreements should be mutually supportive with a view to achieving sustainable development." The protocol is not much concerned with domestic aspects of biotechnology.[65] Its real focus is the regulation of transboundary shipments of living modified organisms (LMOs) and the impact of those LMOs on recipient states. Second, the protocol's orientation, though it does not say so expressly, is to LMOs of the plant world. Plant seeds have the significant feature of being capable of "transferring or replicating genetic material" long after having been detached from their host. They have a post-harvest period during which their categorization as "living modified organisms" is preserved, and they therefore are squarely caught by the protocol. *Transgenic fish, on the other hand, are not "living modified organisms" after having been harvested.* Post-harvest, they cease to be "living," as they no longer have the capability of "transferring or replicating genetic material." They become a non-living commodity that is *outside* the ambit of the Biosafety Protocol. They maneuver around the protocol's net. This makes sense, given the focus of the Biosafety Protocol on protecting biodiversity from unwanted and uncontrolled infiltration by GM organisms. A dead fish is not a threat to ecological balance. If this view of the application of the Biosafety Protocol is correct,[66] from the Canadian perspective it likely exempts the bulk of future transgenic fish production, certainly at least transgenic salmon, since the industry by all accounts would be geared to the domestic grow-out of fish to market maturity, rather than just the development and production (and transboundary shipment) of eggs, fry and smolt.

To be clear, though, the Biosafety Protocol would have full application to the export of living transgenic fish, regardless of the stage of maturity, from Canada to another country.[67] The protocol is also applicable should Canada be designated as the "Party of import" of living transgenic fish.[68]

Notwithstanding that the protocol may have limited relevance to the internal rules regarding the rearing and handling of transgenic fish, Canadian regulators will still have to be cognizant of its framework and requirements. Article 2(1) obligates each "Party" to the protocol to "take necessary and appropriate legal, administrative and other measures to implement its obligations under this Protocol." Beyond this overarching legal duty, practicalities will probably dictate a substantial degree of conformity between specific protocol obligations (assuming eventual ratification) and Canada's domestic regulatory regime.[69]

Consumer choice

Embedded in many of the debates regarding GM food products, both inside and outside Canada, are concerns about the consuming public having full, accurate and readily available information about those foods. The concerns often seem to be vocalized in demands for mandatory labeling of GM foods. Labeling, however, is only one aspect of the "consumer choice" side of GM food regulation. Indeed, labels come only at the end of a much larger process of scrutinizing a particular GM food for health and environmental safety. Both the *CBAC Report* and the *Royal Society Report* put substantial emphasis on regulatory transparency and public involvement in the GM food approval process, and on improved information dispersal to support consumer choice.[70] These areas of emphasis seem to reflect international trends. The *CBAC Report* comments that "[i]nternationally, biotechnology regulatory systems are evolving toward increased transparency, often with enhanced opportunities for public input."[71]

Public notification and request for comments procedures are in place, although they vary somewhat, in the United States, Australia and New Zealand.[72] The situation in the United States is noteworthy because it seems to be somewhat skewed in favor of the proprietary interests of biotech developers. The Food and Drug Administration (FDA) in the United States has control of both the food safety and the environmental implications of commercialization of transgenic animals.[73] The FDA's review process has attracted some criticism for being excessively "closed."[74] At the other end of the spectrum is the European Union, which has more of a focus on openness and public involvement, and less on protecting commercial rights. By way of example, the European Parliament, in February 2001, "adopted a directive concerning the deliberate environmental release of GMOs that requires assessment reports to be made public and that the public be given an opportunity to comment before the field trials and market approval."[75]

Notwithstanding the differences in current national approaches, at least in principle, there does not seem to be too much debate at the international level on the value, vis-à-vis GM foods, of increasing the role of citizen participation in policy formulation, of regularizing and making more transparent the functions of government, and of improving public awareness and the quantity and quality of information made available. These are all viewed as worthwhile objectives, with details of implementation to be worked out state by state depending on cultural idiosyncrasies, funding availability and similar factors. It is labeling that stirs up the real dust when one assesses "consumer choice" issues related to GM foods.

As of mid-2002, thirty-five states had mandatory labeling laws in place for GM foods.[76] These states included China, Australia, Japan, Norway, Switzerland and the fifteen states then comprising the European Union.[77] The addition of China, as of 1 July 2002,[78] to the list of states requiring compulsory labeling of GM food and food products is notable because of the large Chinese population and the resulting implications for its trading partners.[79]

While some ENGOs declare that labeling of GM foods "is fast becoming the *de facto* international standard,"[80] this in fact is something of an exaggeration. There exists powerful international opposition to the drive to force mandatory labeling for all GM foods and food products. The Greenpeace organization points to the United States as the ringleader of the anti-labeling countries. There is no doubt current US policy stands staunchly opposed to mandatory labeling. In response to a draft document (tentatively called *Proposed Draft Guidelines for the Labeling of Foods and Food Ingredients Obtained through Certain Techniques of Genetic Modification/Genetic Engineering*) circulated for a 2002 meeting of the Codex Committee on Food Labeling (CCFL), the United States went on the record as saying:

> [F]oods derived from biotechnology are not inherently less safe than other foods ... [and] the United States strongly believes that the Committee should hold in abeyance any further discussion on mandatory process-based labeling until more comprehensive information is available regarding the implications of such labeling, particularly information relating to the costs and impact on international trade.[81]

The position of the United States on mandatory labeling is supported by Argentina and Canada.[82] Those three countries account for 96 percent of the world's transgenic crop production.[83] Some environmental interest groups assert that the US/Argentina/Canada triumvirate opposes mandatory labeling primarily to protect domestic producers/exporters of GM foods.[84] Of course there is truth in this. There are real, significant costs and trade implications associated with mandatory labeling. Most biotech foods currently in commercial production are of the first-generation variety and have been on the market for years. What is the point, the triumvirate asks, of forcing producers to spend large amounts of money to relabel? Not only that, the warning label perception would be likely to drive down consumption of explicitly identified GM foods. The common position of the United States, Argentina and Canada is not bereft of logic. In the absence of actual health and environmental risks, they say, there is no compelling reason to mark GM foods, and several good reasons (all commercial) not to.

There are clearly deep divergences of opinion in the international forum in connection with the need for mandatory labeling of GM foods. They reflect the complexities of the larger GM food debate. Once a GM food product is assessed and determined to be safe (both for human consumption and for the environment), it may be asked why then differentiate between that food and its non-GM equivalent? The answer is framed firmly as a "right to know." Labeling advocates suggest that any adverse implications of GM foods, notwithstanding the most rigorous contemporary testing, may only be identifiable after a long period of time. Consumers who want to avoid those possible implications altogether, it is argued, have the right to do so. There are also members of the public who oppose GM foods on ethical, religious or similar "conscientious objector" grounds. They, too, labeling proponents say, have the right to choose not to consume GM foods.

The consumer choice regulatory issues *vis-à-vis* GM foods are essentially reducible to a classic contest between commercial interests, on the one hand, and the right of the consuming public to complete information so as to enable thoroughly informed decisions, on the other. The more expansive the information requirements, the higher the costs and the lower the profits of the GM food producers. And those GM food producers are, at present, concentrated in a very small number of countries, led by the United States. This has caused a split in the international community between those states favoring full consumer information (including mandatory labeling) and those in opposition. The latter camp includes the GM food producers, such as the United States, as well as countries heavily dependent on the GM producing states. The former camp includes the non-GM food producers, who have the luxury of being able to promote consumer rights without having to face serious adverse economic consequences from taking such a policy position.[85] Consumer mobilization, often sparked by efforts of ENGOs, also plays a part in the policy-making decisions of states.[86]

The GM food discourse in the international community highlights two significant conclusions relative to issues of consumer choice. First, there is currently gridlock within the Codex Alimentarius Commission as to how to proceed with mandatory labeling. This situation is likely to continue for the foreseeable future. The CCFL[87] "operates by consensus according to general Codex rules that remain surprisingly undefined."[88] The United States, Argentina, Canada and others have taken advantage of this situation to drag out labeling discussions year after year. There is no reason to believe this will soon end. The United States at present has little incentive to change tactics. The CCFL process permits, almost encourages, interminable procedural maneuvering. Even if that process was amended so as to permit CCFL to formulate recommendations on labeling of GM foods, and the Codex Alimentarius Commission accepted those recommendations, it would not necessarily mean that uniform global labeling practices would follow. No enforcement mechanisms exist for Codex standards. States and non-state entities are free to adopt them, or not. Yet the standards promulgated by Codex, as indicated previously in this chapter, do have substantial persuasive force in the international community, and may be adverted to in trade disputes. There exists, as a consequence, a strong incentive on the part of GM food-producing states to defer for as long as possible the day when Codex becomes directly seized of the issue. That certainly seems to be the strategy of the United States and like-minded states as they continue to hamstring the CCFL process.

The second conclusion about the consumer choice issues highlighted by the contemporary international GM food discourse is that there is still insufficient overall concern about GM foods to bump that discourse to the next level of intensity. First-generation GM food products have been with us for a substantial period of time, and have so far presented themselves as nothing other than benign. No broad-based cause for concern therefore exists. This may change when more advanced GM food products, including transgenic fish, near market entry. This approaching wave of next-generation GM foods is one of the factors prompting the many studies, internationally and domestically, of regulatory issues. The level of debate over mandatory labeling and other consumer choice issues will undoubtedly heighten as the next-generation GM food wave grows nearer. For the present, however, governments generally seem to have ample leeway to formulate domestic policies to accommodate their respective states' commercial interests with "softer" demands for enhanced consumer information and input.

Domestic regulatory structures and procedures

Existing regime

Canada's existing structures and procedures relating to food regulation involve several federal departments and agencies, each with its own general area of responsibility (although there is some overlap). There is presently no separate and distinct responsible body or procedural approach for GM food products. Health Canada occupies the central role in assuring the safety of Canada's food supply[89] and has charge of labeling matters relating to health and safety issues.[90] The Canadian Food Inspection Agency (CFIA)[91] and Environment Canada also play key roles in the regulation of food.

The current human food safety assessment process used by Health Canada apparently is unique. The *CBAC Report* states that Canada "is the only country where regulatory oversight is triggered by 'novelty' rather than 'process'."[92] By "novelty" is meant that a particular food product has novel characteristics compared to existing approved food products. The fact that those novel characteristics are the result of genetic modification is irrelevant. Process, in other words, does not matter. The Canadian approach is therefore anomalous. In contrast to this, "some form of 'process-triggered' regulation is the rule in all other countries that have developed regulatory systems for GM foods."[93] Because the initial stage of the Canadian approach to food regulation is fairly sweeping, regulators have invoked "unique terminology and definitions."[94]

Health Canada's regulatory jurisdiction regarding food safety and related matters comes from the *Food and Drugs Act*[95] and the regulations made thereunder. The three primary foci for food assessments, whether GM or non-GM foods, are toxicity, allergenicity and nutritional content.[96] GM foods fall under the *Novel Food Regulations*.[97] These regulations establish "important background criteria,"[98] but the "more instructive document is that entitled *Guidelines for the Safety Assessment of Novel Foods* (Health Canada,

1994)."[99] The Guidelines expressly adopt a threshold test based on comparative analysis. The Canadian approach to the safety assessment of a novel food product therefore starts (and may end) by examining certain attributes of that product against "those of its traditional counterpart." The technical jargon for this is "substantial equivalence." It is identified as lying "[a]t the heart of Health Canada's safety assessment process,"[100] and it comes in for considerable scrutiny and commentary in both the *CBAC Report* and the *Royal Society Report*.[101]

Environment Canada, under the *Canadian Environmental Protection Act 1999*[102] (CEPA), maintains "overall responsibility for performing environmental risk assessments of new substances manufactured or imported into Canada, including organisms produced through biotechnology."[103] If, however, an equivalent environmental assessment is mandated by other legislation, Environment Canada is relieved of its responsibility under CEPA. This is how CFIA became responsible for assessing GM plants and their effects on the environment and biodiversity. In a similar usurpation of Environment Canada's role as the overseer of environmental assessments, Fisheries and Oceans Canada (DFO) is in the midst of developing new regulations under the *Fisheries Act*[104] to permit it to conduct assessments of all "transgenic aquatic organisms."[105]

Environment Canada administers the CEPA regulations (the *New Substances Notification (NSN) Regulations*[106]) that pertain to GM organisms. The NSN Regulations require that "all 'new' substances, including products of biotechnology, are reported and assessed for their potential to adversely affect human health and the environment before being manufactured in Canada or imported across its borders."[107]

The NSN Regulations "call for information to be provided by the proponent about many aspects of the modified organism's biological and ecological niche, and concerning potential or actual environmental impacts of its unconfined release."[108] The *Royal Society Report* assesses the "information requirements as listed in the CEPA regulations [as] quite substantial."[109] However, it qualifies that comment further on by concluding, "based on interviews with Environment Canada officials ... [that] the CEPA Regulations have no explicit data requirements for information pertaining to the potential effects on conservation and biodiversity posed by GM animals."[110] The Royal Society Expert Panel categorized this as "a significant weakness in the current legislation"[111] that leaves the "existing regulatory framework ... ill-prepared, from an environmental safety perspective, for imminent applications for the approval of transgenic animals for commercial production."[112]

As things currently stand in Canada, then, any application for the environmental release of transgenic fish would be processed by Environment Canada under the NSN Regulations, such as they are. But until a release into the environment is actually *intended*, the existing environmental protection regulations do not kick in.

As has already been mentioned, DFO is drafting new regulations under the *Fisheries Act* specific to "transgenic aquatic organisms." That work was still ongoing at the time of writing this chapter.[113] The work follows two major regulatory initiatives undertaken by DFO in the late 1990s. They are the: (1) the *National Code on Introductions and Transfers of Aquatic Organisms*[114] (DFO National Code); and (2) the *Draft Policy on Research with, and Rearing of, Transgenic Aquatic Organisms* (DFO Draft Policy). The DFO National Code was disseminated in January 2002. However, that final version appears to have been made inapplicable to transgenic aquatic organisms.[115] And indications are that the DFO Draft Policy has been left to wither, as present effort is directed to completion of the *Fisheries Act* regulations.[116] The DFO Draft Policy was, in any event, intended only "to be used on an interim basis, until specific Regulations are enacted."[117] The motivation for the quick preparation of the DFO Draft Policy, according to the *Royal Society Report*, was the expectation of imminent application for approval of production of transgenic fish, which was premised on the application filed in the United States in early 2000.[118] Given its status as the precursor to the *Fisheries Act* regulations-in-progress, the DFO Draft Policy retains some relevance despite its current dormant state.[119] The primary criticism of the DFO Draft Policy leveled by the Royal Society Expert Panel was in respect of DFO's recommendations therein in support of the possible use of sterilization techniques to protect the environment from transgenic fish.[120]

The domestic regime in place in respect of consumer choice aspects of Canadian food regulation has already been touched on. It is intertwined with food safety, reduces for the most part to labeling matters, and is in the hands of Health Canada and CFIA.

So how do transgenic fish fit into the existing Canadian regulatory regime, especially in respect of food safety, environmental protection and consumer choice? The answer is relatively straightforward. At this point, little in the way of special arrangements has been made for transgenic fish. The new *Fisheries Act* regulations will address the environmental implications associated with research on and rearing of transgenic aquatic organisms. Precisely how that is done remains to be seen. Aside from this activity specific to transgenic fish in the area of environmental protection, no other aspect of the Canadian regulatory regime applicable to GM foods differs materially for transgenic fish. That regime is geared to plants and seeds. But relatively little tweaking likely would be necessary for the current regime to accommodate food products made from transgenic fish. Such products would have to undergo the same food safety and nutrition assessments, based on whatever standards are ultimately accepted for GM foods generally, as other GM foods, and would also be affected by the same consumer choice issues. As has already been alluded to, however, the current Canadian regulatory regime is likely headed for substantial modification.

The current GM food discourse in Canada: what is being said and how might it affect regulation of transgenic fish?

With attitudes evolving and initiatives developing around GM foods at the international level, as well as domestically, the big question for Canadian transgenic fish producers is: how will all this affect existing regulatory structures and procedures, and to what extent can and will those structures and procedures, in turn, be adapted to accommodate commercial production of transgenic fish? The previous section of this chapter, for example, made reference to new *Fisheries Act* regulations being drafted for transgenic aquatic organisms. The precise shape those regulations take will reflect the level of Canada's commitment to safeguard the environment from such organisms. The discussion in this section returns to each of the regulatory areas of food safety, environmental protection and consumer choice to offer some specific comments on what may be in store for transgenic fish farmers in Canada.

Food safety

The issue of GM food safety is a deceptively complex one. At its heart is the question: "How safe is 'safe'?" The answer to that ultimately depends on the perspective of the poser of the question. As indicated in the first part of this chapter, there is no international consensus to use for guidance. Viewpoints span the spectrum, ranging from the FAO, which endorses "substantial equivalence"[121] and generally promotes the careful use of GM foods to help address the immediate needs of global hunger, to some developed states that use a broad brush in categorizing GM foods as inherently risky and unnecessary.[122] Canada, as an active member of the international community, must factor in these viewpoints in shaping its own approach to regulating the safety aspects of GM foods.

Canada is a "major importer and exporter of primary, intermediate and final food products,"[123] and the gradual infiltration of GM products into the global food supply therefore has significant ramifications for Canadian producers and consumers. As the world continues to shrink, Canada cannot, even if it wanted to, insulate its internal policies from those developing outside its borders. The astute tactic is therefore to use all reasonable means to affect what is evolving on the international scene. There appears to be broad recognition and acceptance of this approach in government. The indications are everywhere. A good example is Canada's very active involvement in the efforts of the Codex Alimentarius Commission to develop internationally accepted standards for the safety and identification of GM foods. In the labeling context, the CBAC has noted the importance of taking this type of internationalist approach.[124] Indeed, the significance of aligning Canada's domestic GM food policies with those of the rest of the world is a recurring theme in both the *CBAC Report* and the *Royal Society Report*. International trends, approaches and initiatives are continually referred to and examined for relevance to Canada.

The emergence of the precautionary principle is perhaps the most note-worthy of those international developments.[125] The *CBAC Report* and the *Royal Society Report* advocate the employment of considerable caution in Canada's development of standards of assessment for "next-generation" GM foods.[126] The *Royal Society Report* recommends "the precautionary regulatory assumption that, in general, new technologies should not be presumed safe unless there is a reliable scientific basis for considering them safe."[127] The *CBAC Report* recommends that "regulatory authorities take a precautionary approach to all stages of development and commercialization of a GM food (laboratory research, confined field trials, pre-market risk assessment and post-market surveillance . . .)."[128] Both reports support the maintenance and strengthening of a risk-based approach to the assessment of GM foods based on rigorous scientific testing and evaluation, and pre-scheduled reviews of product approval decisions.[129] The Royal Society Expert Panel recommends that CFIA "develop detailed guidelines describing the approval process for transgenic animals intended for . . . food production."[130] All of this suggests more demanding and regimented assessment procedures for new GM foods than have been applied to "first-generation" products.[131] Also, the reports recommend that the developers of biotech foods bear the burden (and cost) of "carry[ing] out the full range of tests necessary to demonstrate reliably that they do not pose unacceptable risks."[132]

Transgenic fish producers will be caught by this toughened future regime. Canadian policy-makers face difficult choices ahead.[133] There will be a natural predilection to look to broadly accepted international standards. The Codex Alimentarius Commission will be key. Canada undoubtedly will continue its strong participation in the work of Codex, and will probably adopt standards for GM foods as sanctioned by that organization. The reasons for this prediction are twofold. First, given the competing mandates of Codex itself (protection of human food safety and promotion of fair, free-flowing international trade and, indirectly through FAO, of world food security), the GM food standards it advocates are likely to strike a reasonable balance. For Canada, faced with similar pressures, the path of least resis-tance, and the one easiest to justify to domestic parties with interests in the regulation of GM foods, may be one that runs parallel to that of Codex. Second, as a leading global producer and exporter of GM foods, Canada will be strongly influenced to go with safety standards that are at least as rigor-ous as those of the Codex Alimentarius Commission. That strategy will probably keep Canada "on side" in respect of its own international trade commitments, and provide to it serious ammunition in any trade disputes that may arise with states taking extreme protectionist stances.

Environmental protection

The great uncertainty regarding the environmental effects of transgenic fish has led to a relatively high degree of consensus among all interested parties on the need to take meaningful steps to protect marine ecosystems. Fish obviously live in an environment very different from, say, that of a plant such as canola. They also reproduce differently, self-propel and are carnivorous. A host of unique environmental issues consequently come into play for fish. This prompted the only recommendations in either the *CBAC Report* or the *Royal Society Report* directed specifically at transgenic fish.[134]

The central recommendation of the *Royal Society Report*, at least insofar as transgenic fish are concerned, is that it would be "*prudent* and *precautionary* to impose a moratorium on the rearing of GM fish in aquatic facilities."[135] This recommendation is based on: (1) "the paucity of scientific data and information pertaining to the environmental consequences of genetic and ecological interactions between cultured and wild fish"; (2) "the difficulty . . . in being able to use laboratory research to predict environmental consequences reliably"; and (3) "the unpredictable nature of complex pleiotropic effects of gene insertions."

In addition to the moratorium, the four other substantive recommendations of the *Royal Society Report* relating directly to transgenic fish, all of an environmental protection nature, are that: (1) "[a]pproval for commercial production of transgenic fish be conditional on the rearing of fish in land-based facilities only"; (2) "[r]eliable assessment of the potential environmental risks posed by transgenic fish can only be addressed by comprehensive research programs devoted to the study of interactions between wild and cultured fish"; (3) "[p]otential risks to the environment posed by transgenic fish must be assessed not just case-by-case, but also on a population-by-population basis"; and (4) "[i]dentification of pleiotropic, or secondary, effects on the phenotype resulting from the insertion of single gene constructs be a research priority."[136]

Canada's existing international commitments, both moral and legal, serve only to reinforce the positions vocalized by the Royal Society Expert Panel. It would be very difficult for Canada to adopt anything but an environmentally protectionist position on transgenic fish. To do otherwise would run foul of the obligations encapsulated in the Biodiversity Convention, the Biosafety Protocol (assuming ratification), the ICES Code, NASCO Guidelines and FAO Code. From the perspective of transgenic fish developers, much energy must surely now be focused on creating efficient and foolproof sterility techniques. To the extent those developers can offer 100 percent sterile GM smolt and fry to rearing facilities, many of the environmental concerns unique to transgenic fish melt away. Guaranteed sterility also puts back into play the possibility of raising transgenic fish in ocean-based nets or cages, a more economically attractive alternative to fish farmers.[137]

It will be interesting to see what emerges in the new *Fisheries Act* regulations. If those regulations should take anything but an ultra-cautious approach to protecting the environment, the ire of many will be aroused and Canada's commitment to its international obligations will be called into question.[138]

Consumer choice

Both the CBAC and the Royal Society Expert Panel grappled extensively with the new reality of how best to uphold the consuming public's right to full information and choice when it comes to GM food products. One of the four themes of the *CBAC Report* is "Information and Consumer Choice." The *Royal Society Report* similarly includes a discussion on the pros and cons of mandatory labeling. Indeed, the whole "consumer choice" issue, for practical purposes, very much boils down to a matter of labeling. To label or not to label is the question. The *CBAC Report* suggests that it would be premature for Canada to adopt a mandatory labeling scheme prior to "an agreed-upon Canadian standard [being] developed and tested."[139] The report also reiterates that, whether voluntary or mandatory, a "single internationally accepted standard is highly desirable and perhaps essential in the longer run."[140] The *CBAC Report* recommends that Canada "establish a voluntary labeling system for foods with GM content based on a set of clear labeling criteria, derived from a broadly accepted standard."[141] The report further recommends enhancement of Canada's "continuing effort, in concert with other countries, to develop a harmonized approach to labeling in regard to GM foods."[142] The Royal Society Expert Panel also supported a voluntary system of GM food labeling.[143]

The labeling debate is not simply a matter of having the "right to know," as some environmental groups (such as Greenpeace Canada) argue.[144] It is intimately connected to both trade and food safety issues. On the one hand, detailed labels have the potential to obviate some food safety issues. For example, a GM food product to which a small section of the population may be allergic could be properly identified as such and still allowed to enjoy broad market distribution (similar to how nut products are currently handled). On the other hand, to the extent that Canada's labeling laws are out of step with those of its major trading partners, serious economic woes will certainly ensue. In this respect, the position of the United States is critical. It is difficult to imagine Canada adopting labeling rules different than the rules of the United States.

The CBAC recommended that any voluntary labeling system be subjected to review in five years to determine whether it has in fact given consumers an adequate degree of choice.[145] If it has not, other approaches, including mandatory labeling, would then be assessed. The Canadian Council of Grocery Distributors and the Canadian General Standards Board are apparently in the process of together developing a Canadian standard for the voluntary labeling of GM foods.[146]

All things considered, it seems quite probable that Canada will see a voluntary labeling system in place for GM foods within the not-too-distant future. Mandatory labeling is liable not to happen for many years, if ever, unless there is some drastic and unforeseen occurrence, such as the United States changing its stance or the Codex Alimentarius Commission deciding to support mandatory labels (and, given CCFL politics, the latter is extremely unlikely without the former). There is no good reason to believe that the approach to labeling for transgenic fish will be any different than that generally applicable to GM foods. As transgenic fish will be a new product, its producers will have no recourse to the argument that they cannot absorb the costs of relabeling. In fact, it can be anticipated that GM fish farmers may, from the outset, embrace labeling as a means of pre-emptively defusing at least one of the controversies associated with their product.

Conclusion

The one thing that can be said with certainty about the existing regulatory regime for GM foods in Canada is that it will see many significant changes in the next few years. Technological advances, if nothing else, will compel change. Other forces are at play, though, besides technology. There is growing consumer awareness at home and GM conservatism abroad. There is the reality of economic gain and the potential of enhanced world food security (forces sometimes, oddly enough, running on parallel courses in the GM food debate). There are concerns about ethics, religion and globalization. There is a broad commitment to ensure that human-made organisms do not run amok in the natural world, causing perhaps irreparable damage. Canadian policy-makers must sort through all this and come up with a plan that strikes a balance between the protection and promotion of multiple domestic interests and the honoring of multiple international commitments. The task is a formidable one.

Transgenic fish, whose technology is just emerging from the pipeline, find themselves in the middle of the GM food fray. Such fish, or products made from them, will be treated with at least as much caution and trepidation as "next-generation" GM foods derived from plants. A Canadian producer of transgenic fish can therefore anticipate encountering difficulties on many fronts – so much so, in fact, that it seems doubtful such a product will land on grocery shelves anytime soon.

The main impetus to commercial production of GM fish is economic: the promise of hefty profits. Lined up in opposition are the various non-economic factors discussed in this chapter. To the extent the economic incentive is strong, some or all non-economic factors may be mitigated or overcome. It is like bowling: the larger and more powerful the economic ball, the more readily non-economic pins can be scattered. That seems to be the way of the modern regulatory state. Humans follow their innovative urges (often tied in with profit-seeking) far and wide, usually reined in only by laws designed to buffer their fellows and the (shared) environment from the worst consequences of those urges. If it were not for those urges, of course, most of us would still be leading "short and brutish" lives. Improvement of the human condition is a laudable and never-ending quest, but care and consideration are necessary counterbalances – particularly when pursuit of profit is the central motive underlying a specific "improvement" effort.

The primary economic attribute of transgenic fish is their extraordinarily enhanced growth rates.[147] The raw profit potential is enormous. That expedited growth, though, does not necessarily translate into pure profit. Faster-growing fish bring significantly higher costs: fish feed, labor and transportation, to name just some. And then there are the capital start-up costs. Land-based rearing facilities, if mandated, do not come cheap (even if government helps out, which is the Canadian way). The cost of research and development has to be recouped, and the costs associated with obtaining regulatory approvals must be offset. When the product goes to market, chances are it will have a "GM" label on it. Sales may, as a result, not be as robust as hoped. And the price fetched for GM fish, compared to the "natural" counterpart, is likely to be lower, perhaps substantially so. Then there are the costs and headaches of dealing with the export of such a product to recipient states that may be just as happy not to have it. Ultimately, the economic "ball" represented by transgenic fish may not be large enough, or have sufficient steam behind it, to knock over many (or any) non-economic "pins."

There are two big "wild cards" for the nascent Canadian transgenic fish industry. One is essentially technological, the other geopolitical. On the technology side, it has been noted that GM research has the potential to make fish that have positive attributes other than enhanced growth. The growth-enhanced salmon already developed apparently are more efficient feeders than their non-GM cousins.[148] This has important environmental implications, since most fishmeal is made from small wild fish without much market appeal. It is generally accepted that it takes more than a pound of wild fish, reduced to feed, to produce every pound of farmed fish. That raises "concerns that aquaculture might fail to yield a net gain of fish protein for the world."[149] There is obviously something inherently disturbing about this. If it is accurate, it means the fish farm that is often viewed as friendly to the wild fishery is actually directly contributing to its ongoing depletion. GM fish, if they are more efficient at feed conversion than non-GM fish, could help reverse this negative impact on wild stocks (assuming that overall production remains constant, which, of course, is not what GM fish producers have in mind – expedited growth cycles are supposed to bring large increases in the *total amount* of farmed fish brought to market).

Aside from the feed conversion factor, GM technology could result in fish that are in several ways truly more benign *vis-à-vis* the natural environment. A GM fish could be made that efficiently uses grain-based fishmeal, thus resulting in a very positive impact on wild fish. A GM fish could be made that is disease resistant, thus reducing or eliminating the need to feed antibiotics to farmed fish and thereby rendering them more "organic." Similarly, a more "organic" fish would be one that is genetically programmed to handle pests (such as sea lice), bringing about a reduction or the elimination of the application of pesticides to farmed fish. The right combination of genetic tinkering could, hypothetically, create a fish that is something of an environmental "superhero." If such a scenario should unfold, it would surely strengthen the position of transgenic fish proponents.[150]

On the geopolitical side, the key word is "America." For Canadian GM fish producers, the United States has the potential to add much critical mass to the economic "ball," and to propel it down the lane with considerable force.[151] The size of the US market, the apparent lack of widespread aversion to GM foods on the part of US consumers (unlike those in Europe) and the existing degree of integration of the two economies make the US consuming public the logical target of the Canadian-based GM fish producer. After all, Americans already account for the bulk of Canada's annual aquaculture exports.[152] And the FDA application submitted in 2000[153] means that US approval may predate any Canadian approval eventually sought. In fact, with a green light for transgenic fish in the United States, full Canadian approval perhaps becomes secondary. Transgenic fish-rearing facilities in Canada could service the US market, turn a healthy profit and be disinclined to get caught up in a Canadian system that is not as "producer-friendly" as the American one. That, however, would generate even more pressure for Canadian regulatory authorities to align themselves with the US approach. Such pressure would become particularly acute with the passage of time and no evidence of significant adverse effects by transgenic fish on human health or the environment.[154]

The US situation therefore seems to be central both to Canada's future transgenic fish production industry and to the tone and content of its evolving regulatory regime for transgenic fish.[155] The Canadian market is not large enough to overcome the many issues attached to GM fish. Other potentially lucrative markets, such as those of Europe, are presently too problematic. The regulatory path in Canada is strewn with numerous impediments. Policy-makers can equivocate around "precaution" indefinitely, and may do so. Only political will can clear the path. Political will is most often fueled by economics. Without a strong push from the economic side, which in all likelihood must be an American push, Canadian-produced transgenic fish probably will be swimming in many circles for years to come.

This may be seen as a good or bad thing, depending on one's perspective and the relative weight one assigns to economic versus non-economic factors. If a guess had to be ventured, it would be that economics (perhaps strengthened by GM food's promise to enhance global food security and any future technological breakthroughs) ultimately will come out victorious, and transgenic fish will leap into the North American food supply. A finding of adverse health implications would stop this process in its tracks. Otherwise, faith in human ability to self-regulate and control risks likely will prevail. Of course, history shows that humankind is not particularly adept at self-regulation. The ecosystem threats possibly presented by transgenic fish represent an area of focused concern. Regulation can never fully corral human greed, ignorance and carelessness. With only a few transgenic fish perhaps having the capacity to cause serious, irreversible disruption in the world's oceans, it seems that environmental protection may in the long run be the one significant piece of the regulatory puzzle that proves the most difficult to fit snugly into place. Canadians, who at least in principle tend to give environmental issues relatively high priority, will likely not be easily deflected from the general sentiment that meaningfully protective steps are essential.

Notes

Research support for this chapter was provided by AquaNet, a Centres of Excellence Network for Aquaculture in Canada, based at Memorial University of Newfoundland and funded by the Natural Sciences and Engineering Research Council of Canada and the Social Sciences and Humanities Research Council of Canada through Industry Canada. I would like to thank Professor David VanderZwaag for his helpful comments on and criticisms of the work that forms the basis of this chapter. A modified version of this chapter recently appeared as an article in (2004) 1 *Macquarie Journal of International and Comparative Environmental Law*.

1 "The International Service for the Acquisition of Agri-biotech Applications estimates that the global area of transgenic crops in 2001 was about 53 million hectares, a more than 30-fold increase since 1996, grown by 5.5 million farmers in 13 countries." Canadian Biotechnology Advisory Committee, Report to the Government of Canada Biotechnology Ministerial Coordinating Committee, *Improving the Regulation of Genetically Modified Foods in Canada* (Ottawa, August 2002) at 5. Online. Available http://cbac-cccb.ca/epic/internet/incbac-cccb.nsf/vwapj/cbac_report_e.pdf/$FILE/cbac_report_e.pdf (accessed 6 March 2004) [*CBAC Report*].

2 *Ibid.* at 34.

3 "With the expected availability of genomic information for many species in the next few years, the floodgates of genetic modifications could open and release on the market an unprecedented variety of genetically enhanced products." Royal Society of Canada, Expert Panel Report on the Future of Food Biotechnology, *Elements of Precaution: Recommendations for the Regulation of Food Biotechnology in Canada* (Ottawa, January 2001) at 14. Online. Available http://www.rsc.ca/foodbiotechnology/GmreportEN.pdf (date accessed: 06 March 2004) [*Royal Society Report*].

4 *Ibid.*

5 *CBAC Report*, *supra* note 1 at 24–25.

6 There are variations on the definition of "transgenic." The following is taken from the "Definitions" section of Canada's *National Code on Introductions and Transfers of Aquatic Organisms* (Ottawa: Department of Fisheries and Oceans, January 2002). Online. Available http://www.dfo-mpo.gc.ca/science/aquaculture/code/prelim_e.htm (1 December 2003) [DFO National Code]:

> *Transgenic organisms* – Organisms bearing within their DNA, copies of novel genetic constructs introduced through recombinant DNA technology. This includes novel genetic constructs within species as well as interspecies transfers. Such organisms are usually (but not always) produced by micro-injection of DNA into newly fertilized eggs."

7 *CBAC Report*, *supra* note 1 at 25.

8 *Royal Society Report, supra* note 3 at 25, 26. There exist various novel gene constructs that have been, or probably will be, the subject of finfish research.
Genetically superior fish could have characteristics such as improved growth
rates, feed conversion efficiencies, disease resistance, cold and freeze resistance,
tolerance to low oxygen levels and the ability to utilize low-cost or non-animal
protein diets. *Ibid.* at 27.

9 *Ibid.* at 27.

10 *Ibid.*

11 *Ibid.* Contrary to common perception, however, the GM Atlantic salmon at
maturity is more or less the same size as a mature non-GM Atlantic salmon.
The allure of enhanced growth rate is that the fish can be brought to market
size more quickly, thus allowing for a significant increase in production within
a fixed time frame. *Ibid.*

12 The "first application to be made in North America for the commercial production of a transgenic fish ... was made in early 2000 in the United States
[to the Food and Drug Administration]." *Royal Society Report, supra* note 3 at
27. As of July 2003, that application remained active, with assessment work
ongoing and no decisions having been made. Also as of July 2003, no parallel
application had been submitted in Canada (personal communication, 11 July
2003, from E. Entis, CEO of AQUA Bounty Farms Inc., *infra* note 24). So far
as the author is aware, in early 2005 the US application continued active, but
undetermined, and no application had been made in Canada.

13 The CBAC is a creation of the government of Canada. It is a body composed of
independent experts whose mandate is to assist the federal government in the
formulation of public policy on various biotechnology matters. Advice from
the CBAC flows to the Biotechnology Ministerial Coordinating Committee,
which consists of the federal ministers of Industry, Agriculture and Agri-Food,
Health, Environment, Fisheries and Oceans, Natural Resources, and International Trade. See the CBAC website. Online. Available http://www.cbac-
cccb.ca (accessed 5 March 2004).

14 The Royal Society of Canada, The Canadian Academy of the Sciences and
Humanities, is a national body of distinguished scholars and scientists. As of
November 2003, it consisted of about 1,800 Fellows. The Royal Society's
main goal is the promotion of learning and research in the natural and social
sciences and in the humanities. One of the functions of the Royal Society is to
provide independent expert advice, typically to government, on public policy
issues through its program of Expert Panel reports. See the Royal Society
website. Online. Available http://www.rsc.ca (accessed 5 March 2004).

15 Article 2 of the *United Nations Framework Convention on Biodiversity*, 5 June
1992, UN Doc. UNEP/bio.Div/N7-INC.514, reprinted in (1994) 31 *International Legal Materials* 818. Online. Available http://www.biodiv.org/conven-
tion/articles.asp (accessed 9 March 2004) [Biodiversity Convention], defines
"biotechnology" as "any technological application that uses biological systems,
living organisms, or derivatives thereof, to make or modify products or
processes for specific use."

16 "Action Plan of the Government of Canada in Response to the Royal Society of
Canada Expert Panel Report" (Ottawa, 23 November 2001). Online. Available
http://www.hc-sc.gc.ca/english/pdf/RSC_response.pdf (accessed 9 March 2004)
[Action Plan].

17 "Progress Report – Action Plan of the Government of Canada in Response to
the Royal Society of Canada Expert Panel Report" (Ottawa, January 2002).
Online. Available http://www.hc-sc.gc.ca/english/pdf/royalsociety/progress_
report.pdf (accessed 9 March 2004) [Progress Report No. 1]; "Progress Report
– Action Plan of the Government of Canada in response to the Royal Society of

Canada Expert Panel Report" (Ottawa, May 2002). Online. Available http://www.hc-sc.gc.ca/english/pdf/royalsociety/progress_report_may.pdf (accessed 9 March 2004) [Progress Report No. 2]; "Progress Report – Action Plan of the Government of Canada in response to the Royal Society of Canada Expert Panel Report" (Ottawa, December 2002). Online. Available http://www.hc-sc.gc.ca/english/pdf/royalsociety/DecProgReportE.pdf (accessed 9 March 2004) [Progress Report No. 3]; "Progress Report – Action Plan of the Government of Canada in Response to the Royal Society of Canada Expert Panel Report" (Ottawa, June 2003). Online. Available http://www.hc-sc.gc.ca/english/protection/royalsociety/progress_report_june.html (accessed 9 March 2004) [Progress Report No. 4]; "Progress Report – Action Plan of the Government of Canada in response to the Royal Society of Canada Expert Panel Report" (Ottawa, December 2003). Online. Available http://www.hc-sc.gc.ca/english/protection/royalsociety/progress_report_december2003.html (accessed 29 January 2005) [Progress Report No. 5]; "Progress Report – Action Plan of the Government of Canada in Response to the Royal Society of Canada Expert Panel Report" (Ottawa, August 2004). Online. Available http://www.hc-sc.gc.ca/english/protection/royalsociety/progress_report_august_2004.html (accessed 29 January 2005) [Progress Report No. 6].

18 In Canada, the GM food controversy has manifested itself in several ways. The *CBAC Report, supra* note 1 at 11, chronicles the developments on the domestic front during the period of its project. These included a Private Member's Bill (Bill C-287) to amend the *Food and Drugs Act*, R.S.C. 1985, c. F-27, to require the labeling of GM foods. The bill was defeated in the House of Commons. There have also been numerous developments on the international front that have helped to foment, and have partially driven, the domestic debate.

19 "The plant biotechnology products now under development will present a wider range of novel traits and will be more complex than the current products. The GM foods commercialized to date involve primarily single-gene insertions, whereas the products being developed involve the introduction of multi-gene traits that either produce entirely new metabolic pathways or significantly alter existing ones. This will make the prediction and assessment of side effects more difficult." (*CBAC Report, supra* note 1 at 10).

20 "At present, a relatively small number of companies hold the vast majority of plant biotechnology patents and an increasing share of the GM food market. This market dominance and concentration of economic power is seen by some as a source of diminished self-sufficiency in food production and a threat to the sovereignty of some underdeveloped countries. Others regard this industrial structure as a necessity due to the time and expense involved in developing products from the research stage through regulatory approval." *Ibid.* at 44.

21 "Biotechnology provides powerful tools for the sustainable development of agriculture, fisheries and forestry, as well as the food industry. When appropriately integrated with other technologies for the production of food, agricultural products and services, biotechnology can be of significant assistance in meeting the needs of an expanding and increasingly urbanized population in the next millennium." United Nations Food and Agriculture Organization, "FAO Statement on Biotechnology" (Rome, 15 March 2000). Online. Available http://www.fao.org/biotech/stat.asp (accessed 9 March 2004) [FAO Biotechnology Statement].

22 Much of the increased public awareness of GM foods and food products and the issues related to them has been catalyzed and focused by environmental non-governmental organizations such as Greenpeace Canada. Greenpeace Canada in the early 2000s has showcased GM foods as one of its top issues. It has launched a website to inform consumers of the products on their grocery

shelves that contain GM ingredients. Online. Available http://www.green-peace.ca (accessed 9 March 2004).

23 The food safety, environmental protection and consumer choice aspects of the regulation of GM foods are, in the view of this author, "tier one" regulatory issues. There are a number of "softer" issues, such as the socio-economic and ethical implications of food biotechnology. These, however, tend to garner less attention in the GM food debate, especially in the international forum. These issues are consequently not ascribed much direct focus in this chapter. They do, however, come up tangentially from time to time, particularly in the discussions of "consumer choice" aspects of GM food regulation.

The other "tier one" regulatory issue relative to transgenic fish is trade. Trade matters are closely intertwined with food safety and consumer choice (especially labeling) issues. A state that wishes to export a GM food product must be cognizant of potential barriers and restrictions. To the extent that an exporter has a product barred from entry to another state, or has conditions and/or restrictions imposed on that export, international trade rules and regimes come into play. Similar considerations apply to a state wishing to bar or restrict the entry into it of GM food products. With different states taking varying positions on GM foods (and sometimes on types of GM foods), there arises significant potential for abusive application of trade restrictions. A state may move to bar or restrict entry of a GM food product on the ostensible basis of alleged human health and/or safety concerns. Given the many unknowns regarding GM foods, however, such a move could as easily be rooted in disguised protectionism as in a genuine concern about health and safety implications.

For an excellent discussion of aquaculture and the multilateral trade regime, please refer to Chapter 11 of this book, contributed by T. L. McDorman and T. Ström. These authors provide a detailed outline of the framework pertaining to the international flow of trade in aquacultural products (which has equal application to GM aquacultural products).

24 See, for example, J. van Aken, "Genetically Engineered Fish: Swimming against the Tide of Reason" (January 2000) at 3. Online. Available http://www.greenpeace.ca/e/campaign/gmo/documents/swimming_12_00.pdf (accessed 4 December 2003) [Swimming against the Tide], wherein a company called A/F Protein Inc. is identified as "leading the race to commercialize growth enhanced GE [genetically engineered] fish." The website of A/F Protein Inc. (now called AQUA Bounty Farms Inc.) indicates that it is head-quartered in Waltham, Massachusetts, that it is the parent of A/F Protein Canada Inc. (now called AQUA Bounty Canada Inc.), based in St. John's, Newfoundland, and that A/F Protein Canada Inc. operates a research facility in Fortune Bay, Prince Edward Island, under the business style "Aqua Bounty Farms" [A/F Protein Inc., AQUA Bounty Farms Inc., A/F Protein Canada Inc., AQUA Bounty Canada Inc. and Aqua Bounty Farms – collectively, "Aqua Bounty Farms"].

25 "The [aquaculture] industry reported total operating revenues of [CDN]$704.5 million in 2001. . . . Of that, finfish, mostly salmon, accounted for $602.0 million, or almost 90 percent of total sales. . . . During the last decade, the export market has consistently expanded, driven in large part by demand for salmon in the United States. In 2001, the value of aquaculture exports, which totaled $444.3 million, increased 17 percent from the previous year, more than triple 1992 levels." Statistics Canada–Agriculture Division, "Aquaculture Statistics, 2001" (October 2002) at 2. Online. Available http://www.statcan.ca/english/freepub/23-222-XIE/23-222-XIE01000.pdf (accessed 9 March 2004) [Aquaculture Statistics]. The predicted growth of

Canada's aquaculture industry is premised on several factors, including: (1) the availability of large marine areas for exploitation by commercial aquaculture interests; (2) the willingness of small, often isolated communities to embrace aquaculture production facilities, particularly in light of decreased participation in traditional fisheries and the paucity of other localized economic opportunities; (3) the proximity of a huge, increasingly health-conscious US market; and (4) active encouragement, sometimes manifesting itself as undisguised promotion, of aquaculture by both federal and provincial governments.

At the federal level in Canada, aquaculture is promoted by the Office for the Commissioner of Aquaculture Development, which falls under the auspices of the Department of Fisheries and Oceans. Aquaculture is also promoted at the provincial level in most provinces. For example, in Nova Scotia it is championed by the Department of Agriculture and Fisheries. Even a cursory review of the websites for these two government departments (http://www.gov. ns.ca/nsaf/aquaculture/home.htm and http://www.ocad-bcda.gc.ca/ehome. html, respectively) discloses that economic development of the industry occupies the central and paramount policy position for these particular departments. The Nova Scotia Department's website states, for instance, that "[a]quaculture offers one of the best opportunities for economic development in coastal areas ... [and] [t]he government strongly supports the development of this sustainable long term industry." "Regional Aquaculture Development Advisory Committees (RADACs) – Background Information." Online. Available http://www.gov.ns.ca/nsaf/aquaculture/radac/index.htm (accessed 9 March 2004).

26 *CBAC Report, supra* note 1 at 9.

27 The Codex Alimentarius Commission was created in 1963 by the FAO and WHO to "develop food standards, guidelines and related texts such as codes of practice under the Joint FAO/WHO Food Standards Programme." The stated primary purposes of the Joint FAO/WHO Food Standards Programme are "protecting health of the consumers and ensuring fair trade practices in the food trade, and promoting coordination of all food standards work undertaken by international governmental and non-governmental organizations." Online. Available http://www.codexalimentarius.net (accessed 9 March 2004).

28 *CBAC Report, supra* note 1 at 9. Lesser-known international bodies include the Secretariat of the International Plant Protection Convention and the International Office of Epizootics (also known as the World Animal Health Organization). *Ibid.* Additionally, there are bilateral and regional initiatives. These include, by way of example, "the network on plant biotechnology for Latin America and the Caribbean (REDBIO), which involves 33 countries." FAO Biotechnology Statement, *supra* note 21 at 3.

29 *Ibid.*

30 *Ibid.* Other commentators, such as Greenpeace Canada, express a more cynical view of Canada's activities at the international level: "Canada is one of the world's worst countries in blocking action on labeling GE [genetically engineered] food. ... Canada, the US and Argentina are determined to force the world to eat GE food without knowing it – because they grow 96 percent of the world's GE crops." Greenpeace Canada, "Greenpeace in Halifax for Global Meeting on Genetically Engineered Foods" (Halifax, 6 May 2002). Online. Available http://www.greenpeace.ca (accessed 9 March 2004) [Halifax Codex Meeting].

31 "Recent reports issued include the CODEX report on the development of internationally accepted principles of risk analysis, and guidelines for the safety assessment of foods derived from modern biotechnology. Reports by the Organization for Economic Cooperation and Development (OECD), Food and

Agriculture Organization (FAO) and the World Health Organization (WHO) addressed food and safety nutritional factors, and allergenicity. Reports have also been published by national scientific bodies, such as the Royal Society of London and the US National Academy of Sciences in the United States, as well as other national and multinational activities, which supplement international consensus-building efforts." *Ibid.* at 12 [footnotes deleted].

32 FAO Press Release, "FAO Stresses Potential of Biotechnology but Calls for Caution" (17 March 2000). Online. Available http://www.fao.org/WAICENT/ OIS/PRESS_NE/PRESSENG/2000/pren0017.htm (accessed 9 March 2004).

33 *Ibid.*

34 "The Codex Alimentarius Committee on Food Labeling is a United Nations committee which ... held its first meeting in 1996 with a mandate to establish guidelines for the labeling of foods derived from biotechnology." Greenpeace Canada, "What Is Codex Alimentarius?" (Halifax, NS, 6 May 2002). Online. Available http://www.greenpeace.ca/e/campaign/gmo/index.php (accessed 16 March 2003).

35 Codex Alimentarius Commission, "FAO/WHO Food Safety Assessments of Foods Derived from Biotechnology." Online. Available http://www.codexalimentarius.net/biotech.stm (accessed 9 March 2004).

36 Codex Alimentarius Commission, "Biotechnology and Food Safety: The Codex Ad Hoc Intergovernmental Task Force on Foods Derived from Biotechnology." Online. Available http://www.fao.org/es/ESN/food/risk_biotech_taskforce_en.stm (accessed 28 January 2005).

37 *Ibid.*

38 Codex Alimentarius Commission, "Biotechnology and Food Safety: FAO/WHO Work on the Safety Evaluation of GM Foods." Online. Available http://www.fao.org/es/ESN/food/risk_biotech_en.stm   (accessed   9   March 2004).

39 Codex Alimentarius Commission, "Biotechnology and Food Safety: FAO/WHO Expert Consultations." Online. Available http://www.fao.org/es/ ESN/food/risk_biotech_consultations_en.stm (accessed 9 March 2004).

40 FAO Biotechnology Statement, *supra* note 21 at 3–4.

41 Van Aken, *supra* note 24 at 1.

42 *Ibid.* at 2.

43 *Royal Society Report*, *supra* note 3 at 156.

> The effect of genetic interactions on the viability and persistence of wild fish populations will depend on the degree to which individuals are adapted to their local environment, on the genetic differentiation between wild and cultured individuals, on the probability and magnitude of outbreeding depression (i.e. a fitness reduction in hybrids from matings between individuals from two genetically distinct populations), and on the size of potentially affected wild populations relative to their carrying capacities.
>
> (*Ibid.* at 152)

44 *Ibid.* at 87.

45 An "estimated 32,000 to 86,000 farmed Atlantic salmon escaped from netpens [into British Columbia waters] between January and September 2000." *Ibid.* at 151.

46 Rio Declaration on Environment and Development, 13 June 1992, UN Doc.A/CONF.151/5/Rev.1, reprinted in (1992) 31 *International Legal Materials* 874. Online. Available http://www.unep.org/unep/rio.htm (accessed 9 March 2004) [Rio Declaration]. Principle 15 states:

> In order to protect the environment, the precautionary approach shall be

widely applied by States according to their capabilities. Where there are threats of serious or irreversible damage, lack of full scientific certainty shall not be used as a reason for postponing cost-effective measures to prevent environmental degradation.

47 According to A. Jiordan and T. O'Riordan, "The Precautionary Principle in Contemporary Environmental Policy and Politics," in C. Raffensperger and J. A. Tickner, eds., *Protecting Public Health and the Environment* (Washington, DC, and Covelo, CA: Island Press, 1999) at 19–21 ["Precautionary Principle in Contemporary Policy/Politics"], the precautionary principle had its genesis in the former West Germany in the early 1970s, picked up some impetus from the 1972 UN Conference on the Human Environment held in Stockholm, gathered force in Europe in the 1980s, and by 1993 had been "accepted as a fundamental guiding principle of EU environmental policy."

48 D. Freestone and E. Hey, "Origins and Development of the Precautionary Principle," in D. Freestone and E. Hay, eds., *The Precautionary Principle and International Law: The Challenge of Implementation* (The Hague: Kluwer Law International, 1996) at 41.

49 See, for example, the dissenting opinion of Sir Geoffrey Palmer in *New Zealand v. France* [1995] I.C.J. 288, wherein he stated, at 413, that "the precautionary principle . . . may now be a principle of customary international law relating to the environment."

50 Jiordan and O'Riordan, *supra* note 47 at 22:

> [T]he Precautionary Principle still has neither a commonly accepted definition nor a set of criteria to guide its implementation. . . . While it is applauded as a "good thing", no one is quite sure about what it really means, or how it might be implemented. Advocates foresee precaution developing into "*the* fundamental principle of environmental protection policy at [all] scales." Skeptics, however, claim its popularity derives from its vagueness; that it fails to bind anyone to anything or resolve any of the deep dilemmas that characterize modern environmental policy-making. [emphasis in original] [references deleted]

One excellent commentary that makes a strong argument in favor of the use of the precautionary principle to harmonize the "two fundamentally different conceptions of GM technology" ("these conceptions [being] the Frankenstein and the Better Living through Chemistry narratives") is J. S. Applegate, "The Prometheus Principle: Using the Precautionary Principle to Harmonize the Regulation of Genetically Modified Organisms" (2001) 9 *Indiana Journal of Global Legal Studies* 207 at 208.

51 NASCO was established under the Convention for the Conservation of Salmon in the North Atlantic Ocean. Online. Available http://www.nasco.int/convention.htm (accessed 9 March 2004) [NASCO Convention], which entered into force on 1 October 1983. NASCO has as its object the "conservation, restoration, enhancement and rational management of salmon stocks . . . [in] the [North] Atlantic Ocean. "About NASCO." Online. Available http://www.nasco.int/about.htm (accessed 9 March 2004). The parties to the Convention are Canada, Denmark (in respect of the Faroe Islands and Greenland), the European Union, Iceland, Norway, the Russian Federation and the United States.

52 ICES is an organization the modern incarnation of which was established in 1964 by the Convention for the International Council for the Exploration of the Sea, 12 September 1964. Online. Available http://www.ices.dk/aboutus/convention.asp (accessed 9 March 2004) [ICES Convention]. The primary

objective of the ICES is to coordinate and promote marine research in the North Atlantic. "About Us – What Do We Do?" Online. Available http://www.ices.dk/aboutus/aboutus.asp (accessed 9 March 2004). ICES has nineteen member countries, including Canada and the United States. *Ibid.*

53   NASCO Guidelines for Action on Transgenic Salmon. Online. Available http://www.nasco.int/pdf/nasco_cnl_03_57.pdf (accessed 9 March 2004) [NASCO Guidelines].

54   *Ibid.*

55   The 1995 Code has been updated; see Code of Practice on the Introductions and Transfers of Marine Organisms 2003. Online. Available http://www.ices.dk/reports/general/2003/Codemarineintroductions2003.pdf (accessed 5 April 2004) [ICES Code].

56   *Ibid.* at para. (c) of Section V.

57   *Supra* note 3 at 166.

58   Code of Conduct for Responsible Fisheries. Online. Available http://www.fao.org/fi/agreem/codecond/ficonde.asp#9 (accessed 9 March 2004) [FAO Code]. The FAO Code is a non-mandatory instrument that sets out various principles and standards applicable to the conservation, management and development of global fisheries. *Ibid.* at Preface. It is intended to provide a "framework for national and international efforts to ensure sustainable exploitation of aquatic living resources in harmony with the environment." *Ibid.*

59   "States should conserve genetic diversity and maintain integrity of aquatic communities and ecosystems by appropriate management. In particular, efforts should be undertaken to minimize the harmful effects of introducing non-native species or genetically altered stocks used for aquaculture including culture-based fisheries into waters, especially where there is a significant potential for the spread of such non-native species or genetically altered stocks into waters under the jurisdiction of other States as well as waters under the jurisdiction of the State of origin. States should, whenever possible, promote steps to minimize adverse genetic, disease and other effects of escaped farm fish on wild stocks."

60   Biodiversity Convention, *supra* note 15.

61   *Cartagena Protocol on Biosafety to the Convention on Biological Diversity*, 29 January 2000. Online. Available http://www.biodiv.org/biosafety/protocol.asp (accessed 25 March 2004) [Biosafety Protocol, Cartagena Protocol or simply protocol].

62   The Biodiversity Convention was ratified by Canada in 1992, and as of September 2005 had 188 parties. The Biosafety Protocol entered into force on 11 September 2003, ninety days after receipt of the fiftieth instrument of ratification (Article 37(1)). Although Canada signed the protocol on 19 April 2001, as of 1 January 2005 Canada had not ratified the protocol. The discussion of the protocol in this chapter assumes its ultimate ratification by Canada. The website of the Office of Biotechnology of the Canadian Food Inspection Agency suggests that Canada is still considering whether or not to ratify the protocol. Online. Available http://www.inspection.gc.ca/english/sci/biotech/enviro/protoce.shtmldate (accessed 28 January 2005).

63   "In accordance with the precautionary approach contained in Principle 15 of the Rio Declaration on the Environment and Development, the objective of this Protocol is to contribute to ensuring an adequate level of protection in the field of the safe transfer, handling and use of living modified organisms resulting from modern biotechnology that may have adverse effects on the conservation and sustainable use of biological diversity, taking also into account risks to human health, and specifically focusing on transboundary movements."

64   Biosafety Protocol, *supra* note 61 at Article 4. Paragraphs (g) and (h) of Article

3 provide the following key definitions: "'[l]iving modified organism' means any living organism that possesses a novel combination of genetic material obtained through the use of modern biotechnology"; "'[l]iving organism' means any biological entity capable of transferring or replicating genetic material, including sterile organisms, viruses and viroids."

65 Articles 1 and 4 do, though, speak of the protocol applying generally to the "use" and "handling" of LMOs in the broad context of the "conservation and sustainable use of biological diversity." Parties to the protocol would therefore, ostensibly, have to ensure conformity of their internal rules and practices with the requirements of the protocol. The language used in Article 2(2) in fact is broader than that of Articles 1 and 4. Article 2(2) says that "Parties shall ensure that the *development*, handling, transport, use, transfer and *release* of any living modified organisms are undertaken in a manner that prevents or reduces the risks to biological diversity, taking also into account risks to human health" (emphasis added). The use of the word "reduces," however, renders this provision somewhat weaker than it otherwise would have been.

66 It is supported by the comments of K. Buechle in "The Great, Global Promise of Genetically Modified Organisms: Overcoming Fear, Misconceptions, and the Cartagena Protocol on Biosafety," (2001) 9 *Indiana Journal of Global Legal Studies* 283 at 286, 287:

> The Protocol only deals with LMOs, not processed LMOs. As article 3(h) makes clear, a living organism is one that can replicate genetic material. Once an LMO has been processed, it is no longer capable of replicating genetic information ... "GM product" or "GM food" may refer to an LMO or to a processed food product that would not fall under the [Biosafety Protocol] LMO definition. [footnotes deleted]

67 If the purpose of the export is to achieve the "intentional introduction into the environment of the Party of import," then the "advance informed agreement (AIA) procedure," referred to in Article 7(1) and described in Articles 8, 9, 10 and 12, becomes applicable. The AIA procedure is a fairly onerous one and calls for the written prior consent of the state to which the genetically modified product is being exported. Most exporters of LMOs will, one assumes, seek to avoid application of the AIA procedure whenever possible. For transgenic fish, the way around it is to stipulate that the fish are "destined for contained use undertaken in accordance with the standards of the Party of import," per Article 6(2), in which case the AIA procedure is deemed not to apply. Outside the AIA procedure, the protocol is not nearly so onerous. Shipments of LMOs "destined for contained use" would mandate adherence to certain notification and identification requirements, but would not need the advance written agreement of the state of import. What is or is not "destined for contained use" may be open to some interpretive creativity. First, the word "destined" has an element of subjectivity to it, as well as a temporal limitation. An LMO transported from one country to another, even with the genuine intention of having it kept fully and securely contained, may at a later point in time be removed from that containment. Second, the wording of Article 6(2) leaves itself open to exploitation by unscrupulous parties who structure an LMO shipment to be for "contained use," knowing full well that the LMO, once shipment is complete, may be contained inadequately or not at all. Finally, the protocol's definition of "contained use" is itself somewhat malleable.

68 There seems to be ample protective scope within the protocol to fend off any unwanted incursions of transgenic fish. Article 10 sets out the AIA decision procedure. Article 11 deals with LMOs intended for direct use as food or feed, or for processing. The "food/feed" referred to in Article 11 must either be, or

contain, an LMO. Examples would be GM grain, whether intended for direct or processed human consumption, or for livestock feed, and GM animals shipped alive across national borders to be butchered at the place of destination. Articles 10(6) and 11(8) contain precautionary principle language that leaves much discretion with the "Party of import." Article 26 of the protocol also contains express wording permitting a state, in "reaching a decision on import under this Protocol or its domestic measures implementing this Protocol," to take into account "socio-economic considerations" (so long as they are consistent with other international obligations of the state). To the extent that Articles 10 and/or 11 are somehow found to be insufficiently flexible to support a state's decision on an import, which seems unlikely, Article 26 would be available to offer justification for that decision. Such a decision could, therefore, under the terms of the protocol, be taken *either* on "scientific" *or* on "non-scientific" grounds (or both).

69 The necessity for "contained use" of certain LMOs, as discussed in note 67, is a good example of this. Strong containment requirements for LMOs, whether those LMOs are sterile or not (according to the definition), represent one means of minimizing risks from LMOs, no matter what their source.

70 Recommendations 4.1, 4.9, 6.1 and 7.2 of the *Royal Society Report, supra* note 3, all concern ways of enhancing transparency and the provision of assessment information to the public. Regulations 2.1–2.10 and 6.1 to 6.3 of the *CBAC Report, supra* note 1, all concern increasing public involvement in the decision-making process and the transparency of that process, as well as the improvement of information flow to consumers to support their purchasing choices.

71 *CBAC Report, supra* note 1 at 20.

72 *Ibid.* at 20–21.

73 Pew Initiative on Food and Biotechnology, *Future Fish: Issues in Science and Regulation of Transgenic Fish* (Washington, DC, January 2003) at 46. Online. Available http://pewagbiotech.org/research/fish/ (accessed 9 March 2004) [*Future Fish*].

74 "Only after the FDA grants a new animal drug application may it release [data and] information [related to the application]. Even then, it makes public only portions of the file. The practical effect is that even after the FDA approves an application, much of the information submitted as part of that application, or any earlier underlying information from an investigational application, remains undisclosed. This closed process may have merit for protecting the intellectual property involved in the traditional human or animal drug approval. When applied to transgenic animals, however, it blocks public consideration of, and input into, a range of policy issues that go beyond technical and scientific considerations of safety. In particular, questions about what constitutes an acceptable level of environmental risk are at least as much policy questions as they are scientific questions. . . . This lack of transparency and public participation in the new animal drug approval process significantly challenges an agency hoping to retain public confidence in its decision-making process." *Ibid.* at 52.

75 Council Directive 2001/18/EC, repealing Council Directive 90/220/EEC, as cited in *CBAC Report, supra* note 1 at 21.

76 Greenpeace Canada, Halifax Codex Meeting, *supra* note 30.

77 Greenpeace Canada, "China Gets Mandatory GE Labeling while Canadians Still Guess What They're Eating" (Hong Kong, 1 July 2002). Online. Available http://www.greenpeace.ca (accessed 9 March 2004) ["China Gets Mandatory Labeling"].

78 *Ibid.*

79 In the world of food labeling politics, there is a great deal that hinges on crit-

ical mass. With a market as big as China's, mandatory label identification of GM products means that some ripple effect is unavoidable. Exporters doing significant business in China will be compelled to change the labels on their products destined for that market, and the prohibitive costs associated with running different product lines (in terms of labeling) might then prompt a decision to use GM labels even on products shipped to states without mandatory labeling legislation (or consumed in the state of production). To the extent that a particular state's food exporters have to make their product labels conform to the requirements of recipient states, the government of that state must consider the relative costs and benefits of adopting similar requirements. Greenpeace Canada, *ibid.*, reported that, in connection with the Chinese move, Thailand, Malaysia, Indonesia and the Philippines were all discussing the compulsory labeling of GM foods and food products.

80 *Ibid.*
81 Quoted in B. Kneen, "CODEX and the Politics of Labeling" (16 April 2002 – article prepared for CropChoice) at 4. Online. Available http://www.greenpeace.ca (accessed 15 November 2002).
82 *Ibid.* at 3.
83 *Ibid.*
84 Greenpeace, in particular, sees the issue in much starker terms. It assumes, until presented with convincing proof to the contrary, that GM foods are potentially dangerous. This approach runs contrary to the stated US position, which stresses the "inherent safety" of approved biotech foods. Groups in favor of mandatory labeling consider US-led opposition to labeling to be nothing short of a conscious strategy to permeate the world with GM organisms such that, eventually, there would be no practical means of differentiating between GM and non-GM foods. This, goes the theory, not only saves GM food producers costs in the short term, but insulates them from liability in the long term should their products turn out to have adverse consequences (*ibid.* at 1 and 7):

> [T]he strategy of the biotech industry and its government partners, particularly in the US and Canada, has been to contaminate the global food system as fast and as extensively as possible and then say, it is too late to label. . . . The biotech industry is desperate to kill the call for traceability (of which Europe is the major advocate). The industry simply does not want to take responsibility for its products or to accept liability.

85 One exception to this is China, which accounts for 3 percent of worldwide transgenic crops. *CBAC Report, supra* note 1 at 5.
86 The campaign of Greenpeace Canada offers a study in progress of how this works (*supra* note 22). Despite Greenpeace Canada's activities, however, there seems to be little evidence so far of any significant increase in the level of public concern over GM foods, at least not to the extent of forcing mandatory labeling.
87 Canada, it should be noted, is "both host country and chair" of the CCFL. Kneen, *supra* note 81 at 1.
88 *Ibid.*
89 Health Canada is solely responsible for assessing the safety of foods for human consumption, including GM foods and other novel foods, and for allowing them to be sold in Canada. It is responsible for implementing the provisions of the *Food and Drugs Act* that relate to public health, safety and nutrition; for establishing policies and standards for the safety and nutritional quality of foods sold in Canada; and for assessing the effectiveness of CFIA activities related to food safety. *CBAC Report, supra* note 1 at 8.

90 Health Canada is responsible for labeling in respect of health and safety matters, while the CFIA "handles general food labeling policies and regulations not related to health and safety." *Ibid.*

91 The thrust of the CFIA's efforts, at the risk of oversimplification, is on the farm. The CFIA looks after

> regulating GM plants, assessing their impact on the environment and biodiversity, including the possibility of gene flow and impact on non-target organisms, and is responsible for ensuring livestock feed safety, including feed composition, toxicology, nutrition and dietary exposure. . . . CFIA operates under the powers of the Seeds Act, the Plant Protection Act, the Feeds Act, the Fertilizer Act, and the Health of Animals Act. It also shares some responsibilities with Environment Canada under the Canadian Environmental Protection Act (CEPA), and with Health Canada under the Pest Control Products Act (PCPA) and the Food and Drugs Act.
>
> *Royal Society Report, supra* note 3 at 35.

The CFIA is said to be the first government agency typically encountered by a person or corporation wanting to introduce a new GM crop plant. Much of the CFIA's work involves assessing the environmental impacts of GM plants. This is done in large part through confined field trials.

92 *Ibid.* at 5.

93 *Ibid.* at 6. The *CBAC Report* also notes that the Biosafety Protocol's definition of LMO emphasizes process. *Ibid.*

94 *Ibid.*

> Rather than referring to GM plants or GM foods, the guidelines and regulations refer to plants with novel traits and novel foods, respectively. The regulations define a novel food as any food that does not have a history of safe use as a food, or has been manufactured or packaged in a way not previously applied to that food and that causes a significant change in the food's properties. *A third category of novel foods is GM foods, including foods derived from mutagenesis.* [emphasis added].
>
> *Ibid.*

95 R.S.C. 1985, c. F-27, as amended.

96 *Royal Society Report, supra* note 3 at 44–86.

97 *Food and Drugs Regulations*, Division 28, C.R.C., c. 870.

98 *Royal Society Report, supra* note 3 at 37.

99 *Ibid.*

> These guidelines (as opposed to regulations) specify that a guiding principle in the safety assessment is based on a "comparison of molecular, compositional and nutritional data for the modified organism to those of its traditional counterpart." They suggest that data should be provided on dietary exposure, nutrient composition, anti-nutrients, and nutrient bio-availability. If concerns still remain following this analysis, "toxicity studies would be required as necessary, on the whole food, food constituent or specific component in question." Finally, using data supplied by the applicant, Environment Canada and Health Canada consult together to decide whether a product is "toxic" to the environment and human health."
>
> *Ibid.*

100 *CBAC Report, supra* note 1 at 8.

101 The "substantial equivalence" concept apparently was "formulated by the [Organisation for Economic Co-operation and Development (OECD)] in 1993

[and] was the result of consultations with some 60 experts from 19 countries on methods to assess the safety of GM foods." *Ibid.* at 25.

102 S.C. 1999, c. 33, as amended.

103 *CBAC Report, supra* note 1 at 7.

104 R.S.C. 1985, c. F-14, as amended.

105 *CBAC Report, supra* note 1 at 7. To the extent that transgenic fish remain at the research and development stage, all work done

> within a contained setting such as a laboratory or greenhouse is not currently subject to regulatory oversight and authorization in Canada ... [although the] Canadian Institutes of Health Research have guidelines designed to prevent the environmental release of GMOs [and] [m]ost research institutions – both public and private – have their own codes of conduct and oversight committees.
>
> *Ibid.*

106 SOR/94–260 [*NSN Regulations*].

107 *CBAC Report, supra* note 1 at 9.

108 *Royal Society Report, supra* note 3 at 38.

109 *Ibid.* at 39.

110 *Ibid.* at 165.

111 *Ibid.*

112 *Ibid.*

113 Personal communication, 21 November 2003, from I. Price, DFO, Ottawa. Ms. Price advised that the "hard science" base for the risk assessment process relative to transgenic aquatic organisms was still being created. As of early 2005, no new *Fisheries Act* regulations had been made regarding transgenic aquatic organisms and, so far as the author is aware, no draft regulations had been released to the public.

114 *Supra* note 6.

115 This notwithstanding the fact that early versions of the DFO National Code, including the one critiqued by the Royal Society Expert Panel, covered GM fish. Evidently DFO had a late change of heart, or a realization that the DFO Draft Policy or pending Fisheries Act regulations would be the better place to deal with transgenic aquatic organisms. Section 1.1.4 of the final version of the DFO National Code states that "[i]ssues related to ... *transgenic aquatic organisms* are not covered by this Code" (emphasis in original).

116 Personal communication, 2 December 2002, from L. Stewart, Communications Adviser, Office of the Commissioner for Aquaculture Development, Department of Fisheries and Oceans, Ottawa: "I understand ... that the policy document 'Research with, and Rearing of, Transgenic Aquatic Organisms' is no longer in use as it is out of date and ... we are moving to regulations instead."

117 *Royal Society Report, supra* note 3 at 163.

118 *Supra* note 12.

119 DFO will not, however, release the DFO Draft Policy for public viewing. Its content can therefore only be discerned, and then not in any comprehensive fashion, through comments offered by the Royal Society Expert Panel.

120 The Royal Society Expert Panel compared the DFO Draft Policy position against the ICES Code and NASCO Guidelines, as well as making its own assessment of the practical risks of using sterility techniques as the sole means of insulating the environment from transgenic fish. The *Royal Society Report* expresses skepticism about the "mitigative utility of rendering GM fish sterile in aquatic facilities" (at 170) largely because "existing techniques for effecting sterility are not 100 percent effective" (at 166), but also because of the

"considerable uncertainty associated with … the consequences of ecological interactions between [sterilized fish] and wild fish" (at 160). Its conclusion is that sterilization is an insufficient means, at this point in time, of ensuring environmental protection from transgenic fish.

121   The debate around the virtues and shortcomings of substantial equivalence symbolizes much of the larger debate about the degree of scrutiny appropriate for GM foods. Substantial equivalence essentially stands for the proposition that if a GM food product looks, feels, tastes and acts the same as its non-GM counterpart, no more need be done. It is a relatively simple procedure that appeals both to commercial producers of GM food products and to international agencies focused on global food security issues. The *CBAC Report*, *supra* note 1 at 54 and accompanying text, cites several joint FAO/WHO reports, and an OECD report, which apparently endorse substantial equivalence as an assessment tool. The *CBAC Report*, at 26, states that "[t]he most recent FAO/WHO joint expert consultation on GM foods from plants concluded that the proper application of substantial equivalence contributes to a robust safety assessment framework and that it is the best strategy for safety assurance currently available" (footnote deleted). The problem with substantial equivalence as an assessment technique is that it is perceived by some as "subjective, inconsistent and pseudo-scientific" (*CBAC Report* at 25 – footnotes deleted), in effect a shortcut endorsed by those who, for whatever reasons, want to avoid the application of "hard science" in assessing GM foods. This criticism led the Royal Society, in its Recommendation 8.1, to reject the use of substantial equivalence "as a decision threshold to exempt new GM products from rigorous safety assessments" and, in its Recommendation 7.1, to advocate that "testing should replace the current regulatory reliance on 'substantial equivalence' as a design threshold." The CBAC does not come down quite so hard on the concept of "substantial equivalence." The *CBAC Report* states, at 27, that, "notwithstanding its inherent limitations … substantial equivalence remains a useful approach to structuring the environmental and food safety assessment of GM foods and crops."

122   For example, in Norway not only must a proposal to produce GM fish guarantee *no* adverse effect on the environment, but any such production must also be able to show tangible positive results. See s. 10 of *The Act Relating to the Production and Use of Genetically Modified Organisms (Gene Technology Act)*, Act No. 38 of 2 April 1993.

123   *CBAC Report, supra* note 1 at 41.

124   "The development of an international labeling standard, accepted by all of Canada's trading partners, is the surest way to obviate the negative consequences of mandatory labeling while providing meaningful consumer choice. It would be highly advantageous for Canada to actively promote the development of such a standard." *Ibid.* at 42.

125   Recognition of the growing international development and application of the precautionary principle prompted the federal government recently to complete a broad interdepartmental consultation process regarding the principles that should underpin the application of the precautionary principle by Canadian regulators. Those consultations generated a "Discussion Document" in September 2001. Privy Council Office, 2001, *A Canadian Perspective on the Precautionary Approach/Principle*, Discussion Document. Online. Available http://www.pco-bcp.gc.ca/raoics-srdc/docs/precaution/Discussion/discussion_e.htm (accessed 9 March 2004) [Discussion Document on Precaution]. The Discussion Document on Precaution concludes that the "precautionary approach/precautionary principle is a distinctive approach within risk management that primarily affects the devel-

opment of options and the decision phases, and is ultimately guided by judgment, based on values and priorities" (at 4) (references deleted). Vague as this is, the Discussion Document on Precaution does offer some comments that suggest development of a pragmatic Canadian approach to the precautionary principle. One comment (at 5), recognizes that it is "legitimate for decisions to be guided by society's chosen level of protection against risk." This suggests the possibility of broad commercialization of GM foods, so long as risk assessment has been done to a level that is acceptable to society generally, and/or that identification standards are adopted that provide consumers with a real ability to choose (and so avoid potential risks). Another series of comments repeatedly stresses the need to apply a *"sufficiently* sound or credible scientific basis" (emphasis added) (*ibid.* at 10–11) in the decision-making procedures relating to GM foods. This suggests an inclination to promote a scientific approach to assessment of GM foods that is reasonable, but not perversely stringent. Again, though, where the precise lines are to be drawn is open to debate. The *Discussion Document on Precaution* also makes a point of carefully emphasizing that the precautionary principle is an international phenomenon that is constantly evolving and, importantly, affects Canada's vital interests (economic and otherwise). Canada should, it says, therefore make it a high priority to try to shape the precautionary principle to ensure it fits Canadian reality, including in the area of GM foods.

126 The *CBAC Report, supra* note 1 at 5, notes the relatively large number of "first-generation" GM approvals to the time of the report:

> To date, Health Canada has authorized 52 novel (42 transgenic) foods for marketing in Canada, and the Canadian Food Inspection Agency (CFIA) has authorized 39 plants with novel traits (31 transgenic) for unconfined environmental release. Forty GM crops (31 transgenic) have been approved for use in livestock feeds, including some not grown in Canada, such as cotton.

The CBAC also noted, however, at 27, that it "found no evidence to indicate that substantial equivalence has been used as a decision threshold to exempt GM foods from appropriate regulatory oversight." The implication from this is that assessment of first-generation GM products has been, overall, adequate – but the level of assessment for next-generation products will have to be intensified.

127 *Royal Society Report, supra* note 3, Recommendation 8.1.

128 *CBAC Report, supra* note 1, Recommendation 3.2.

129 Recommendations 3.1 and 3.6, and *Royal Society Report, supra* note 3, Recommendation 7.1.

130 *Royal Society Report, supra* note 3, Recommendation 5.1.

131 The Action Plan issued by the Canadian government in response to the *Royal Society Report* confirms the support of the government for a precautionary approach to biotech food consistent with Principle 15 of the Rio Declaration. One wonders what else the federal government could possibly have said. Of course it is cautious about approving products that might possibly hurt Canadians and/or their environment. But is that traditional caution equivalent to application of the "precautionary principle"? That is the core question.

132 See, for example, *Royal Society Report, supra* note 3, Recommendation 8.2.

133 Because of the divergent interests involved in the GM food debate, and the fact that government is pulled in multiple directions, many of the recommendations of the CBAC and the Royal Society Expert Panel focused on independence, transparency and accountability in the future regulation of GM foods. Two of the *CBAC Report*'s eight categories of recommendations, for instance,

are "Structure, organization and operation of the federal food regulatory system" and "Transparency and public involvement." Regardless of what assessment standards are ultimately adopted, the reports emphasize that there must be absolute public confidence that those standards be applied fully and consistently in every case.

134   The *CBAC Report*, *supra* note 1, has little in it pertaining directly to transgenic fish. It is, overall, very much plant and crop orientated. It also followed the *Royal Society Report*, temporally speaking, and the CBAC therefore consciously endeavored not to duplicate what had been covered in the *Royal Society Report*. One CBAC Recommendation (3.4), *supra* note 1, may have implications for transgenic fish research facilities. It encourages "government [to] undertake a study to evaluate the effectiveness of existing guidelines covering experimental work with genetically modified organisms in laboratories and greenhouses ... with a view to determining the need for national guidelines or statutory measures."

135   *Royal Society Report*, *supra* note 3, Recommendation 6.13 (emphasis added).

136   *Ibid.* Recommendations 6.14–6.17.

137   It does not, however, completely clear the way for ocean-based facilities. Environmental concerns still exist regarding sterile fish released into natural ecosystems:

> Even if sterility could be assured, release of triploid fish into the environment presents certain hazards. Triploids of some species, while sterile, still have enough sex hormones in their bloodstream to enter into normal courtship and spawning behaviour. Escaped sterile triploid fish could interfere with the reproduction of wild relatives by mating with fertile wild adults. The most severe consequence would be reproductive interference of declining, threatened, or endangered species.
>
> (*Future Fish*, *supra* note 73 at 27–28)

138   The Government of Canada Action Plan in Response to the *Royal Society Report*, *supra* note 16, indicates agreement with the expressed need to keep reproductively capable transgenic fish and transgenic aquatic organisms in secure land-based facilities.

139   *CBAC Report*, *supra* note 1 at 42.

140   *Ibid.*

141   *Ibid.* at Recommendation 7.1.

142   *Ibid.* at Recommendation 7.4.

143   *Royal Society Report*, *supra* note 3 at 226. Only where there are identified health risks or significant nutritional changes does the Royal Society Expert Panel recommend compulsory labels.

144   "[R]easons why consumers might choose to consume or avoid GM foods include perceived or potential health risks or benefits, perceived or potential environmental risks or benefits, a fundamental ethical opposition to genetic modification of any kind, religious beliefs, food quality and price, broader societal concerns (such as globalization, food security issues and concentration of corporate power), and lack of confidence in the regulatory system." *CBAC Report*, *supra* note 1 at 38.

145   *Ibid.* at Recommendation 7.2.

146   *Ibid.* at note 95 and accompanying text.

147   Genetic engineering can be used to alter any number of traits. In fish, however, the focus of research has been on growth enhancement and cold-water tolerance. To the extent that transgenic fish are restricted to controlled land-based facilities, the ability to tolerate cold water becomes irrelevant. It is, in any event, the trait of greatly enhanced growth that has generated most of

the interest in transgenic fish and pushed them to the verge of commercialization.

148 *Future Fish, supra* note 73 at 7.

149 *Ibid.* at 16.

150 Some of the possible future advantages of transgenic aquaculture are discussed in *Future Fish, ibid.* at 34. Perhaps the single most important non-economic factor relative to a Canadian transgenic fish industry is the perfection of a practical and effective mass sterilization technique. Such a technique not only would neutralize much (though not all – see *supra* note 137) of the concern on the environmental side, thus removing from the lane one of the stickier non-economic "pins," but at the same time would significantly reduce production costs (by enabling the use of ocean cages/nets), thus increasing the size and velocity of the economic "ball." Sterilization therefore has the potential to effect a tipping of the regulatory scales for transgenic fish.

151 In one sense, therefore, the Canadian situation may be a somewhat anomalous one and perhaps may not provide to other states quite as much guidance and instruction as was intimated at the beginning of this chapter.

152 Aquaculture Statistics, *supra* note 25.

153 *Supra* note 12.

154 The regulatory harmonization between the two countries also would make it difficult for Canada to stand on the sidelines while the United States proceeded with commercial production of GM fish. If those fish are deemed safe for consumption by US citizens, Canadian regulators would be hard-pressed to hold to the position that Canadians should have a higher or different standard of safety applied to them. Cross-border shopping would, practically, make it impossible to prevent at least some transgenic fish being eaten by Canadians.

155 The US situation itself is extremely interesting in the way it seems to be evolving. *Future Fish, supra* note 73, for instance, is quite critical of the regulatory regime that has emerged in the United States to deal with transgenic fish. The core concern seems to be that there exists some built-in bias in favor of rights of property and profit. That is, the big business interests pushing GM fish may bulldoze over other "lesser" interests. One of the recommendations of *Future Fish* is, therefore, increased public participation and accountability regarding the US regulatory approval process.

Part V

Comparative national legal approaches

14 Offshore marine aquaculture in US federal waters

Picking up the pieces and painting a picture

Jeremy Firestone

Introduction

The state of mariculture policy development in the United States as of December 2004 is aptly described in the following passages:

> [A]quaculture in the United States lacked coherent support and direction from the federal government. Poor coordination, lack of leadership, and inadequate financial support have traditionally characterized programs related to aquaculture.[1]

> [N]o formal framework exists to govern the leasing and development of private commercial aquaculture activities in public waters. A predictable and orderly process for ensuring a fair return to the operator and to the public for the use of public resources is necessary to the development of marine aquaculture. It is recommended that Congress create a legal framework to foster appropriate development, to anticipate potential conflicts over proposed uses, to assess potential environmental impacts of marine aquaculture, to develop appropriate mitigation measures for unavoidable impacts, and to assign fair public and private rents and returns on such operations.[2]

> [T]he absence of a well-defined and efficient policy framework which fulfills public trust responsibilities in public waters while offering a predicable review, permitting, leasing, and monitoring process ... hinder[s] the development of this industry.[3]

Perhaps surprisingly (or unsurprisingly?), the three passages quoted were written respectively in 1978, 1992 and 2000. By sharing their sentiments, I intend to imply neither that a framework for mariculture in the United States is non-existent nor that the existing framework is unworthy of study, but rather, only that it is poorly defined. Thus, one goal of this chapter is to review the present framework for mariculture in the United States, focusing on legislative, administrative and judicial developments. In addition, ongoing developments in the United States allow us to sketch out where mariculture policy in the United States may be heading should it eventually mature beyond its current embryonic state. Specifically, an ongoing effort by a research team to devise an operational framework for US mariculture in conjunction with the work of two Ocean Commissions[4] – one sponsored by the US government, the other by a non-governmental organization (NGO) – provides a roadmap for mariculture policy development.

Background

Aquaculture production is increasing around the globe (excluding Africa and the countries of the former Soviet Union),[5] and now represents more than 28 percent of the total global seafood supply.[6] In the United States, aquaculture production has increased steadily over the course of the past decade and, in the near term, is projected to continue to grow, both absolutely and as a percentage of seafood consumption.[7] The United States Department of Agriculture (USDA) documented more than 4,000 aquaculture farms in United States in 1996, the vast majority of which were producing fish for food.[8]

Aquaculture is practiced as a commercial venture in all fifty US states and in US territories, commonwealths and freely associated states, from the farming of Atlantic salmon off the Maine coast, to mussels in Washington's Puget Sound, to catfish in Mississippi, to alligators in Louisiana, and to giant clams in the Federated States of Micronesia. While as many as thirty species are commonly aquatically farmed in the United States, fewer than ten of these comprise the bulk of US aquaculture food production: catfish, crawfish, hybrid striped bass, salmon, tilapia, trout and various mollusks.[9]

The annual value of US aquaculture now approaches a market value of US$1 billion and about 500,000 metric tons.[10] Although the increase in North American production has not been as rapid as the overall increase in global production,[11] US consumption patterns influence global production: while the United States ranks eleventh in aquaculture production, it ranks third in seafood consumption.[12] As a result, the US fish trade deficit stands at US$7.1 billion.[13]

The trend towards increased aquaculture production in North America is more pronounced if one focuses solely on mariculture production. From 1988 to 1997, North American mariculture production increased from 45,000 to 209,000 metric tons, an increase of more than 450 percent.[14] As of 1997, mariculture production in North America accounted for more than 40 percent of the total North American aquaculture production. The prime mariculture species reared in the United States are cupped oysters, hard clams and Atlantic salmon.[15]

In the United States, state and federal government advocacy has played and will continue to play an important role in aquaculture development. In the United States, as in other nations, aquaculture is viewed not only as an answer to the seafood shortage, but, in addition, as an attractive revenue stream and employment base. In the State of Maine, for example, where the US Atlantic salmon mariculture industry is based, Atlantic salmon mariculture alone generates annual revenues in the neighborhood of US$18 million and provides nearly 700 jobs in two Maine counties.[16] Given that economic footprint, it is perhaps not surprising that the US Department of Commerce is promoting a fivefold increase in US aquaculture production by year 2025.[17]

Yet, mariculture will have difficulty developing further in the United States without successfully negotiating a variety of environmental, social, political and technological obstacles that affect both its public perception and economic viability. For example, mariculture facilities discharge effluents that are difficult to quantify, monitor and control, and can provoke high-profile conflicts over use of ocean space. Moreover, one of the most lucrative segments of US aquaculture – Atlantic salmon mariculture – rests high in the food chain and thus requires environmentally damaging accessory fisheries.

The current regulatory framework for mariculture in the United States is characterized by a patchwork of state and federal laws that was not conceived of with mariculture in mind. Indeed, for the most part it was not envisioned that ocean space would be an engine of economic development. For example, the US Environmental Protection Agency (EPA) regulates the discharge of aquaculture facility effluents, the US Army Corps of Engineers (Army Corps) regulates the placement of any obstruction in US navigable waters, while the National Oceanic and Atmospheric Administration (NOAA), which is part of the Department of Commerce, has asserted jurisdiction over aquaculture, given the potential for aquaculture operations to negatively impact wild fish stocks.[18] Interestingly, no agency has authority to lease ocean space for the purposes of aquaculture, although the Minerals Management Service (MMS), which is housed within the Department of the Interior, has comparable authority to lease ocean space for the purposes of oil and gas exploration.[19]

The lack of a coherent regulatory framework for mariculture development in the United States is a major constraint on its expansion, as both regulators and culturists lack a clear blueprint on how to navigate the present maze. Given the inevitability of aquaculture development in the United States and the potential environmental and other problems associated with expanded mariculture, the most important questions currently surrounding mariculture in the United States are: In which areas of the sea will development[20] occur? And under what regulatory framework?[21]

Mariculture and federalism in the US system

As an initial matter, it is useful to have an understanding of how federalism impacts natural resource management in ocean space over which the United

States asserts jurisdiction. As is well known, the United States has yet to ratify the *United Nations Convention on the Law of the Sea* (UNCLOS). Nonetheless, customary international law likely supports US claims to a twelve-nautical mile territorial sea, a twelve- to 200-nautical mile exclusive economic zone (EEZ), and other maritime zones recognized by UNCLOS. In addition, by presidential proclamation, the United States extended its territorial sea to twelve nautical miles and, as early as 1976, legislatively extended its fishery zone to 200 miles.[22]

The international maritime zones do not dictate the state role in natural resource management. Rather, for most purposes the *Submerged Lands Act* (SLA)[23] controls, having granted title to the states to submerged lands extending three miles from the low-water mark and provided the states with control over natural resources within that three-mile belt. Somewhat similarly, state jurisdiction under the *Federal Water Pollution Control Act* (a.k.a. the Clean Water Act, CWA)[24] is limited to three nautical miles, yet the CWA definition of state jurisdiction is not entirely consistent with the SLA.[25] For these reasons, I refer to those areas where the states predominate as state offshore waters[26] and those areas (generally 3–200 miles) where the federal government predominates as federal offshore waters.

To the extent that mariculture operations are conducted in federal offshore waters, they may affect states in a couple of ways. First, they may require onshore infrastructure, such as staging areas and base yards, to support offshore operations. Second, escapes, should they occur, may impact state fisheries. Under the *Coastal Zone Management Act* (CZMA)[27] and regulations promulgated pursuant thereto,[28] to the extent that a proposed mariculture operation will affect a land or water use or natural resource of a state's coastal zone, the applicant is required to certify that the operation "complies with and will be conducted in a manner consistent with the [state coastal] management program." If a state objects to the consistency certification, an applicant cannot proceed further, although under limited circumstances it may appeal to the Secretary of Commerce.[29] Nonetheless, in most cases, inconsistency might be difficult to find, particularly if a state authorizes aquaculture in state offshore waters. However, because Alaska and California do not permit finfish aquaculture in their respective coastal zones, they may be able to use the consistency provisions of the CZMA to bar commercial finfish mariculture in federal offshore waters off their coasts.[30]

Major legislative mariculture developments

In 1980, Congress enacted the principal aquaculture-specific US legislation, the *National Aquaculture Act* (NAA).[31] In the NAA, Congress declared it to be "in the national interest, and ... the national policy, to encourage the development of aquaculture in the United States."[32] While Congress recognized that the primary responsibility to develop aquaculture in the United States fell to the private sector, it also noted that three federal agencies in particular – the Departments of Agriculture, Commerce and Interior – had a role to play in the industry's advancement. Yet Congress left each agency's role somewhat ill defined. Interestingly, some six months earlier, the Secretaries of those same departments recognized that their efforts related to aquaculture development could be mutually reinforcing.[33] They had thus entered into an interagency agreement that specified their respective roles and responsibilities.[34] The interagency agreement required all three departments to designate an aquaculture coordinator. And while it recognized that some issues would cut across jurisdictional lines and designated responsibilities, for the most part Agriculture would focus on freshwater research and support activities, Commerce on marine, estuarine and anadromous species, and Interior on freshwater finfish research.[35]

In an attempt to address the fact that aquaculture development in the United States had been "inhibited by many scientific, economic, legal, and production factors,"[36] Congress also established an interagency coordinating body – the Joint Subcommittee on Aquaculture (JSA) – within the Office of Science and Technology Policy. The JSA is made up of a number of cabinet- and subcabinet-level members, including the Secretaries of Agriculture, the Interior, Commerce, Energy, and Health and Human Services, as well as the EPA Administrator and the Chief of the Army Corps. Although initially the chair of the JSA rotated, in 1985 Congress passed the *National Aquaculture Improvement Act* (Public Law No. 99-198), establishing the Secretary of Agriculture as its permanent chair.

The primary mission of the JSA is "to increase the overall effectiveness and productivity of Federal aquaculture research, technology transfer, and assistance programs."[37] The means chosen by Congress for the JSA to implement that mission was the development of a National Aquaculture Development Plan. As noted by Cicin-Sain *et al.*,[38] the plan, which covered about thirty programs in twelve federal agencies, and was completed in 1983,

> created a National Aquatic Information Center, and a network of Regional Aquaculture Centers. The plan also identified the major problems facing the industry: inadequate credit, diffused legal jurisdiction, lack of management information, lack of supportive government policies, and lack of reliable supplies of feed stocks. To date, inadequate resources have been directed towards addressing these issues, and they remain concerns for the industry today.

Despite these efforts at improving interagency coordination, in 1995 the Office of Technology Assessment noted: "[D]espite a long history of debate over Federal agency roles in aquaculture, and establishment of a coordinating body, specific agency roles and responsibilities remain unclear."[39]

In 1996, the National Aquaculture Development Plan was revised in draft form,[40] but that draft has yet to be finalized. The draft amended plan calls for federal action related to (1) research and technology development and transfer; (2) education and training; (3) information and data collection, dissemination and exchange; (4) sustainability and environmental compatibility; (5) aquatic animal health; (6) seafood safety and quality assurance; (7) financial services and incentives; (8) marketing and international trade; (9) coordination and partnerships; and (10) improvements in the federal regulatory framework.

One other development at the national level that is notable, although not mariculture specific, was the adoption of the *Oceans Act 2000* (Public Law No. 106-256), which established the US Commission on Ocean Policy. Congress assigned the commission the task of developing a national ocean policy and to report its findings to the President and to Congress. This was a tall order, given the historic sector-by-sector (e.g. oil and gas, fisheries, and marine mammals) management of ocean and coastal resources in the United States and lack of coordinated planning in federal offshore waters, as well as a number of potential new uses of these waters that would need to be considered in addition to mariculture, including wind farming and bioprospecting.[41] The US Commission on Ocean Policy released its final report, which will be discussed later in the chapter, in September 2004.

Administrative developments related to aquaculture

Regulation as a fishery

The *Magnuson–Stevens Fishery Conservation and Management Act* ("*Magnuson–Stevens Act*") governs fishing in federal offshore waters.[42] This Act confers on the NOAA, the National Marine Fisheries Service (NMFS) and a series of regional fishery management councils (FMCs) the power to regulate capture fisheries. There is no explicit authorization in the *Magnuson–Stevens Act* to regulate aquaculture, and the legislative history of the Act is silent on that question. While the *Magnuson–Stevens Act* defines fishing to include "catching, taking, or harvesting of fish,"[43] it does not define "harvesting." It does, however, define "United States harvested fish" as "fish caught, taken, or harvested by vessels of the United States within any fishery regulated under this chapter."[44] Because "harvesting" connotes bringing in a crop, the NOAA asserted in 1993 that it has jurisdiction over mariculture in those instances when a mariculture facility includes a vessel.[45] The NOAA's position was apparently in response to a contrary position advocated by the US Department of Justice, which had asserted that *Magnuson–Stevens Act* jurisdiction was limited to "naturally occurring stocks."[46]

The NOAA position could be read as asserting jurisdiction in any instance when a maricultured specimen is removed from a facility and placed on a vessel. Because the *Magnuson–Stevens Act* imposes restrictions on the total allowable catch, seasons and ownership, the NOAA's interpretation may amount to a substantial impediment to development of aquaculture in federal offshore waters.[47]

Regulation of traditional effluents by the EPA

The CWA prohibits the discharge of a pollutant from a point source to US waters without a permit.[48] These permits require compliance with technology-based effluent limitations and state water quality standards. To the extent a person wishes to discharge effluent to the territorial sea,[49] contiguous zone or ocean, s. 403 of the CWA[50] requires that the discharge also must be in compliance with any ocean discharge criterion adopted by the US EPA.[51] These criteria are primarily directed at protecting the ecological health of the marine environment.[52]

More specific to aquaculture, an implementing regulation[53] requires an aquaculture operation (referred to as "Concentrated Aquatic Animal Production Facility," CAAPF) to obtain a permit if it discharges on at least thirty days in a year and exceeds certain production limits.[54] In addition, the EPA, on a case-by-case basis, may otherwise require a permit after finding that an aquaculture operation is a significant contributor of water pollution. The CAAPF rule is often confused with another discharge regulation related to aquaculture,[55] which regulates the discharge of pollutants *into* an aquaculture project. This latter provision, however, has no bearing on discharge *from* aquaculture facilities.

The EPA and the regulation of escapes

Recently, EPA took a cautious step[56] towards distinguishing which farmed fish should be regulated as pollutants and which should not. Specifically, in proposed aquaculture effluent guidelines, EPA stated that persons operating net-pen systems above a specified production threshold (greater than the threshold for regulation of conventional effluents) must "develop and implement [best management] practices [BMPs] to minimize the potential [unintended] escape of non-native species."[57] These practices – such as installing double netting in a net-pen operation – would be embodied in a non-native species escapement plan.[58] In its final rule, however, the EPA decided against including this provision. Instead, the final rule includes a "narrative effluent limitation that requires facilities to implement operational controls," with the goal of assuring that those facilities are properly maintained.[59]

Despite the EPA's decision not to include any direct measures to limit escapes, there are several interesting aspects to its abandoned proposal. First, nothing in the final rule suggests that the EPA concluded that it did not have authority to regulate escapes; the EPA has thus implicitly determined that an escape constitutes the "discharge" of a "pollutant" from a point source. Previously, the EPA had only gone so far as to indicate its "understanding" of the judicial interpretation of the term "biological materials."[60]

Second, the EPA's proposal was limited to the escape of non-native fish rather than any farmed fish, despite the fact that subtle genetic distinctions may exist between wild and captive-bred populations.[61] Third, by relying on BMPs, the draft effluent guidelines failed to take into account the potential biological impact of an escape. Thus, whether or not wild populations potentially affected are endangered was immaterial. Fourth, the EPA indicated that it was considering banning the intentional release of farmed fish, which can occur if, for example, the fish are not "growing rapidly enough to justify continued feeding."[62]

Code of conduct

The NOAA has recently taken steps to establish guidelines for aquaculture, with the publication of a draft code of conduct for mariculture operations.[63] While potentially far-reaching in some respects, unlike the EPA's proposed effluent guidelines, the proposed code of conduct is "soft" law; that is, even if the code of conduct is finalized, compliance will be voluntary. In pertinent part, the draft code calls for (1) mariculture operations to be guided by "precautionary" and "adaptive" management and BMPs; (2) adoption of siting criteria as well as monitoring, assessment and enforcement sufficient to ensure the conservation of genetic diversity, the maintenance of the functional integrity of ecosystems, and the protection of endangered species; (3) regulation appropriate to the status of the organism as native, non-indigenous or genetically altered; (4) minimization of disease incidence; and (5) escape prevention, combined with remedial action to address significant escape incidents.

Relationship between aquaculture and wild populations

Expert agencies and an expert panel of scientists in the United States recently examined the relationship between aquaculture and wild populations in a particularly compelling milieu: the farming of non-native stocks of Atlantic salmon in Maine, which is also home to endangered populations of wild Atlantic salmon. In particular, a joint review of the status of wild Atlantic salmon stocks by the National Marine Fisheries Service (NMFS) and the Fish and Wildlife Service (FWS) (jointly referred to as the "Services") concluded that despite 128 years of stocking, "hatchery fish have not substantially introgressed with the remnant populations and genomes" of the Gulf of Maine population segment.[64] At the same time, three of the factors cited in support of the endangered species listing decision were (a) the presence of a "large number of aquaculture hatchery origin juveniles" in the Pleasant River; (b) the possibility of disease transmission from cultured to wild stocks; and (c) the discovery of escaped farmed salmon in five Maine rivers.[65]

The Services' demarcation of wild Atlantic salmon found in the seven Maine rivers as a distinct population segment (DPS) led to a call for the

National Academies of Science (NAS) National Research Council to review that finding.[66] In the course of its review, which was supportive of the DPS finding, the National Research Council's Committee on Atlantic Salmon in Maine concluded that farmed Atlantic salmon differ in genetic makeup from the Gulf of Maine DPS as a result of the use of non-native strains, selection by breeders (for growth rate, fat content, disease resistance and delayed maturity) and "inadvertent selection by the novel environment (e.g. reduced fright response, disease resistance, and altered aggressive behaviors)."[67] It did note, however, that those "same traits may not be adaptive in the wild."[68]

Recent judicial developments regarding the regulation of mariculture facilities[69]

In *US Public Interest Research Group (USPIRG)* v. *Atlantic Salmon, Inc.*,[70] a federal district court (trial-level) found that various materials added by the Atlantic salmon mariculture operations to the waters of the United States were "pollutants" within the meaning of the CWA.[71] The CWA defines the term "pollutant" as "dredged spoil, solid waste, incinerator residue, sewage, garbage, sewage sludge, munitions, chemical wastes, biological materials, radioactive materials, heat, wrecked or discarded equipment, rock, sand, cellar dirt and industrial, municipal, and agricultural waste discharged into water."[72] Specifically, the court[73] held that (1) salmon feces and urine "constitute 'biological materials' or 'agricultural wastes'"; (2) the uneaten pigments canthaxanthin and astaxanthin, and the antibiotic oxytetracycline, which "flow [. . .] out of the pens or fall [. . .] through the net pens . . . are subsumed in the category of 'chemical wastes'"; (3) cypermethrin, which is used to kill sea lice, and the chemicals Finquel and Parasite-S, "released into the water after their use," are included "within the category of 'chemical wastes'"; (4) copper, a component of an antifoulant that is applied to the nets to reduce marine growth, is specifically listed by the EPA as a toxic pollutant;[74] and (5) fish that "do not naturally occur in the water, such as non-North American salmon," fall within the term "biological materials"[75] and are therefore "pollutants" under the Act. Because the defendant did not have a CWA permit allowing it to discharge those pollutants, the court found it to be in violation of that Act.

Although the concern with possible genetic pollution of wild stock from mariculture escapees was well articulated in a number of fora before the USPIRG cases, the court did not mention possible genetic pollution of native stocks by escapees as a basis for its decision. Nor did it rely on the endangered status of wild Atlantic salmon stocks. Rather, the court focused solely on the fact that the escapees were of different origin than the wild stock and distinguishable from wild stock by external markers.

In a companion case, *USPIRG* v. *Heritage Salmon*, the defendant, Heritage Salmon, settled with USPIRG after a preliminary decision from the court.[76] In pertinent part, the settlement requires Heritage, prior to receiving any applicable permits from the United States or the State of Maine, to: (1) forgo the use of non-North American stocks and transgenic salmonids; (2) limit the stocking densities in its net pens; (3) fallow its salmon farms; and (4) take precautions and institute measures so that cultured fish do not escape.[77] The settlement also requires Heritage to undertake measures to ensure water quality and prevent benthic impacts,[78] as well as to pay a total of US$750,000 to fund wild Atlantic salmon restoration efforts and to reimburse the plaintiffs for the cost of litigation, including attorney fees.[79]

Another case involving the regulation of Atlantic salmon operations under the CWA arose in the State of Washington. That case, *Marine Environmental Consortium* v. *Washington Department of Ecology*, proceeded under state law because Washington had been delegated authority to administer the CWA permit program within its boundaries.[80] The Washington case was a judicial appeal of a decision by the Washington Pollution Control Hearing Board (WPCHB) – a quasi-judicial state administrative body – upholding the issuance of discharge permits. The focus of the administrative proceeding had been on the extent of risk posed to native Pacific salmon by farm-raised Atlantic salmon, particularly the risk that Atlantic salmon might colonize the rivers of Puget Sound. The WPCHB had before it evidence of two escape incidents – 105,000 and 369,000 escapees, respectively in July 1996 and July 1997 – as well as evidence of the presence of at least twelve Atlantic salmon smolts in the Tsitika River. The WPCHB concluded, *inter alia*, that, "while undesirable," the accidental release of Atlantic salmon did "not pose a significant threat to native salmon" nor "degrade water quality," yet at the same time identified substantial evidence in the record to support a finding that "regular and large releases such as those that occurred in 1996 and 1997 could constitute a threat to Pacific salmon."[81] On the issue of spawning, the WPCHB concluded that while there "may have been successful spawning" of escapees, there was "no evidence to support that Atlantic salmon was 'self-sustaining,'" and thus the WPCHB took a middle ground, ordering the Washington Department of Ecology (WDE) to "take the Tsitika findings fully into account when it considers and reissues" the permits.[82]

The state court, deferring to the administrative interpretation and application of the law by the WPCHB, held that the inadvertent release of Atlantic salmon is neither "pollution" within the meaning of Washington law nor, at current level of escapement, a "nuisance"; nor does it "render [state] waters harmful, detrimental or injurious to salmonid species"; nor violate water quality standards.[83] In other words, to demonstrate that escaped Atlantic salmon are "pollution" under Washington state law, a complainant must establish not only a biological alteration of, or discharge of a substance into, state waters, but, in addition, that such alteration or discharge causes a "nuisance" or otherwise renders the waters "harmful, detrimental or injurious" to native salmonid species. No similar showing of harm is required under federal law. The court also affirmed the WPCHB's findings with respect to the Tsitka River and its decision to require the WDE to modify the permits accordingly.

Another case out of the State of Washington, *Association to Protect Hammersley, Eld, and Totten Inlets* v. *Taylor Resources, Inc.*, addresses CWA regulation of mussel harvesting facilities.[84] In mussel farming, unlike Atlantic salmon farming, which adds fish food and other chemicals to the water, mussels are nurtured exclusively by the nutrients within the water; that is, nothing is added. Nevertheless, as a by-product of their metabolism, mussels generate particulates, feces and pseudo-feces, ammonium, and inorganic phosphate – and mussels shells are released from the nets as well. The US Ninth Circuit Court of Appeals, after concluding that the term "biological materials" is limited to waste products of some human [or industrial] process," held that mussel shells and mussel feces and other by-products, although released into the environment, "come from the natural growth and development of mussels" rather than from the "waste product of a transformative human process," and as such are not regulated by the CWA.[85] The court nonetheless implied that the escape of live fish from mariculture facilities would constitute a discharge of biological materials (and hence the discharge of a pollutant).[86]

Developments outside of government

Numerous studies have addressed policy and legal issues that arise in offshore aquaculture. Many of those studies are catalogued in Cicin-Sain *et al.*'s 2000 report.[87] The major shortcomings identified in the current regulatory regime by those studies include (1) redundant regulations; (2) ill-defined property rights; (3) the failure to reconcile conflicting uses and to mediate among conflicting users of ocean space; and (4) questionable assertions of jurisdiction over aquaculture by a number of federal agencies.[88]

In 2000, the Center for the Study of Marine Policy at the University of Delaware (with a team of internal and external ocean and policy law specialists as well as aquaculture scientists and an aquaculture industry member) completed a study and issued a report that began to lay the groundwork for a policy framework for aquaculture in federal offshore waters.[89] The research team, like those before it, identified several gaps and problems that were hindering mariculture development in federal offshore waters: (1) the long-heard critique that aquaculture lacked a well-defined policy framework; (2) concern over environmental impacts; and (3) the often-seen clash between private rights and public expectations; that is, finding the appropriate balance between the need of aquaculturalists to gain a sufficient degree of exclusivity in the ocean space and the desires of other users.

Drawing lessons from nineteen existing studies of US aquaculture policy, six case studies, experiences in twenty-two coastal states and territories and eight other countries (Australia, Canada, Chile, Ireland, Japan, New Zealand, Norway and the United Kingdom), and the efforts of international organizations such as the Food and Agriculture Organization (FAO), the report[90] set forth a broad policy framework that was to be guided by the following policy criteria:

- encourages responsible mariculture in federal offshore waters;
- promotes a decision-making process that is efficient, coordinated and predictable;
- employs a precautionary approach;
- addresses the use of native and non-native species separately;
- is consistent with existing laws and agency responsibilities;
- is equitable and fair to offshore aquaculture and to other US users of federal offshore waters;
- is consistent, to the maximum extent possible, with the coastal, water, environmental and aquaculture policies of adjacent coastal states;
- is consistent with US obligations under international agreements;
- fits within the context of an overall framework for sustainable use/non-use of federal offshore waters;
- produces a fair return to the public for the use of federal ocean space;
- is conducted in a transparent manner with opportunities for public involvement; and
- is adaptive and promotes opportunities for innovation, data collection and learning.

The desire to flesh out a more specific policy framework for mariculture in federal offshore waters led Cicin-Sain and her collaborators to reconstitute themselves and to expand their research team in order to bring to bear additional expertise to the question. The federal offshore waters aquaculture policy project research team now includes a wider range of perspectives and comprises of voices from industry, academia, coastal states, the environmental community, and aquaculture coordinators from the US Sea Grant program, which specializes in education, training and outreach on ocean and coastal matters.[91] The goals of the present effort are twofold. The first goal is to operationalize and build on the earlier framework by providing specific recommendations on the management structure that needs to be constituted in the federal government in order to make mariculture development in federal offshore waters feasible and to devise a detailed operational framework for the actual siting, leasing, permitting and environmental review of mariculture facilities, as well as monitoring and enforcement related thereto, including the drafting of proposed legislation.

The second, related goal is to build consensus among national and regional stakeholders on specific aspects to include in such a framework through the convening of national and regional workshops. In September 2002, the research team convened a national workshop with federal officials from agencies that have regulatory responsibilities over or related to the development of aquaculture in federal offshore waters, staff of Congressional members or committees likely to have jurisdiction over aquaculture, US Commission on Ocean Policy staff, and other stakeholders to receive their insights and feedback on some of the approaches then being considered by the research team. Feedback also was sought regionally in the three regions that are most active in the development of aquaculture in *state* offshore waters: the Gulf of Mexico, New England and the Pacific coast/Hawaii.

Although the federal offshore waters aquaculture policy project has not issued a final report, a May 2003 letter, with attached preliminary recommendations, to the US Commission on Ocean Policy from the team sheds considerable light.[92] It is apparent from the letter that the endeavor has been a difficult balancing act between attempting to specify a policy framework specific to mariculture on the one hand, and placing mariculture within the broader, multi-use governance scheme for federal offshore waters that has been enunciated by the US Commission on Ocean Policy on the other hand. Indeed, the bigger challenge is to

> design an approach for the management of new uses of federal offshore waters that encourages sustainable use, and that creates linkages between new and already existing uses of the ocean, such as oil and gas operations, fishing, habitat and ecosystem preservation, transportation and recreation, among others.[93]

The approach to be advocated can be broken into six constituent parts.

1 *New administrative functionalities.* The team recommends the establishment of a new high-level Office of Offshore Aquaculture (OOA) within the NOAA, preferably as a component of a larger, higher-level Office of Ocean Management, which would have as its charge oversight of the "development of all new uses of federal waters" and "assessing, planning for, and reporting on ocean and coastal areas, their conditions and their uses."[94] It is envisioned that the OOA will work closely with the JSA in "developing and implementing the offshore aquaculture planning, research and outreach components of the National Aquaculture Development Plan."[95]

2 *Systematic mapping and assessment of features, resources and uses of federal offshore waters.* This effort would be undertaken to evaluate the suitability of areas for aquaculture development. Although this work would primarily fall to the federal government, it also would "involve region-specific assessment efforts by regional councils, States, or others."[96] The following options for mariculture development in federal offshore waters have been developed:[97]

   • Sites that the federal government would "pre-permit" for marine aquaculture. Under this option, the federal government would secure necessary general permits to address state and federal regulatory requirements.

- The federal government would designate areas for offshore aquaculture, taking into consideration impacts to the marine environment and potential conflicts with other uses and users.
- The creation of marine aquaculture parks that would "provide initial infrastructure, environmental assessment information, and designated areas for pilot, research, and longer-term commercial projects."[98]
- Map-based marine zoning of federal offshore waters, with areas zoned for exclusive or multiple use, depending on site characteristics and existing and desired uses.
- In the interim, until the above measures are a reality, leases would be issued on a case-by-case basis for sites that would likely be identified by aquaculture firms.

Because of the potential for cumulative and/or synergistic effects of mariculture facilities, OOA also would need to develop spacing guidance.[99]

3   *Leasing authority.* To successfully operate in federal ocean waters, a culturist needs to exercise sufficient control over the site. Creating a leasing authority in NOAA OOA would convey some degree of exclusivity while protecting the public's interests in the resource by setting forth detailed rights and responsibilities of the culturalist, including the obligation to pay rent (and royalties, as appropriate) and to ensure proper closure of the facility. A lease would need to address the following elements: (a) eligibility; (b) delineation of the three-dimensional space to be leased; (c) duration; (d) degree of exclusivity awarded; (e) compensation by the lessee (user pay); (f) environmental monitoring; (g) access and information provision; (h) health and safety; (i) lease transferability; (j) performance bonds to ensure site restoration; and (k) abandonment, revocation and termination.[100] In addition, because leases would be the prime vehicle through which to regulate ongoing aquaculture operations, each lease would need to establish operational parameters that would place limits on operations (e.g. stocking densities and feed rates) and/or to rely on environmental discharge parameters, leaving it to the aquaculture firm to devise the most efficient means to manage its operations given discharge limitations. Four types of leases are being considered:

- *Experimental.* For scientific research and to further development of gear or techniques.
- *Short-term (or interim).* To facilitate development of a facility's operational plan.
- *Long-term (or standard).* For culturists with fully developed operational plans.
- *Emergency.* To provide facility operators with the capability to respond rapidly to and temporarily relocate their facilities in the event that conditions (such as extreme weather events) warrant relocation.[101]

4 *Joint leasing and permitting.* NOAA OOA should create a multi-purpose lease/permit application to guide review by all relevant federal and state agencies and to facilitate public input.[102] Some agencies, such as the EPA and the Army Corps have to consider whether or not to issue permits; other federal agencies, such as the NMFS and the FWS, must be consulted regarding potential impacts on essential fish habitat, endangered species and marine mammals, while affected states need to be brought into the loop to analyze whether the proposed facility operations would be consistent with their respective coastal zone management plans. In addition to completing the application form, applicants would be required to submit a detailed operational plan that would include an environmental characterization of the area proposed to be leased, a discharge monitoring plan, an emergency response plan, an escape response plan and an abandonment/closure plan.[103] The "purpose of the joint permitting/leasing system is to simplify the existing jumble of fragmented permitting processes and unclear authority and environmental review procedures, and to attach, as appropriate, detailed conditions and ongoing monitoring to the conduct of aquaculture operations."[104]

5 *Guidelines, principles and procedures for environmental review and monitoring.* The OOA should conduct or direct environmental reviews and associated monitoring, with the extent of review required tailored to each stage of the process (planning, leasing, monitoring). As the research team notes, in order to ensure an effective environmental review process, "Congress should clarify and confirm" that the *National Environmental Policy Act* (NEPA)[105] "applies to activities in the territorial sea, in the EEZ, and on the continental shelf."[106] Seven guiding principles for environmental review of aquaculture in federal offshore waters have been identified: (1) ecological sustainability; (2) the precautionary approach; (3) environmental carrying capacity; (4) environmental impact assessment; (5) adaptive management; (6) transparency/public participation; and (7) promotion of integrated aquaculture, including polyculture.[107] In addition, an eighth principle, "polluter pays," informs the conclusion that culturists should be responsible for addressing natural resource damage in the event of significant adverse impacts to the marine environment.[108]

6 *Operation, monitoring, compliance and enforcement.* BMPs and codes of conduct should guide offshore aquaculture operations.[109] In addition, the NOAA OOA needs to establish procedures, practices and requirements to ensure that aquaculture activities do not exceed the carrying capacity of the environment and that the site is returned to its pre-leased state at project termination.[110] For example, performance bonds to ensure removal of aquaculture structures, site remediation, and implementation of monitoring for ecological health, operational safety and compliance with the terms of leases and permits are required.[111] The

federal government must dedicate sufficient financial and personnel resources to monitor the offshore aquaculture industry, to ensure its compliance with leases, permits and requirements of law, and to bring the government's enforcement authority to bear in the event of violations.[112] In regard to the last item, additional enforcement authority is needed to complement the proposed new leasing authority. Although the executive branch could implement some of the actions detailed above – that is, without congressional action, such as creation of the OOA – other actions, such as leasing authority, would need specific legislative authorization.

A month after Cicin-Sain and her colleagues had outlined their preliminary thoughts on marine aquaculture governance, the Pew Oceans Commission weighed in with its final report: *America's Living Oceans: Charting a Course for Sea Change.* In pertinent part, it recommends significant change in the manner in which US oceans are governed. If adopted, the changes would significantly impact the manner in which the federal government regulates marine aquaculture. The report also makes specific recommendations regarding offshore aquaculture. Turning first to issues of governance, the Pew Commission recommends that Congress enact legislation providing a "unifying set of principles and standards for governance," including ecological sustainability, maintenance of marine biological diversity and healthy marine ecosystems, and precautionary action in favor of conservation.[113] To accomplish those objectives, the Pew Commission advocates implementing ocean ecosystem governance at the regional level and adopting regional ocean governance plans that employ zoning to reduce user conflicts and that provide indicators of ecosystem health, measurable goals and enforceable policies.[114] The Pew Commission calls on Congress to consolidate federal programs related to oceans and the atmosphere in a single independent agency and to establish a national ocean policy council within the White House.[115] Turning to mariculture-specific concerns, the Pew Commission recommends that Congress enact legislation establishing aquaculture siting, design and operation criteria as well as standards that minimize adverse impacts, promote species that are not overly dependent on fishmeal and fish oil, and limit mariculture in federal offshore waters to native stocks.[116]

US Commission on Ocean Policy

In its September 2004 report – *An Ocean Blueprint for the 21st Century Final Report of the U.S. Commission on Ocean Policy* – the US Commission on Ocean Policy makes recommendations of relevance to marine aquaculture that are broadly consistent with those put forth by the Pew Commission. To begin with, the US Commission on Ocean Policy advocates management that is participatory and based on the best available science and information; that is adaptive, ecosystem-based and protective of marine biodiversity; that recognizes ocean–land–atmospheric connections and multiple uses; and that at its core is sustainable and reflective of the notion of stewardship.[117] To meet this management objective, it calls, in pertinent part, for the establishment of a National Ocean Council within the Executive Office of the President to ensure that ocean issues receive high-level government attention,[118] the adoption of legislation that would clarify NOAA's structure and align its mission along the principle of ecosystem-based management, and the consolidation of ocean programs and functions.[119] The US Commission on Ocean Policy also seeks to devolve ocean governance to the regional level.[120]

Underscoring the importance that the US Commission on Ocean Policy attributed to marine aquaculture, it devotes an entire chapter of its report to this subject.[121] The Commission recommends that NOAA be designated the lead federal agency for marine aquaculture and that NOAA establish an Office of Sustainable Marine Aquaculture.[122] The new office would then be charged with establishing a leasing and permitting regulatory regime to streamline government review and to enhance coordination with other ocean uses.[123] The Commission recommends as well that aquaculturalists pay the government a "reasonable portion of the resource rent generated from marine aquaculture projects that rely on ocean resources held in the public trust" and post a performance bond to assure performance and proper removal of the aquaculture facility.[124]

Conclusion

Although the quotations that opened this chapter give one pause – a consistent quarter-century clarion call for clarity and consistency has gone unanswered – several recent developments may provide the impetus for action. To begin with, the findings of the US Commission on Ocean Policy should place the issue of ocean management squarely on Congress's plate. Moreover, the operational framework for aquaculture in federal offshore waters should complement the work of the Commission on Ocean Policy. In addition, recent lawsuits against aquaculture firms under the CWA may mobilize a constituency for change within the industry. And finally, aquaculture may be able to piggyback the recent activity in another emerging use of federal offshore waters, wind farming – a use that may find a more receptive ear in Congress and in the White House.

Assuming that administrative and legislative action occurs in the aquaculture realm, what will that development mean to the aquaculture community? Although a move from state to federal offshore waters may pose some costs to the aquaculture industry in terms of higher operating costs and increased risk of harm from weather events, such a move is likely to have compensating benefits. To begin with, whatever regulatory framework is eventually adopted for federal offshore waters, it will provide the industry with a regulatory environment that is more consistent than presently exists as it shifts or expands operations from one state jurisdiction to another, up and down the Atlantic and Pacific coasts and along the Gulf coast. Indeed, states have such diverse mariculture rules as a blanket prohibition on finfish farming in Alaska, a submerged lands lease from the Department of Agriculture and approval by the State Cabinet in Florida, and aquaculture siting on a local and county level in Texas and Washington, respectively.[125] Moreover, unlike in state offshore waters, where both state and federal law apply, in federal offshore waters only federal law applies, and thus, to a greater extent, mariculture firms that operate in federal offshore waters can avoid potentially conflicting and duplicative laws and regulations.

For example, in state offshore waters, in addition to any state laws that may be applicable, the discharge prohibitions of the CWA would be applicable, as would federal requirements related to the use of pesticides (the *Federal Insecticide, Fungicide, and Rodenticide Act*, FIFRA)[126] and antibiotics (the *Federal Food, Drug, and Cosmetic Act*)[127] the placement of an obstruction in the navigable waters (the *Rivers and Harbors Act*, RHA)[128] and the prohibitions against taking an endangered species (the *Endangered Species Act*)[129] or marine mammals (the *Marine Mammal Protection Act*).[130] In contrast, in federal offshore waters, state regulation, for the most part, is likely to have minimal effect. For example, s. 401 of the CWA,[131] which would otherwise require a federal permit applicant to receive certification from a state that its discharge is in compliance with state water quality standards, is not applicable in federal offshore waters because the jurisdictional reach of the states under s. 401 extends only three miles offshore.[132]

It is true, however, that even if mariculture operations occur in federal waters, some state laws and programs may still have bite. For example, s. 307 of the CZMA[133] requires federal permit applicants to obtain state certification that any permitted action that will affect land uses, water uses or natural resources of a state's coastal zone is consistent with that state's coastal zone management plan. States also, for example, could require a permit to transport live fish through state jurisdictional waters.[134] Finally, states may participate in permit and lease development through the environmental evaluation and public participation process mandated by the NEPA.

The further mariculture operations move offshore, the more likely it is they will be able to avoid conflicts with other uses (e.g., the conflicts between wild and cultured salmon stocks and among mariculture, commercial fishing, oil and gas, and preservation) and other users and interested parties. Moving further offshore also should ameliorate some of mariculture's ecological impacts. For example, the capacity of the ocean to assimilate nutrient loadings from mariculture operations should be greatly enhanced as one moves from near-shore to the wide expanse of federal offshore waters. Thus, when, and if, a regulatory framework is put in place for federal offshore aquaculture, firms will have to weigh the benefits of avoiding state regulation against the costs of moving operations to federal offshore waters.

Postscript

In 2005, an NOAA-drafted bill, the National Offshore Aquaculture Act of 2005 was introduced in the U.S. Senate, as bill S.1195. Shortly thereafter, the Congress passed the Energy Policy Act of 2005, which at section 388 (to be codified at 43 U.S.C. § 1337) authorizes the MMS to enter into leases for offshore renewable energy production and for marine-related activities associated with an offshore energy facility. Although this authority arguably includes leasing for aquaculture, at present MMS is only considering allowing platforms to be converted to aquaculture use provided that another agency such as the NOAA approves the underlying activity. 70 Federal Register 77346 (30 December 2005). The Cicin-Sain, *et al.*, report on an offshore aquaculture framework, *supra* n. 4, was released to the public in early 2006 and follows the approach outlined above. Finally, the Pew Charitable Trusts and Woods Hole Oceanographic Institution established Marine Aquaculture Task Force in July 2005; the Task Force is expected to complete its work in early 2007 (http://www.whoi.edu/sbl/liteSite.do?litesiteid=2790).

Notes

The author wishes to acknowledge Biliana Cicin-Sain and the other members of the Offshore Marine Aquaculture Policy Project and Robert Barber for advancing his thinking on these matters, and Brooks Bowen for his careful eye. Portions of this chapter are adapted from Jeremy Firestone and Robert Barber, "Fish as Pollutants: Limitations of and Crosscurrents in Law, Science, Management, and Policy" (2003) 78 *Washington Law Review* 693–756.

1 National Research Council, Commission on Natural Resources, Committee on Aquaculture Board on Agriculture and Renewable Resources, *Aquaculture in the United States: Constraints and Opportunities* (Washington, DC: National Academy Press, 1978), quoted in National Marine Fisheries Service (NMFS), *The Rationale for a New Initiative in Marine Aquaculture* (Washington, DC: NOAA, September 2002) at 5. Online. Available http://www.nmfs.noaa.gov/trade/AQ/AQWPPrint.pdf (accessed 26 February 2004).

2 National Research Council, Committee on Assessment of Technology and Opportunities for Marine Aquaculture in the United States, Marine Board, Commission on Engineering and Technical Systems, *Marine Aquaculture: Opportunities for Growth Committee* (Washington, DC: National Academy Press, 1992) at 7. Online. Available http://search.nap.edu/books/0309046750/html/ (accessed 26 February 2004).

3 Biliana Cicin-Sain, S. M. Bunsick, R. DeVoe, T. Eichenberg, J. Ewart, H. Halvorson, R. W. Knecht and R. Rheault, *Development of a Policy Framework for Offshore Marine Aquaculture in the 3–200 Mile US Ocean Zone* (Newark, DE: Center for the Study of Marine Policy, August 2000). Online. Available http://darc.cms.udel.edu/sgeez/sgeez1.html (accessed 26 February 2004).

4 See Biliana Cicin-Sain, Susan M. Bunsick, John Corbin, M. Richard DeVoe, Tim Eichenberg, John Ewart, Jeremy Firestone, Kristen Fletcher, Harlyn O. Halvorson, Tony MacDonald, Ralph Rayburn, Robert B. Rheault and Boyce Thorne-Miller, *Towards an Operational Framework for Offshore Aquaculture in the United States*, Center for the Study of Marine Policy. See: Delaware

Aquaculture Resource Center (DARC) website, http://darc.cms.udel. edu/sgeez/ about.html (accessed 26 February 2004); the US Commission on Ocean Policy website, http://www.oceancommission.gov (accessed 26 February 2004); and the Pew Oceans Commission website, http://www.pewoceans.org (accessed 26 February 2004). The Pew Oceans Commission released its final report, *America's Living Oceans: Charting a Course for Sea Change* (2003), on 4 June 2003. Online. Available http://www.pewoceans.org/oceans/downloads/ oceans_report.pdf (accessed 26 February 2004). The US Commission on Ocean Policy issued its final 800-page report, *An Ocean Blueprint for the 21st Century: Final Report of the U.S. Commission on Ocean Policy* (2004) on 20 September 2004. Pre-publication copy, online. Available http://www.oceancommission. gov/documents/prepub_report/welcome.html (accessed 23 December 2004). Chapter 22 of that report addresses marine aquaculture.

5 Sena S. De Silva, "A Global Perspective of Aquaculture in the New Millennium," in R. P. Subasinghe, P. Bueno, M. J. Phillips, C. Hough, S. E. McGladdery and J. R. Arthur, eds., *Aquaculture in the Third Millennium* (Bangkok: NACA; and Rome: FAO, 2001) at 431–459.

6 *Ibid.*

7 Rebecca J. Goldburg, Matthew S. Elliott and Rosamond L. Naylor, *Marine Aquaculture in the United States: Environmental Impacts and Policy Options* (Arlington, VA: Pew Oceans Commission, 2001). Online. Available http://www.pewoceans.org/reports/137PEWAquacultureF.pdf (accessed 26 February 2004). Aquaculture is the fastest-growing food-producing sector in the United States. NMFS, *supra* note 1, App. II at 30.

8 United States Department of Agriculture (USDA), *1996 Census of Aquaculture*, AC97-SP3 (Washington, DC: USDA, 1997).

9 United States Congress, Office of Technology Assessment (OTA), *Current Status of Federal Involvement in US Aquaculture*, OTA-BP-ENV-170 (Washington, DC: OTA, 1995).

10 NMFS, *supra* note 1 at 18–19 and App. II at 29.

11 Goldburg *et al.*, *supra* note 7 at 2. In North America, aquaculture production increased an average of 3.6 percent per year from 1984 to 2001. Global aquaculture production has increased by 9 percent per year over the same time frame. See Paul G. Olin, "Current Status of Aquaculture in North America in Aquaculture in the Third Millennium," at 377–396, in *Aquaculture in the Third Millennium*, *supra* note 5.

12 Goldburg *et al.*, *supra* note 7 at 2. Americans consume 20.9 kilograms of fish and shellfish per year and spend US$54.4 billion per year on fishery products, a little over two-thirds of which is at food establishments. NMFS, *supra* note 1, App. I at 27.

13 NMFS, *supra* note 1 at 28.

14 United Nations Food and Agriculture Organization (FAO), "Long-Term Outlook: Some Plausible Structural Changes in Production and Demand" (Chapter 4), *State of World Fisheries and Aquaculture (SOFIA)* (Rome: FAO, 2000).

15 NMFS, *supra* note 1, App. II at 30.

16 *Ibid.*

17 See Goldburg *et al.*, *supra* note 7 at 2.

18 *Clean Water Act*, U.S.C., v. 33, s. 1342 and 1343; *Rivers and Harbors Act* (RHA), U.S.C., v. 33, s. 403; and letter from James W. Brennan, Acting General Counsel, NOAA, to Robert Blumberg, Bureau of Oceans and International Environmental and Scientific Affairs, US Department of State, *Re: American Norwegian Fish Farm, Inc.* (1 February 1993); and attached memorandum from Jay S. Johnson, Deputy General Counsel, and Margaret F. Hayes,

Assistant General Counsel for Fisheries, to James W. Brennan, *Regulation of Aquaculture in the EEZ* (1 February 1993).

19 *Outer Continental Shelf Lands Act* (OCSLA), U.S.C., v. 43, ss. 1331ff.
20 A number of other potential uses of federal waters are being discussed, with wind farming receiving most of the attention.
21 The future is now. In July 2003, NOAA/NMFS published a notice of an application for an exemption from fish permitting to authorize a twenty-four-month study of the feasibility of culturing cobia, mahi-mahi, greater amberjack, Florida pompano, red snapper and cubera snapper in net-cages 33 miles off the coast of Florida. NOAA, Fisheries of the Caribbean, Gulf of Mexico, and South Atlantic; Reef Fish Fishery of the Gulf of Mexico; Coastal Migratory Pelagic Resources of the Gulf of Mexico and South Atlantic; Exempted Fishing Permit, F.R., v. 68 at 44745–44747 (30 July 2003).
22 Proclamation 5928, 27 December 1988; *Magnuson–Stevens Fishery Conservation and Management Act*, U.S.C., v. 16, ss. 1851ff.
23 43 U.S.C., s. 1301.
24 U.S.C. 1997, v. 33, s. 1362(8).
25 Litigation pursuant to the SLA resulted in Texas and Florida (Gulf Coast only) being awarded title to submerged lands extending 3 marine leagues (about 9 nautical miles) because those were the boundaries as they existed at the time of statehood. See, for example, *United States* v. *Florida*, 420 U.S. 531 (1975). The CWA recognizes no exceptions. See *NRDC* v. *EPA*, 863 F.2d 1420 (9th Cir. 1988).
26 With notable exceptions (Florida, Hawaii and Maine), most states, like the federal government, lack a comprehensive aquaculture framework. Pew Oceans Commission, *supra* note 4 at 78.
27 U.S.C. 1997, v. 16, ss. 1451–1465.
28 C.F.R. 2002, v. 15, s. 930.57(a).
29 C.F.R. 2002, v. 15, s. 930.63(e) and 930.130(e).
30 A.S., s. 16.40.210; California Senate Bill 245 (approved 12 October 2003) (prohibiting the spawning, incubation or cultivation of any species of finfish belonging to the family Salmonidae as well as transgenic fish species and any exotic species of finfish). Online. Available http://www.leginfo.ca.gov/pub/bill/sen/sb_0201-0250/sb_245_bill_20031012_chaptered.html (accessed 25 February 2004).
31 U.S.C. 1997, v. 16, ss. 2801–2810.
32 For an in-depth history of aquaculture development, see Robert Roy Strickney, *Aquaculture in the United States: A Historical Survey* (New York: John Wiley, 1996).
33 See OTA, *supra* note 9.
34 *Interagency Agreement among Department of Agriculture, Department of Commerce, Department of the Interior, Subject: Designation of Areas of Responsibility in Aquaculture* (Washington, DC: USDA, USDOC and USDOI, 1980).
35 *Ibid.* at Appendix D.
36 U.S.C. 1997, v. 16, s. 2801(a)(7).
37 See mission statement. Online. Available http://ag.ansc.purdue.edu/aquanic/jsa/mission.htm (26 February 2004).
38 *Supra* note 3 at 64.
39 OTA, *supra* note 9, Foreword at v.
40 Online. Available http://aquanic.org/publicat/govagen/usda/dnadp.htm (accessed 26 February 2004).
41 For a discussion of offshore wind in addition to mariculture, see Jeremy Firestone, Willett Kempton, Andrew Krueger and Christy E. Loper, "Regulating Offshore Wind Power and Aquaculture: Messages from Land and Sea," (2004) 14(1) *Cornell Journal of Law and Public Policy*, 71–111.

42 U.S.C., v. 16, ss. 1851ff.
43 U.S.C., v. 16, s. 1802 (15).
44 U.S.C., v. 16, s. 1802 (42).
45 See Johnson, *supra* note 18.
46 See Brennan, *supra* note 18.
47 NOAA/NMFS may now see the difficulties that its position has generated. It now cites the enactment of laws designed to "manage wild-stock fisheries or natural resources without consideration for" the national aquaculture industry as a constraint on the industry. NMFS, *supra* note 1 at 22.
48 U.S.C., v. 33, ss. 1311, 1342 and 1362(12); *EPA* v. *California ex rel. State Water Resources Control Board*, 426 U.S. 200, 205 (1976); *International Paper* v. *Ouellette*, 479 U.S. 481, 489 (1987).
49 The territorial sea is defined for the purposes of the CWA as that within 3 nautical miles of the shore (U.S.C., v. 33, s. 1362(8)). The CWA has not been amended to bring it into line with President Reagan's proclamation that extended the territorial sea to 12 nautical miles, Proclamation 5928 (27 December 1988), or Article 3 of the *United Nations Convention on Law of the Sea*, which provides the same. *UN Convention on the Law of the Sea*, 10 December 1982, 21 *International Legal Materials* 1261 (entered into force 16 November 1994), Article 3. Online. Available http://www.un.org/Depts/los/ convention_agreements/convention_overview_convention.htm (accessed 26 February 2004). For a view of the federal government's present view of jurisdictional boundaries, see: Department of Homeland Security, Coast Guard, Territorial Seas, Navigable Waters, and Jurisdiction, F.R., v. 68 at 42595–42602 (18 July 2003).
50 U.S.C., v. 33, s. 1343.
51 Regulations implementing section 403 are found at C.F.R., v. 40, s. 125.120–125.124 (2003). See also Ocean Discharge Criteria, 45 F.R., v. 45 at 65, 942 (3 October 1980) (to be codified at C.F.R., v. 40, pt. 125). In the waning days of the Clinton administration, the EPA announced a proposed revision to Ocean Discharge Criteria, but did not publish the rule prior to the beginning of the Bush administration. On 24 January 2001, the Bush administration "withdrew" the proposed revision in order to provide the new EPA Administrator with an opportunity to review it. See National Legal Center for the Public Interest, *Judicial Legislative Watch Report*, March 2001, Vol. 22(4) at 1, 5. Online. Available http://www.nlcpi.org/books/pdf/jlwr_march01.pdf (accessed 1 March 2004). To date, no further action has been taken on the proposed rule.
52 C.F.R., v. 40, s. 125.122.
53 C.F.R., v. 40, s. 122.124.
54 Facilities that grow or hold cold-water fish or other aquatic animals and that produce less than 9,090 harvest weight kilograms (approximately 20,000 pounds) in a year *or* that feed less than 2,272 kilograms (approximately 5,000 pounds) of food during a calendar month are excluded, as are those that grow or hold warm-water fish or other aquatic animals and that produce less than 45,454 harvest weight kilograms (approximately 100,000 pounds) in a year *or* that employ closed ponds that discharge only during periods of excess runoff. Regulations, *supra* note 51.
55 C.F.R., v. 40, s. 122.125.
56 EPA underscored its ambivalence to address this "potential area of concern" with its statement that it was "considering whether it should establish national requirements for net pens systems at all." US EPA, *Effluent Limitation Guidelines and New Source Performance Standards for the Concentrated Aquatic Animal Production Point Source Category*; Proposed Rule, F.R., v. 67 at 57872, 57901 (12 September 2002), to be codified at C.F.R., v. 40, pt. 451.

57 *Ibid.* at 57928.
58 US EPA, *Effluent Limitation Guidelines and New Source Performance Standards for the Concentrated Aquatic Animal Production Point Source Category*; Final Rule, F.R., v. 67 at 51892–51930 (23 August 2004), to be codified at C.F.R., v. 40, pt. 451.
59 *Ibid.* at 51911.
60 US EPA, *Draft Aquatic Nuisance Species in Ballast Water Discharges: Issues and Options* (Washington, DC: US EPA, 10 September 2001) at 31. Online. Available http://www.epa.gov/owow/invasive_species/ballast_report/ (accessed 26 February 2004).
61 F.R. v. 67 at 57925, to be codified at C.F.R., v. 40, s. 451.2(k). The proposed effluent guidelines explicitly exempt "species raised for stocking by public agencies" from the definition of "non-native aquatic animal species" regardless of whether that stocking is for the purpose of endangered species enhancement, commercial enhancement or sport/recreational fishing. *Ibid.*
62 F.R., v. 67 at 57887.
63 NMFS, *A Code of Conduct for Responsible Aquaculture Development in the US Exclusive Economic Zone* (Washington, DC: NMFS, 23 August 2002). Online. Available http://www.nmfs.noaa.gov/aquaculture.htm (accessed 26 February 2004). Other efforts by the federal government to improve the environmental performance of aquaculture operations in the United States include: (1) US EPA, *Draft Guidance for Aquatic Animal Production Facilities to Assist in Reducing the Discharge of Pollutants*, EPA-821-B-02-002 (Washington, DC: US EPA, August 2002), Online. Available http://epa.gov/guide/aquaculture/guidance/complete.pdf (accessed 26 February 2004) (setting forth best management practices and attempts to assist aquaculturalists in meeting the proposed effluent guidelines); and (2) Atlantic States Marine Fisheries Commission (ASMFC), *Guidance Relative to Development of Responsible Aquaculture Activities in Atlantic Coast States*, Special Report No. 76 (Washington, DC: ASMFC, November 2002).
64 Fish and Wildlife Service, Department of the Interior and National Oceanic and Atmospheric Administration, Department of Commerce, *Endangered and Threatened Species; Final Endangered Status for a Distinct Population Segment of Anadromous Atlantic Salmon* (Salmo salar) *in the Gulf of Maine*, F.R., v. 65 at 69459, 69460, 69465 (17 November 2000).
65 *Ibid.* at 69464, 69471 and 69478.
66 The *Endangered Species Act*, U.S.C., v. 16, s. 1532(16), regulates species, subspecies and distinct population segments.
67 Committee on Atlantic Salmon in Maine, Board on Environmental Studies and Toxicology, Ocean Studies Board, Natural Research Council, *Genetic Status of Atlantic Salmon in Maine: Interim Report* (Washington, DC: NAS, 2000) at 20.
68 *Ibid.*
69 For a more detailed discussion, see Jeremy Firestone and Robert Barber, "Fish as Pollutants: Limitations of and Crosscurrents in Law, Science, Management, and Policy," (2003) 78 *Washington Law Review* 693–756.
70 215 F. Supp. 2d 239 (D. Maine 2002).
71 The court made similar findings involving different pollutants in two unpublished companion cases.
72 33 U.S.C. 1362(6).
73 215 F. Supp 2d. at 243–244 and 247–248.
74 C.F.R., v. 40, s. 401.15(22).
75 For an argument that aquaculture escapees regardless of origin (and releases of stocked fish for recreational fisheries) are "pollutants," see Firestone and Barber, *supra* note 69.

76 Civ. No. 00-150-B-S (D. Maine 2002).

77 Consent Decree and Order, paras. 17, 18, 22 and 23.

78 *Ibid.* at paras. 31–53.

79 *Ibid.* at paras. 54 and 57.

80 No. 99-2-00797-0 (Wash. Sup. Ct. 1 December 2000).

81 *Ibid.* at 4, paras. I.VIII and I.IX.

82 *Ibid.* at 7, para. I.XIII.

83 *Ibid.* at paras. II.VIII and II.X.

84 (APHETI) 299 F.3d 1007 (9th Cir. 2002).

85 299 F.3d at 1016–1019.

86 *Ibid.* at 1017.

87 Cicin-Sain *et al., supra* note 3, Chapter 2.

88 See, for example, Tim Eichenberg and Barbara Vestal, "Improving the Legal Framework for Marine Aquaculture: The Role of Water Quality Laws and the Public Trust Doctrine," (1992) 2 *Territorial Sea Journal* 339; and Alison Rieser, "Defining the Federal Role in Offshore Aquaculture: Should It Feature Delegation to the States?" (1997) 2 *Ocean and Coastal Law Journal* 209–234. The latter article grew out of a 1996 symposium "Open Ocean Aquaculture" and was published with other articles on aquaculture in a special 1997 issue of the *Ocean and Coastal Law Journal*.

89 See Cicin-Sain *et al., supra* note 3.

90 *Ibid.* at 139.

91 The author is a member of the expanded research team.

92 Letter from Dr. Biliana Cicin-Sain, Principal Investigator, and Members of Offshore Marine Aquaculture Policy Project to Admiral James D. Watkins, Chair, US Commission on Ocean Policy, and members of the US Commission on Ocean Policy, 19 May 2003.

93 *Ibid.* at 2.

94 *Ibid.*

95 Addendum to Letter to US Commission on Ocean Policy – *19 May 2003, ibid.* at 2, Detailed Draft Recommendation No. 3.

96 *Ibid.* at 2.

97 Addendum, *supra* note 95, Detailed Draft Recommendation No. 7.

98 *Ibid.*

99 *Ibid.* at Detailed Draft Recommendation No. 33.

100 For a discussion of the leasing needs of aquaculture firms and the responsibilities that would then attach, see Richard M. DeVoe, "Regulation and Permitting," in R. R. Stickney, ed., *Encyclopedia of Aquaculture* (New York: John Wiley, 2000) at 744–760.

101 Addendum, *supra* note 95, Detailed Draft Recommendation No. 11.

102 Letter to US Commission on Ocean Policy – *19 May 2003, supra* note 92 at 2; Addendum, *supra* note 95, Detailed Draft Recommendation No. 12.

103 Addendum, *supra* note 95, Detailed Draft Recommendation No. 13.

104 Letter to US Commission on Ocean Policy – *19 May 2003, supra* note 92 at 2.

105 U.S.C., v. 42, s. 4332.

106 Addendum, *supra* note 95, Detailed Draft Recommendation No. 19. Recently, the Bush II Administration argued otherwise; however, a federal court found that the NEPA applies in the EEZ. See *Natural Resources Defense Council* v. *United States Department of the Navy*, CV-01-07781 CAS (RZx), (C.D. CA 17 September 2002) (opinion by Christina A. Snyder) (despite the presumption against extraterritoriality, the NEPA applies in the EEZ).

107 Addendum, *supra* note 95, Detailed Draft Recommendation No. 20.

108 *Ibid.* at Detailed Draft Recommendation Nos. 24 and 34.

109 *Ibid.* at Detailed Draft Recommendation No. 29.
110 *Ibid.* at Detailed Draft Recommendation No. 33; Letter to US Commission on Ocean Policy – *19 May 2003, supra* note 92 at 2.
111 *Ibid.*
112 Addendum, *supra* note 95, Detailed Draft Recommendation No. 28.
113 Pew Oceans Commission, *supra* note 4 at 102.
114 *Ibid.* at 103–105.
115 *Ibid.* at 107–108.
116 *Ibid.* at 126–127.
117 US Commission on Ocean Policy, *supra* note 4, Chapter 3, at 32–33. In response, on 17 December 2004 the President established a Cabinet-level panel on ocean policy and released a 40-page Ocean Action Plan. Press Briefing by Chairman of the Council on Environmental Quality, James Connaughton on US Oceans Action Plan (17 December 2004). Online. Available http://www.whitehouse.gov/news/releases/2004/12/20041217-12.html (accessed 23 December 2004); Executive Order: Committee on Ocean Policy (17 December 2004). Online. Available http://www.whitehouse.gov/news/releases/2004/12/20041217-5.html (accessed 23 December 2004).
118 *Ibid.* at 48–49.
119 *Ibid.* at 73–80.
120 *Ibid.* at 57–60.
121 *Ibid.* at Chapter 22. It notes that the United States presently has a US$7 billion "seafood trade deficit." *Ibid.* at 285.
122 *Ibid.* at 288.
123 *Ibid.* at 288–289.
124 *Ibid.* at 289.
125 See Goldburg *et al.*, *supra* note 7.
126 U.S.C., v. 7, ss. 136ff.
127 U.S.C., v. 21, ss. 301ff.
128 U.S.C., v. 33, s. 403.
129 U.S.C., v. 16, s. 1538.
130 U.S.C., v. 16, s. 1372.
131 U.S.C., v. 33, s. 1341.
132 U.S.C., v. 33, s. 1362(8); *NRDC* v. *EPA*, 863 F.2d 1420 (9th Cir. 1988).
133 U.S.C., v. 16, s. 1456.
134 See, for example, A.A.C., v. 5, s. 41.005; C.G.S., ss. 26–57.

15 Australian aquaculture
Opportunities and challenges

Marcus Haward

Introduction

Australia has responsibility for the fourth largest maritime jurisdiction in the world. The Australian exclusive economic zone (EEZ) and claimable continental shelf is 16 million km², extending from tropical to Antarctic waters. While the Australian EEZ is not highly productive on a world scale, it nonetheless supports a number of commercially lucrative fisheries, including tuna and billfish, high-value shellfish and crustaceans, and increasingly important mariculture of salmonids and southern bluefin tuna.[1] Australian fisheries have experienced a period of impressive growth in the recent past, driven by significant developments in aquaculture. The gross value of Australian seafood production in 2002–2003 was A\$2.3 billion[2] and is forecast to reach A\$5 billion by 2020. It is expected that aquaculture developments will furnish the vast majority of this economic growth. Aquaculture production has trebled in the decade to 2002–2003, and the value of this production more than doubled in the same period. Exports of aquaculture have grown in value from A\$246.7 million in 1991–1992 to A743.5 million in 2002–2003[3] (see Figures 15.1 and 15.2). Moreover, aquaculture and associated processing are vital rural industries, sustaining regional communities around the coastline.[4]

This chapter begins with a brief overview of the development of aquaculture in Australia, followed by an outline of the legal and policy framework governing Australian fisheries and aquaculture operations. Jurisdiction, and resultant interaction between the Commonwealth and state governments,[5] is a key to this framework. This chapter focuses on the opportunities and challenges facing aquaculture operations in Australia as identified in the development of a national Aquaculture Industry Action Agenda. The Aquaculture Industry Action Agenda aims to achieve a vision of Australia "being the world's most globally competitive aquaculture producer," with an industry mission of "total commitment to economic, social and environmental benefits from aquaculture."

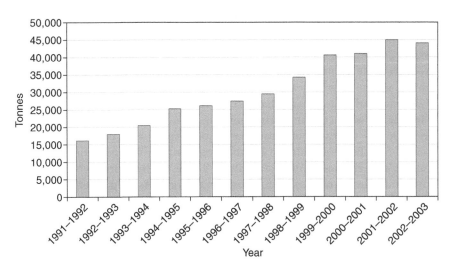

Figure 15.1 Australian aquaculture production (sources: Australian Bureau of Agricultural and Resource Economics (ABARE); *Australian Fisheries Statistics 2003*; *Australian Fisheries Statistics 1994*).

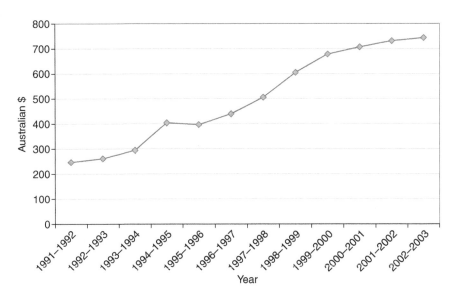

Figure 15.2 Australian aquaculture: value (sources: as Figure 15.1).

Australian aquaculture: developments

Australian aquaculture began with the "culturing" of shellfish, most notably oysters and mussels, in the latter decades of the nineteenth century. Pearl production developed in the 1950s[6] using Japanese-developed techniques. Finfish aquaculture increased in significance with commercial development of Atlantic salmon cage aquaculture in the mid-1980s and successful cage "farming" of wild-caught southern bluefin tuna in the 1990s.[7] Finfish operations are the mainstay of Australian aquaculture, with new species being trialed. Shellfish aquaculture is also developing, with operations based on scallops and abalone being established. These shellfish products are aimed both at domestic markets and for export to north Asian markets. The Australian aquaculture industry is diverse, comprising approximately 3,000 individual businesses[8] and fifty aquaculture associations and councils. The National Aquaculture Council represents 60 percent of the aquaculture industry by value.[9] The lack of a national peak industry body has been identified as a current constraint, with such a body seen as being able to "improve the industry's contribution to aquaculture policy and program development."[10]

The development of finfish aquaculture in Australia began with the successful introduction of Atlantic salmon aquaculture into Tasmania. This introduction occurred as the state pursued an aggressive developmental strategy for its fisheries in the late 1970s, seeking out new opportunities and niche markets. Salmon that had been imported into Australia as a sport fish in the 1960s (into lakes associated with major hydroelectric developments in the state of New South Wales) provided the original source of stock. This had the advantage of quarantining the original stock from problems associated with transferring wild-caught fish into farming operations. Tasmania had a long history of work with salmonids. Trout and trout eggs were imported from England into the state in the mid-1880s to support an angling industry in highland rivers. These fish provided the base of trout hatchery material and led to regular introductions of fry into Tasmanian rivers and lakes.

Atlantic salmon aquaculture in Australia is centered on Tasmanian waters. The operations are based on cages located in estuaries and inshore areas, with related onshore facilities close to the cages. As in other states, management of aquaculture in Tasmania (including tenure and lease of sites, fisheries husbandry and processing, environmental controls, occupational health and safety, and marine farm planning) is regulated by state legislation and regulations.[11] In 2002–2003, the farm-gate value of Atlantic salmon was approximately A\$108 million, with more than 14,000 tonnes produced for both the domestic and export markets.[12] Profitability of Tasmanian operations has been affected by climatic conditions (higher than average summer water temperatures impact on the reproductive fitness of brood stock[13]) and global oversupply of fish. This led to a further rationalization of the industry with the merger of Tasmania's two largest salmon producers in February 2003. This occurred when Tassal, the state's largest salmon producer, was in receivership, with debts of A$40 million. The merged company traded out of financial difficulty, and the new entity had a successful stock market launch on November 2003.

The "farming" of southern bluefin tuna has developed from an experimental farm introduced in 1991–1992. A number of factors have contributed to the development of this aquaculture operation. Substantial reductions in quotas of wild fish caught between the three major harvesting countries, Japan, Australia and New Zealand,[14] and reduced profitability of Australian operations encouraged innovative approaches to returning the fishery to viability. Initial experiments in Japan had indicated that bluefin tuna (a fast-swimming, highly migratory species in the wild) could adapt to cage conditions.[15] Active support from Australian industry encouraged Japanese interests to establish a larger-scale experimental cage operation in the relatively sheltered waters off Port Lincoln, South Australia.[16] This farm attracted funding and technical support from the Japanese Overseas Fishery Cooperation Foundation in a collaborative venture with the Tuna Boat Owners Association of Australia.

Initial success and "proving up" of the farming operations led to a major commercial focus of these operations from the mid-1990s. Australian operators gradually shifted operations from fishing to farming, supported by their close links with Japanese importers. Australia targeted product for Japanese supermarkets, which in the 1980s began to displace the fishmonger as the point of sale of fisheries products.[17] At present, over 95 percent of the Australian catch[18] of southern bluefin tuna is placed in cages and fattened for processing and export to Japan. The Australian quota of 5,265 tonnes of southern bluefin tuna is grown out to over 8,000 tonnes of high-grade sashimi tuna.[19] The tuna farming operation has been commercially successful, with a gross value of production exceeding A$250 million in 2000–2001.[20] The Australian operations have attracted considerable worldwide interest. Australian technology is now utilized by, and Australian interests are now linked to, tuna farms in Asia, Europe and North America.

The development of tuna farming has raised several issues. A major storm in 1996 saw major losses of stock as a result of cages being located in shallow water. As a result, cages are rotated more regularly and moved into deeper water if weather conditions warrant. Ongoing environmental monitoring is undertaken around and in the cages. Research is continuing on feeds (caged southern bluefin tuna are fed pilchards and other seafood, with frozen pilchards imported from Chile[21]). In Tasmania, there has been criticism of the environmental impacts of the tuna farming operations, and court action over siting and operations of cage-based aquaculture.[22]

While tuna and salmon operations dominate the value of Australian aquaculture, considerable efforts are being made with abalone. Shore-based propagation of abalone is undertaken in Tasmania, Victoria and South

Australia, with the farmed product aimed to fill a niche market for small fish in North Asian markets. An innovative approach has been the use of a converted freighter as an "at sea" abalone farm, reducing the costs associated with land-based abalone farming.[23]

Australian fisheries and aquaculture management: the legal and policy framework

Jurisdiction over Australian fisheries is shared between the Commonwealth and the states.[24] Commonwealth power is based on s. 51(x) of the Constitution ("fisheries in Australian waters beyond territorial limits") and state legislation through provisions of each state's constitution.[25] Each of the states and the mainland territories has a coastline and interests in coastal management.[26] The question of jurisdiction over fisheries increased in salience from the late 1940s, with jurisdiction divided by a boundary established by the three-nautical mile (nm) territorial sea. Resolution of jurisdictional conflicts in the 1970s saw this boundary entrenched by the Offshore Constitutional Settlement (OCS).[27] The OCS established a series of "agreed arrangements" affecting management of a variety of marine resource sectors, including fisheries. In effect, these arrangements entrenched the state or territory jurisdiction over what became known as "state waters" – waters within a line drawn three nm offshore of the territorial sea baseline. The OCS enables arrangements to be established to "streamline" management of fisheries that transcended the three nm boundary.[28]

The OCS has remained a major policy framework for Australian fisheries and aquaculture, with aquaculture operations managed by the states. It includes a range of environmental and planning instruments, as well as local government planning controls. Despite this direct involvement by the states and local government, the Commonwealth retains important influence in Australian aquaculture. The Commonwealth controls quarantine (including the importation of aquaculture feeds), regulates biosafety and gene technology, and increasingly provides environmental oversight of state fisheries. In addition, the Commonwealth provides export market certification, is concerned with the control and management of aquatic animal diseases, and supports research and development, economic forecasts and trade information.[29]

Management of Australian fisheries underwent substantial changes in the 1990s. A major turning point in the administration and management of Australian (and particularly Commonwealth) fisheries occurred in early 1985, when the Australian Fisheries Conference was held in Canberra. Prior to this conference, the Senate Standing Committee on Trade and Commerce undertook a major review of the Australian fishing industry.[30] This report, presented to Parliament in 1982, made a series of recommendations related to the development of the Australian fishing industry. The recommendations of the Fisheries Conference and the Archer Report heralded a new era in fisheries management. The 1980s ended with the release of a major Commonwealth policy statement, *New Directions for Commonwealth Fisheries Management in the 1990s*.[31] This new era was, as noted earlier, also marked by increasing interest in aquaculture operations, following the successful commercialization of sea-caged Atlantic salmon operations in Tasmania.

The *New Directions* statement provided the basis of major legislative and administrative changes implemented by the Commonwealth government in 1991. These included the development of a statutory body, the Australian Fisheries Management Authority (AFMA), to assume management responsibilities previously undertaken by the Australian Fisheries Service within the Department of Primary Industry and Energy. Government acceptance of the focus of the *New Directions* statement led to the development of a package of laws in 1991 that, with amendments, provide the basis for current management of Commonwealth fisheries.[32]

Developments in broader Commonwealth government policies, primarily through the implementation of *Australia's Oceans Policy*,[33] reinforce the requirement that fisheries policy and management (including aquaculture) need to be reorientated towards an ecosystem-based approach.[34] One of the more significant challenges affecting Australian fisheries policy and management has been the increasing external scrutiny of management.[35] This is reflected in the growing impact of Commonwealth environmental legislation on fisheries management, development of sustainability indicators, and the extension in 1999 of schedule 4 of the *Wildlife Protection (Regulation of Exports and Imports) Act 1982* to fisheries. This scrutiny has extended to state-administered aquaculture operations, with several cases being brought before state courts over impacts and operations of aquaculture operations.

The *Environment Protection and Biodiversity Conservation Act 1999* (EPBC Act) entered into force in July 2000 and gives domestic effect to the Biodiversity Convention and the Jakarta Mandate.[36] Prior to the development of this legislation, these commitments were met by the National Biological Diversity Strategy, which, *inter alia*, endorsed the adoption of ecologically sustainable fishing practices to achieve conservation of biological diversity in fisheries. The EPBC Act provides protection for matters of national environmental importance, including World Heritage properties, Ramsar wetlands of international importance, nationally threatened animal and plant species and ecological communities, internationally protected migratory species, and Commonwealth marine areas.[37] The EPBC Act protects the environment on Commonwealth land and regulates actions of Commonwealth departments and agencies that may have a significant impact on the environment. It provides for a streamlined system for environmental assessment and project approval through the establishment of bilateral agreements with the states and territories.

The EPBC Act promotes ecologically sustainable development and establishes an integrated approach to biological conservation. It establishes the Australian whale sanctuary and contains provision for public comment and consultation. The EPBC Act replaces the following Commonwealth legislation: the *Environment Protection (Impact of Proposals) Act 1974*; the *Endangered Species Protection Act 1992*; the *National Parks and Wildlife Conservation Act 1975*; the *World Heritage Properties Conservation Act 1983*; and the *Whale Protection Act 1980*. The EPBC Act gives the Department of Environment and Heritage a major role in reviewing management arrangements established after July 2000. It also provides statutory support for management of by-catch in Australian fisheries.[38]

In 2000, a review of Commonwealth fisheries policy was announced. This review recognized that Australian fisheries have undergone significant changes since the release of *New Directions*, and that a new policy statement was needed.[39] The increased importance and future potential of aquaculture was highlighted as one "change driver" influencing the development of a new policy statement. While progress was made in developing the statement in 1999–2000, the federal election of 2001 and appointment of a new minister responsible for fisheries after the re-election of the Liberal coalition government postponed the completion of the policy review. *Looking to the Future: A Review of Commonwealth Fisheries Policy*[40] was released in June 2003. This report noted the "changing policy environment" since the release of the *New Directions* statement in 1989. One change that was identified was the "rapidly growing aquaculture industry."[41] The report noted that while "aquaculture in Australia occurs almost exclusively in state and territory waters,"[42] "there will be an emerging requirement for the aquaculture industry to gain access to Commonwealth waters to cultivate and farm marine species."[43] The failure of the Commonwealth's *Fisheries Management Act* to address aquaculture activities was seen to create uncertainty.[44]

As was indicated earlier, the state and territory governments have responsibility for aquaculture in Australia, as these operations occur within the three-nm state waters zone. Australian aquaculture operations generally take place within state waters – that is, the area from low-water mark to three nm offshore within the twelve-nm territorial sea, with the states having primary responsibility for management and regulation of aquaculture. The states administer aquaculture through a variety of legislation, such as a section with a broader fisheries Act as in Victoria,[45] or dedicated marine farming or aquaculture legislation as in Tasmania.[46] The development of aquaculture operations in the 1980s and 1990s was one impetus for legislative reviews of state fisheries; the other was the need to accommodate post-Brundtland Report[47] principles such as ecologically sustainable development and precautionary approaches to management, which had been established as broad policy parameters at both the Commonwealth and state levels of government.

Local government has an important role in aquaculture development: it is responsible for much of the terrestrial land-use planning and approvals for land-based infrastructure for marine farming operations. Aquaculture operations have focused considerable community attention on uses of the coastal zone, and siting, or planned expansion, of operations has rarely occurred without controversy over loss of access or amenity.

Notwithstanding the lack of formal Commonwealth jurisdiction within state waters, Commonwealth–state relations form the framework for ongoing development of Australian aquaculture. Indeed, there is currently no legal recognition for aquaculture in Commonwealth waters, as aquaculture in not recognized in the *Fisheries Management Act 1991*.[48] Clearly, as industry seeks to extend operations into Commonwealth waters, such lacunae need to be addressed. The Commonwealth government has, however, been active in fostering a national perspective on aquaculture developments. This interest has been derived from a number of directions, including the benefits of regional cooperation with Asian countries within the Network of Aquaculture Centres in Asia-Pacific (NACA).[49] The Commonwealth also has direct responsibility in areas such as quarantine, biosafety and regulation of the use of gene technology that directly impact on aquaculture.[50]

Commonwealth, the states and intergovernmental relations

Intergovernmental interaction is a significant feature of Australian federalism. While the states retain significant jurisdiction, the Commonwealth influences a range of policy areas though the process of judicial review, reinforcing Commonwealth jurisdiction in relation to the domestic implementation of international treaties, and its fiscal dominance gives it considerable power to allocate funds to the states.[51] These developments have increased the policy reach of the Commonwealth but have not occurred in a zero-sum manner. Increased influence by the Commonwealth has been matched by increased activity by the states and a resultant increase in intergovernmental institutions and processes. These institutions and processes are centered on ministerial councils comprising relevant Commonwealth, state and territory ministers, supported by committees of senior officials from each government. These institutions have been described as "lubricating the federal system"[52] and are seen to provide an important "interface" between Australian governments, enabling joint action in a vast number of policy areas.[53]

The importance of collaboration between Australian governments over aquaculture development has been a common theme from the mid-1980s. In 1988, the Australian Science and Technology Council's report *Casting the Net*[54] recommended preparation of an overview of the national status and future potential of aquaculture. The 1989 report of the Review Committee on Marine Industries, Science and Technology, *Oceans of Wealth*,[55] recommended undertaking "a study of government roles in the promotion of effective development of aquaculture" and that "each state should make a detailed review of the impact of the government regulatory framework on aquaculture."[56] An intergovernmental Working Group on Aquaculture

(WGA), with members from the states, the Northern Territory and the Commonwealth, were commissioned by the Standing Committee on Fisheries and Aquaculture (SCFA) to respond to these reports. The WGA response was approved by the SCFA and Ministerial Council in 1993, and formed the *National Strategy on Aquaculture in Australia*,[57] which was released in March 1994.

The *National Strategy* outlined ten goals that would "overcome the current constraints on the aquaculture industry" and "create an environment in which the aquaculture can capitalize on its advantages."[58] Goal 3 was to provide a "coordinated government framework to support industry development."[59] A further example of the importance of intergovernmental relations, together with industry involvement, was the collaborative effort to establish a national strategic plan for aquatic animal health (AQUAPLAN) in 1999. AQUAPLAN, a five-year plan, concluded its first iteration in 2003.[60] The need for long-term funding of this initiative has been recognized by industry and is currently being addressed as part of a broader government–industry agenda.

Government–industry relations

Unlike other primary industry sectors in Australia, fisheries and aquaculture lack strong and united peak industry bodies. One key challenge is to create effective industry bodies to engage with governments and help advance industry's interests. The diversity and spread of Australian aquaculture, while providing significant opportunities, poses a major challenge and reinforces the need for a strong industry body.

The aquaculture industry is a major player in regional Australia, making a significant contribution to regional economic and community development. Aquaculture operations have provided employment and can provide further economic multipliers in areas that have suffered significant economic decline. Aquaculture is also seen as providing considerable benefits to Australia's indigenous communities.[61] A major spin-off from the tuna farming and salmon aquaculture in Port Lincoln and southern Tasmania has been the range of small and medium-sized enterprises (SMEs) that have developed to support aquaculture operations. Initiatives such as the Kimberly Aquaculture Aboriginal Corporation (KAAC), established in 2002 with an operation focusing on trochus shell aquaculture but with opportunities to develop prawn and pearl farming in the future, are seen as providing similar benefits to aboriginal and islander communities.[62] Development of community-based and/or commercial aquaculture projects could support a host of related training and infrastructure development requirements that are central to the well-being and survival of these communities.[63]

Governments (both Commonwealth and state) have recognized the benefits of a sustainable aquaculture industry and have moved to ensure that ongoing development takes the form of a strategic partnership with industry. This model recognizes that industry is driven by commercial realities but operates within an environment shaped or bounded by government policies. This includes an important commitment to ecologically sustainable use of marine resources. The development of such a strategic partnership – the Aquaculture Industry Action Agenda – between 1999 and 2002 is a major achievement. The agenda identifies both the opportunities and the challenges facing Australian aquaculture.

The future: the Aquaculture Industry Action agenda

The Aquaculture Industry Action Agenda, comprising an industry-initiated and government-supported vision, mission statement and outline policy framework, arose from a national aquaculture workshop held in August 1999 that identified the opportunities and "impediments" to continued growth in this industry sector.[64] The workshop "delivered a clear commitment and provided a sound basis for the development of a national action plan for the Australian aquaculture industry" and "developed and endorsed . . . a vision (including a growth target) and mission statements for the Australian aquaculture industry."[65] The Agenda's vision is that "[b]y 2010 a vibrant and rapidly growing Australian aquaculture industry will achieve $2.5 billion in sales by being the world's most efficient aquaculture producer."[66] Its mission is "total commitment to economic, social and environmental benefits from aquaculture."[67]

The workshop noted the importance of industry coordination and organization, promotion of industry, recognition of the significance of environmental issues, markets and marketing, and research and development.[68] One outcome was an agreement among industry representatives to form a new peak industry body – the National Aquaculture Council (NAC) – to replace the Australian Aquaculture Forum. It was determined that NAC would be industry driven and open to organizations, businesses and individuals involved in aquaculture.[69]

In 2000, the Commonwealth government announced that it would work with industry to achieve industry's vision. An expert group, the National Aquaculture Development Committee, with a majority of industry members, was established to assist in the development of an agenda for action. The committee's terms of reference included direct reference to the industry vision. As part of the consultation process associated with the development of the action agenda, the National Aquaculture Development Committee released a discussion paper on June 2001 "to stimulate discussion on a range of issues during the development of the aquaculture Action Agenda."[70]

The Action Agenda is based on maximizing what Australian industry sees as its competitive advantages:

its ability to produce a wide range of warm and cold water species enabling us to access a wide range of niche markets. As a small producer by world standards, Australia relies on being able to target niche, high-value markets with a wide range of premium-quality produce.[71]

The National Aquaculture Development Committee noted that Australian aquaculture benefits from a "positive business environment, stable government, a strong regulatory framework for ecologically sustainable development, fish health, a highly skilled workforce, and world-class scientists and educational institutions."[72]

The report to government and industry by the National Aquaculture Development Committee included a summary of proposed initiatives and recommended actions.[73] These actions included a national aquaculture policy statement; implementing an industry-driven action agenda; growing the industry within an ecologically sustainable framework; investing for growth; promoting aquaculture products in Australia and globally; tackling research and innovation challenges; making the most of education, training and workplace opportunities; and creating an industry for all Australians.[74] In December 2002, the Commonwealth government announced that it supported the direction of the Action Agenda[75] and that it was progressing to implementing the recommendations from the National Aquaculture Development Committee. This included the release of a National Aquaculture Policy Statement and best-practice principles for the development and management of aquaculture. The National Aquaculture Policy Statement was released on 25 July 2003,[76] as part of a strategic direction and commitments arising from the Aquaculture Industry Action Agenda.

The Aquaculture Industry Action Agenda is seen as "an important step in [the] direction of ensuring responsible environmental management for long-term success."[77] The sustainable aquaculture independent working group reporting to the Prime Minister's Science, Engineering and Innovation Council (PMSEIC) noted that in addition to the critical issues of the industry's need for a sustained and focused research and development effort, two major issues stand out for joint consideration by government and industry: scale and marketing, and environment and sustainability.[78] Addressing these issues provides challenges as well as opportunities for Australian aquaculture. The current focus on niche markets for high-value products needs to be addressed and a strategy needs to be developed to expand markets and to address increasing competition in these new markets. The need to adopt a focused marketing strategy is also identified as a challenge, with the suggestion to borrow from the lessons learned from the highly successful model developed by the Australia wine industry.[79]

Conclusion

Australian aquaculture operations are significant, and targeted for continued expansion over the next decade. This planned expansion raises a number of opportunities but also challenges. The development and implementation of a National Aquaculture Policy, together with ongoing implementation of the Aquaculture Industry Action Agenda, are important elements in ensuring that opportunities are met and challenges confronted. The challenges facing governments include issues related to jurisdiction (e.g. property rights in aquaculture operations in Commonwealth waters) and ensuring appropriate policy settings to guide development. For industry, the challenge is to maintain commitment to the world's best practices and to ensure effective industry linkages and communication. Conflicts over siting and operations are likely to continue, and there will be increased importance placed on environmental management in the sector. Growth in the involvement of Indigenous peoples in aquaculture provides great opportunities for these communities. To adopt an industry-based metaphor, Australian aquaculture has succeeded in ensuring "grow-out" and has achieved much in the past decade. The challenge to ensure an ecologically sustainable industry remains a priority for industry and governments.

Notes

1 A. Caton, ed., *Fishery Status Reports 2000–01* (Canberra: Bureau of Rural Sciences, 2002).
2 Australian Bureau of Agricultural and Resource Economics (ABARE), *Australian Fisheries Statistics 2003* (Canberra: ABARE and Fisheries Research and Development Corporation (FRDC), 2004) at 1.
3 *Ibid.* at 32.
4 Agriculture, Fisheries and Forestry Australia (AFFA), *Fisheries and Forestry Industries Division Operational Plan 1999–2000* (Canberra: AFFA, 1999).
5 Australia is a federal state with division of powers and responsibilities based on a Constitution written on the US model and enacted in 1901. The federal or national government is termed the Commonwealth and the subnational divisions are known as states. The merging of a US-style federal compact with responsible government has provided tensions for most of the past century. The Constitution divides power between the Commonwealth or federal government and the states. In formal terms, the Commonwealth has a number of enumerated heads of power, with residual powers falling to the states. In practice, the Commonwealth exerts considerable influence over the states because of its significant financial power and the process of judicial review, which has seen its power to enact international treaties into domestic law impact on a range of areas that were formerly the responsibility of the states.
6 Department of Agriculture, Fisheries and Forestry – Australia (AFFA), *Australian Aquaculture*. Online. Available http://www.affa.gov.au (accessed 5 March 2004).
7 Caton, *supra* note 1.
8 National Aquaculture Development Committee (NADC), *Aquaculture Industry Action Agenda: Report to Government and Industry* (Canberra: Aquaculture Action Agenda Taskforce, Department of Agriculture, Fisheries and Forests – Australia, July 2002) at 19.
9 *Ibid.*
10 *Ibid.*
11 In Tasmania, the *Marine Farm Planning Act 1995* provides an overarching legislative framework. This Act, developed as part of a wide-ranging post-UNCED

reform of Tasmanian environmental and resource management legislation, includes a commitment to ecologically sustainable development and encompasses the precautionary approach as key policy parameters.

12 ABARE, *supra* note 2 at 32. About 15 percent of this production was exported. NADC, *supra* note 8, at 12.

13 Prime Minister's Science Engineering and Innovation Council (PMSEIC), Eighth Meeting, 31 May 2002, Agenda Item 4, "Sustainable Aquaculture," at 16.

14 Management of the southern bluefin tuna industry was conducted under a voluntary trilateral arrangement between the countries that was formalized in the early 1990s with the negotiation of the Commission for the Conservation of Southern Bluefin Tuna (CCSBT). See A. Bergin and M. Haward, "The Southern Bluefin Tuna Fishery: Recent Developments in International Management," (1994) 18 *Marine Policy* 289–309.

15 A. Bergin and M. Haward, *Future Trends in Japanese Tuna Fisheries* (Canberra: Department of Politics, University College, University of New South Wales, June 1991).

16 Port Lincoln had become the center of southern bluefin tuna operations in Australia, with restructuring of operations leading to operators in Port Lincoln holding a majority of the Australian quota.

17 S. Williams, "Japan Enters a New Marketing Revolution," (March 1994) *Australian Fisheries* 27–31.

18 The Australian southern bluefin tuna quota is regulated under catch limits set by the CCSBT. See: A. Bergin and M. Haward, *Japan's Tuna Fishing Industry: A Setting Sun or New Dawn?* (Commack, NY: Nova Science Publishers, 1996); M. Haward and A. Bergin, "The Political Economy of Japanese Distant Water Tuna Fisheries," (2001) 25 *Marine Policy* 91–101.

19 PMSEIC, *supra* note 13 at 28.

20 Caton, *supra* note 1.

21 These imports were one ground used by Canada to challenge the Tasmanian government's ban on imports of salmon, and the Australian government's import risk assessment in the WTO *Australian Salmon* case. See World Trade Organization WT/DS 18/R and WT/DS18/AB/R (November 1998).

22 Legal challenges have been mounted over tuna farming developments in South Australia that have been appealed to the State Supreme Court. See, for example, *Tuna Boat Owners of South Australia Inc* v. *Dac and Ano* [2000] S.A.S.C. 238.

23 "Floating Abalone Farm Commences First Full Year of Operations," (2003) 25 *Professional Fisherman* 10.

24 It is a concurrent power; s. 51(x) of the Commonwealth Constitution gives the Commonwealth a rather confusing "head of power," having power over "fisheries in Australian water beyond territorial limits," while the respective state constitutions provide the basis for state laws.

25 D. R. Rothwell and M. Haward, "Federal and International Perspectives on Australia's Maritime Claims," (1996) 20 *Marine Policy* 29–46.

26 The land-locked Australian Capital Territory (the location of Canberra, the federal capital city) administers the Jervis Bay Territory on the south coast of the state of New South Wales.

27 M. Haward, "The Australian Offshore Constitutional Settlement," (1989) 13 *Marine Policy* 334–348.

28 Following agreement between the Commonwealth and the state, OCS arrangements could be made to ensure that only one government manages fish stocks.

29 AFFA, *supra* note 6.

30 Australia, Senate Standing Committee on Trade and Commerce, *Development of the Australian Fishing Industry* (Canberra: Australian Government Publishing Service, 1982) (Senator Brian Archer, Chairman).

31 Australia, *New Directions for Commonwealth Fisheries Management in the 1990s: A Government Policy Statement* (Canberra: Australian Government Publishing Service, 1989).

32 This legislation includes the *Fisheries Administration Act 1991*, which established the Australian Fisheries Management Authority and the Fishing Industry Policy Council. This Act also established management advisory committees (MACs) and delimited their functions. In addition to the *Fisheries Administration Act 1991* and the *Fisheries Management Act 1991*, other legislation included the *Fisheries Agreements (Payments) Act 1991*, the *Fishing Legislation (Consequential Provisions) Act 1991*, the *Fishing Levy Act 1991*, the *Foreign Fishing Licences Act 1991* and the *Statutory Fishing Charge Act 1991*. See M. Haward, "The Commonwealth in Australian Fisheries Management: 1955–1995," (1995) *Australasian Journal of Natural Resources Law and Policy* 313–325.

33 Commonwealth of Australia, *Australia's Ocean Policy* (Canberra: Environment Australia, 1998).

34 See, generally, A. Bergin and M. Haward, "Australia's New Oceans Policy," (1999) 14 *International Journal of Marine and Coastal Law* 387–398.

35 *Ibid.* See also E. Foster and M. Haward, "Integrated Management Councils: A Conceptual Model for Ocean Policy Conflict Management in Australia," (2003) 46(6–7) *Ocean and Coastal Management* 547–563; T. Potts and M. Haward, "Sustainability Indicator Systems and Australian Fisheries Management," (2001) 117 *Maritime Studies* 1–10.

36 *Convention on Biological Diversity*, done at Rio de Janeiro, 5 June 1992, entered force 29 December 1993, (1992) 31 *International Legal Materials* 818; *Jakarta Mandate on Marine and Coastal Biological Diversity 1995* (Recommendations on Scientific, Technical and Technological Aspects of the Conservation and Sustainable Use of Marine and Coastal Biological Diversity) (1995) Doc. UNEP/CBD/COP/2/5. Online. Available http://www.oceanlaw.net/texts/jakarta.htm (accessed 5 March 2004).

37 EPBC 1999, s. 24 defines the Commonwealth marine area as:

    (a) any waters of the sea inside the seaward boundary of the exclusive economic zone, except:

        i   waters, rights in respect of which have been vested in a state by section 4 of the *Coastal Waters (State Title) Act 1980* on in the Northern Territory by section 4 of the *Coastal Waters (Northern Territory Title) Act 1980*; and

        ii waters within the limits of a State or Northern Territory;

    (b) the seabed under waters covered by paragraph (a);

    (c) airspace over the waters covered by paragraph (a);

    (d) any waters over the continental shelf except:

        i   waters, rights in respect of which have been vested in a state by section 4 of the *Coastal Waters (State Title) Act 1980* on in the Northern Territory by section 4 of the *Coastal Waters (Northern Territory Title) Act 1980*; and

        ii waters within the limits of a State or Northern Territory;

        iii waters covered by paragraph (a);

    (e) any seabed under waters covered by paragraph (d);

    (f) any airspace over the waters covered by paragraph (d).

38 EPBC 1999, Part 13, Division 4 refers to "listed marine species" that prohibits by-catch of species with high conservation values as determined by scientific advice to the minister.

39 W. Truss, "Federal Fisheries Review to Consult widely," *Media Release* AFFA 002/11WT (20 November 2000).

40 Commonwealth Department of Agriculture, Fisheries and Forestry – Australia, *Looking to the Future: A Review of Commonwealth Fisheries Policy* (Canberra: Commonwealth Department of Agriculture, Fisheries and Forestry – Australia, June 2003).

41 *Ibid.* at 3.

42 *Ibid.* at 28.

43 *Ibid.*

44 *Ibid.*

45 *Fisheries Act* (Vic.) 1995.

46 *Marine Farm Planning Act* (Tas.) 1995.

47 World Commission on Environment and Development, *Our Common Future*, Australian edition (Melbourne: Oxford University Press, 1987).

48 NADC, *supra* note 8 at 18.

49 The Commonwealth Parliament was informed that "membership of NACA will provide Australia with the opportunity to contribute to the sustainable development of aquaculture in Australia and through the Asian region. It will allow Australia to forge strong links with Asian-Pacific aquaculture and fisheries policy makers." *National Interest Analysis – Agreement on the Network of Aquaculture Centres in Asia and the Pacific*, done at Bangkok, 8 January 1988, 21 October 1997.

50 L. Galli, *Genetic Modification in Aquaculture: A Review of Potential Benefits and Risks* (Canberra: Bureau of Rural Sciences, 2002) at 24.

51 There is insufficient space here to detail the complexities of federal financial relations in Australia. The Commonwealth, through its power to collect income taxes, raises more revenue than it needs for its own purposes, and the states that lack significant growth taxes suffer considerable financial imbalances.

52 R. Wettenhall, "Intergovernmental Agencies: Lubricating a Federal System," (April 1985) *Current Affairs Bulletin* 28–35. See also G. B. Pollard, *Managing the Interface: Intergovernmental Affairs Agencies in Canada* (Kingston: Institute of Intergovernmental Relations, Queen's University, 1986).

53 R. J. K. Chapman, "Public Policy, Federalism, Intergovernmental Relations: The Federal Factor," (1990) 20 *Publius* 69–84.

54 Australian Science and Technology Council, *Casting the Net: Post Harvest Technologies and Opportunities in the Fishing Industry* (Canberra: Australian Government Publishing Service, 1988).

55 Review Committee on Marine Industries, Science and Technology, *Oceans of Wealth* (Canberra: AGPS, 1989).

56 *Ibid.* at 31.

57 Standing Committee on Fisheries and Aquaculture (Working Group on Aquaculture), *National Strategy on Aquaculture in Australia* (Canberra: Commonwealth of Australia, 1993).

58 *Ibid.* at 3.

59 *Ibid.* at 4.

60 Aquaplan concluded in late 2002. A new work plan and focus was developed by the Commonwealth government's Aquatic Animal Health Committee, which had its first meeting in February 2003. The new work plan included results from Aquaplan and results from subcommittees established by the Aquatic Animal Health Committee.

61 C. Lee, "Aquaculture: An Opportunity for Australia's Indigenous Communities," (2002) *Western Fisheries* 34–37.

62 *Ibid.* at 35.

63 *Ibid.* at 37.

64 NADC, *supra* note 8 at 7.
65 National Aquaculture Workshop, August 1999, *Changing Direction Workshop: Facilitator's Report*, prepared by Denis Hussey, ACIL Consulting. Online. Available http://www.affa.gov.au (accessed 5 March 2004).
66 *Ibid.*
67 *Ibid.*
68 *Ibid.*
69 *Ibid.*
70 National Aquaculture Development Committee, *Aquaculture Industry Action Agenda: Discussion Paper* (Canberra: AFFA, June 2001).
71 NADC, *supra* note 8 at 9.
72 *Ibid.*
73 *Ibid.* at 35.
74 *Ibid.* at 35–38.
75 *Media Release*, "Plans for Future of Aquaculture Revealed" (Canberra: AFFA, 13 December 2002).
76 Online. Available http://www.affa.gov.au/corporate_docs/publications/pdf/fisheries/fishupdate.pdf (accessed 25 March 2004).
77 PMSEIC, *supra* note 13 at 20.
78 *Ibid.*
79 *Ibid.* at 22.

16 New Zealand mariculture
Unfairly challenged?

Hamish Rennie

Introduction

New Zealand's aquacultural development is dominated by marine farming (mariculture), especially shellfish farming. Since 1991, it has developed within the context of world-leading, but substantively different, governance regimes for fisheries and coastal resources. The rapid growth of mussel farming and the associated "race for space" have posed considerable challenges to attempts to develop integrated coastal management regimes and, in 2001, resulted in the imposition of a national moratorium on all applications for new farm sites. The challenges to its development have come primarily from fishery rights holders and recreational and environmental communities. The situation has been clouded further by controversial claims to the foreshore and seabed by the Indigenous people, the Maori.[1] The combination of these situational factors makes New Zealand's mariculture experience of particular interest to nations seeking to implement new aquaculture policy or integrated coastal management regimes.

In comparing the New Zealand experience with that of other countries, it is essential to recognize its relatively straightforward system of government. Technically a monarchy with a Governor-General appointed by the Queen of England, New Zealand is, in all respects, essentially governed by a democratically elected House of Representatives. There are no state legislatures to debate jurisdictional issues, nor is there a formal constitution, separate presidential office or upper house of any type. There are, however, two forms of local government authorities – regional councils and territorial authorities – that have planning and by-law making capacity, but no ability to make laws. The twelve regional councils are watershed-based elected authorities that have assumed much greater importance in resource management since 1991. Their planning and resource management authority extends to the twelve-nautical mile boundary of New Zealand's territorial sea. The seventy-five territorial authorities are the elected municipal and district councils, and they have retained significant powers in determining land use within their boundaries, which generally end at the mean high-water line of spring tides.

The legitimacy of the New Zealand government is underpinned by the

1840 Treaty of Waitangi between chiefs of most of the Maori tribes and the Crown of England. Key aspects of the treaty are that it extended to Maori the rights of British citizenship, and thereby the basis for recognition of customary and aboriginal rights under British common law,[2] and it guaranteed to Maori their rights to their "treasures," specifically including fisheries.

Any ancestral blood link to Maori enables one to claim to be "Maori," and 14.7 percent of the four million people living in New Zealand claimed that they were Maori at the most recent census. Interracial marriage is common, the proportion of Maori is expected to continue to grow, and there has been a cultural resurgence since the 1970s. This has been reinforced by significant legal victories,[3] reports from the Waitangi Tribunal[4] and a process set in place for the government, representing the Crown, to negotiate settlements with individual tribes for past and present grievances. Although many Maori still live in the rural areas where they have traditionally retained land ownership to significant areas often held in common title, they do not live on "reservations" such as those found in North America. Maori predominate in the lower socio-economic statistics, but in other respects form an integral part of the larger New Zealand society.

This chapter begins with a brief overview of the significant political changes in New Zealand over the past thirty years. This is followed by an overview of the development of aquaculture and an outline of the legal and policy frameworks governing New Zealand marine farming since the 1970s.[5] The chapter concludes with a discussion of the industry response, the moratorium and government proposals, and Maori responses to these reforms. Ideological changes in government policy and associated devolved mechanisms of policy implementation are the keys to interpreting the frameworks. The focus is on the principles underlying the interactions between the post-1991 governance regimes and the industry, and why that led to the moratorium and the new governance regime being established in the mid-2000s. An examination of these developments suggests that marine farmers are being unfairly discriminated against.

Political context

Despite being essentially an island nation, New Zealand had a relatively small inshore fishery until its 1978 declaration of a twelve-nautical mile territorial sea and 200-nautical mile exclusive economic zone (Figure 16.1). Government subsidies aided a boom in fishing industry development leading to concerns over the sustainability of fish stocks and the subsequent imposition of the quota management system (QMS), discussed below. The 1970s also saw New Zealand join the Organisation for Economic Co-operation and Development (OECD) and the beginning of the end of its special trading relationships with Britain as the latter joined the European Community. The potential loss of British market access was a major blow to New Zealand's primary production-dependent economy and this was

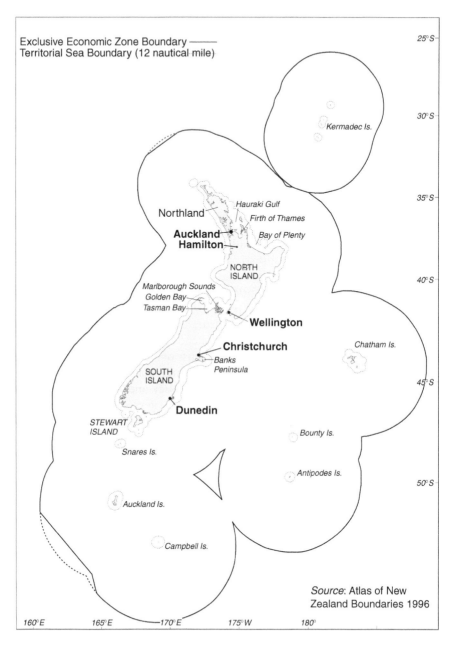

Exclusive Economic Zone Boundary ———
Territorial Sea Boundary (12 nautical mile)

Kermadec Is.

Hauraki Gulf
Northland
Firth of Thames
Auckland
Bay of Plenty
Hamilton

NORTH
ISLAND

Marlborough Sounds
Golden Bay
Tasman Bay
Wellington

Christchurch
Chatham Is.
*Banks
Peninsula*

SOUTH
ISLAND

Dunedin

*STEWART
ISLAND*
Bounty Is.

Snares Is.
Antipodes Is.

Auckland Is.

Campbell Is.

25°S
30°S
35°S
40°S
45°S
50°S

160°E   165°E   170°E   175°W   180°

Source: Atlas of New
Zealand Boundaries 1996

Figure 16.1 New Zealand and places mentioned in the text (source: *Jan Kelly and Brian Marshall, Atlas of New Zealand Boundaries 1996* (Auckland: Auckland University Press)).

exacerbated by the 1970s oil crisis. The latter emphasized the cost of transporting New Zealand's commodities to its major international markets.

The well-established welfare state meant a high level of individual taxation and associated redistribution of wealth. Central government paid for such things as health care and education, and was the major employer in the economy, especially in the infrastructure and development areas. It provided many subsidies to the primary and other sectors, and protected domestic industries with tariffs against imports. In response to the worsening economic and energy outlook, New Zealand's government embarked on several large, primarily energy-based projects to boost and diversify economic development.

The incoming Labour government of 1984, however, found itself in a significant monetary crisis. The energy projects had largely failed to deliver the expected returns and had galvanized a strong environmental movement. The new government's response was to initiate across-the-board neoliberal free-market reforms of the state, the country and the resource management systems.[6] Among the (usually explicit) underlying ideological principles were the following:

- Governments should not distort the operation of the free market by providing subsidies or direction (that is, they should not "pick winners").
- Governments should not be involved in business (this principle resulted in a sell-off of state-owned infrastructure, such as railways and ports, and its construction agencies).
- Landowners should be free to develop their property as they see fit, provided the environmental impacts are addressed (that is, the principles of "the polluter pays" and internalizing external costs were to apply).
- Decision-making should be made by the consumer, and in terms of governance this meant that resource management decision-making should be devolved to local authorities more responsive to the local electors.
- Decision-making should be efficient, transparent and accountable.

The resultant upheavals in the New Zealand economy were accompanied by a resurgent Maori culture seeking redress for a variety of grievances, many of which related to control over natural resources and land. Under the Labour Party, as a small nation still largely dependent on commodity exports, New Zealand strengthened its staunchly multilateralist international position, seeing the United Nations as a key to removing trade barriers on commodities. New Zealand played a leading role in international negotiations leading to the establishment of the World Trade Organization and was also active in Agenda 21 and subsequent international agreements. In many of these negotiations, it aligned itself with Canada, Australia and the United States.

New Zealand claimed the high moral ground in international negotiations on the basis of having removed its own subsidies to, especially, the primary sector. The subsidies that it had removed included the hidden subsidies involved in cleaning up environmental negatives caused by resource users, such as water pollution and overfishing. Its primary legislative tools for achieving this were the *Resource Management Act 1991* and, for fisheries, the QMS, introduced progressively since 1984. Subsequent governments have maintained these approaches.

By the new millennium, income taxation had been substantially reduced, but a tax on goods and services was introduced. Many formerly protected industries have closed. The economy is much more diverse, and attention is being placed on niche marketing of value added products such as wine and mussels. The political rhetoric has changed from one of equality in outcomes to equality in opportunities. This is linked to following a "free-market" approach, with some affirmative action to ensure equality of opportunity for those who have less fortunate starts. There is little apparent electoral support for a return to the pre-1984 era, with the main, and most minor, elected political parties now considered to be somewhat right of center. In summary, there has been a significant change in the ideological positioning of New Zealand politically and socio-culturally since 1984. Marine aquaculture development reflects these changes in the larger socio-political structure.

New Zealand mariculture development

Aquaculture can be categorized into marine (mariculture, which involves farming flora or fauna in the sea or intertidal areas) and freshwater (land-based, which includes freshwater farming and farming in facilities built on land and using either salt or fresh water, but excludes ponds that are naturally flushed by salt water). New Zealand has had legislation enabling aquaculture for over 130 years, and Maori have claimed an even longer tradition of cultivating marine species.[7] Trout and salmon were successfully introduced over 100 years ago for sustenance and recreational fishing, and have been augmented by hatcheries. However, commercial salmon and land-based paua farming have only developed, and at a very modest level, in the past twenty years.[8]

Despite the more than 15,000 kilometers of diverse coastline spread across 20 degrees of latitude, very high water quality and good year-round growing temperatures, marine aquaculture did not become established as a significant industry in New Zealand until the latter part of the twentieth century. The lack of growth before the 1960s reflected the distance to significant markets and the cost of fuel. The development of appropriate low-cost farming technology and management practices, the high-quality environment and the establishment of unique niche markets for its mussel industry have enabled New Zealand to overcome its relative isolation.

Some have argued, however, that the rate of growth of aquaculture over the past forty years has been significantly constrained by its enabling legislation.[9] The drive for new aquaculture legislation came from a Parliamentary

Fishing Industry Committee in the early 1960s to enable would-be marine farmers to acquire secure property rights without the need of a special Act of Parliament for each farm.[10] Previously, only temporary permits had been obtainable owing to constraints in the *Harbours Act 1950*.[11] The *Rock Oyster Farming Act 1964* (ROFA) was the harbinger of modern marine farming legislation, but it was constrained to enabling leases of up to 2 hectares of foreshore and seabed for farming only the native rock oyster (*Crassostrea glomerata*). In the late 1960s, partly as an industry-led attempt to establish trout farming, but primarily as a government drive for regional development and because of a failing mussel fishing industry, ROFA was replaced with the generic *Marine Farming Act 1968* (MFA68). MFA68 continued the established regulatory practice limiting the size of a lease to no larger than 2 hectares, but farmers were allowed to own more than one lease from 1968 onwards. The criteria for reaching decisions gave considerable discretion to the minister.

Rock oyster farming was fostered through the 1960s by a government seeking to develop a new industry to compete in the export market. It employed an Australian farmer to act as an extension officer, and he introduced the intertidal, rack-based oyster farming system. The accidental arrival in 1971 of the Pacific oyster (*Crassostrea gigas*) was initially treated as an invasive threat to the industry, but subsequently became the dominant species farmed. Initial experiments with oyster hatchery production of seed were abandoned as non-commercial, and the industry continues to rely on seed caught on-site.

The overexploitation and complete collapse of commercial mussel fishing in the 1960s led to experiments in commercial mussel farming in 1965.[12] The industry is based on the endemic green-lipped mussel (*Perna canaliculus*) and commenced in former mussel fishing areas in the ria-like Hauraki Gulf and, almost simultaneously, in the drowned river valleys of the Marlborough Sounds. The Spanish raft system used initially was soon abandoned in favor of the aesthetically more pleasing Japanese longline system introduced by the Fishing Industry Board in 1974.

The mussel industry relies on spat caught on grow-out sites, on special spat catching sites and, especially, spat harvested from seaweed cast up on beaches. The industry's development has been greatly aided by extension services provided by industry organizations, including cooperative spat catching sites. The mussel meat product's early (1980) acceptance by the US and European regulatory agencies was crucial in its development.[13] As with oysters, there is no need for a depuration period, and the industry has developed sophisticated water quality monitoring and associated farm management systems. The mussels reach harvestable size in 12–18 months.

Recreation interests strongly, and ultimately successfully, opposed commercial trout farming. Because salmon farming was linked with trout farming, salmon were prohibited from being marine farmed commercially until the two were delinked in 1983. Pacific salmon (*Oncorhyncus tshawytscha*)

Table 16.1 Production and value of New Zealand marine aquaculture, 2001 (FAO data)

Species	Production (metric tonnes)	Value (thousand US dollars)
Green-lipped mussels	64,000	65,917
Pacific (Chinook) salmon	7,887	22,944
Pacific oysters	3,500	5,709
Total	75,387	94,570

sea-cage farming commenced in 1982 on an experimental license in Stewart Island by BP and subsequently spread to the Marlborough Sounds and Banks Peninsula. Despite considerably expanding its production, its relatively high cost and the ready international competition from Australia and Chile for the Northern Hemisphere's "off-season" market has constrained the salmon farming industry's growth (Table 16.1).

There have been experiments with farming other species, ranging from finfish to seaweeds, rock lobsters and seahorses, and for a range of markets, but the commercial industry remains essentially mussel, oyster and salmon farming. There is, however, an important qualifier when comparing mariculture in New Zealand with that overseas. Scallop enhancement, such as that conducted in the Nelson–Tasman region of the upper South Island, does not involve structures, containment or control of the reseeded scallops and is not classified as either sea-ranching or aquaculture under New Zealand law. As will be discussed later, this has posed significant difficulties for the aquaculture industry.

Post-1970 legislative and policy frameworks

The passage of the *Marine Farming Act 1971* (MFA71) paved the way for the expansion of mariculture in New Zealand. The major changes introduced by MFA71 were the provisions for a licensing system to complement the leasing system, further refinement of criteria to be considered when issuing a license or lease, greater freedom in the size of farms allowed, and provision for the creation of marine farm plans.

The provisions on size were altered to provide the controlling authority with discretion over the total farm size, which was to be "no greater" than it considered the applicant could "successfully develop ... within the term of the lease or licence."[14] Environmental impact assessments have been required to be carried out by the licensing authority before issuing a license or lease since 1974.[15]

Licenses could be issued for up to fourteen years, and had a right of renewal.[16] An applicant would apply to the Ministry of Agriculture and Fisheries[17] (MAF) for the license. The ministry would then seek concurrence from the agencies responsible for navigation, physical structures, and the use and occupation of space. In addition, fish-feeding required discharge permits under the *Water and Soil Conservation Act 1967*, but this was overlooked until the late 1980s. Interestingly, neither the applicant nor the ministry was required to specifically notify adjacent landowners and Maori tribes of proposals to develop marine farms.

There were no provisions for appeal of a MAF decision, except on points of law. The criteria on which decisions could be made effectively gave precedence to navigation, commercial fishing, mining, existing or proposed recreational or scientific use, the adjacent landowners' activities and the public interest. Whereas leases provided the holders with the exclusive right to the area, licenses provided only the right to undertake the mariculture activity within that area. The public could continue to use and pass through licensed areas. As a general policy, no leases for new areas were issued after MFA71.

Marine farm plans tended to create zones in which marine farming was *prohibited*, rather than create zones *for* marine farming. By the early 1980s, the minimum processing time for farm applications was of the order of thirty months.[18] Often the marine farm plans merely supported the outcomes of consultative community planning exercises undertaken by local authorities planning other coastal activities under other legislation.

The *Town and Country Planning Act 1977* introduced provisions enabling the establishment of "regional" councils on a voluntary basis and giving them the responsibility for regional planning schemes covering the marine area out to the boundaries of the territorial sea. Specified maritime planning authorities could also prepare maritime planning schemes. Only a few areas took up these opportunities, but those that did included rules in their plans that addressed marine farming, usually by effectively prohibiting it either generally or from selected places. Prohibitions were usually justified on the grounds that farms would interfere with navigation, recreational use, amenity values, or Maori fishing reserves and special places.

In 1986, the government introduced the QMS for twenty-seven commercial species. The system is based on individual transferable quotas (ITQs). ITQ holders have the "right to harvest" a *proportion* of the total allowable commercial catch (TACC) of a particular species from a particular quota management *area*. The actual tonnage may vary from year to year, depending on the TACC set. The TACC is based on an annual stock assessment process that estimates the maximum sustainable yield for a species. The ITQ is held *in perpetuity* and can be freely traded on the market. There has been some consolidation of ITQ holdings into the hands of major fishing companies since its introduction. Most major fishing companies and processors have significant marine farm investments.

In the Tasman and Golden Bay areas, the government passed special legislation enabling the quota holders of the scallop stock in the area to undertake cooperative scallop enhancement and rotational fishing practices. These have proved very effective in sustaining the fishery.[19] Spat catching for these scallop operations includes using moored structures essentially indistinguishable from mussel spat collectors and requiring similar approvals to those for marine farming, but with tighter restrictions on the period covered by the license.

Unlike the fishing industry-based process used to introduce the QMS, the *Resource Management Act 1991* (RMA) represented the culmination of several years of extensive public consultation and reviews of resource management legislation. A desperate rearguard action by the Ministry for Agriculture and Fisheries and the fishing industry kept fisheries out of the RMA's control.[20] The RMA did, however, replace the other legislation that governed the gaining of marine farming permits.

The key aspects of the RMA reflected the general principles underlying the "New Right" agenda. The purpose of the RMA was the "sustainable management" of natural and physical resources. It set out a number of matters of national importance and other matters that had to be taken into account or had regard to by decision-makers working under the Act. It also set in place a framework for policy and planning at national, regional and territorial authority levels. The central government could produce national policy statements, and a national policy statement on the coastal environment was mandatory. Regional councils had to produce regional policy statements for their entire region, and regional coastal plans for the area from mean high water at spring tides to the twelve-nautical mile limit. This area is known as the "coastal marine area." Regional councils could also produce regional plans on other resources for which they were given authority, for example water and air. Territorial authorities had to produce district plans addressing land use for their territories. Each lower tier of government plan or policy could not be inconsistent with a higher level.

The RMA also established a single process for granting permits to do things that were discretionary under the plans. Decisions were to be made at the relevant local authority level, depending on the nature of the resources being affected. Decision-making was to occur within specified time frames, although if appeals were lodged against decisions, then the Environment Court[21] was not constrained by the same time constraints. Ministers could only "call in" or have final decision-making roles in particular cases of national significance. Otherwise, they and their agencies were subject to the RMA processes.

Importantly, the new RMA adopted the principle of "effects-based" planning. Rules in plans were to be based not on activities, but on environmental effects. Thus, rules could not be made that prohibited the activity of marine farming, but rules and zones could be made in plans on the basis of thresholds of effects. Therefore, any discharges of specified contaminants above specified rates might be prohibited.

Every activity, consequently, required some form of biophysical environmental impact assessment. In contrast to the previous regime, the costs for these are borne by the person wishing to carry out the activity. If the activity has effects that cross a threshold set out in plans or policy documents prepared under the RMA, then the permit process comes into operation. The maximum term of a permit is thirty-five years, with no right of renewal or preference, and the permits are transferable, but not divisible. There are no size restrictions, but if public access is effectively excluded from areas of ten hectares or greater, or effectively restricted from areas of fifty hectares or greater, the application is heard by the regional council, but the final decision is made by the Minister of Conservation on the council's recommendation.[22] The criteria the minister considers do not differ from those of the Environment Court or the regional or territorial authorities, and do not allow consideration of trade competition (e.g. with fisheries). The process is thus transparent and subject to judicial review, thereby providing public accountability.

The RMA enabled people to carry out their activities provided they addressed the effects their activities had on the environment. The market was left to address the social and moral issues. This meant that socially or morally incompatible uses could be located adjacent to each other unless opponents could demonstrate that the biophysical effects of the activity were such as to warrant the activity being prevented. Under the RMA, if a hotel could be built in a particular location, say next to a school, then so too could a casino or brothel, because those activities would have essentially the same biophysical effects as the hotel. A church, however, might have greater difficulty, as the activities undertaken might have significant noise effects during bell-ringing or choir practice.

Seabed, foreshore and water column space in the coastal marine area were not privately owned, but were administered by the government on the behalf of the public. The occupation of space is an effect on the environment in that it affects the capacity of the space to be occupied by another activity. The RMA was therefore the means by which to gain the right to occupy space in the coastal marine area.

The RMA epitomized the post-1984 neoliberal policy principles of successive governments. Most importantly, governments at all levels were not to pick winners through their zoning decisions; they could only decide the effects that they wished to avoid. Decision-making had been devolved away from the central government, clear criteria and transparent processes were in place, and it was up to potential developers to invest in the technology to avoid, remedy or mitigate their effects if they wished to proceed.

Marine farmers found themselves at the interface between two entirely different pieces of legislation. They needed a coastal permit to occupy the marine space and have the marine aquaculture facilities and, because fisheries were outside the RMA, they needed a "marine farm permit"[23] to hold and harvest the fish. This has become known as the dual permit system. The marine farming permit is issued only after a coastal permit has been obtained from the appropriate regional council.[24] The Ministry of Fisheries makes decisions on the marine farm permit based on the "undue adverse effects" on commercial fishing and fisheries resources.

The precautionary principle

The relative lack of scientific knowledge of marine resources and processes meant that a "precautionary approach" was adopted as a guiding principle in the RMA. All activities with environmental effects (except fishing) are prohibited in the marine area unless otherwise permitted either by a rule in a regional coastal plan or by a coastal permit obtained from the regional council. The courts have found that the precautionary principle is implicit in the RMA's overall effects-based approach. Each effect of a proposed activity, such as a marine farm, is assessed qualitatively or, preferably, quantitatively in terms of the probability of occurrence and the force of the impacts. Risk assessment is employed, and New Zealand and international standards have been drawn on in court decisions on the appropriate processes to be used, notably in the assessment of potential impacts of different forms of farming on marine mammals.[25] Implementation of the *Convention on Biological Diversity* has resulted in an added emphasis on the maintenance of indigenous coastal diversity.[26] The introduction of non-locally endemic species (e.g. Pacific oysters) from other parts of New Zealand has also been prevented in some instances.[27]

In addition, the *Biosecurity Act 1993* and the *Hazardous Substances and New Organisms Act 1996* (HSNO) control the management of pests and the introduction of non-indigenous species and genetically modified organisms (GMOs). The HSNO has assumed particular importance as a consequence of a heated national debate over genetic engineering and the field testing of GMOs. It requires that anyone intending to import, develop, field-test or release into New Zealand new organisms (including GMOs) must obtain a permit from the Environmental Risk Management Authority (ERMA). ERMA must "take into account the need for caution in managing adverse effects where there is scientific and technical uncertainty about those effects."[28]

Industry response to the different regimes

The industry has grown rapidly regardless of the regimes, but the RMA freed the regulatory system immensely. The new coastal plans being developed did not have prohibitions against marine farming, and areas previously prohibited to marine farming were now open at the discretion of the local authorities. Even before the new RMA plans were developed, however, applications were lodged for areas where the old plans had not put prohibitions in place. Such areas were often without prohibitions, primarily because of assumptions that the sites were not technologically or financially viable and were not under significant pressure. Alternatively, there was an expectation that there would be such opposition to the use of some areas that marine farms would not be allowed.

Marine farmers were concerned that the new regime would make it harder to get permits. They were concerned that local authorities would be less likely to approve farms than MAF had been. Moreover, the Minister of Conservation was given the role of approving regional coastal plans and significant developments. The Department of Conservation had been established in 1987 with the explicit purpose of advocating for conservation of the environment,[29] and by 1991 had become the major government department opposing marine farm applications.[30] Under the RMA, marine farmers would bear the costs of the application process rather than pay the nominal fee previously charged by MAF, and they would have to consult directly with the community and potential opponents.

Consequently, immediately prior to the commencement of the RMA, many applications were lodged, either to add approval for a wider range of species to be farmed on existing approved licenses, or to extend existing sites, or to obtain new sites. The race for space was most intense in the Marlborough Sounds, and allocation was based on first in, first served.[31]

In the Golden and Tasman Bays, issues became especially heated when several mussel farmers applied for spat catching permits to extensive areas in which the Tasman and Golden Bay scallop quota holders had been operating the scallop enhancement scheme. The local council's initial proposed regional coastal plan suggested rules that would effectively push marine farms out beyond 3 nautical miles to protect the environment and meet the navigational concerns of recreational boaters. The resulting multifaceted legal action involved a three-month court hearing, many millions of dollars and a suicide. It became clear that the quota holders did not have rights to occupy space as they held no coastal permits or marine farming permits. Therefore, their argument against the applications was tenuous under the RMA. The court proposed a solution whereby it identified large aquaculture management areas (AMAs) in the coastal plan, and aquaculture would be prohibited outside those areas.[32]

The proposed new plans gained weight in decision-making processes for individual marine farm applications in other regions, depending on the stage to which the plans had advanced through public processes and appeals.[33] Entrepreneurial site developers and existing marine farmers recognized the opportunities to obtain sites in new places and at unprecedented scales. Applications were made for several sites of almost fifty hectares, more than ten times larger than the average marine farm site prior to the RMA. As investors became familiar with the RMA, the requirement for the Minister of Conservation and local authorities to operate in accordance with the criteria specified in the RMA became apparent. Developers invested in new technology and developed proposals for offshore farms of up to 10,000 hectares.[34]

It has been the delays and imprecision of the criteria considered by the Ministry of Fisheries that has proved the most significant impediment to developments under the new regime. The problem for the Ministry of Fisheries lies partly in its lack of administrative resources for these types of developments, but primarily in the imprecision of the legislation that requires it to consider intersector, or trade, competition. Specifically, the Ministry is requiring much more extensive ecological assessments than previously, and is wrestling with the implications of the test for undue adverse effect on the commercial fishing sector.[35] This has caused many in the industry now to indicate a preference for all marine farm permitting decisions to be made entirely under the RMA and by local authorities.

Government moratorium and proposals

In response to the increased demand, and especially to the Tasman–Golden Bay case, in November 2001 the government imposed a national moratorium on new applications. The primary initial purpose of the moratorium was to provide regional councils with time to put specific AMAs in their plans and for the government to pass legislation implementing the proposed aquaculture reforms.[36] The proposed reforms include only allowing new applications for marine aquaculture within AMAs. Moreover, these areas would be identified through community consultation processes (appealable to the courts) and agreement from the Ministry of Fisheries. The local fish quota holders would also effectively have a right to veto some areas where they were considered by the ministry to be "unduly adversely affected" unless suitable arrangements (e.g. financial compensation) could be reached.

It is proposed that once AMAs are established, they will be developed progressively, allowing monitoring of the environmental effects in the process. The default mechanism for allocating space will be by tendering. Once the regional council has approved a coastal permit for a farm, the farmer will register it with the Ministry of Fisheries. That registration will be automatic and replace the requirement to obtain a marine farming permit for holding and harvesting marine flora and fauna. A one-stop shop will have been created, because the effects on fishers will have been addressed through the creation of the AMA.

The legitimacy of these proposals has been thrown into doubt as a result of court action by various Maori tribes claiming ownership of the foreshore and seabed. Their initial success in the Court of Appeal created considerable wider, continuing political turmoil.

Maori response

In the 1980s, Maori tribes successfully challenged the implementation of the QMS and the allocation of ITQ on the basis that their common law customary right to these resources had never been extinguished by legislation.[37] Subsequent negotiations ended with the 1992 "Sealords Deal."[38] Aquaculture and freshwater fishing were not included in the settlement.

Among other things, the deal guaranteed Maori up to 20 percent of the quota allocated for commercial fish stocks brought into the QMS and half the ownership of Sealords, the largest fishing company in New Zealand. The quota and the Sealords shares are held by the Treaty of Waitangi Fisheries Commission.[39] The Commission, and Maori as tribal entities or as individuals, have also acquired quota through their own endeavors, and Sealords has continued to expand. The actual Maori ownership of ITQ therefore probably exceeds, but does not replace, that required to be provided by the government.

The RMA developed concurrently with the aftermath of the court cases and key Waitangi Tribunal reports into grievances, and shows their imprint. It requires that decision-makers consult with Maori; take into account the (undefined) "principles" of the Treaty of Waitangi; have regard to kaitiakitanga (guardianship); and recognize and provide for the relationship of Maori and their culture and traditions with their ancestral lands, water, *wahi tapu* (special places) and other *taonga* (treasures). This potentially provides Maori with considerable influence in decision-making. Some marine farm developers have established partnerships with Maori tribes or individuals to smooth the application process, and many Maori-owned organizations and individuals have invested directly in marine farm ownership. Sealord and the Commission have also become major players in marine farming, and Sealord has recently helped create a benchmark standard for organic mussel farming, believed to be a world first.[40]

Despite these provisions, some Maori were concerned over the government's new aquaculture proposals and discriminatory difficulties they believed they had encountered from local authorities in trying to gain marine farming rights. In 2003, they lodged claims against the proposed aquaculture law reforms, and subsequently against the government's foreshore and seabed ownership proposals, in both the Waitangi Tribunal and the courts. The government and others opposed these claims. The Court of Appeal overturned previous understandings of the unowned nature of the territorial sea and found that Maori might still hold customary rights to parts of the foreshore and seabed.[41]

This threatened to undermine the presumption in the aquaculture proposals that the foreshore and seabed were not privately owned. The government responded with a set of policy proposals that, if they became law, would remove Maori customary ownership rights, but provide Maori with a veto right over any proposed use of the water column, seabed or foreshore if and where they could establish customary title. Moreover, Maori would be able to gain an *in perpetuity* development right for such areas that would override the thirty-five-year limitation on permits, but would still need to obtain coastal permits. Provision might also be made for a preferential allocation to Maori of a proportion of the sites within an AMA.[42] Whether this would sufficiently compensate Maori for removing their existing customary rights would probably depend on the location and nature of the customary rights lost.

The government's proposals provoked considerable heated debate throughout the nation. Two Waitangi Tribunal reports have found that the proposals breach both the Treaty of Waitangi and principles of fairness in international and domestic law.[43]

Unfairly treated?

It is quite apparent that marine farmers too are being unfairly treated in the development and use of marine resources in New Zealand. The RMA regime placed them on a level playing field with all other marine users except fishers. The property rights regime created by the ITQ system for fishers was implemented long after the establishment of aquaculture as a viable industry, and it was initiated because of the inability of fishers to sustain their fisheries. Marine farming, on the other hand, has been seen as a sustainable alternative to unsustainable fishing. Why should it be seen as an *alternative* when it is arguably an equally valid industry, especially given the lengthy historical period of aquaculture activities claimed by Maori?

Moreover, the moratorium on aquaculture developments has not been matched with a similar moratorium on the development of marinas or other activities that might occupy space that could have been used by aquaculture. In addition, the government has continued to provide quota rights to new species, and these ITQ holders may thereby gain a veto or compensation right over proposed AMAs. This aspect of the proposals also provides the fishing industry with an arguably unwarranted advantage in gaining permission for its own marine farms and may effectively exclude the type of non-fishing people who initiated and developed much of New Zealand's aquaculture industry.

There is also an inherent unfairness in that the aquaculture industry has responded to the challenge put to it by government to invest in technology that addresses its environmental effects. Its subsequent growth should be seen as a credit to the regime. The new proposals, however, essentially restrict aquaculture to the technologies and species that have already been established in New Zealand. The AMAs and associated prohibitions will remove the incentives for the industry to farm new species and to develop new technologies. Other industries, such as marine tourism developments, will be allowed to proceed unencumbered by equivalent "tourism management areas" and prohibitions. Efficient and environmentally sustainable uses of marine space will be prevented, and less efficient and sustainable activities might well take up the space.

Most galling of all for marine farmers is that they must undertake a thorough environmental impact assessment of the sites sought, but the same does not apply to fishers. An AMA, or a marine farmer, may therefore be prevented from taking up a site above a particular benthic community, only to see a scallop dredger destroy the same community the next day. Similarly, while marine farmers are unlikely to be permitted to establish farms using species not previously present in the area,[44] recreational or commercial fishers may easily discard unwanted by-catch in areas where those species have not previously been found. Discharges from recreational fishing boats and others may also create lower water quality in particular areas than would aquaculture activities.

Conclusion

New Zealand's experience in mariculture development suggests that an integrated system of allocating space to all activities is essential for the achievement of efficient and environmentally friendly development. Such concepts are fundamental in the language of integrated coastal management, and it has been the failure to draw fisheries into that system that has become a fundamental constraint on marine farming development. Moreover, attempts to implement new systems of allocating space should not be embarked upon before considering fully the existing rights to that space. In issuing *in perpetuity* ITQ for fisheries, the New Zealand government had to revisit its presumption of its ability to ride over Maori rights, and we have seen a repeat performance with its proposals for the aquaculture sector. While it has implemented such principles as precaution, transparency, accountability, the maintenance of public access, public ownership of marine space, and "polluter pays" for aquaculture and other activities governed under the RMA, it has failed to implement all of these in the ITQ approach to fisheries.

This in turn places other users, especially marine farming, at an unfair disadvantage relative to fishers. Aquaculture has demonstrated that when it is treated on a level playing field, fairly with respect to other marine users, then it has the capacity to respond with innovative marine farming technology and management practices that are environmentally sustainable. Seeking special provisions for aquaculture, such as AMAs, is more likely to disadvantage aquaculture than advantage it. The recent proposals for aquaculture in New Zealand represent a subtle change from the *laissez-faire*, free-market neoliberalist ideals of the 1980s reforms for coastal management, and a return to a more interventionist central government approach to picking individual winners. The winners here are not the aquaculture industry or Maori, but the ITQ holders – fortuitous beneficiaries of the same neoliberal reforms now being undermined.

Policy-makers, however, need to bear in mind that if they remove all consideration of socio-cultural matters, then they may find the industry becomes dominated by those with the capital to invest in the technology and management practices that minimize their environmental effects. Fewer capital resources may mean less environmentally sustainable outcomes. The industry may become inaccessible for small entrepreneurs and sustenance or subsistence farmers, and become the domain of those able to enlist major capital resources. Whether such an inequitable outcome of ownership distribution is preferable to accepting adverse environmental effects may be a luxury only able to be afforded by developed and affluent countries.

Postscript

In November and December 2004, the New Zealand government rushed into law controversial new statutes vesting ownership of the foreshore and seabed in the Crown and implementing the new aquaculture regime. As anticipated in this chapter, the new aquaculture legislation ended the moratorium on aquaculture, but has prohibited aquaculture development unless it occurs within an aquaculture management area (AMA) identified as such in a regional coastal plan prepared by the relevant local government under the *Resource Management Act 1991*. An AMA cannot be established if the Chief Executive of the Ministry of Fisheries considers that it would unduly adversely restrict or displace recreational or customary fishing. If it would unduly restrict or displace commercial fishing, then a reservation is placed on it, and those seeking to obtain space for aquaculture activities within the AMA must have succeeded in reaching an "aquaculture agreement" with those who own 90 percent of the quota of fish for that area. This essentially provides a veto capability for large ITQ holders. The preferred mechanism for allocating space to those who have aquaculture agreements is to tender "authorizations." Once authorizations have been allocated, then the authorization holders can apply for coastal permits, which are considered in terms of environmental effects. Before tendering, 20 percent of the available space in the AMA is to be provided to a Maori aquaculture trustee to allocate to *iwi*. The space to be provided to Maori is to be representative of the available space and to be of an economically viable size. The allocation of this 20 percent does not prevent Maori from tendering for more space.

Notes

The support of the University of Waikato and of Derek Johnson and SISWO's Centre for Maritime Research (MARE) in Amsterdam is gratefully acknowledged. The map used in this chapter was provided by Max Oulton, Waikato University's cartographer.

1 Maori are a Polynesian people who arrived in New Zealand at least 800 years ago and are in all respects treated as indigenous to New Zealand. Maori organizational structures are based on the family (*whanau*), the extended family (*hapu*) and affiliations to tribes (*iwi*).
2 P. McHugh, *The Maori Magna Carta: New Zealand Law and the Treaty of Waitangi* (Auckland: Oxford University Press, 1991).
3 *Te Weehi* v. *Regional Fisheries Officer* [1986] 1 NZLR 680, 19 August 1986, M662/85 High Court, Williamson J; *New Zealand Maori Council* v. *Attorney-General* [1987] 1 NZLR 641, 29 June 1987, CA54/87, Court of Appeal, Cooke P; *New Zealand Maori Council* v. *Attorney-General*, 8 October 1987, Greig J, CP553/87, High Court – Wellington.
4 The Waitangi Tribunal was established by the government to hear claims of grievances against the Crown, both past and present, in terms of breaches of the Treaty of Waitangi. It is quasi-judicial and produces reports with findings in relation to whether there have been any breaches, and makes recommendations as to possible ways to resolve these breaches. Following such reports, and some-

times preceding them, the government usually enters into negotiations with affected parties to redress any breaches.

5 The analysis of marine farming development and associated legislation, policy and plans draws extensively from my doctoral thesis, H. Rennie, "A Geography of Marine Farming Rights in New Zealand: Some Rubbings of Patterns on the Face of the Sea," Geography Department, Waikato University, 2002.

6 For a detailed critical discussion of the reforms, see J. Kelsey, *Rolling Back the State: Privatisation of Power in Aotearoa/New Zealand* (Wellington: Bridget Williams Books, 1993); A. Memon, *Keeping New Zealand Green: Recent Environmental Reforms* (Dunedin: University of Otago Press, 1993); and R. Le Heron and E. Pawson, eds., *Changing Places: New Zealand in the Nineties* (Auckland: Longman Paul, 1996).

7 The first legislation providing for aquaculture was the *Oyster Fisheries Act 1866*, which enabled landowners to establish artificial oyster beds on the shore adjacent to their land where there were no natural beds. Maori claims are described in *Ahu Moana: The Aquaculture and Marine Farming Report (Wai 953)* (Wellington: Waitangi Tribunal, 2003).

8 Commercial aquaculture is orientated to producing flora and fauna for sale, not necessarily to maximize profit or for food, but does not include releasing fish for general recreational fishing (cf. N. Hishamunda and N. Ridler, "Sustainable Commercial Aquaculture: A Survey of Administrative Procedures and Legal Frameworks," (2003) 7(3 and 4) *Aquaculture Economics and Management* 168–169).

9 Ministry of Fisheries and Ministry for the Environment, *Aquaculture: Join the Discussion. Public Consultation on the Future Management of Aquaculture* (Wellington: Ministry of Fisheries, 2000).

10 Fishing Industry Committee, "Report of the Fishing Industry Committee" I.14A, *Appendices to the Journals of the House of Representatives* 1962. The committee also recommended the establishment of the Fishing Industry Board and that the commercial farming of trout and salmon be permitted.

11 Fishing Industry Committee, "Report of the Fishing Industry Committee" I.14A, *Appendices to the Journals of the House of Representatives* 1971 and Fisheries Committee, *Report of the Fisheries Committee to the National Development Conference May 1969 N.D.C.6* (Wellington: National Development Conference, 1969).

12 The first commercial crop was harvested in 1968.

13 Previous growth had been sparked by interest in the potential market for dried mussel powder exports.

14 *Marine Farming Act 1971*, s. 3.

15 Required by the *Environmental Protection and Enhancement Procedures* created as an internally binding government Cabinet memorandum in 1974.

16 Whether this right implies an automatic renewal if all operating conditions are met has been a point of debate between industry and government, but has not been tested in the courts. The government maintains that there has not been a right of automatic renewal since 1993, despite all renewal applications having been approved (see discussion in *Ahu Moana, supra* note 7).

17 Marine farming was initially administered by the Department of Marine, but was shifted in the 1970s to the more export-orientated Ministry of Agriculture and Fisheries before the establishment of the Ministry of Fisheries in 1995. In each agency, aquaculture has been of relatively minor importance in the overall administrative responsibilities and priorities.

18 P. Tortell, "Environment," in P. Smith and J. Taylor, eds., *Prospects for Snapper Farming and Reseeding in New Zealand*, Fisheries Research Division Occasional Publication No. 37 (Wellington: New Zealand Ministry of Agriculture and Fisheries, 1982), 33–38.

19 Areas were reseeded one year, left fallow for a year, and then dredged the third year in a crop rotation style of enhancement deliberately kept outside the RMA and marine farm permit system. This was considered a model for how the ITQ system should work. See M. Arbuckle and K. Drummond, "Evolution of Self-Governance within a Harvesting System Governed by Individual Transferable Quota," in R. Shotton, ed., *FishRights99 Conference: Use of Property Rights in Fisheries Management*, Vol. 2 (Rome: FAO, 2000), 370–382.

20 See H. Rennie, "The Coastal Environment," in A. Memon and H. Perkins, eds., *Environmental Planning in New Zealand* (Palmerston North, NZ: Dunmore Press, 1993).

21 A specialist court established to hear cases under the Act.

22 An oyster rack farm may effectively exclude public access, but a mussel farm suspended from longlines, especially if the bulk of the array is suspended at 10 meters below the surface, is more likely to only restrict public access.

23 Under the *Fisheries Act 1983* as amended in 1993.

24 Or the Minister of Conservation, if the scale of the effects of the proposed activity exceeds thresholds, described in regional coastal plans, redefining the activity as being a matter of "national significance."

25 *Clifford Bay Marine Farms Ltd* v. *Marlborough District Council*. Environment Court, 27 September 2003 (C131/2003), Jackson J. Spatial and temporal scales must be considered.

26 Principle 11 of *The New Zealand Coastal Policy Statement 1994*, the only national policy statement so far prepared under the RMA by the central government to guide local authorities.

27 *Greensill* v. *Waikato Regional Council*. Planning Tribunal, 6 March 1995 (W17/95). Treadwell J; and *Pigeon Bay Aquaculture Ltd* v. *Canterbury Regional Council*. Environment Court, 18 March 1999 (C32/85), Jackson J.

28 HSNO, s. 7. See also *Bleakley* v. *Environmental Risk Management Authority* 2001, 3 NZLR 213 (HC).

29 However, it also had inherited the responsibilities for administering the *Harbours Act 1950*, which governed the placement of structures in the coastal marine area. Its concurrence was required for *Marine Farming Act* leases and licenses, and its decision criterion under the *Harbours Act* was "the public interest." Concurrence was required from the Ministry of Transport for navigational issues.

30 The potential conflict of interest between its "conservation advocacy" role and its "public interest licensing" role led to a review of the appropriateness of the combination prior to enactment of the RMA, but the review team was sufficiently convinced of the ability of the department to juggle the two roles that it has retained the oversight of coastal structures and coastal planning for the central government under the RMA.

31 *Fleetwing Farms Ltd.* v. *Marlborough District Council* [1997] N.Z.R.M.A. 385. The race for space grew so intense that a two-year Marlborough region moratorium on applications was declared by the Minister of Conservation under the RMA in 1996. This was supposed to be followed by a tendering process, but this was never implemented and the moratorium ended.

32 *Golden Bay Marine Farmers* v. *Tasman District Council*. Environment Court, 27 April 2001 (W42/2001), Kenderdine J.

33 More than ten years after the RMA was enacted, most regional councils have not completed the sections of their plans relevant to aquaculture, which has meant guidance being primarily taken from case law and a variety of soon-to-be-replaced plans (see H. Rennie, "Coastal Fisheries and Marine Planning in Transition," in P. Memon and H. Perkins, eds., *Environmental Planning and Management in New Zealand* (Palmerston North, NZ: Dunmore Press, 2000).

34 An example of these new developments is that in the Bay of Plenty. The applicant comprised a consortium of three companies: an *iwi* and a fish processor holding 40 percent each, and a professional site development company holding 20 percent. The site approved by the regional council comprised some 4,500 hectares between 3 and 7 nautical miles offshore and comprising an array of mussel longlines suspended 10 meters beneath the water surface. At the time of writing, the marine farming permit to go with the site was still under consideration by the Ministry of Fisheries, which was trying to work through the implications for the fishing sector.

35 The key test case found that the ministry had total discretion in terms of both the nature of environmental assessment and the level of undue adverse effect, but noted that undue adverse effect did not mean no adverse effect and that adverse effects were to be expected. *Tasman Bay Amateur Marine Fishers* v. *Ministry of Fisheries*. High Court, 28 April 1999 (CP 25/97), Gallen J.

36 It was expected that the legislation will be introduced into Parliament in late 2004 . See the postscript to this chapter.

37 *Te Weehi* v. *Regional Fisheries Officer* [1986] 1 N.Z.L.R. 680; *Ngai Tahu Maori Trust Board and others* v. *Attorney-General*, 2 November 1987, oral judgment of Greig J regarding CP 559/87, CP610/87, CP614/87, High Court, Wellington.

38 *Treaty of Waitangi (Fisheries Claims) Settlement Act 1992*.

39 More commonly known as *Te Ohu Kai Moana*, comprising a well-qualified, minister-appointed Maori board that manages the assets until such time as a mechanism has been determined to effectively distribute them to tribes for the benefit of all Maori. The distribution of assets was approved in 2004.

40 "Organic Mussel Venture Creates World Benchmark," (2004) 72(5) *Tangaroa*. New Zealand's organic certifying agency, Bio-Gro, is to present the model to the International Federation of Organic Agriculture Movements as a proposed world standard for aquaculture.

41 *Ngati Apa and others* v. *Attorney-General* [2003] 3 N.Z.L.R. 643.

42 In 2004, the High Court confirmed that regional councils currently do not have the power under existing law to give preferential rights to Maori for areas within AMAs. *Hauraki Maori Trust Board* v. *Waikato RC* High Court, Auckland, 4 April 2004 (CIV485-0999), Randerson J.

43 *Ahu Moana, supra* note 7, and *Report on the Crown's Foreshore and Seabed Policy (Wai 1071)* (Wellington: Waitangi Tribunal, 2004).

44 *Greensill* v. *Waikato Regional Council*. Planning Tribunal, 6 March 1995 (W17/95), Treadwell J; and *Pigeon Bay Aquaculture Ltd* v. *Canterbury Regional Council*. Environment Court, 18 March 1999 (C32/85), Jackson J.

17 Conclusion

Towards sustainable aquaculture through principled access and operations

Arthur Hanson

Introduction

Aquaculture faces similar challenges in many parts of the world. While the range of responses is quite varied, it can be expected that both operational approaches and policies will continue to evolve, but likely with some degree of convergence. Certainly, what has become everyday practice now is quite different from practice a decade ago, and we are likely to see changes of even greater magnitude in the coming years. Cultivation of more species will present new operational needs and policy tensions.

For reasons of both social and environmental conflicts and economic opportunity, aquaculture law and policy need to be cast into a context of sustainable development. Sustainable development certainly can be ambiguous, but it may well be the best concept we have. Yet there is no concurrence on the definition of sustainable aquaculture. And, looking ahead, our vision of sustainability by 2020–2030 may be quite different from that of today. Importantly, it will be characterized by transition to a "biological economy." A biological economy represents an outcome where innovation produces very significant economic costs and benefits based on our increased knowledge of biological factors such as genetics and ecology. In practical terms, we will spend a lot of money trying to maintain ecological integrity, and gain benefits from applications of biotechnology or other advanced innovations. Issues such as trading rules and consumer acceptance will be critical; there is no guarantee that even superior methods of production will be automatically accepted, in either rich countries or poor ones. That is the lesson of the recent past.

Aquaculture today is on the cutting edge of this new, poorly recognized shift to a new economy where sustainable development decisions prevail. These decisions will be based on science and technology, policy, law and regulations, economic incentives, and participatory approaches. We do not have a good "institutional roadmap" of these for our management of ocean use generally. And, while such a roadmap appears to be emerging for the future of what we hope can be defined as sustainable aquaculture, it is still very incomplete and open to debate.

Thus, no country has yet fully implemented a longer-term vision of sustainable ocean use and development, or a satisfying sustainable aquaculture program as part of it. Contributors to this volume have drawn out possibilities of win–win strategies that would help to address problems faced by the aquaculture industry as it continues along an expansionary, but arguably unsustainable, pathway, identifying controls that would address environmental, community and other issues.

However, we are stuck with designing the ship while sailing it. Since aquaculture, and indeed many other ocean uses, continue to grow at a rapid rate, we cannot presume that we have the luxury of time in dry dock to complete our transition in law, policy and other needs. Instead, we must presume that, having launched our efforts in somewhat stormy waters, the building process must take place at sea, and possibly with hurricane-force winds ahead. The job facing those truly interested in the future of this industry must be to build a ship that can truly sail into the wind. These nautical metaphors may be crude, but they underline that a decade of serious and difficult work lies ahead in all parts of the world, and certainly in Canada.

We likely have not even identified all the challenges. We can expect surprises, and therefore the "learning by doing" approach makes a lot of sense. That will not be enough, however, since sustainable aquaculture development will depend upon trust-building of various sorts, and science upon which some of the "big breaks" for the future are dependent. Adaptive planning and management, which treats development as an experiment, with full participation of those involved directly, plus stakeholders, is an important means for preparing for uncertain times, and for introducing new ideas. So far there are few rigorous applications of this approach to policy formulation and implementation with respect to aquaculture.

Principled access and operations

The key principles guiding access to and operations of aquaculture need to be strengthened in law and regulation.

Access

"*Social license*" involves the building of public confidence and legitimacy of aquaculture. One issue is the role of coastal communities and riparian landowners – how much power should they have that is enshrined in law? Social license is also, in part, in the hands of international and national stakeholders and consumers. Their role and influence is increasing but often based on campaigns and choices, not necessarily supported by either policies or laws. However, international legal regimes, including trade laws, are further legitimizing their presence in debates over the right to operate.

Subsidiarity refers to the local role in aquatic use planning. The principle is that decisions about access should be taken at the lowest level that is feasible rather than its always being assumed that they should be taken at the national or provincial level. This approach requires local-level capacity-building, local institutional development, impact assessment of policies and practices on local economies, and strong decision support systems for local levels. In Norway, local sea-use planning – done in advance of need – has reduced conflict between ocean users.

Operations

Canada and other countries can learn much from international experiences such as the US Operational Framework project mentioned in Chapter 14. Operations-related issues for law and policy include the following:

Waste control

Waste control is a key concern that is already receiving much attention. Indeed, some question whether cage-based culture will continue to be the most viable choice, or whether land-based operations with advanced waste treatment will eventually become more common. Much has already been accomplished over the past decade on the issue of waste monitoring and reduction. The assimilative capacity of receiving waters likely will be reflected in the development and application of future operational standards.

Health

Health issues are characterized by complex approval processes, the possibility of international non-tariff trade barriers, and the potential for consumer scares. Further attention to operational standards related to human health may be expected, certainly in response to biotechnology introductions, contaminants, etc., and, more generally, to reducing risk not only for human consumption but also concerning both the risk of health to the aquacultured species and diseases or pests that might affect wild stocks and other species living in the surrounding waters.

Ecological integrity

The concept of ecological integrity, refers to the health of the whole ecosystem. It is enshrined in law at present as the objective in management of Canada's national parks system. It will probably emerge as an operational concern for marine ecosystems that have been altered dramatically through fisheries, pollution and other uses. If it does, it will have important implications for aquaculture, since it will be a defined objective that will have to be addressed through operational procedures that may be considerably different from today's, and may require considerably more monitoring and development of different indicators.

Three main ecological concerns associated with aquaculture are (1) genetic mixing of farmed and wild stocks and the introduction of exotic species or genetically modified fish to marine waters; (2) the spread of disease and parasites to wild stocks, for example sea lice in British Columbian waters; and (3) the degradation of benthic habitats. We can ask ourselves whether or not the law adequately regulates for ecological integrity – and, beyond that, whether we know how to deal with this subject in policy and operational terms. So far, the answer appears to be a qualified "no."

Timely regulatory and communications action

Systems for timely regulatory and communications action are not keeping pace with the current and anticipated rate of aquaculture development. This is an important finding that is not limited to Canada. Unless this situation changes, constraints on the right to operate, and the possibility of future campaigns against various types of aquaculture, are inevitable. However, considerable effort has been expended at all levels of government and within the industry on addressing the need for better communication of the changes occurring in how aquaculture is conducted, and on actual development of new regulatory frameworks.

Law and regulation: problems and challenges

Without question, the legal and regulatory regime for aquaculture is complex. *Problems* in law and regulation relate to legal pluralism, overlapping permitting, limited capacity to implement, the need for greater incentives for cohesion within sectors of the industry, and adequate funding to address legal and administrative development and reform. In Canada, regulation of aquaculture raises numerous interjurisdictional and constitutional issues.

While the national aquaculture law and policy regime in Canada has "hardly left port," and significant gaps exist in provincial laws, there are *opportunities* to use legal and regulatory tools to foster sustainable aquaculture. Current laws can be flexible, shaped by both regulatory change and use of administrative approaches. International standards and precedents indicate that more coherent approaches are possible. There is some move to incorporate market-based and voluntary approaches to facilitate aquaculture operations, and these will become even more important internationally during the years ahead. These voluntary approaches frequently are the preliminaries to more binding laws and policies.

Law and legislation need to be based on smart regulation with a mix of command and control and market-based incentives, fostering partnerships and cooperation to make the system work. An enabling framework should make it easier, rather than more difficult, for the industry to meet specified objectives. And a framework should be comprehensive enough to address all major needs accepted as valid by governments and society. Elements of such a framework, and key questions to ponder, include:

- *Fairness.* Is aquaculture being asked to meet a higher level of regulation than other ocean users?
- *Unified approach.* Is a national Aquaculture Act or an integrated approach, linking existing law and regulations and filling the gaps, a useful approach?
- *Coordination with other ocean uses.* Does Canada's *Oceans Act* adequately address the needs for an overarching ocean users' law and policy framework?
- *Adaptive management.* What are the implications of adaptive resource management for developing new laws and regulations, and for enforcing existing measures?
- *International trade.* As international trade law and policies shift to a focus on sustainable development, what are the implications for a trade-dependent industry such as aquaculture?

Operational tools: precaution, risk management and "sound science"

Precaution, risk management and "sound science" are concepts that can greatly influence aquaculture law and policy and their implications for environmental protection, economic opportunities, social and community well-being, national sustainability branding, aquaculture industrial security and development, investment, and even political careers. The questions are whether these three concepts need to be addressed as a package, and how they can be made operational.

As articulated in the Rio Declaration, the *precautionary principle* states that lack of conclusive scientific evidence does not justify inaction, particularly when the consequences of inaction may be devastating or when the costs of action are negligible. The principle is enshrined in Canadian legislation (e.g., the *Oceans Act* and the *Canadian Environmental Protection Act*), and guiding principles are in place across the federal government. Unfortunately, it is a principle that is difficult to implement. The precautionary principle/approach will not go away; Canada needs to learn how to make it work.

Questions regarding practical application relate to the link between *risk assessment* and the precautionary principle, which is not very clear in operational terms, its application in setting reasonable standards, and the implications of legislated implementation against weaker statements of policy and administrative procedure. If risk assessment is enshrined as a specific tool for aquaculture policies and laws, it could prove to be a costly, technical business if not applied appropriately.

"Sound science" is not always a meaningful term for law. It is not clear whom it serves, and how it may be linked to adaptive management. Rather, the question is how international science, national scientific efforts and traditional knowledge can be reconciled in addressing precaution and risk management.

Policy and planning tools

Planning is the proactive link to vision. However, within governments it is difficult to establish planning frameworks when there are no or few current applications awaiting approval. There are numerous planning tools available, including geographic information systems (GIS), zoning and participatory mechanisms. At present, land-use planning for the sea does not work well, since zoning is incomplete, and the land-based principles do not encompass the multidimensional aspect of the marine environment (e.g. "sour bottom," water flows creating conditions for algal blooms, movements of sea lice). In general, planning has not caught up to the pressures of aquaculture development anywhere, although Norway appears to be in a better position than most. Integrated planning for ocean use is now in vogue, but will it do the job? There are very practical limits to it, and we need to link integrated planning more effectively to law, economics, and local ecological and community needs.

Environmental impact assessment (EIA) is an internationally applied policy and planning tool that can be useful in assessing the impact of aquaculture operations in the marine environment. Specific issues relating to its application include the level of detail (screening versus in-depth "class screening"), consideration of integrated developments and cumulative impacts, the link to other regulatory processes such as food safety, avoiding excess burden of multiple jurisdictions/assessments, and integration of internationally acceptable standards. Strategic impact assessment of policies, including those for sustainable aquaculture, could be useful.

"Learning by doing" implies that the first round of management applications will not get it right, but with increased experience, improved science and stakeholder input, over time sustainable aquaculture management will improve. Adaptive management requires a commitment to experimentation and learning from the results. This approach should contribute to improved environmental impact assessment, precaution and risk management applications, and the approach should be supported in laws and policies. At this time, adaptive management is a technique not generally mentioned in laws.

Market-based tools include certification of environmental management systems, codes of practice, and industry-based environmental monitoring. All require adherence to high standards. The Scottish Assurance Scheme and the Marine Stewardship Council offer examples of such approaches that are being adopted around the world. Certification will become increasingly important, and economists and lawyers need to talk more so that market instruments can be used more effectively.

Despite their best efforts, regulators, practitioners and the general public continue to have disputes over various aspects of aquaculture development and regulation. A range of alternative dispute resolution mechanisms can be incorporated in legislation, though it is not clear how successfully they can be applied in comprehensive planning/zoning. Conflict can be reduced through consultation *before* development occurs and through the establishment of partnerships. Legislation can explicitly incorporate dispute resolution procedures, as is the case in Nova Scotia. In order to work, funding needs to be available to bring together the full range of users and stakeholders, and to provide the necessary information. Adaptive management may help by leading to decentralized decision-making, leading to reduced levels of conflict. In order to deconstruct conflict, we need to recognize that different types of conflicts (e.g. data, interest, structural, value or relationship conflicts) will need an array of responses. What policies and laws are needed to resolve each of these types of conflicts?

Policy-makers must also address aquaculture industry concerns over the complex and confusing maze of regulations, and also by inadequate regulations, governing aquaculture operations. Industry representatives want reassurance that there is political will to address outstanding issues, and that law and policy research is founded in good science. To this end, people in the aquaculture industry need to be engaged by government regulators and invited to participate in the policy-making process.

Aboriginal concerns and rights

Some First Nations perceive law as a mechanism to further colonize. They see "sharing, reciprocity and respect" as the principles underlying utilization and development of aquaculture resources. Traditional and spiritual concerns are fundamental, and these do not translate well into legal perspectives. To address social needs, aquaculture needs to link to the "real economies" of First Nations communities.

Given the limited consultation of First Nations during the development of current aquaculture regimes, some Aboriginal people are calling for full-scale public review. There is a confirmed duty to consult, guided by law, but some variability emerges in court tests. Consultation on resource issues will have to increase and be meaningful. Domestic law has been used to resist aquaculture; international law and United Nations forums may be used in the future.

As the evolution towards greater recognition of First Nations' rights continues, based particularly on court decisions, legal precedents have emerged to govern First Nations' participation in the development of aquaculture law and policy. So far, the rights being tested primarily relate to upholding of treaty obligations, land rights, and hunting and fishing. Future tests may be expected to encompass offshore waters, the seabed, the foreshore, riparian rights, freshwater access and the protection of ecological integrity. Aquaculture operations will cross over several types of rights (land, freshwater, ocean space).

Conclusion

The Conference of the Parties to the Convention on Biological Diversity in Decision VII/5 on Marine and Coastal Biodiversity[1] emphasized many of the research challenges still confronting mariculture that complicate the further formulation of laws and policies. Those research needs include, among others:[2]

- development of criteria for judging the seriousness of biodiversity effects of mariculture operations;
- development of criteria for when environmental impact assessments should be required;
- studies related to impacts of mariculture on genetic, species and ecosystem diversity, including the genetic effects of biotechnology developments and the effects of genetic pollution from farmed populations on wild populations;
- comparative studies on legislative, economic and financial mechanisms for regulating mariculture activities;
- development of quantitative and qualitative criteria for assessing mariculture impacts on the environment, including social and cultural impacts.

As the chapters in this volume have documented, the strengthening of aquaculture law and policy is an ongoing challenge in light of the many conflicting interests that must be satisfied and the negative views and political inertias that need to be faced. Key sustainability principles emerging from international law, such as precaution, the ecosystem approach, public participation, integration, and intergenerational and intragenerational equity, offer hope in the global search for sustainable seas and healthy coastal communities. However, best practices are still at the frontier of being realized.

Notes

1 Online. Available htpp://www.biodiv.org/decisions/?dec=VII/5.
2 *Ibid.* at Appendix 5 (Research and monitoring priorities associated with Programme Element 4: Mariculture).

Index